International Conference on Extraction, Transport, Storage and Processing of Hydrocarbons & Minerals (ETSaP 2020)

IOP Conference Series: Materials Science and Engineering
Volume 952

Tyumen, Russia
24 – 25 August 2020

Editor:

Rasul A. Mamadaliev

ISBN: 978-1-7138-4780-9
ISSN: 1757-8981

Printed from e-media with permission by:

Curran Associates, Inc.
57 Morehouse Lane
Red Hook, NY 12571

Some format issues inherent in the e-media version may also appear in this print version.

Institute of Physics
Dirac House, Temple Back
Bristol BS1 6BE UK

Phone: 44 1 17 929 7481
Fax: 44 1 17 920 0979

techtracking@iop.org

Additional copies of this publication are available from:

Curran Associates, Inc.
57 Morehouse Lane
Red Hook, NY 12571 USA
Phone: 845-758-0400
Fax: 845-758-2633
Email: curran@proceedings.com
Web: www.proceedings.com

TABLE OF CONTENTS

IOP Conf. Series: Materials Science and Engineering **952** (2020) 011001 doi:10.1088/1757-899X/952/1/011001

PREFACE

The International Conference on Extraction, Transport, Storage and Processing of Hydrocarbons and Minerals (ETSaP 2020) successfully took place in Tyumen, Russia on 19–20 August 2019. ETSaP 2020 was held through the cooperative efforts of Tyumen industrial University, North-Caucasian Institute of Mining and Metallurgy (State Technological University) (Vladikavkaz).

This year, the conference has covered a wide range of subject matters, and the geography has expanded greatly. The participants presented the results of their research on minerals mining and hydrocarbons extraction, as well as on transportation, storage and processing of minerals and hydrocarbons. Moreover, issues of infrastructure development in oil-producing and mining regions as well as responsible resource extraction and environment-friendly industry were discussed.

Two days of the conference became the time of intense learning for all the participants and surely gave them the impetus to further research. We are pleased to present to our readers the latest research on extraction and processing of hydrocarbons and minerals. After the careful reviewing, 72 papers were selected for publication that present the most valuable and significant research results.

We would like to thank all the participants of ETSaP 2020 as well as the organizing committee for their efforts in holding the conference. We also express out deep gratitude to the reviewers and editors for their invaluable contribution to publication of the ETSaP 2020 proceedings.

IOP Conf. Series: Materials Science and Engineering **952** (2020) 011002 doi:10.1088/1757-899X/952/1/011002

Peer review declaration

All papers published in this volume of *IOP Conference Series: Materials Science and Engineering* have been peer reviewed through processes administered by the Editors. Reviews were conducted by expert referees to the professional and scientific standards expected of a proceedings journal published by IOP Publishing.

Type of peer review: Single-blind peer review by two members of the editorial team and post-acceptance language and style check (by an invited linguist).

- **Conference submission management system:** submissions were received and handled via e-mail
- **Number of submissions received:** 114
- **Number of submissions sent for review:** 87
- **Number of submissions accepted:** 72
- **Acceptance Rate (Number of Submissions Accepted / Number of Submissions Received X 100):** 65%
- **Average number of reviews per paper:** 2
- **Total number of reviewers involved:** 8
- **Any additional info on review process:** all the papers were reviewed by the volume editor and by the session editor, notifications of acceptance and invitation were send after review of thesis (not all reported papers were accepted and included in the volume).
- **Contact person for queries:** Rasul A. Mamadaliev, senior lecturer of department of mechanical engineering, Tyumen Industrial University, corresponding editor, Mamadaliev_it@mail.ru

IOP Conf. Series: Materials Science and Engineering **952** (2020) 012001 doi:10.1088/1757-899X/952/1/012001

Reduction of well stimulation period after hydraulic fracturing

A E Verisokin[1], T A Gun'kina[1], V A Vasil'yev [1], A I Shchekin[1], D Yu Serdyukov[2]

[1] Institute of Oil and Gas, NCFU, 1, Pushkin Str., Stavropol, 355009, Russia
[2] NGDU Rechitsaneft, RUP PO Belorusneft, 43 Lenin Str., Rechitsa, 247500, Belarus

E-mail: verisokin.aleksandr@mail.ru

Abstract. This article discusses the problems associated with the removal of proppant, mechanical particles of the rock and undecomposed fluid from the hydraulic fracture into the well. The necessity of developing a device for washing the perforation interval has been identified and justified. It is shown that hardened and viscoelastic gel particles fall into perforations and adhere to the walls of channels and wells. Subsequently, these zones of gel deposition reduce permeability in perforation channels. The author has developed the design of a device for decolmation (declogging), which reduces the time for well development after hydraulic fracturing. To increase the productivity of the considered wells, it is necessary to flush them with the presented device in the mode of maximum approximation to deposits. To carry particles better to the surface, it is proposed to use foam systems generated by the jet pump, which is described in this article.

1. Introduction

Today, more and more attention is paid to the development of hard-to-recover reserves in the world. Such reserves include oil in sufficiently dense rocks, which lie in poorly permeable reservoirs. At the same time, emphasis is placed on the introduction of new technologies for the intensification of production, which have successfully been used in oil fields with highly permeable reservoirs. One such effective technique is hydraulic fracturing (Fracking). During the hydraulic fracturing operation, a branched drainage system is formed in the bottom-hole zone of the reservoir by supplying the hydraulic fluid under high pressure [1]. The development of many well-known hydrocarbon deposits with permeable reservoirs of less than 5 - 10 * 10^{-3} μm^2 shows that hydraulic fracturing is almost the only method that can be used to develop such reserves. Additional requirements are imposed on the operation of low-permeable reservoirs, which in some cases casts doubt on the economic feasibility of hydraulic fracturing and significantly increases the risk of failure of the operation.

A significant part of fracking operations is carried out in our country. Currently in Russia more than 50% of wells undergo hydraulic fracturing. For example, in the fields of Western Siberia, more than 25 thousand hydraulic fractures have been carried out. Every year, the importance of hydraulic fracturing, both in the world and Russian practice, will continue to increase. In turn, the technology and technique of the hydraulic fracturing method will be more complicated. Nevertheless, despite the huge amount of research on hydraulic fracturing, the technology of this operation has many unresolved scientific problems. Presently, in Western Siberian fields, in 80% of the cases, there are complications in well development after hydraulic fracturing. The main part of production wells operates under conditions of formation of "products" of hydraulic fracturing at the bottom with partial or complete overlapping of the perforation interval. Eliminating gel residues from the well is a daunting task.

According to the results of field observations, it was proved that the proppant backflow from the fracture to the well is maximum at the stage of well development. This necessitates the development of new progressive technical means for improving the methods of well development after hydraulic fracturing.

2. Materials and methods

The research materials compiled the results of hydraulic fracturing in the fields of Western Siberia in reservoirs of medium and low permeability. To solve the problem of blocking perforation channels with an undecomposed gel and reducing the development time after hydraulic fracturing, numerous theoretical and analytical studies were carried out to develop a set of devices. A literature review and patent analysis on the issue under study were carried out. Criticism of analogues has been carried out and more advanced devices have been developed for flushing perforation channels and perforation intervals of wells from "silt" carried out from fractured and destroyed proppant.

3. Research results

The removal of proppant from the well leads to a breakdown of the electric centrifugal pump, as well as to a decrease in well productivity. The image of the wellbore in the perforation interval after mudding with an undecomposed gel with proppant particles is shown in Figure 1.

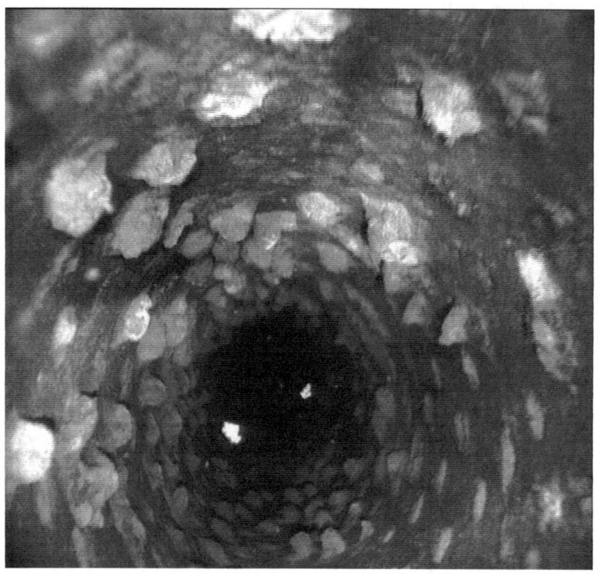

Figure 1. Wellbore in the perforation interval after mudding

It should also be noted that in case of poor cementing of the well, during hydraulic fracturing, part of the proppant enters the voids in the cement ring, and during development it is taken out. In addition to gel sedimentation on the walls of the well, this chemical liquid also settles in perforation channels, clogging them.

Oil and gas industry experts believe that some of the carrier fluid does not collapse even under the influence of modern destructors - additives that contribute to a controlled decrease in fracture fluid viscosity and its removal from the hydraulic fracture [2-5]. If the viscoelastic "islands" remain inside the crack for a long time, then they harden. In the future, these sections of the fracture will have a reduced permeability and adversely affect the productivity of the well after hydraulic fracturing [6-8].

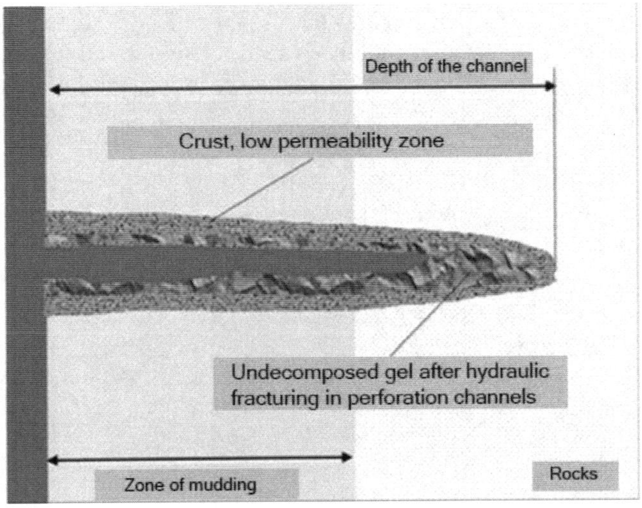

Figure 2. Mudding of perforation channels with undecomposed hydraulic fracturing gel

Solid and viscoelastic particles of the destructors also enter the perforation channels and adhere to the walls of the well. These deposits reduce permeability in perforated channels. To improve the productivity of such wells, flushing should be applied with the highest proximity to complications. You can get rid of such zones with the help of a device for reclamation of wells, which is shown in Figure 3. The device consists of a casing 1 with a cylindrical axial channel 2, an adapter 3 closed on top with a blind bottom 4 with a conical protrusion 5, which goes into a cylindrical axial channel 2 in the conical axial channel 6, the angle formed is equal to the angle of inclination of the conical protrusion 5.

In the body of the adapter 3 above the blind bottom 4, several tangential channels 7 are made, hydraulically connecting the axial channel 8 of the adapter 3 with the cylindrical axial channel 2 of the casing 1 (Figure 3).

A bore is made at the lower end of the casing 1, in which a support seat 9 is installed. Inside the axial channels, cylindrical 2 and conical 6, hollow pipes 10 are provided, equipped with an annular protrusion 11, on which a thrust sleeve 12 is mounted, which is installed with the possibility of contact with the inner surfaces of the conical 6 and cylindrical 2 axial channels of the casing 1. The upper end of the hollow pipe 10 is located freely between the inner wall of the cylindrical axial channel 2 of the casing 1 and the conical protrusion 5. At the lower end of the hollow pipe 10 an end valve 13 with a spherical surface is mounted, resting on the counter spherical surface in the support seat 9. At the same time, the end valve 13 is provided with a supply tube 14 passing through the axial channel of the support seat 9 and connected to the sleeve 15 covering the cylinder 16. In the axial channel of the cylinder 16, a drainage tube 17 is installed, provided with an annular protrusion 18, in the axial channel 19 of which a nozzle 20 is installed. An annular chamber 21 is formed between the drainage tube 17 and the inner surface of the cylinder 16, in which a spring 22 is placed, which compresses an annular protrusion 18 of the drainage tube 17 to the end of the nut 23 connected with the cylinder 16.

An inner annular bore 24 is made in the sleeve 15, which is connected by the bypass hole 25 with the axial channel 26 of the supply tube 14 and the holes 27 in the body of the cylinder 16 with its axial channel 28 between the nut 23 and the annular protrusion 18 of the drain tube 17. In the its body there are radial holes 29 for hydraulic connection of the axial channel 28 of the cylinder 16 with the axial channel 19 of the drainage tube 17 located in the initial position under the nut 23. The cylinder 16 is fixed in the sleeve 15 with a lock nut 30. The axial channel 19 of the drainage tube 17 extends outside the cylinder 16 and 31. In the closed plug adapter, body 3 is threaded for connection to the coiled tubing 32 of the coiled tubing unit.

The annular bore 24 in the sleeve 15 is protected from aggressive media by the seals 33 and 34. The annular gap between the drain pipe 17 and the cylinder16 is insulated by a seal 35, which is mounted on the annular projection 18. The annular gap between the nut 23 and the drain pipe 17 is closed by the seal 36. The principle of operation of the device is described in [9].

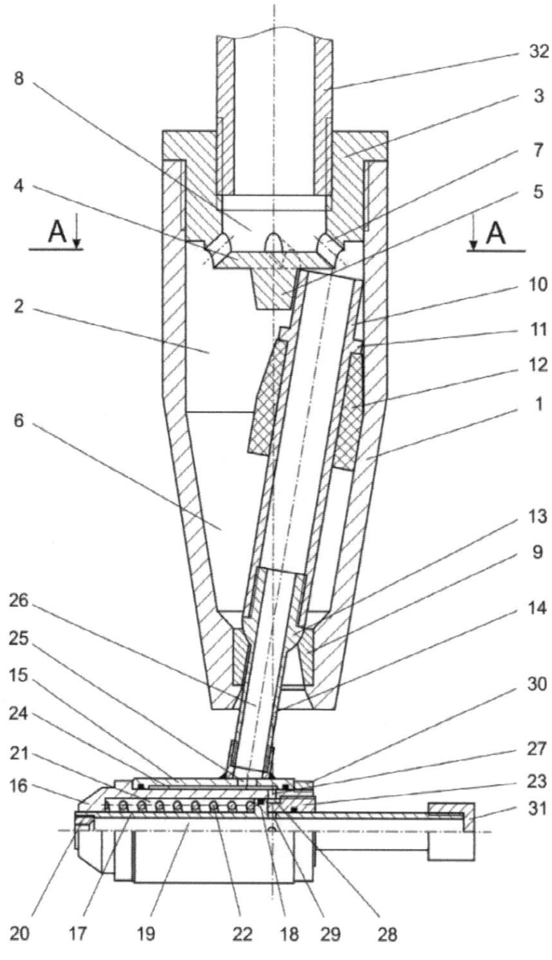

Figure 3. The design of the device in the context, in the initial position of the parts (longitudinal section)

The use of this device will increase the effect of the working agent jet on deposits by using its total flow rate, which is directed through the nozzle, as well as isolate the internal cavity of the device from the ingress of formation fluid and particles of the destroyed proppant at the end of the flushing process.

The arrangement of the device in the well is shown in Figure 4. The technological feature of this device is that it can be lowered through tubing using a coiled tubing installation.

Technical characteristics of the flushed wells are shown in table 1.

In the fields of Western Siberia, flexible tubings (FT) are increasingly used to eliminate the "products" of hydraulic fracturing in the wellbore that fall to the bottom. It is proposed to use foam systems for flushing the well.

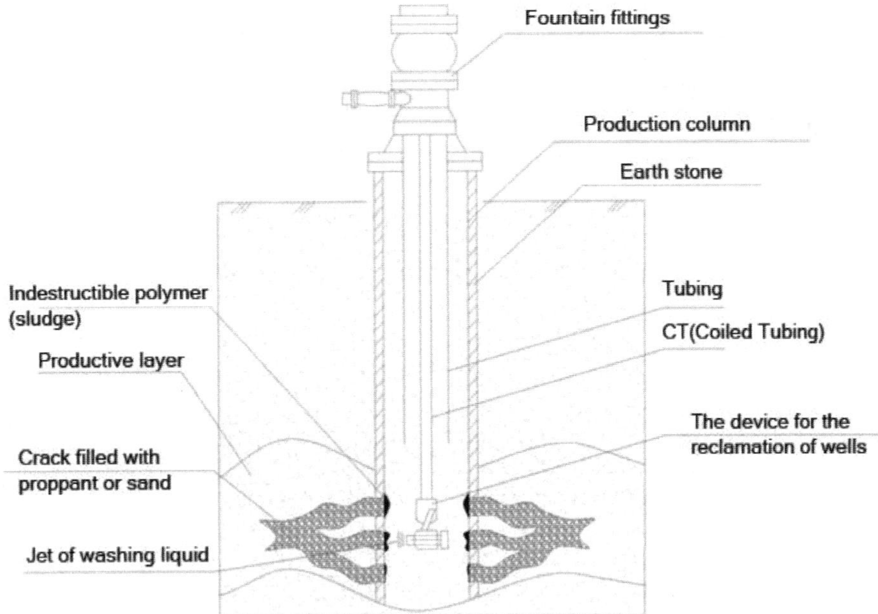

Figure 4. The layout of the device in the well

Table 1. Technical characteristics of the device

Liquid flow through the layout l / min	70 – 110
Working fluid pressure, not more than, MPa	66
Maximum layout diameter, mm	70
Working fluid	Water
Diameters of flexible continuous tubing (FCT), mm	50,8 и 25,4
Operating temperature of the installation, not more than 0˚C	145

The use of foam systems is due to the following advantages:

- when technological fluids get into the formation, reservoir properties of the bottom-hole formation zone (BHF) are preserved;
- the use of foam systems allows you to quickly adjust the pressure at the bottom of the well, so you can quickly switch from repression to depression;
- the foam has excellent bearing capacity, which improves the cleaning of the perforation interval from particles of gel and proppant after hydraulic fracturing.

However, the use of foam systems as a circulating agent in a well is accompanied by increased complexity. Special equipment must be used to create foam. For better foam generation, the author has developed a jet pump design. The scheme of the inkjet apparatus is shown in Figure 5. A detailed principle of the pump is given in [9].

4. Conclusion

1. In oil fields, large volumes of proppant, mechanical particles, and undecomposed hydraulic fracturing fluid exit the well. Some of the carrier fluid is not disintegrated even under the influence of a destructor. A mixture of mechanical particles, proppant fragments with an undecomposed gel forms

a difficult-to-remove mixture in the perforation channels and on the walls of the well - "sludge". Subsequently, these hardened residues reduce the permeability in the perforation channels. To improve the productivity of such wells, it is necessary to flush them with the closest possible complication. You can get rid of such complications with the help of the proposed set of devices for well development after hydraulic fracturing.

2. A device for well reclamation has been developed: Patent No. 2651869, Application No. 2017109500.

3. A jet pump has been developed for better foam generation: Patent No. 2643882, Application No. 2017114222.

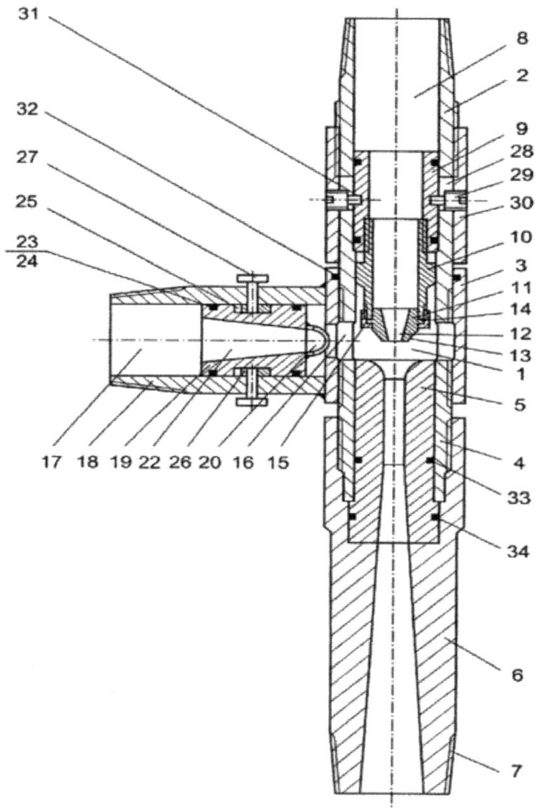

Figure 5. Jet pump: 1 - receiving chamber, 2 - passive medium, 3 - connecting sleeve; 4 - connected to the casing, 5 - mixing chamber, 6 - diffuser, 7 - connecting thread, 8 - axial channel, 9 – nipple, 10 - nozzle holder, 11 - nozzle, 12 - conical axial channel, 13 - curved surface, 14 - annular protrusion 15 - union nut, 16 - tangential channel, 17 - axial channel, 18 – pipe, 19 - figured sleeve, 20 - flat tip, 21 - slit gap, 22 - axial channel, 23 - annular grooves, 24 - sealing rings , 25 - annular groove, 26 - cams, 27 - screws , 28 - two longitudinal grooves, 29 - fingers, 30 - cylinder, 31 - annular groove, 32, 33, 34

References

[1] Verisokin A, Vasil'yev V and Gun'kina T 2019 Packer design research used in hydraulic fracturing *IOP Conf. Series: Earth and Environmental Science* **378** 7

[2] Zinchenko I 2007 *The use of hydraulic fracturing for stimulation of inflow in gas condensate wells of the Yamburgskoye field and prospects for the application of the method in the process of further development of deposits* (Moscow - IRC Gazprom LLC)

[3] Anderson J, Simpson M and Bosinski P 2003 Producing Natural Gas from Coal. *Oilfield Review, Autumn* **9** 23-30

[4] Kanevskaya R 1998 *Foreign and domestic experience of hydraulic fracturing* (Moscow – VNIIOENG)

[5] Keybal A and Keybal A 2009 Reasons for proppant backflow to the wellbore after hydraulic fracturing of the reservoir *Drilling and Oil* **11** 48-52

[6] Merkulov A 2007 Impulse technologies for stimulation and hydraulic fracturing (Part I) *Oil industry* **9** 127-129

[7] Sudo R 2007 Development of low-permeable formations in the fields of RITEK OJSC using hydraulic fracturing *Oil industry* **3** 48-50

[8] Fakhretdinov R and Brovchuk A 2007 The results of the use of hydraulic fracturing for the development of the southern licensed territory of the Priobskoye oil field *Oil industry* **3** 44-47

[9] Verisokin A 2018 The proppant test procedure for hydraulic fracturing *Science and Technology in the gas industry* **2** 62-69

IOP Conf. Series: Materials Science and Engineering **952** (2020) 012002 doi:10.1088/1757-899X/952/1/012002

Non-stationary mass transfer of oil-water mixture in reservoirs with a system of horizontal wells

I N Zhizhimontov[1], Yu D Zemenkov[2] and A B Shabarov[1,2]

[1] Tyumen State University, 6, Volodarskogo str., Tyumen, 625000, Russia
[2] Tyumen Industrial University, 38, Volodarskogo str., Tyumen, 625000, Russia

E-mail: inzhizhimontov@yandex.ru

Abstract. The paper presents a physical and mathematical model of quasi-three-dimensional non-stationary filtration of a water-oil mixture in a system of horizontal wells and the corresponding numerical calculation method. Based on them, a computer program was created, designed to solve a number of scientific and production problems. The developed model and program are aimed at theoretical and numerical study of the processes of mass transfer of the oil-water mixture in the field of analysis, maintenance and development of natural formations with high-viscosity oil deposits. The results of numerical experiments for calculating the two-phase flow of an incompressible fluid in a reservoir are given as an example of a real field, located in the north of Western Siberia. The paper proposes an approach to discretization of the computational domain, based on the construction of a dynamic computational grid in a natural semi-fixed coordinate system. For the first time, the problem of calculating two-phase mass transfer using a semi-fixed natural grid and an analytical expression for the flow rates of phases in current streams of a found shape has been solved. Based on the proposed model, calculation and parametric studies of various modes of field development with a system of parallel horizontal wells were performed.

1. Introduction

In recent decades, physical and mathematical modeling has been the main tool for making decisions on the analysis, maintenance and development of oil and gas fields.

The correct solution of the filtration problem, taking into account the specifics of the deposits in question, allows predicting hydrocarbon production in the long term, localizing residual hydrocarbon reserves, and evaluating the effectiveness of measures to maintain reservoir pressure and other measures to increase oil recovery.

The first steps in the development of the theory of filtration were laid by the French engineer A. Darcy, who published an experimental law in 1865, which corresponded the pressure gradient to the linear filtration rate [9]. This work is considered the beginning of the theory of filtration. To date, many works by domestic and foreign authors are devoted to the development of the theoretical foundations of filtration. The works of many scientists, such as R.I. Nigmatulin [3], G.N. Barenblatt [1], K.S. Basniev [2], studied complex two-phase and three-phase flows in a reservoir and wells.

Taking into account the latest achievements in the field of applied mathematics and a constant increase in the productivity of computing technologies, the modern idea of modeling the processes of developing hydrocarbon deposits is reduced to solving a detailed three-dimensional problem [10, 11].

IOP Conf. Series: Materials Science and Engineering **952** (2020) 012002 doi:10.1088/1757-899X/952/1/012002

Despite this, the practice of using hydrodynamic models shows that the capabilities of simulators are far from always meeting the needs that arise when solving important production problems associated with field development [6]. In particular, this applies to modeling development processes in complex unconventional fields. An urgent problem is the construction of computational models and computer programs for calculating the parameters of the development of viscous oil fields, especially near horizontal wells.

2. Physico-mathematical model of mass transfer

The system of equations for calculating the non-stationary filtration of the oil-water mixture includes [1, 2, 3, 7, 8]:

- phase continuity equation:

$$\frac{\partial(m\, S_i\, \rho_i{}^0)}{\partial\tau} + div(\rho_i{}^0\, \vec{w_i}) = 0, \qquad (1)$$

- phase equation (generalized Darcy equation):

$$\vec{w_i} = \frac{k_0 f_i}{\mu_i}\, grad(P), \qquad (2)$$

- equation of balance of internal energy of a mixture:

$$\frac{\partial(\rho U)}{\partial\tau} + div\,(\vec{q} + \rho\, U\, \vec{w_i}) = 0, \qquad (3)$$

where τ is time; i = 1, 2; oil parameters correspond to the index i = 1; for water i = 2; m is porosity; S_i is the saturation by the i-th phase(considering $S_1 + S_2 = 1$); $\rho_i{}^0$ is true density; k_0 is absolute permeability; f_i is relative phase permeability; μ_i is dynamic viscosity; $\rho U = (\rho_1 c_1 + \rho_2 c_2 + \rho_{rock} c_{rock})T$ is specific internal energy of the mixture, $\vec{q}$ is heat flux vector due to heat conduction (λ).

In equations (1–3), the unknowns are velocity ($\vec{w_i}$), pressure (P), temperature (T), and saturation (S_i). For the system of equations (1–3), the initial (at the time τ = 0) distribution of parameters in the flow region is set, as well as the conditions on the boundary of the region.

3. Problem statement

The paper considers quasi-three-dimensional isothermal non-stationary processes that occur during the operation of an oil field with a system of horizontal injection and production wells. The system "injection well – reservoir – production well" is considered. The scheme of the problem under consideration is presented in Figure 1.

The following assumptions are used in the work:

1. The three-dimensional problem is reduced to a combination of three interrelated tasks:

1) quasi-one-dimensional problem of changing pressure, flow rate of water (steam) and temperature (in the case of solving the thermal problem) in the horizontal section of the injection well;

2) the set N_x of two-dimensional two-phase filtering problems in vertical layers of $\Delta z_i = L_z/N_z$, where $i = \overline{1, N_z}$

3) quasi-one-dimensional problem of changing pressure, temperature and flow rates of oil and water (at this stage we restrict ourselves to solving the two-phase filtration problem) in the horizontal section of the well.

The oil-water mixture flows in horizontal sections of production wells along the Z-axis $0 \leq Z < L_z$; like in injection wells, hot water or other coolant flows along the Z axis, while Z varies from $Z = L_z$ to $Z = 0$. The projection of the filtration rate along the Z axis (W_z) is assumed to be significantly smaller compared to W_x and W_y, thus, equality (4) holds:

$$W = \sqrt{W_x^2 + W_y^2} \qquad (4)$$

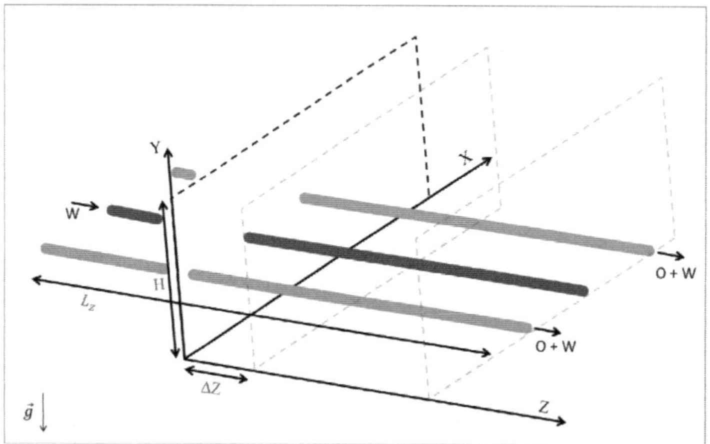

Figure 1. Schematic representation of a quasi-three-dimensional problem in a system of parallel horizontal wells

Legend in Figure 1: H is the reservoir thickness, L_z is the length of the horizontal section of the well, ΔZ are the dimensions of the model cell along the Z axis, W is water, O + W is the water-oil mixture.

2. A quasistationary approximation is considered with a known distribution of parameters at time t^n and certain external influences over time Δt, parameters at time $t^{n+1} = t^n + \Delta t$, where $n = \overline{1, N_t}$.

The development system (Figure 1), which is used in practice, is described by solving a three-dimensional non-stationary problem of heat and mass transfer in a system of horizontal sections of an injection well - formation - horizontal section of a producing well.

In this work, aimed at improving methods for calculating hydrodynamic parameters, it is assumed that the temperature field in the region of the filtration flow is known.

4. Calculation algorithm

The following iterative method for solving the problem under the above assumptions is proposed, which includes the following blocks:

1) Formation of a computational grid in the reservoir area and in cross sections in wells.

2) Setting the initial data for modeling:

- geometric dimensions: diameter of the injection (d_i) and production wells (d_p); distance between horizontal sections of wells (L_x); effective length of the horizontal section of wells (L_z);
- initial reservoir parameter distributions: porosity (m), absolute (K_o) and relative phase permeability ($f_{o,w}(S_w)$), water or oil saturation ($S_o = 1 - S_w$), pressure (P_0) and temperature (P_0) of the reservoir at the initial moment of time (t^0);
- initial pressure in the injection (P_i) and production (P_p) well or rates ($G_{o,w}$) and temperature (T_i) in the injection well;
- generalized data on the thermophysical and rheological properties of fluids and rocks: true densities $\rho^0_{o,w,rock}(T, P)$, dynamic viscosities $\mu_{o,w}(T, P)$, thermal conductivity and heat capacity coefficients $\lambda_{o,w}(T)$, $C_{o,w}(T)$;
- calculation parameters: time step (Δt), number of sections (N_z), number of streamlines and equipotential surfaces, error during iterations over the flows (ξ_G).

3) The initial distributions of pressure $P_{i,p}(x, t$ and temperature $T_i(x, t)$ in the horizontal section of the injection and production wells along the length $0 \le x \le L_x$ are calculated; the pressure distribution is expressed from the equation of balance of momentum [7] and is described by equation (5):

$$P_i(x) = P_i(0) + a_k\rho[\frac{v(0)^2 - v(x)^2}{2} + g\big(z(0) - z(x)\big) + (l' - l_{fr})], \quad (5)$$

where $\frac{v(0)^2-v(x)^2}{2}$ is the change in kinetic energy; $g\big(z(0)-z(x)\big)$ work in the field of gravity; $l' = -(v - v_x)\frac{G}{\rho S}$ is specific work associated with mass transfer through the lateral surface S; $l_{fr} = \lambda_{fr}\frac{\Delta x}{d}\frac{v^2}{2}$ is work of friction forces according to the Darcy–Weisbach equation.

4) The method of complex source potentials [4, 5, 7] determines the initial position of streamlines and equipotential surfaces in the planes $Z_i = const$ $(i = \overline{1, N_z})$, control volumes are formed at the points of intersection of the lines. Thus, we use the natural coordinate system $(\bar{l}, \bar{l_1})$ [4, 5], where $l_i = const$ corresponds to the lines of equal potential $\varphi_i = const$ determined in the calculation process and the streamlines $\psi_j = const$. The integrated potential for modeling the flow, taking into account the geometric shape of the object of study (Figure 1), taking into account the influence of the boundaries of the reservoir, is presented in the form of a superposition of sources and sinks [7]:

$$w(z) = \frac{Qi}{2\pi} * \ln\left(z - \frac{h}{2}\right) - \frac{Qp}{2\pi} * \ln\left(z + \frac{h}{2}\right) + \frac{Qi}{2\pi} * \ln\left(z - \left(H - \frac{h}{2}\right)\right) -$$

$$-\frac{Qp}{2\pi} * \ln\left(z + \left(H - \frac{h}{2}\right)\right) + \frac{Qp}{2\pi} * \ln\left(z - \left(H + \frac{h}{2}\right)\right) + \frac{Qi}{2\pi} * \ln\left(z + \left(H + \frac{h}{2}\right)\right). \quad (6)$$

5) From the system of quasi-one-dimensional equations of fluid motion in horizontal wells [8], we find the change in velocities, pressures, saturations, and temperatures along the length $0 \le x \le L_x$.

The pressure distribution along the flow stream will be expressed as

$$p(l) = p_1 - J_1(\acute{l}) * G. \quad (7)$$

The flow rate of the mixture through the flow stream is determined from (7) taking into account $\acute{l} = L_j$, where L_j is the current stream length, $p = p_2$ and $J_1(\acute{l}) = J_1(L)$:

$$G = \frac{p_1 - p_2}{J_1(\acute{l})}. \quad (8)$$

The definite integral $J_1(\acute{l})$ has the following form:

$$J_1(\acute{l}) = \int_0^{\acute{l}} \frac{dl}{k_0 * S * \left(\frac{\rho_o^0 f_o}{\mu_o} + \frac{\rho_w^0 f_w}{\mu_w}\right)}. \quad (9)$$

The distribution of phase velocities along the flow stream l taking into account (8) and (9):

$$w_o(l) = -\frac{k_o f_o}{\mu_o}\left(-\frac{G}{k_0 F\left(\frac{\rho_o^0 f_o}{\mu_o} + \frac{\rho_w^0 f_w}{\mu_w}\right)}\right) = \frac{G}{F\left(\rho_o^0 + \rho_w^0 \frac{f_w \mu_o}{\mu_w f_o}\right)}, \quad (10)$$

$$w_w(l) = -\frac{k_o f_o}{\mu_w}\left(-\frac{G}{k_0 F\left(\frac{\rho_o^0 f_o}{\mu_o} + \frac{\rho_w^0 f_w}{\mu_w}\right)}\right) = \frac{G}{F\left(\rho_w^0 + \rho_o^0 \frac{f_o \mu_w}{\mu_o f_w}\right)}. \quad (11)$$

Oil and water flow rates are in the section $l = L$ with area $F(L)$:

$$\begin{cases} G_o = w_o(L) * \left(\rho_o^0 * F(L)\right) = \frac{G}{1 + \frac{\rho_w^0 f_w \mu_o}{\rho_o^0 f_o \mu_w}}, \\ G_w = G - G_o. \end{cases} \quad (12)$$

Then the saturation by the i-th phase at time t^{n+1} equals:

$$S_i^{n+1} = S_i^n + \left[\left(\rho_{i1}^0 w_1 F_1\right) - \left(\rho_{i2}^0 w_2 F_2\right)\right]^n \frac{\Delta t}{V * m * \rho_i^0}. \quad (13)$$

6) With the control volumes found in step 4 in the natural coordinate system, the initial pressure distributions in the current streams and phase flows are determined, by analogy with [8], but taking into account the shape of the streamline;

7) Taking into account the parameters found in step 6, the calculated streamlines (step 4) and the parameters of the control volumes in the planes $x_i = const\ i = \overline{1, N_x}$ are refined. Thus, a dynamic natural semi-fixed coordinate system is used to discretize space. To account for flow dynamics and changes in control volumes, lines of equal potentials are fixed, and streamlines are refined taking into account constantly changing parameters of the filtration flow.

8) The iterative procedure according to steps 5 - 7 is repeated until convergence from the conditions:

$$\frac{\Delta G_w{}^{k+1} - \Delta G_w{}^{k}}{G} \leq \xi_G, \quad (10)$$

$$\frac{\Delta G_o{}^{k+1} - \Delta G_o{}^{k}}{G} \leq \xi_G, \quad (11)$$

where k is the iteration number, ξ_G is the permissible error during iterations over the flows in the layers $x_i = const\ (i = \overline{1, N_x})$;

9) The calculation according to steps 5-8 is repeated for time intervals (t^n, t^{n+1}) for $n = \overline{2, N_t}$, where N_t is the number of time steps;

10) Visualization and processing of simulation results;

11) Adaptation of results and calculation and parametric studies.

5. Calculation and parametric studies on the example of a real field with high-viscosity oil

As an example, we used real data from a field in Western Siberia with a similar development system. The relative phase permeability functions were set according to the results of digital stochastic pore-network modeling, a detailed description of the approach is presented in [12]. The initial data for the calculation are summarized in Table 1.

Table 1. Source data for modeling

Parameter	Designation	Value
Reservoir thickness [m]	H_{res}	20
Coefficient of porosity [%]	m	30
Initial oil saturation [%]	S_o	70
Initial reservoir pressure [MPa]	P_0	80
Initial reservoir temperature [°C]	T_0	20
Downhole pressure of injection well [MPa]	P_i	100
Downhole pressure of producing well [MPa]	P_p	60
Length of horizontal section of wells [m]	L_z	500
Diameter of wells [mm]	$d_{p,i}$	169

At the first stage, technological indicators were calculated for various modes of field development (Figure 2).

IOP Conf. Series: Materials Science and Engineering **952** (2020) 012002 doi:10.1088/1757-899X/952/1/012002

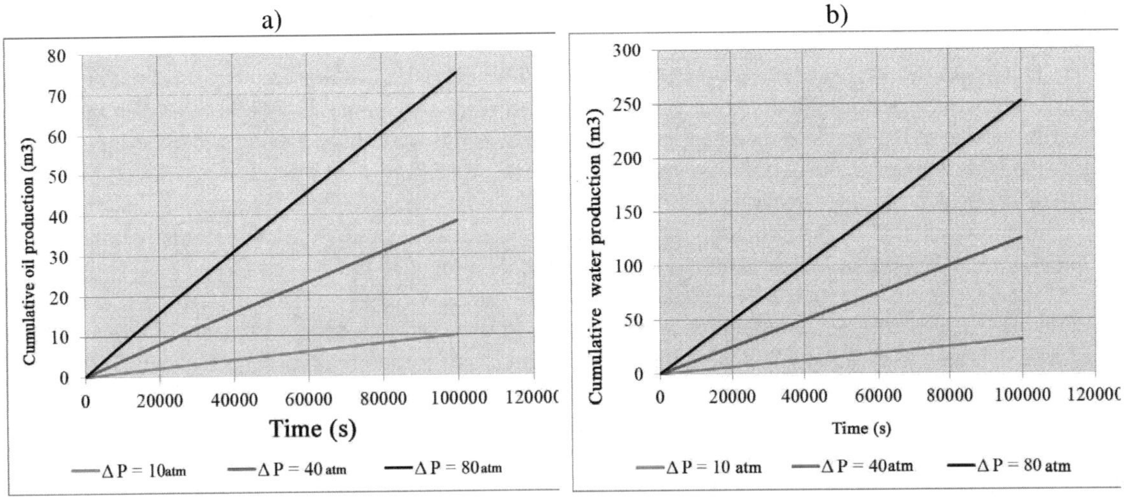

Figure 2. Technological indicators of development in various modes: a) the dynamics of cumulative oil production, b) the dynamics of cumulative water production

At stage two, a series of isothermal calculations was performed with different initial formation temperatures (Figure 3a). Initial formation temperatures were taken equal to 20, 60 and 80 °C. In this series of experiments, the key factor influencing the result was the dependence of fluid viscosity, surface tension, and the change in the relative permeability diagram. The dependence of the dynamic viscosity and surface tension were obtained from reservoir fluid samples, and the patterns of change in the relative permeability diagrams were justified by the proposed digital core research technology [12].

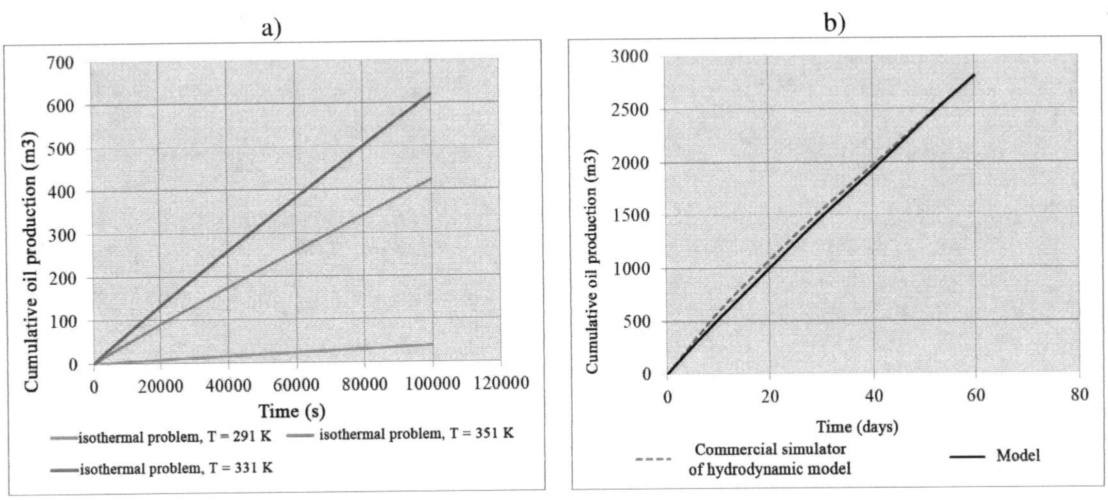

Figure 3. A series of isothermal calculations with different initial formation temperatures (a); comparison of simulation results for cumulative oil production of the developed program and a commercial simulator (b)

IOP Conf. Series: Materials Science and Engineering **952** (2020) 012002 doi:10.1088/1757-899X/952/1/012002

The reliability of the research is provided by the application of the fundamental laws of the mechanics of multiphase systems in conjunction with modern physical and mathematical models and numerical methods for solving problems. The results obtained are substantiated by verification of data with well-known commercial hydrodynamic simulators (Figure 3b).

6. Conclusions

1. The paper for the first time proposed an approach to discretization of the computational domain, based on the construction of a dynamic computational grid in a natural semi-fixed coordinate system.

2. A technique for thinning the shape of current streams as applied to filtering a two-phase mixture with the determination of RPT by the core digital research technology was developed and implemented as a computer program.

3. A new method for calculating quasi-three-dimensional non-stationary isothermal operating modes of a system of parallel horizontal wells has been developed.

4. Based on the proposed model, calculation method, and software implementation of computational algorithms, computational-parametric studies of various development modes with a system of parallel horizontal wells were performed using an example of a real hydrocarbon field.

The proposed calculation methods and their software implementations are the basis for further modeling of heat and mass transfer processes in order to theoretically and numerically study the processes that occur during thermal methods of developing high-viscosity oil fields.

References

[1] Barenblatt G N, Entov V M and Ryzhik V M 1984 *Movement of liquids and gases in natural palms* (Moscow: Nedra)

[2] Basniev K S, Kochina I N and Maksimov V M 1993 *Underground hydromechanics (*Moscow: Nedra)

[3] Nigmatulin R I 1987 *Dynamics of multiphase media (*Moscow: Science)

[4] Sirotkin Ya A 1963 Calculation of the axisymmetric vortex flow of an inviscid compressible fluid in radial turbomachines *News of the USSR Academy of Sciences. Mechanics and mechanical engineering* **3**

[5] Stepanov G Yu 1962 Hydrodynamics of lattices of turbomachines (Moscow)

[6] Stepanov S V 2016 *A set of computing technologies to improve the quality of modeling the development of oil and gas and oil fields*, Dissertation for the degree of Doctor of Technical Sciences, Tyumen.

[7] Shabarov A B 2013 *Hydrodynamics* (Tyumen: Tyumen State University Publishing House)

[8] Shabarov A B 2014 Physico-mathematical model and method for calculating the flow of a gas-condensate mixture in the reservoir *Bulletin of the Tyumen State University. Physics and mathematics* **7**

[9] Darcy N 1856 *Les fontaines publiques de la ville de Dijon: exposition et application des principes à suivre et des formules à employer dans les questions de distribution d'eau ....* (Paris: V. Dalmont)

[10] 2005 *Eclipse Reference guide*. Version 2005A_1

[11] Ivantsov N and Stepanov S 2012 Study of Sharp Drop in Production Rates of Horizontal Wells in Unconsolidated Rock of Heavy Oil Reservoir *SPE 157869, SPE Heavy Oil Conference, Canada*

[12] Zhizhimontov I, Stepanov S and Stepanov A 2019 Development of the Technology for Obtaining Relative Phase Permeabilities at Different Temperatures Based on Digital Stochastic Poro-Network Modeling *SPE 196994, SPE Russian Petroleum Technology Conference*

IOP Conf. Series: Materials Science and Engineering **952** (2020) 012003 doi:10.1088/1757-899X/952/1/012003

Calculation of the thermodynamic characteristics of an upward swirling gas flow under conditions of side wind

A G Obukhov

Tyumen Industrial University, 38, Volodarskogo Str., Tyumen, 625000, Russia

E-mail: agobukhov@inbox.ru

Abstract. The paper presents the results of numerical modeling of lateral wind effects on three-dimensional unsteady air flows in an upward swirling stream of an artificially created tornado after entering a stationary mode. Under the corresponding initial and boundary conditions, the solutions of the complete system of Navier-Stokes equations in the computational domain in the form of a rectangular parallelepiped are numerically obtained using the explicit difference scheme. At fixed points in time, the gas-dynamic characteristics of air flows are calculated under the action of gravity and Coriolis, as well as taking into account the constant horizontal directional wind speed. Numerical calculations showed that the resulting density, temperature, and pressure gradients lead to significant and asymmetric changes in air flows at different points in time and at different points in the computational domain. The total displacement of the vortex in the direction of the wind, the deformation and increase in the diameter of the vortex, which begin in the bottom part and pass into its vertical part, are established. In the considered range of time changes, the stability of the air vortex under study against wind action and the stable operation of the computational scheme used are noted.

1. Introduction

The application of numerical methods to the study of complex unsteady three-dimensional gas flows as a compressible viscous heat-conducting continuous medium poses interest for two reasons.

Firstly, despite a significant number of publications, a description of extremely complex gas flows in free atmospheric swirling flows and gas flows in pipelines, adequate to nature, cannot yet be considered complete, and especially when it comes to non-stationary three-dimensional flows. Therefore, from a scientific point of view, mathematical and numerical modeling can be considered as the most important and, perhaps, the only tool for studying this type of flows.

Secondly, the numerical simulation of such flows has a pronounced applied character, since it not only gives recommendations on the destruction or reduction of the destructive effects of natural atmospheric vortices [1], but also allows calculating all the gas-dynamic and energy characteristics of the flows for more efficient operation of gas pipelines.

The main reasons for the emergence and stable functioning of the ascending swirling air flow [2] are the long existence of the ascending air flow and the rotation of the Earth, which, through the action of the Coriolis force, gives significant peripheral speed to air particles in the bottom part of the ascending swirling flow. The fact of the occurrence of a twist and its direction is strictly mathematically proved at the level of the corresponding theorems [3, 4] and confirmed experimentally [5, 6].

Numerous observations of natural upward swirling flows allowed hypothesizing that in the middle-high vertical part of such flows there is a boundary separating the external resting air from air moving in an upward swirling flow. In gas dynamics, such a boundary is called a contact surface [7, 8]. Therefore, to create a stable upward swirling flow in laboratory conditions, an airtight vertical cylindrical pipe with an exhaust fan directing air through the pipe from the bottom up is used as the contact surface. Numerical calculations of such a gas flow made it possible to give concrete suggestions and recommendations on the possible conduct of a large-scale experiment on swirling large masses of air - an artificial tornado.

The numerical results obtained earlier [9–12] provide a good basis for further research in the gas-dynamic theory of destructive atmospheric vortices. In particular, it seems very interesting to study the effect of external wind action on the ascending swirling air flow. The aim of this work is to numerically simulate the functioning of an artificial tornado under conditions of side wind impact on it, as well as the calculation of its thermodynamic characteristics.

2. Mathematical model

As the mathematical model describing complex three-dimensional unsteady flows of a compressible viscous heat-conducting gas (such as atmospheric air), the complete system of Navier-Stokes equations was used, which in dimensionless variables—taking into account the action of gravity and Coriolis in vector form—has the following form [4]:

$$\begin{cases} \rho_t + \mathbf{V}\cdot\nabla\rho + \rho\,\mathrm{div}\,\mathbf{V} = 0, \\ \mathbf{V}_t + (\mathbf{V}\cdot\nabla)\mathbf{V} + \dfrac{T}{\gamma\rho}\nabla\rho + \dfrac{1}{\gamma}\nabla T = \mathbf{g} - 2\mathbf{\Omega}\times\mathbf{V} + \dfrac{\mu_0}{\rho}\left[\dfrac{1}{4}\nabla(\mathrm{div}\,\mathbf{V}) + \dfrac{3}{4}\Delta\mathbf{V}\right], \\ T_t + \mathbf{V}\cdot\nabla T + (\gamma-1)T\,\mathrm{div}\,\mathbf{V} = \dfrac{\kappa_0}{\rho}\Delta T + \dfrac{\mu_0\gamma(\gamma-1)}{2\rho}\Big\{\Big[(u_x - v_y)^2 + \\ + (u_x - w_z)^2 + (v_y - w_z)^2\Big] + \dfrac{3}{2}\Big[(u_y + v_x)^2 + (u_z + w_x)^2 + (v_z + w_y)^2\Big]\Big\}. \end{cases} \tag{1}$$

In system (1):
t is time;
x, y, z are Cartesian coordinates;
ρ is gas density;
$\mathbf{V} = (u, v, w)$ is gas velocity vector with projections on corresponding Cartesian axes;
T is gas temperature;
$\mathbf{g} = (0, 0, -g)$ is gravitational vector;
$\gamma = 1.4$ is polytropic indicator for air;
$-2\mathbf{\Omega}\times\mathbf{V} = (av - bw, -au, bu)$ is Coriolis force acceleration vector;
$a = 2\Omega\sin\psi,\ b = 2\Omega\cos\psi,\ \Omega = |\mathbf{\Omega}|;$
$\mathbf{\Omega}$ is Earth angular velocity vector;
ψ is the latitude of point O, the origin of the Cartesian coordinate system rotating with the Earth;
$\mu_0 = 0.001$, $\kappa_0 \approx 1.46\mu_0$ are constant values of dimensionless viscosity and thermal conductivity coefficients.

The computational domain (Figure 1) is a rectangular parallelepiped with side lengths $x^0 = 1$, $y^0 = 1$ and $z^0 = 0.04$ along axes Ox, Oy and Oz, respectively, which is filled with a three-dimensional grid of nodes of intersection of three families of planes $x = x_i$, $y = y_j$, $z = z_k$, where $x_i = i\cdot\Delta x$, $y_j = j\cdot\Delta y$, $z_k = k\cdot\Delta z$, $0 \le i \le L$, $0 \le j \le M$, $0 \le k \le N$. Difference steps in three spatial variables $\Delta x = x^0 / L$, $\Delta y = y^0 / M$, $\Delta z = z^0 / N$, $L = 200$, $M = 200$, $N = 20$.

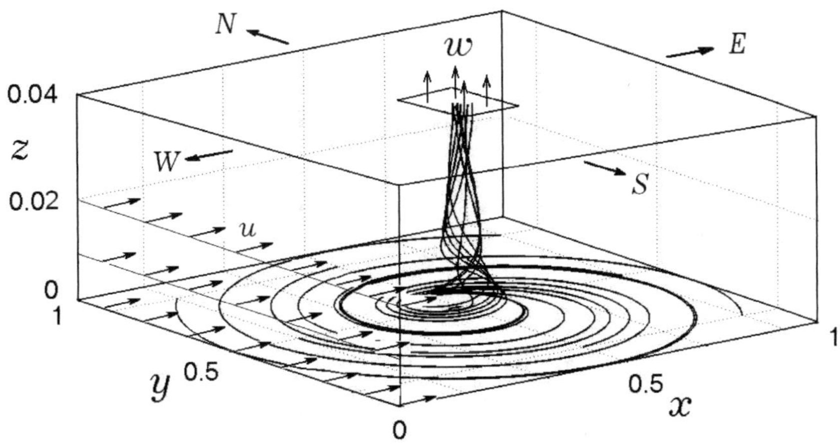

Figure 1. Calculation domain

The initial conditions in all internal nodes of the computational domain are the values of the five functions sought that characterize the air flow in an artificially created tornado at the time $t = t_s$ it enters the stationary mode of operation [12]:

$$u = u(x, y, z, t_s),\ v = v(x, y, z, t_s),\ w = w(x, y, z, t_s),\ T = T(x, y, z, t_s),\ \rho = \rho(x, y, z, t_s).$$

The boundary conditions on the faces of the computational domain are set as follows.

The density on the four side faces of the calculated parallelepiped $x = 0$, $x = x^0$, $y = 0$, $y = y^0$, is taken equal to the values from the stationary distribution of the density of atmospheric air [4]

$$\rho\big|_{x=0,\,x=x^0} = \rho_0(x, y, z), \qquad \rho\big|_{y=0,\,y=y^0} = \rho_0(x, y, z),$$

$$\rho_0(z) = (1 - kz)^{\nu-1},\quad k = \frac{lx_{00}}{T_{00}},\quad l = 0.0065\ \text{К/м},\quad x_{00} = 50\ \text{м},\quad T_{00} = 288^o\text{K},\quad \nu = \frac{\gamma g}{k}.$$

The density on the lower $z = 0$ and upper $z = z^0$ faces obeys the continuity of the flow. This means that the density values at the boundaries of the region are carried out by linear extrapolation along the normal to the given boundary surface from the inner part of the computational domain [4].

For the temperature on all lateral faces, values are set from the stationary distribution of atmospheric air temperature

$$T\big|_{x=0,\,x=x^0} = T_0(x, y, z), \qquad T\big|_{y=0,\,y=y^0} = T_0(x, y, z), \qquad T_0(z) = 1 - kz.$$

The temperature on the lower $z = 0$ and upper $z = z^0$ faces corresponds to the condition of symmetry. In this case, the temperature values are calculated from the condition that their derivative is normal to the given face.

On the upper and lower faces, the conditions of non-leakage are set. In this case, the third component of the velocity is equal to zero $w\big|_{z=0,\,z=z^0} = 0$, and the first and second components of the velocity vector are determined from the symmetry condition, that is, they are calculated from the condition that their derivative to normal of the given face is zero. In addition, vertical speed $w = 0.03$ is set through a square hole with the size of 0.1×0.1 which is in the center of the upper face and simulates vertical air blowing at a speed of 10 m/s through a pipe with a diameter of 5 m.

For the lateral faces of the computational domain, for all velocity vector components that are normal to the faces, continuity conditions are set, while two tangential velocity components are calculated by the condition of symmetry. Additionally, on the left (western) face $x = 0$, with

$0 < z < z^0 / 2$ the normal component of the velocity is set equal to the wind speed $u = 0.03$. Thus, a horizontally directed wind action is simulated on an ascending swirling stream of an artificially created tornado.

The calculation of the three-dimensional unsteady flow is carried out according to an explicit difference scheme by switching from the next n–th time layer to the next time layer $n+1$ with a constant given step Δt. The values of all the desired functions are calculated at all the internal points of the rectangular parallelepiped. After that, the values of the desired functions are determined at all internal points of each of the six faces and twelve edges.

The calculations were carried out with the following input parameters: scaled dimensional values of density, speed, distance, and time are respectively $\rho_{00} = 1.2928$ kg/m^3, $u_{00} = 333$ m/s, $x_{00} = 50$ m, $t_{00} = x_{00} / u_{00} = 0.15$ s. Difference steps in three spatial variables $\Delta x = \Delta y = 0.005$, $\Delta z = 0.002$, while time step $\Delta t = 0.001$.

3. Calculation results

Figures 2-3 show graphs of the density function $\rho(x, y)$ for a height $z = 0.2$ m and two fixed points in time $t_1 = 1$ s, $t_2 = 10$ s. Along axes Ox and Oy are the numbers of nodes of the computational grid, along axis Oz is the gas density in dimensionless quantities.

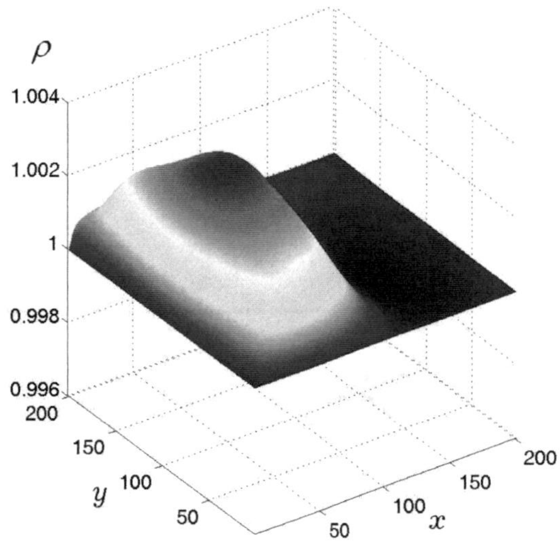

Figure 2. Density for $t_1 = 1$ s

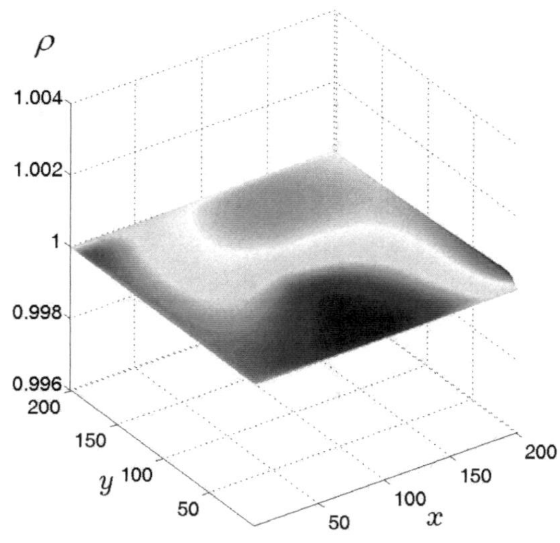

Figure 3. Density for $t_2 = 10$ s

The behavior of the gas density at the indicated height, as follows from the calculations, can be characterized as follows. By the estimated time $t_1 = 1$ s, the density in the western part of the computational domain increases to 1.004 (1.29797 kg/m^3 in dimensional values), while in the eastern part the region of reduced density is characteristic of the stationary state of the ascending swirl flow (Figure 2). Then, with a decrease in its values, the region of increased density propagates along the peripheral regions, covering the center of the vortex. Gradually, it begins to deform with a simultaneous decrease in its size, giving way to an increasing region of lower values (Figure 3).

Figures 4-5 show the temperature distribution as graphs of the functions of two variables at the same height and at the same determined points in time.

IOP Conf. Series: Materials Science and Engineering **952** (2020) 012003 doi:10.1088/1757-899X/952/1/012003

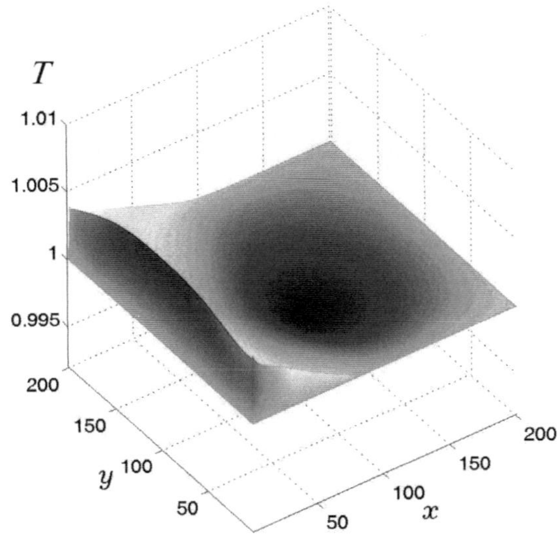

Figure 4. Temperature for $t_1 = 1$ s

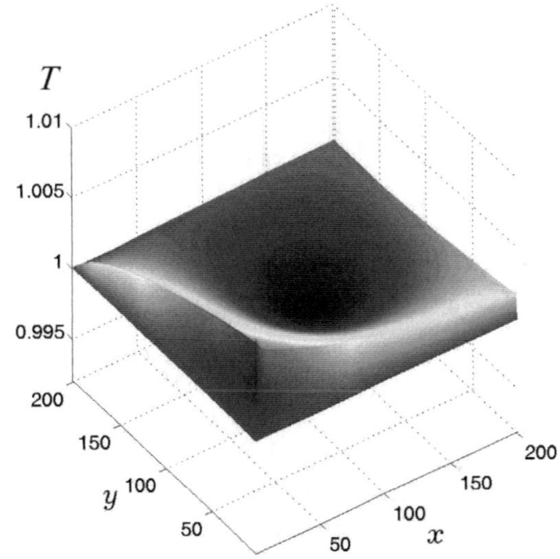

Figure 5. Temperature for $t_2 = 10$ s

At the beginning of the wind action (Figure 4), the temperature on the left (western) face of the computational domain is 1.005 ($289.44^{\prime}$ K in dimensional values). At subsequent time instants, the elevated temperatures begin to cover the funnel-shaped central region of the lower temperature, where the bulk of the vortex is located. Subsequently, elevated temperatures are concentrated near the western $x = 0$ and southern $y = 0$ faces of the computational domain (Figure 5). It should be noted that the excess of temperature over the scale value for all calculated moments of time is about $1.5^{\prime}$ K. Nevertheless, the visualization program allows recording such a temperature difference and its dynamics.

Figures 6-7 show the calculated pressure distributions as graphs of the functions of two variables corresponding to the same height and the same fixed points in time. The behavior of pressure is similar to the behavior of density and temperature.

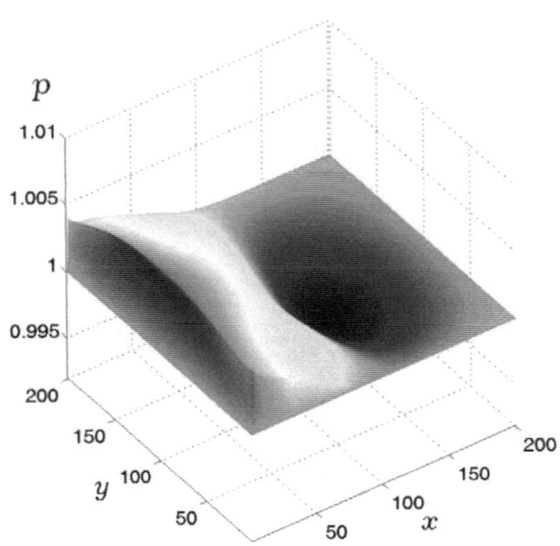

Figure 6. Pressure for $t_1 = 1$ s

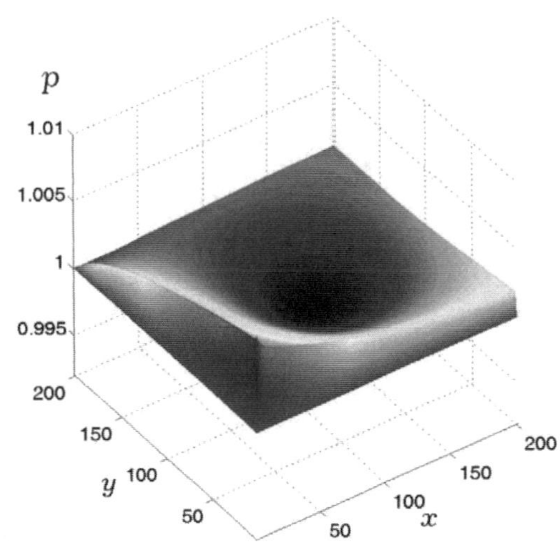

Figure 7. Pressure for $t_2 = 10$ s

Calculations showed that the resulting density, temperature, and pressure gradients lead to significant and asymmetric changes in air flows at different points in time and at different points in the computational domain. There is a general displacement of the vortex and its deformation in the direction of the wind. In the considered range of time changes, the stability of the air vortex under study against wind action and the stable operation of the computational scheme used are noted.

4. Conclusions

A numerical simulation of the lateral wind effect on three-dimensional unsteady air flows in an upward swirling stream of an artificially created tornado in a stationary mode of operation was performed. Using an explicit difference scheme, numerically obtained solutions of the complete system of Navier-Stokes equations in the computational domain in the form of a rectangular parallelepiped. The thermodynamic characteristics of currents are calculated at fixed times under the action of gravity and Coriolis, as well as when constant horizontal wind speed is taken into account.

Numerical calculations showed that the resulting density, temperature, and pressure gradients lead to corresponding changes in air flows at different times and at different points in the computational domain. The displacement of the vortex in the direction of the wind, as well as its uneven deformation, is established.

Refereneces

[1] 2011 Russian Federation: Decree of the President of the Russian Federation *No. 899 On the approval of priority areas for the development of science, technology and technology in the Russian Federation and the list of critical technologies*

[2] Bautin S P 2008 *Tornado and Coriolis Strength* (Novosibirsk: Nauka)

[3] Bautin S P 2009 *The characteristic Cauchy problem and its applications in gas dynamics* (Novosibirsk: Nauka)

[4] Bautin S P, Deryabin S L, Krutova I Yu, and Obukhov A G 2017 *Destructive atmospheric vortices and the rotation of the Earth around its axis* (Yekaterinburg: UrGUPS)

[5] Bautin K V, Bautin S P and Makarov V N 2013 Experimental confirmation of the possibility of creating an air flow swirling by the Coriolis force *Bulletin of Ural State Transport University* **2 (18)** 27–33

[6] Bautin S P and Makarov V V 2016 Creating a stream of air swirling by the Coriolis force using a two-meter diameter pipe *Bulletin of Ural State Transport University* **4 (32)** 39–45

[7] Ovsyannikov L V 2003 *Lectures on the basics of gas dynamics* (M.; Izhevsk: Institute of Computer Research)

[8] Loytsyansky L G 1970 *Mechanics of fluid and gas* (M.: Nauka)

[9] Bautin S P and Obukhov A G 2013 Mathematical simulation of the near-bottom section of an ascending twisting flow *High Temperature* **51 (4)** 509-512

[10] Volkov R E and Obukhov A G 2016 Parallel calculations in studies of the dependence of the gas-dynamic parameters of the ascending swirling gas flow on the blowing rate *News of Higher Education Institutions. Oil and gas* **1** 92–97

[11] Volkov R E and Obukhov A G 2016 A method of parallelizing an algorithm for numerically solving a complete system of Navier-Stokes equations *Bulletin of higher education. Oil and gas* **2** 92–98

[12] Volkov R E and Obukhov A G 2016 The study of the dependence on the blowing rate of the parameters of the ascending swirl flow B*ulletin of the Tyumen State University. Physical and mathematical modeling. Oil, gas, energy* **2 (3)** 24–34

IOP Conf. Series: Materials Science and Engineering **952** (2020) 012004 doi:10.1088/1757-899X/952/1/012004

Risk management of oil and gas transportation facilities based on reliability indicators monitoring

M Yu Zemenkova, Yu D Zemenkov, A A Gladenko and E L Chizhevskaya

Tyumen Industrial University, 38, Volodarskogo Str., Tyumen, 625000, Russia

E-mail: muzemenkova@mail.ru

Abstract. The paper considers the methodological aspects of the analysis of technical risk based on real data on the reliability of objects. Today, there is a problem of assessing the probability of technological events of various types based on statistical data of events at various facilities; however, the current state of automatic control systems for production processes allows preventing an accident, transferring it to the category of incidents. In this case, there is a real distortion of information on statistics of dangerous situations. In modern control systems, information for forecasting and preventing accidents is the most valuable. The paper presents a systematic approach to the creation of new generation technical risk management systems.

1. Introduction

Since 2021, enterprises of the Russian Federation have been expecting changes in the field of environmental policy. “Since 2019, 300 industrial enterprises that have a significant negative impact on the environment should switch to environmentally friendly best available technologies, and from 2021 all enterprises with a high-risk category for the environment should do this,” - V. Putin [1].

On April 20, 2017, President of the Russian Federation V. Putin signed a decree on the Strategy for Ecological Safety of Russia for the period until 2025 [2,4]. The document contains an assessment of the current state of environmental safety, a description of the challenges and threats to environmental safety. The Strategy outlines the main goals and objectives of state policy in this area, assessment and monitoring mechanisms. A strategic planning document in the field of ensuring national security of the Russian Federation that defines the main challenges and threats to environmental safety, goals, objectives and mechanisms for implementing state policy in the field of environmental safety.

The new Environmental Strategy noted that threats to environmental safety remain despite strict policies at enterprises and measures taken to reduce environmental impact levels of chemical, physical, biological and other factors, to prevent natural and man-made emergencies, including emergency situations on dangerous production facilities to adapt industries to adverse climate change.

Effective until 2020, "The concept of improving state policy in the field of industrial safety until 2020" [11], approved by a decision of the Board of the Federal Service for Ecological, Technological and Nuclear Supervision in 2011, determines the necessary condition for differentiating industrial safety measures depending on the degree of risk of accidents and the scale of their consequences is the formation of organizational mechanisms for classifying hazardous production facilities.

IOP Conf. Series: Materials Science and Engineering **952** (2020) 012004 doi:10.1088/1757-899X/952/1/012004

2. Materials and methods

The urgency of the problem of forecasting undesirable events at oil and gas transport facilities is confirmed by systemic accidents and incidents at facilities. Analysis of statistical data indicates the need to create a new generation risk management system based on a modern methodological and instrumental base.

In 2016-2019, significant changes took place in the regulation of industrial safety issues in the oil and gas transportation industry. Modern methods of analysis and reduction of environmental and technological risks are numerous, depending on the type of facilities and many factors, capacious, however, most of them are focused on the use of statistical data on accidents that have occurred. Along with this, along with the concept of "accident", there is the concept of "incident", which is most often encountered in practice.

According to Rostekhnadzor, the statistics of incidents significantly differs from the statistics of accidents, from which the concept of "background" risk acquired special significance (blue, Figs. 1, 2).

The risk of R_A accidents (red, Figures 1, 2) is determined by the following formula.

$$R_A = N/K*1000,$$

N is the number of accidents per year, unit; K is the number of hazardous production facilities

Therefore, the methods and technologies for preventing accidents and monitoring precisely the current technical risks in the industry are becoming especially relevant.

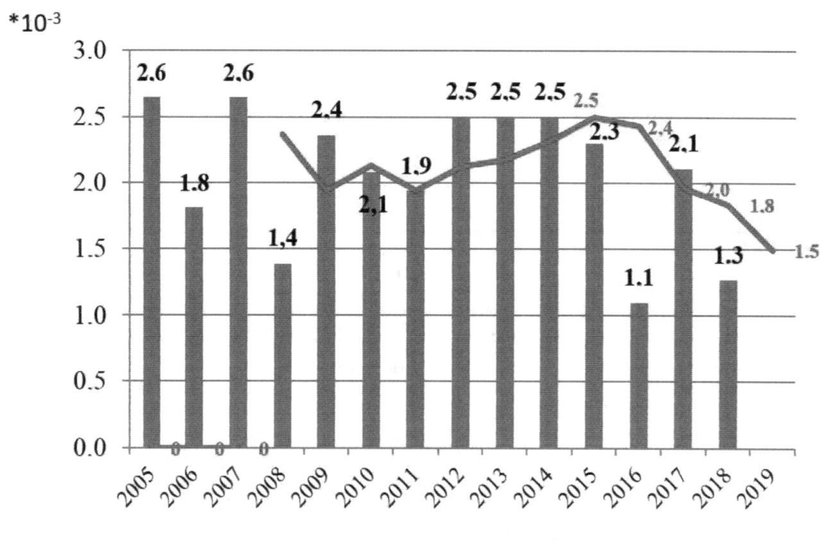

Figure 1. Dynamics of background risks at the facilities of the main pipeline transport of hydrocarbons (318 incidents per 1 accident) [3]

A complex of intellectual technological management of reliability and efficiency of oil and gas systems (CI-TREMS), a fundamentally new technology of intelligent neural network engineering control, forecasting and prevention of emergency situations, incidents, accidents, has been developed at the Department of Hydrocarbon Resources Transport. , optimization and ensuring the effectiveness of technical solutions in the process control of industrial enterprises.

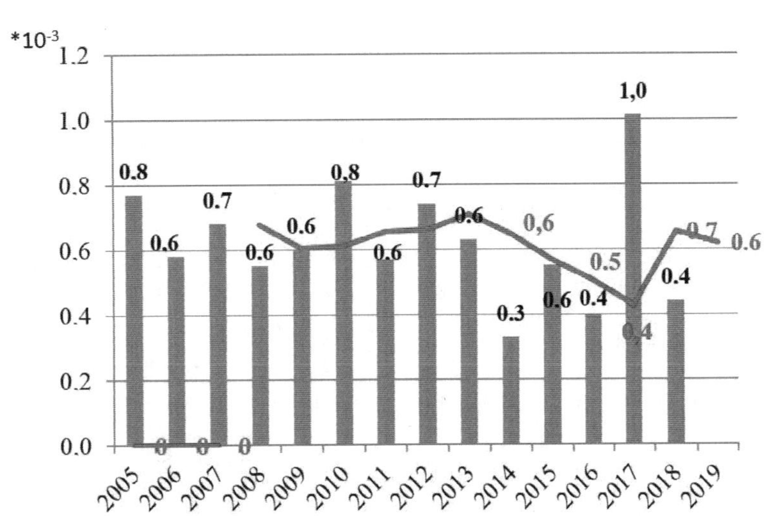

Figure 2. Dynamics of background risks at gas supply facilities (22 incidents per accident) [3]

As a background risk of accidents at hazardous production facilities of the oil and gas complex, it is recommended to use Rostechnadzor data on accidents and injuries expressed in relative values of the risk level of deaths at hazardous production facilities, as well as in estimates of the average expected material damage and the number of accidents in various sectors of the oil and gas complex .

According to [11], for an indicator of the risk of an accident at a hazardous production facility in the oil and gas complex, reliable estimates of the risk of accidents and incidents at the hazardous production facility under consideration or any of the following values of background hazards (background risk of an accident at a hazardous production facility of the oil and gas complex) must be determined:

- ✓ accidents and (or) incidents at a hazardous production facility for which a safety justification is being elaborated;
- ✓ accidents at hazardous production facilities of the oil and gas complex;
- ✓ loss of life in industrial accidents.

Based on modern requirements, for the high-quality monitoring of accident risk indicators at a hazardous production facility in the oil and gas complex and reliable assessment of accident and incident risk values, modern information processing systems based on the processing of online information and databases should be used. Such a system is developed individually and for its creation the authors have methodologically developed the basics. It requires separate functions of the operator-mechanic and operator-technologist for safety, understanding the features of the process.

To create a system in TIU, methodological support was developed using the apparatus of probability theory and statistics, graph models, neural network technologies and machine learning, which allows assessing indicators of the technical condition, reliability and risk analysis of objects of the oil and gas transport system[6-15]. Some features of the device are shown in Table 1 described in more detail in [5].

The current level of development and scale of transport and storage facilities as dangerous opportunities require a qualitative analysis of reliability and technical risks, the possibility of rapid response, therefore, the system should be accessible and provide a qualitative forecast of the state of the systems.

Table 1. Elements of methodological apparatus for monitoring technical risks based on reliability analysis

Methods and their features	Functions	Assessment Features	Tools	Note, novelty
Assessment of technical condition indicators	Elimination of gross system uptime threats	Estimated according to the regulations in the operating mode	Visual inspection, diagnostics	Coefficient method
Statistical analysis of accidents and incidents	Analysis of accidents and incidents and their causes	Based on registered data, quality of databases determines assessment results	Mathematical processing of forms or electronic databases	Must be performed in real time
Assessment techniques using graph models and probabilistic-statistical analysis	Analysis of accident scenarios taking into account current data, assessment of transition probabilities	Using variable transition intensities based on real reliability data	Online incident and accident logging	Variable event intensity
Neural network technology and machine learning	Rapid identification of states and their prediction and prevention	Based on operational information; analysis of a large amount of data using artificial intelligence, identification of states; real-time multivariate analysis	CPCS data arrays	Real-time multivariate analysis
Residual life estimation	Long-term life assessment	Long-term limit state forecast	Diagnostic Tools	Take into account real-time changes
Structural models of reliability analysis	Assessment of potential threats to elements in complex systems, development of re-equipment solutions	Identification of elements of "weak links", the goal is to identify the element	Analysis of technological schemes	Real-time online analysis of mnemonic diagrams, elimination of threats
Factor models for reliability analysis	Assessment of potential operational hazards in complex systems	Identification of the causes of potential threats, the goal is to identify the causes	Analysis of operational information, databases	Real-time online parametric analysis of mnemonic diagrams, elimination of threats
Economic models for assessing damage from consequences of accidents and incidents	Assessment of potential threats in material terms	Material risks, damage are assessed		

IOP Conf. Series: Materials Science and Engineering **952** (2020) 012004 doi:10.1088/1757-899X/952/1/012004

The process of creating and implementing the developed methodology and methodologies in process control systems is very time-consuming and is associated with the following steps:

1) analysis of the technological scheme of the facility, parameters and characteristics, determination of production features of the project and control of parameters and modes;

2) the creation of a diagnostic architecture to assess the technical condition, reliability and safety of the facility and the reliability of technological processes;

3) the justification of the methodological support of the predictive control system;

4) software implementation of models and the creation of a software package;

5) visualization for the perception of information by operational personnel;

6) testing and updating the system at a specific production facility.

As an example, Figures 3, 4 shows a fragment of visualization of a dangerous element of the technical system of the tank farm and pumping station.

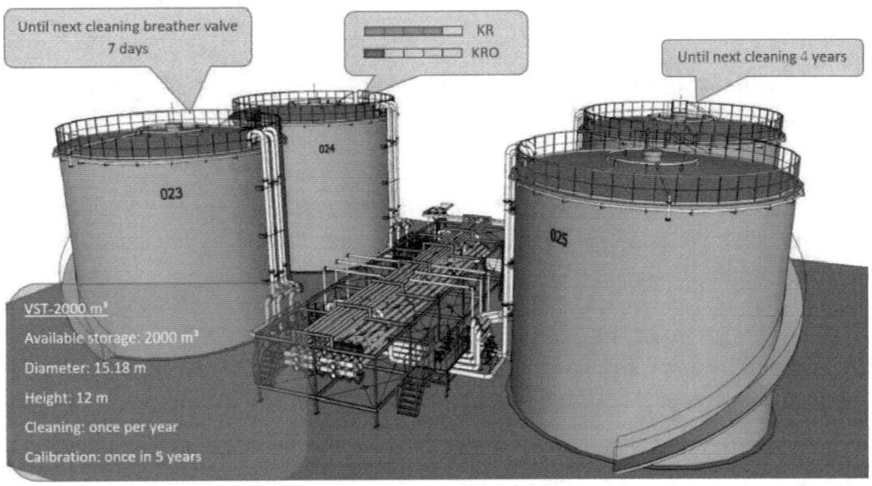

Figure 3. Simulation and visualization of threats using tank farm equipment as an example

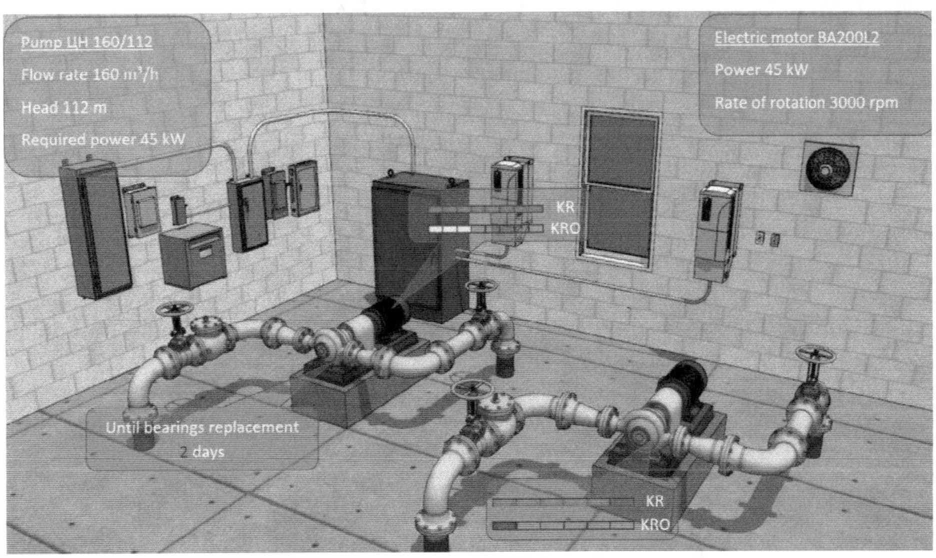

Figure 4. Simulation and visualization of threats on the example of pumping equipment

IOP Conf. Series: Materials Science and Engineering **952** (2020) 012004 doi:10.1088/1757-899X/952/1/012004

3. Conclusion

Using mathematical software systems, a high-quality methodological apparatus, an automated process control system as a database enables on-line analysis of technical risks and allows security specialists and technologists and operators to respond to possible threats and prevent incidents.

References

[1] Russian enterprises will switch to environmentally friendly technologies 2019 Retrieved: https://news.ecoindustry.ru/2018/03/s-2019-goda/

[2] Decree of the President of the Russian Federation 2017 *Environmental Security Strategy of the Russian Federation until 2025* of April 19, No. 176

[3] Federal Service for Ecological, Technological and Nuclear Supervision, available at: http://www.gosnadzor.ru/industrial/oil/lessons/

[4] The concept of improving state policy in the field of industrial safety until 2020, available at: http://docs.cntd.ru/document/902354089.

[5] Zemenkova M Yu 2019 *Methods to reduce technological and environmental risks in the transport and storage of hydrocarbons* (Tyumen: TIU)

[6] Khaikin S 2006 Neural networks: a full course (Moscow: Williams)

[7] Safety Guide "Guidelines for the classification of industrial events in the field of industrial safety at hazardous production facilities of the oil and gas complex" Approved. by order of the Federal Service for Ecological, Technological and Nuclear Supervision of January 24, 2018 N 29 [Electronic resource]. - Access mode: http://sudact.ru/law/prikaz-rostekhnadzora-ot-24012018-n-29-ob/rukovodstvo-po-bezopasnosti-metodicheskie-rekomendatsii/ (accessed: 1.04.2018)

[8] Zemenkova M, Zemenkov Yu, Gladenko A and Podorozhnikov S 2016 Estimation of Emissions During Monitoring of Pipelines in the Dynamic Mode of Operation *MATEC Web Conf.* DOI: 10.1051/matecconf/20168604053

[9] Dudin S, Bahmat G, Zemenkov Yu, Voronin K and Shipovalov A 2017 Strategy for monitoring and ensuring safe operation of Russian gas transportation systems *MATEC Web Conf.* DOI: 10.1051/matecconf/201710606004

[10] Dudin S M, Zemenkov Yu D, Maier A V and Shabarov A B 2016 Research and Design of Thermophysical Gas-Liquid Mixture Parameters in Product Pipelines *IOP Conference Series: Materials Science and Engineering* DOI: 10.1088 / 1757-899X / 154/1/012021

[11] Order of the Federal Service for Ecological, Technological and Nuclear Supervision 2016 No. 349 *Safety Guide "Methodology for establishing the acceptable risk of an accident when substantiating the safety of hazardous production facilities in the oil and gas complex"*

[12] Order of Rostekhnadzor 2016 *Safety Guide "Methodological foundations for hazard analysis and risk assessment of accidents at hazardous production facilities* No. 144

[13] *Stockage et reprise d'huiles légéres*, available at: https://3dwarehouse.sketchup.com/model/97379a6c-f4fa-4cca-942f-0b5c936c2e97/Stockage-et-reprise-dhuiles-l%C3%A9g%C3%A9res

[14] *Pump Station*, available at: https://3dwarehouse.sketchup.com/model/f44232ef-4732-409a-b34b-fa4ca4e37503/Pump-Station

[15] Kulikov A M and Chekardovskiy S M 2018 Improving the safety of operation of tanks with a floating roof in the winter period *IOP Conference Series: Materials Science and Engineering* **445 (1)** 012010 DOI: 10.1088 / 1757-899X / 445/1/012010

IOP Conf. Series: Materials Science and Engineering **952** (2020) 012005 doi:10.1088/1757-899X/952/1/012005

Technology for developing expert modules and DSS comprehensive platform for reliability and risks monitoring in the Oil and Gas Engeneering

M Yu Zemenkova, Yu D Zemenkov, V V Golik and S Belsky

Tyumen Industrial University, 38, Volodarskogo str., Tyumen, 625000, Russia

E-mail: muzemenkova@mail.ru

Abstract. The paper describes the proposed technology for the development of expert modules and DSS for integrated reliability and risk control systems for oil and gas industry, as part of a complex of intelligent technological management of reliability and efficiency of oil and gas systems developed at TIU. All stages of the development of system modules for controlling technical risks, the features of the structure of the decision-making model for DSS, and the components of a static and dynamic expert system for transport and storage of hydrocarbons are considered. The features of creating decision support systems are described, their main functions and decision criteria and their main provisions. As an example, one of the modules of intelligent neural network engineering control for predicting halos of thawing soil in the vicinity of the pipeline and passing communications, and buildings is shown, a model for visualizing processing results is proposed.

1. Introduction

In order to comply with modern environmental safety requirements, for high-quality monitoring of accident risk indicators at a hazardous production facility in the oil and gas complex and for reliable assessment of accident and incident risk values, modern information processing systems based on the processing of online information and databases should be used[1-4].

The complexity of the processes at large-scale objects of transport and storage of hydrocarbons requires the use of modular technologies.

A collaborative intellectual technological reliability and efficiency management of oil and gas systems (CI-TREMS) is developed at the Department of Hydrocarbon Resources Transport, a fundamentally new technology of intelligent neural network engineering control, forecasting and prevention of emergency situations, incidents, accidents, optimization and ensuring the effectiveness of technical solutions in the process control of industrial enterprises.

Consider some aspects related to the proposed technology for the development of the system and its modules. In the development, methods of system analysis dealing with decision-making problems in the analysis of a large amount of information of various nature were used.

2. Materials and methods

The development of an expert system module or a decision-making support system as part of a complex of intelligent technological management of the reliability and efficiency of oil and gas systems includes the following steps:

1. Development of a source database: a list of initial parameters, requirements for data accuracy, a minimum quantity, areas of definition of arguments and functions are determined;
2. Development of tools for collecting the database: modern technologies, instruments, tools, etc. are used, manufacturers and technologies for collecting source data are defined;
3. A system analysis of the object is carried out in the context of the goal of the ES (DSS): there can be various morphological, factorial, structural analysis, analysis of connections, the image of connections in graphic, logical and mathematical forms, neural;
4. Mathematical models, methods of data processing and analysis are substantiated: formulas, dependencies and mathematical methods by which assessment and forecasting are performed;
5. Methods of data analysis, classification, graphs, neural networks, interpretation rules, etc.;
6. The substantiation and calculation of the target criterion (minimum 1) is carried out (based on which a decision is made). Description of possible options and decision-making criteria and methods for their adjustment, decision-making criteria and decision-making technology, description of the method used;
7. Selection and justification of the optimization model and optimization method*; methods: programming, deterministic, random (stochastic), combined, one-dimensional or multidimensional, etc.; optimization model calculation example;
8. Development of the main algorithm of the expert system in the form of block diagrams: 1) algorithm for mathematical calculation; 2) decision making algorithm;
9. Development of software for ES (DSS): screen design; development of a software product that implements a computer model of the system or automates the applied algorithms (calculations, analysis); Python, Excel, Matlab, Matcad, Sharp, Windows forms, and other programs which products can be compatible with operational dispatch control systems; listings, screenshots, description of the program in stages, user manual

Decision making in a complex system is carried out by a technical tool or by a person and is based on a comparison and assessment of options for action. The setting of both the main and particular goals in the system is also usually subject to analysis and research. And again, the main procedure in this case is the decision.

As known, the study of ***decision-making*** procedures and the system organization associated with this is an urgent problem of creating and operating complex systems. We emphasize that all this can be carried out on the basis of specially developed methods, techniques, standard models for organizing the system and making decisions.

Under ***management***, in its most general form, we understand the process of forming the targeted behavior of the system through the information impact produced by a person or device when implemented at a hazardous production facility—an algorithm of action in an emergency or in the face of an incident threat.

The ***decision-making model*** is most often depicted in the form of a circuit with cells, connections between cells and logical transitions. Cells contain specific actions, procedures, which can be very diverse. A joint study of the procedures and their organization follows from the fact that without taking into account the content and features of the cells, creating schemes is impossible. These schemes determine the decision-making strategy in a complex system. It is the study of a coherent set of basic procedures which one should begin solving a specific applied problem with.

The use of a separate software tool is an elementary system analysis procedure. ***As a result, it is the person—the "decision maker" (DM)***—using computer technology who gives an indication of the use of the research results in practice. The functions of the decision maker, the scope of his actions, the separation or not separation from the researchers of the system are problems that in general also apply to the field of system analysis. A variety of counting, controlling, storing, transforming, advising, depicting and other computing techniques is both an integral part of the complex systems themselves, and research.

The specifics of applications of expert systems in comparison with other artificial intelligence systems is as follows. Expert systems are used to solve only difficult practical problems. The expert systems are not inferior to the decisions of a human expert in the quality and effectiveness of solutions. Expert systems solutions have "transparency", i.e. can be explained to the user at a qualitative level. This quality of expert systems is ensured by their ability to reason about their knowledge and conclusions. Expert systems are able to replenish their knowledge in the course of interaction with an expert. It should be noted that at present, the technology of expert systems is used to solve various types of problems (interpretation, prediction, diagnostics, planning, design, control, debugging, instruction, management) in a wide variety of problem areas, such as finance, the oil and gas industry, energy, transport, space, metallurgy, mining, telecommunications, etc.

An ***expert system*** is a software tool that uses expert knowledge to provide highly efficient solutions to informal tasks in a narrow subject area. The basis of ES is a knowledge base (KB) about the subject area, which accumulates in the process of construction and operation of ES. The accumulation and organization of knowledge is the most important property of all ES.

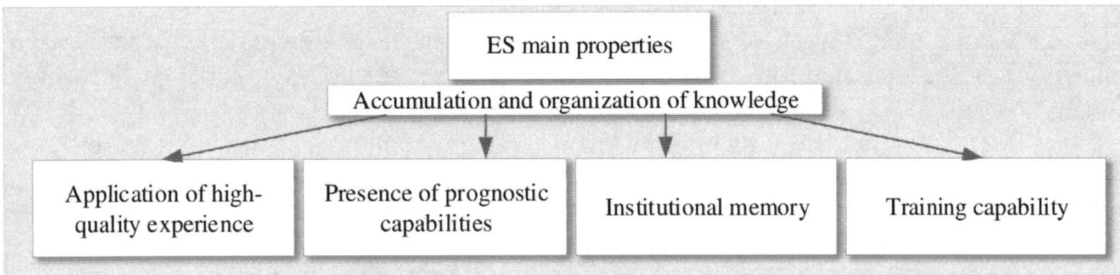

Figure 1. Properties of an expert system

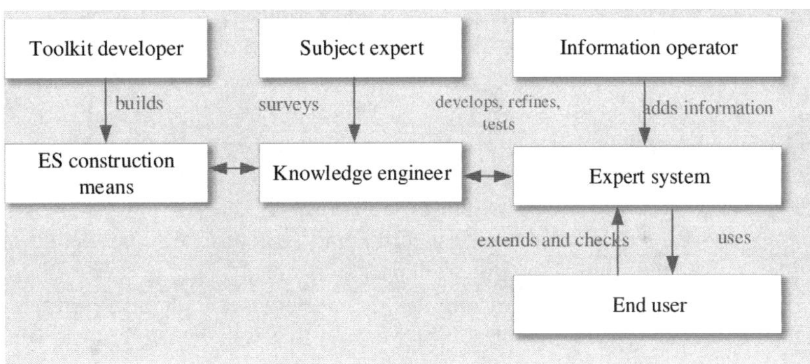

Figure 2. ES functioning structure

A typical static ES consists of the following main components:

- ✓ solver (interpreter);
- ✓ working memory (WM), also called a database (DB);
- ✓ knowledge base (KB);
- ✓ knowledge acquisition components;
- ✓ explanatory component;
- ✓ dialog component.

The distinguished data that the program works with about technical features and operating conditions, any operational and design information is called a knowledge base, while general knowledge about finding solutions to problems is called an output mechanism.

The *knowledge base (KB)* in ES is designed to store long-term data describing the area in question (and not current data), and rules describing the appropriate data transformation of this area. Operational, diagnostic information can be used. The *database (working memory)* is intended for storing the source and intermediate data of the problem currently being solved. This term coincides in name but not in meaning with the term used in information retrieval systems (IRS) and database management systems (DBMS) to refer to all data (primarily long-term) stored in the system.

The *solver*, using the source data from the working memory and knowledge from the knowledge base, forms a sequence of rules that, when applied to the source data, lead to the solution of the problem. The *knowledge acquisition component automates the process of ES filling* with knowledge, carried out by an expert user. The *explanatory component* explains how the system received the solution to the problem (or why it did not receive the solution) and what knowledge it used to make it easier for the expert to test the system and increase the user's confidence in the result. The knowledge base contains facts (data) and rules (or other representations of knowledge) that use these facts as the basis for decision-making.

The output mechanism contains:

✓ an interpreter defining how to apply the rules for deriving new knowledge based on information stored in the knowledge base;

✓ a dispatcher setting the procedure for applying these rules.

Such ESs are called static ESs and have a structure depicted in Figure 2; they are used in those applications where you may not take into account changes in the environment when solving the problem.

The *dialog component* is focused on organizing friendly communication with the user both in the course of solving problems and in the process of acquiring knowledge and explaining the results of work.

Representatives of the following specialties participate in the development of ES:

✓ *expert* in a problem area, the tasks of which will be solved by the ES;

✓ *knowledge engineer is a specialist in ES development* (the technology *used by them* and methods are called knowledge engineering technology (methods));

✓ *programmer* for the development of tools designed to accelerate the development of ES.

It should be noted that the absence of knowledge engineers among the participants in the development (i.e., their replacement by programmers) either leads to failure in the process of creating an ES, or significantly lengthens it.

The expert determines the knowledge (data and rules) that characterize the problem area, ensures the completeness and accuracy of the knowledge entered into ES.

The knowledge engineer helps the expert identify and structure the knowledge necessary for the operation of ES; makes the choice of the toolkit that is most suitable for this problem area, and determines the way of representing knowledge in this toolkit; isolates and programs (by traditional means) standard functions (typical for this problem area) that will be used in the rules introduced by the expert.

The programmer develops the toolkit (if the toolkit is developed anew), containing in the limit all the main components of the ES, and carries out its interfacing with the environment in which it will be used.

The expert system operates in two modes: the mode of acquiring knowledge and the mode of solving the problem (also called the consultation mode or the ES usage mode).

In the mode of knowledge acquisition, communication with ES is carried out (through the knowledge engineer) by an expert. In this mode, the expert, using the knowledge acquisition component, fills the system with knowledge that allows the ES in the solution mode to independently (without an expert) solve problems from the problem area. The expert describes the problem area as a combination of data and rules.

A higher class of applications where it is required to take into account the dynamics of changes in the surrounding world during the execution of applications is called dynamic ES and their generalized structure will look like in Figure 3.

Compared with a static ES, two more components are introduced into a dynamic ES:

✓ subsystem of modeling the external world (external operating conditions of objects);

✓ subsystem of interfacing with the outside world (operating features related to the working conditions of objects).

Dynamic ES communicates with the outside world through a system of controllers and sensors. In addition, the components of the knowledge base and the output mechanism are substantially changed to reflect the temporal logic of events taking place in the real world.

There are platforms that are an object-oriented environment for building and maintaining expert real-time systems designed to monitor, diagnose, optimize, plan and manage a dynamic process.

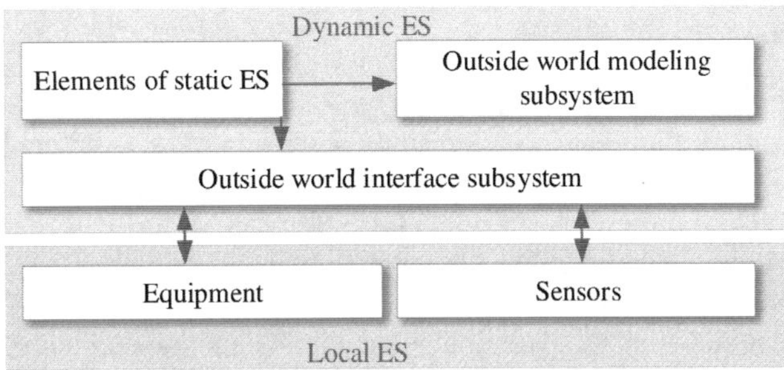

Figure 3. Dynamic ES

Dynamic ESs communicate with the outside world through a system of controllers and sensors. In addition, the components of the knowledge base and the output mechanism are substantially changed to reflect the temporal logic of events taking place in the real world.

Expert systems are designed for high-quality solution of problems in a field defined by developers, in rare cases, in several fields.

Decision Support System (DSS) is a computer-based automated system that, through the collection and analysis of a large amount of information, can influence the decision-making process of an organizational plan, the purpose of which is to help decision-makers in difficult conditions for a complete and objective analysis of subject activity.

Interactive systems allow managers to obtain useful information from primary sources, analyze it, and identify existing business models for solving specific problems.

According to user interaction, three types of DSSs are distinguished:

✓ passive ones help in the decision-making process, but may not put forward a specific proposal;

✓ active ones are directly involved in developing the right solution;

✓ cooperative ones involve interaction of the DSS with the user.

The proposal put forward by the system, can be modified, improved, and then send back to the system for verification by the user.

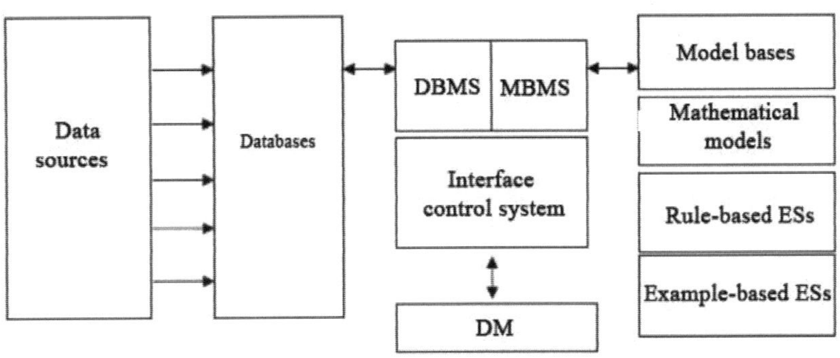

Figure 4. Typical DSS structure

After that, the proposal is again presented to the user, and so on until he approves the decision.

By the method of support, they distinguish:

✓ model-oriented DSSs that use access to statistical or other models in their work;

✓ DSSs based on communications that support the work of two or more users engaged in a common task;

✓ Data-driven DSSs that have access to the organization's time series.

They use in their work not only internal, but also external data; document-oriented DSS manipulate unstructured information contained in various electronic formats; knowledge-oriented DSSs provide specialized factual solutions to problems.

The decision-making criterion is a function that expresses the preferences of the decision-maker (DM) and determines the rule by which an acceptable or optimal solution is chosen.

Based on the results of data analysis, decisions are made. The following acceptance criteria are most often chosen: Sevage, Hurwitz criterion, Hodge-Limon criterion, Germeyer criterion; criteria of accordance to the decision: minimax criterion, Bayes-Laplace criterion, criterion with any evaluative information, the choice of which should be performed, the criterion of products, the minimax composite Bayes-Laplace criterion.

These criteria can be used in turn, and after calculating their values, one has to arbitrarily select some final solution among several options. This allows, firstly, obtaining better insight into all the internal connections of decision-making problems and, secondly, weakening the influence of the subjective factor.

Depending on the degree of unknownness of the upcoming behavior of the initial decision-making parameters, risk conditions are distinguished, in which the probability of occurrence of individual events that affect the final result can be established with some accuracy, and uncertainty conditions, in which, due to the lack of necessary information, such probability cannot be established.

The theory of decision-making under conditions of risk and uncertainty is based on the following starting points:

✓ the decision-making object is clearly determined and the main possible risk factors are known from it.

✓ the indicator that best describes the effectiveness of this decision is selected for the decision-making object.

✓ the indicator characterizing the level of its risk was selected for the decision-making object.

✓ there are a finite number of decision alternatives

✓ there are a finite number of situations in which an event develops under the influence of changes in risk factors. The number of such situations in the decision-making process should be determined in the range from extremely favorable (the most optimistic situation) to extremely unfavorable (the most pessimistic situation).

✓ for each combination of decision-making alternatives and events, a final indicator of the effectiveness of the decision can be determined (the specific value of the amount of net profit, net present value, etc., corresponding to this combination).

✓ for each of the situations under consideration, it is possible or impossible to assess the probability of its implementation. The possibility of making a probability assessment divides the entire system of risk decisions into the previously considered conditions for their justification ("risk conditions" or "uncertainty conditions").

✓ the decision is made according to the best of the considered alternatives.

Figure 5. TPS software interface example

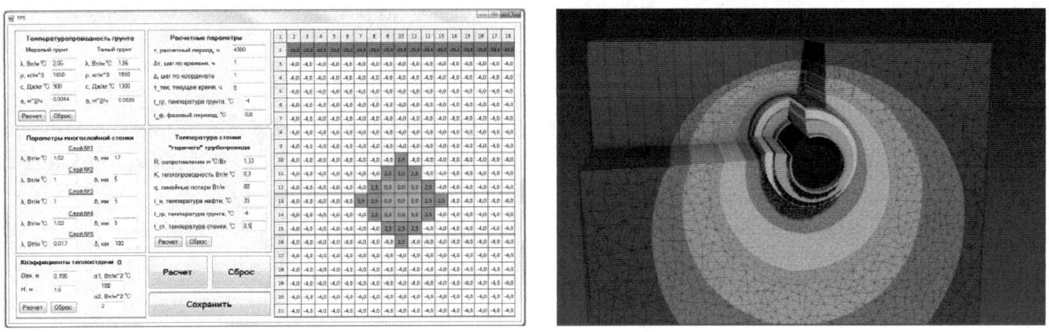

Figure 6. Entering initial conditions in the program interface and graphical representation of the result

The CI-TREMS complex includes a different number of individual modules responsible for various aspects of production.

For example, one of such modules is the TPS program for monitoring the thermal engineering parameters of the soil along the route of the pipeline. This software module can work both autonomously and use the power of modern CAD software for more accurate and in-depth analysis (Figure 5)

The module allows accurately predicting soil temperature fluctuations, simulating the behavior of soil and structures located on it during thawing of the permafrost throughout the entire period of operation of the pipeline, or other structures and facilities under construction and operated in the Arctic region of Russia.

3. Conclusion

In the age of digital technology and the widespread automation of production, such intelligent systems as CI-TREMS [5-14] can reduce the number of emergencies at enterprises in the oil and gas sector, optimize the operation of existing facilities and help with the design of new stations and structures. Work requires a lot of labor to implement in real individual enterprises.

The modules can have simple visualization based on complex models and trends, allowing the operator to make safety decisions to reduce the risks and likelihood of accidents.

Throughout the entire life cycle of an enterprise, equipment or a specific unit, both integrated and individual modules will predict and warn about emergency situations, incidents, accidents. For most experts, it will be indispensable in optimizing and ensuring the effectiveness of technical solutions and managing processes of industrial enterprises in the oil and gas sector.

References

[1] *Russian enterprises will switch to environmentally friendly technologies* 2019 Available at: https://news.ecoindustry.ru/2018/03/s-2019-goda/

[2] Decree of the President of the Russian Federation 2017 *Environmental Security Strategy of the Russian Federation until 2025* No. 176

[3] Federal Service for Ecological, Technological and Nuclear Supervision, available at: http://www.gosnadzor.ru/industrial/oil/lessons/

[4] Boards of the Federal Service for Ecological, Technological and Nuclear Supervision 2011 *The concept of improving state policy in the field of industrial safety until 2020,* available at: http://docs.cntd.ru/document/902354089.

[5] Zemenkova M Yu 2019 *Methods to reduce technological and environmental risks in the transport and storage of hydrocarbons* (Tyumen: TIU)

[6] Federal Service for Ecological, Technological and Nuclear Supervision 2018 *Safety Guide "Guidelines for the classification of industrial events in the field of industrial safety at hazardous production facilities of the oil and gas complex"* No. 29 available at: http://sudact.ru/law/prikaz-rostekhnadzora-ot-24012018-n-29-ob/rukovodstvo-po-bezopasnosti-metodicheskie-rekomendatsii/

[7] Zemenkova M Yu, Kurushina V A, Golik V V and Zemenkov Yu D 2019 *System analysis and monitoring of energy technology complexes. Guidelines for the implementation of the course project for students of the direction 21.04.01 Oil and gas business of the program "Reliability and Safety of Hydrocarbon Resource Transport Facilities" of all forms of training* (Tyumen: Publishing Center BIK, TIU)

[8] Zemenkova M, Zemenkov Yu, Gladenko A and Podorozhnikov S 2016 Estimation of Emissions During Monitoring of Pipelines in the Dynamic Mode of Operation *MATEC Web Conf.* https://doi.org/10.1051/matecconf/20168604053

[9] Dudin S, Bahmat G, Zemenkov Yu, Voronin K and Shipovalov A 2017 Strategy for monitoring and ensuring safe operation of Russian gas transportation systems *MATEC Web Conf.* https://doi.org/10.1051/matecconf/201710606004

[10] Dudin. S M, Zemenkov Yu D, Maier A V and Shabarov A B 2016 Research and Design of Thermophysical Gas-Liquid Mixture Parameters in Product Pipelines *IOP Conference Series: Materials Science and Engineering* DOI: 10.1088 / 1757-899X / 154/1/012021

[11] Pimnev A L, Zemenkova M Y, Zemenkov Y D and Iljyashchenko D P 2016 Mechanical Properties of the Assembly Welded Joint of the Oil Transportation Tank after a Long-Term Service *IOP Conference Series: Materials Science and Engineering* **127 (1)** DOI: 10.1088 / 1757-899X / 127/1/012049

[12] Zemenkova M Y, Shipovalov A N and Zemenkov Y D 2016 Mathematical Modeling of Complex Hydraulic Machinery Systems When Evaluating Reliability Using Graph Theory *IOP Conference Series: Materials Science and Engineering* **127 (1)** DOI: 10.1088 / 1757-899X / 127/1/012056

[13] *Pump Station* 2019, available at: https://3dwarehouse.sketchup.com/model/f44232ef-4732-409a-b34b-fa4ca4e37503/Pump-Station

[14] Chizhevskaya E L, Zemenkova M Y and Mayss N A 2020 Directions to Improve Effectiveness of Management of the Industry Company Personnel *IOP Conf. Series: Earth and Environmental Science* **459** 052093 DOI: 10.1088 / 1755-1315 / 459/5/052093

IOP Conf. Series: Materials Science and Engineering **952** (2020) 012006 doi:10.1088/1757-899X/952/1/012006

Conditions choice for conducting experiments to determine the oil cooling time

A U Yakupov, D A Cherentsov and K S Voronin

Tyumen Industrial University, 38, Volodarskogo str., Tyumen, 625000, Russia

E-mail: ya.yakupov-azamat@yandex.ru

Abstract. The paper considers the option of processing the results of a computer experiment to determine the oil cooling time in a suspended oil pipeline. The authors of the article in previous works proposed a calculation model that allows simulating the process of oil cooling. As part of the design of the experiment, it is necessary to determine the factors affecting the oil cooling time in the pipeline, which will vary in the experiment. Empirical formulas are established, a regression analysis is carried out. The estimates of reproducibility dispersions are calculated. The adequacy hypothesis was tested using the Fisher criterion. Significant regression coefficients were established.

1. Introduction

When operating pipelines through which highly viscous oils are transported, it becomes necessary to determine the temperature of the product [1, 2]. The selected temperature will affect the operating parameters. The state of the piping system and soil are assessed: at high temperatures, a thawing process of permafrost may occur, which can lead to emergency situations [3].

The process of pipeline transportation of oil already demands increased responsibility, which includes non-stationary regimes. Among considered by the authors is an emergency stop of an oil pipeline with a highly viscous product inside the fluidity of which is highly dependent on the temperature of its state. Therefore, when transportation continues after a product reaches a critical temperature, it needs an initial force. A value that may exceed the limit allowable by the strength of the pipeline or the capacity of pumps. Therefore, it is important to determine the oil cooling time to the critical temperature [4-6].

2. Materials and methods

This article describes the experience of using the method of group accounting of arguments to build a mathematical model of parameters that affect the cooling time of oil. The output value may turn out to be independent of one or several input variables presented in Table 1.

In these conditions, the authors consider it appropriate to sort out the set of successively complicated candidate models. To do this, we apply the iterative multi-series algorithm of the argument group accounting method (AGAM) [7]. According to AGAM, the candidate models are tested sequentially by a given criterion; the most frequently used are polynomial support functions in the form of the Kolmogorov-Grabov polynomial.

$$y = b_0 + \sum_{i=1}^{M} b_i x_i + \sum_{i=1}^{M}\sum_{j=1}^{M} b_{ij} x_i x_j + \sum_{i=1}^{M}\sum_{j=1}^{M}\sum_{k=1}^{M} b_{ijk} x_i x_j x_k , \quad (1)$$

where $\bar{x} = (x_1, x_2, \ldots, x_M)$ are the vectors of input variables;
$\bar{b} = (b_1, b_2, \ldots, b_M)$ are the coefficients of the summands.

Table 1. Input Variables

No.	Parameter	Unit
1.	Oil temperature	°C
2.	Specific thermal capacity of oil	J/(kg•K)
3.	Oil viscosity	Pa•s
4.	Thermal conductivity of oil	W/(m•K)
5.	Oil composition	%
6.	Oil pipeline diameter	mm
7.	Insulation thickness	mm
8.	Pipeline Depth	m
9.	Soil heat capacity	J/(kg•K)
10.	Soil thermal conductivity	W/(m•K)
11.	Soil density	kg/m^3
12.	Soil moisture	unit fraction
13.	Soil temperature	°C
14.	Ambient temperature	°C
15.	Wind speed	m/s

The vector of coefficients $\bar{a}$ are found from the training set of values $\bar{x}_n, \bar{y}_n$ $(n = \overline{1, N}, N \geq M)$ using the least squares method.

It is possible to use the combined AGAM algorithm, which does not remain constant, but expands with each new one, complicating the structures of the next form:

$$y = b_0 + b_1 x_i \, , \, i = 1{,}2 \ldots, M.$$

The number of models is selected according to the best criterion.

In the second series, a model of a more complex structure is selected:

$$y = b_0 + b_1 x_i + b_2 x_j, \, i = 1{,}2 \ldots, F; \; j = 1{,}2 \ldots, M; F \leq M.$$

In the third series, models of a more complex form are selected:

$$y = b_0 + b_1 x_i + b_2 x_j + b_3 x_l, \; i = 1{,}2 \ldots, F; \; j = 1{,}2 \ldots, F; l = 1{,}2 \ldots, M.$$

Continue increasing until the value of the minimum criterion decreases.

According to the author, the most convenient to use is the multi-series algorithm [8], in which the iteration rule remains the same for all series. We use in the first series a particular description of the type.

$$y' = b_0' + b_1' x_i + b_2' x_j + b_3' x_i x_j + b_4' x_i^2 + b_5' x_j^2 \, . \quad (2)$$

In the second series:

$$y'' = b_0'' + b_1'' y_i' + b_2'' y_j' + b_3'' y_i' y_j' + b_4'' (y_i')^2 + b_5'' \left(y_j'\right)^2 .$$

In the third series:

$$y''' = b_0''' + b_1'''y_i'' + b_2'''y_j'' + b_3'''y_i''y_j'' + b_4'''(y_i'')^2 + b_5'''(y_j'')^2.$$

The process continues by analogy.

Because it is impossible to find one single model that would have a distinctive external manifestation, one can take a polynomial of any complexity and for each conduct a regression analysis and find the values of the coefficients. And in each equation, the coefficients will be chosen so that the error on the observation interval will be minimal. The iterative multi-series algorithm of the method of group accounting of arguments used has an attribute of external complement, with the help of which it becomes possible to identify a single model. External addition will be the requirement of one result to another. If the result of the first and the requirements of the second do not match, the first will have to be transformed, which is the process of self-organization. To obtain an external supplement in our case, we divide the initial values into the training and test samples. The principle of self-organization of models will be that with the gradual complication of models, the values of internal indicators will gradually decrease, passing through their lows. External indicators will also decrease, passing through their lows, which will make it possible to determine the model of optimal complexity.

A necessary step is to determine the number of values in the sample on which they will be trained, and obtain the coefficients of the arguments. It is proposed that all the studied values of the cooling time n be divided into 3 parts, 2/3 of the parts to be used to obtain the coefficients, and 1/3, to check the results. The standard deviation will be the test criterion.

To build each model, you must perform the following steps on [9]:

a. Create a matrix with input parameter values U'.
b. Create a vector O' in which there will be cooling time values.
c. Enumerate all possible combinations of input variables from the matrix U' and find for each case the values of the coefficients of the polynomials of equation (1).
d. Calculate cooling time values using found coefficients of approximating polynomials.
e. Calculate the standard deviation of the residuals - the differences between the vector O' and the corresponding value of each interpolated polynomial according to the formula:

$$\sigma_{ij} = \left(\frac{1}{n}\sum_{n-1}^{n}\left(Y_n - (y_n')_{ij}\right)^2\right)^{0.5}. \quad (3)$$

Polynomials with the smallest variance values will be used to construct approximating polynomials of the second series.

f. Repeat steps c, d, e for polynomials of the second series.
g. The increase in the level of the series should be performed until the variance of the residuals of the model increases.
h. To confirm the adequacy of the constructed model, it is necessary to calculate the multiple correlation coefficient:

$$R = \sqrt{\frac{\sum_{k=1}^{n}(\hat{y}_k - \bar{y})^2}{\sum_{k=1}^{n}(y_k - \bar{y})^2}} = \sqrt{1 - \frac{\sum_{k=1}^{n}(y_k - \hat{y}_k)^2}{\sum_{k=1}^{n}(y_k - \bar{y})^2}} \quad (4)$$

where $\hat{y}_k$ is the calculated indicator according to the model;

$\bar{y} = \frac{1}{n}\sum_{k=1}^{n} y_k$ is the average value of the indicator;

$\frac{1}{n}\sum_{k=1}^{n}(y_k - \hat{y}_k)^2$ is residual variance;

$\frac{1}{n}\sum_{k=1}^{n}(y_k - \overline{y_k})^2$ is the variance of the effective attribute;

$\frac{1}{n}\sum_{k=1}^{n}(\hat{y}_k - \bar{y})^2$ is factorial variance.

It is carried out after finding the coefficients of the model to assess its suitability. The Fisher criterion is an indicator of suitability, if the calculated value of the criterion does not exceed the tabulated one, then with the corresponding confidence probability the model can be considered adequate.

i. To check the consistency of the obtained dependence with the indicators, it is necessary to calculate the confidence intervals of the regression equations according to the formula:

$$\hat{y}_n \pm t_{0.99}(200-6)\hat{\sigma}\sqrt{1+x_n^T(D^TD)x_n} \ , \tag{5}$$

where D is the factor value matrix;
$t_{0.99}$ is the value of the student distribution quantile for a confidence probability of 0.95;
$\hat{\sigma}$ is standard deviation of random errors.

The significance of the obtained coefficients is estimated. The coefficient is significant in cases where its absolute value is greater than the confidence interval. Information on the significance of the coefficients allows us to consider the possibility of simplifying the subsequent work by eliminating some of the factors; moreover, this information will help in the interpretation of mathematical models.

The above calculations can be represented as an algorithm in Figure 1.

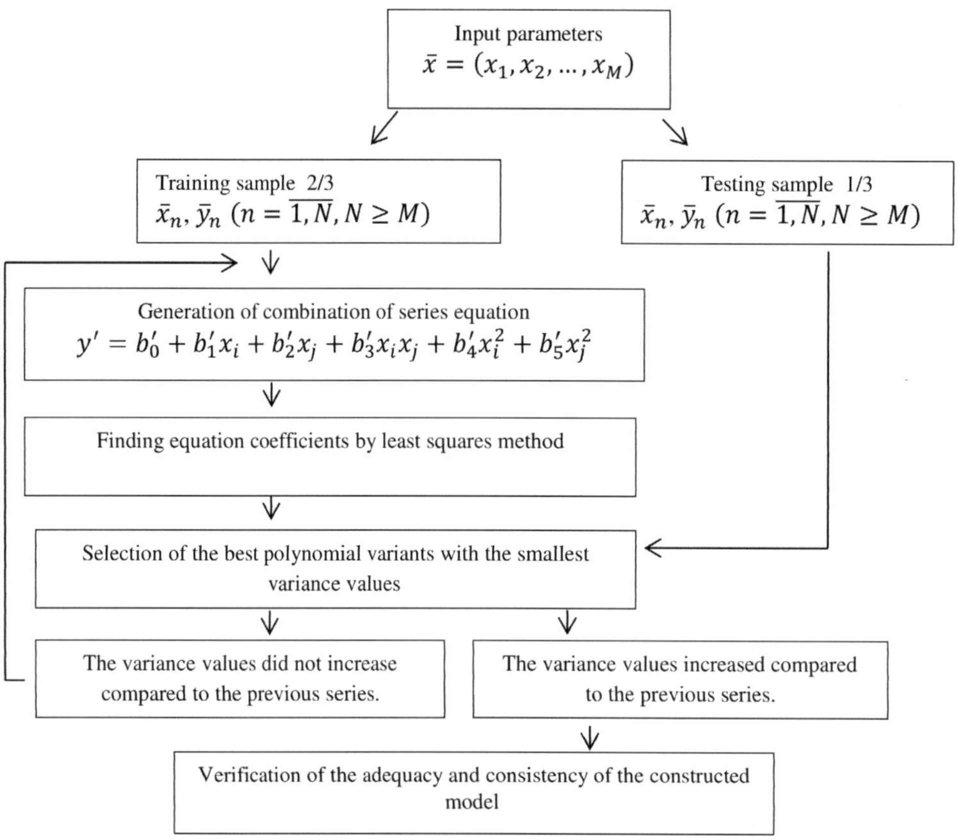

Figure 1. AGAM algorithm

3. Conclusion

The cooling rate of oil can also depend on the conditions of convective heat transfer [10] inside the pipeline, which can have a significant effect, but are not taken into account in this technique. For example, when convective heat transfer is taken into account, it is likely that the angle of inclination of the pipeline will affect the cooling time.

Thus, this technique allows determining the regression equation, evaluating the significance of the parameters, identifying significant parameters for the experiment to determine the cooling time of oil in a suspended underground pipeline.

References

[1] Yakupov A U et al. 2019 Evaluation of the effect of seasonal cooling devices on the time of safe shutdown of the pipeline *News of higher educational institutions. Oil and gas* **3** 120-126

[2] Yakupov A U et al. 2019 Influence of structural features of thermostabilizers on the cooling time of oil in a stopped oil pipeline *News of higher educational institutions. Oil and gas* **6** 140-148

[3] Kvasov I N, Shendaleva E V, Shtengauer O V and Zemenkova M Yu 2019 Prediction of environmental risks in industrial accidents at main and technological oil pipelines *News of higher educational institutions. Oil and gas* **6** 103-117

[4] Voronin K S et al. 2018 Modernization of the individual device for temperature stabilization of the soil *IOP Conf. Ser. Mater. Sci. Eng.* **445 (1)** 012011

[5] Yakupov A U, Voronin K S and Cherentsov D A 2019 Temperature condition of a stopped underground oil pipeline *IOP Conf. Ser. Mater. Sci. Eng.* **663** 012013

[6] Markov E V, Pulnikov S A and Sysoev Yu S 2018 Methodology for calculating the safe stop time of underground pipeline with high pour point oil *IJCIET* 1699-1705

[7] Ivakhnenko A G 1981 *The inductive method of self-organization of models of complex systems* (Kiev: Science. Dumka)

[8] Adler Yu P, Markova E V and Granovsky Yu V 1976 *Planning an experiment in the search for optimal conditions* (Moscow: Nauka)

[9] Zemenkova M Yu and Chekardovsky S M 2016 *Methodology of scientific research in the oil and gas industry. Monograph* (Tyumen: TIU)

[10] Kutateladze S S 1979 *Fundamentals of the theory of mass transfer "Fundamentals of the theory of mass transfer"* (Moscow: Atomizdat)

[11] Zemenkova M Yu, Zemenkov Y D, Pimnev A L and Kurushina E V 2016 System of Controlling the Reliability of Hydraulic Machinery in Oil and Gas Facilities *IOP Conference Series: Materials Science and Engineering* **127 (1)** DOI: 10.1088 / 1757-899X / 127/1/012055

[12] Zemenkova M Y, Shipovalov A N and Zemenkov Y D 2016 Mathematical Modeling of Complex Hydraulic Machinery Systems When Evaluating Reliability Using Graph Theory *IOP Conference Series: Materials Science and Engineering* **127 (1)** DOI: 10.1088 / 1757-899X / 127/1/012056

[13] Chizhevskaya E L, Zemenkova M Y and Mayss N A 2020 Directions to Improve Effectiveness of Management of Management of the Industry Company Personnel *IOP Conf. Series: Earth and Environmental Science* **459** 052093 DOI:10.1088/1755-1315/459/5/052093

Method for improving the liquid hydrocarbon preparation efficiency

A M Kulikov, A A Gladenko and M A Alexandrov

Tyumen Industrial University, 38, Volodarskiy str., Tyumen, 625000, Russia

E-mail: kulikovam@tyuiu.ru

Abstract. According to the "Energy Strategy of Russia for the Period until 2030" and Decree of the Government of the Russian Federation of November 13, 2009 No. 1715-r, oil and gas companies have to use machines and equipment in the production process, as well as methods and processes that minimize energy consumption, and eliminate negative impacts of technological processes on the environment. New methods for the design, construction and operation of oil and gas facilities and improvement of the existing equipment should be developed and implemented.

1. Introduction

The production of gas condensate - a valuable chemical raw material used to produce high-quality types of motor fuels, plastics, synthetic rubbers, fibers and resins - occupies a special place in the activities of oil and gas companies. High rates of natural gas production increase the volume of gas condensate extraction. At the same time, gradual depletion of Cenomanian gas deposits creates the need to develop gas condensate deposits.

According to the Federal State Statistics Service, in 2007-2018, the volume of production of unstable gas condensate increased by more than 90%. In Western Siberia, there are more than half of the Russian gas condensate reserves.

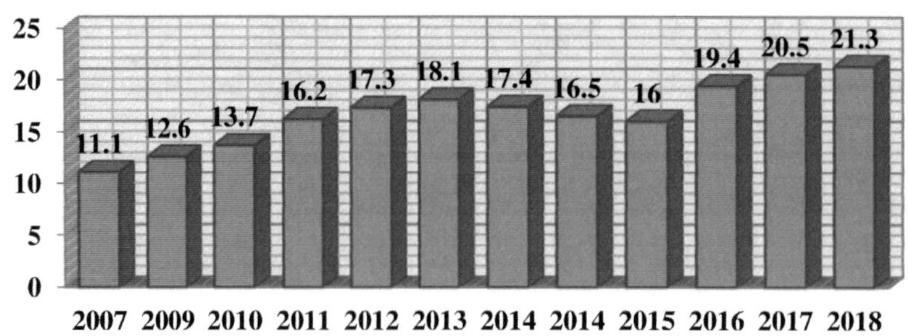

Figure 1. The dynamics of gas condensate production in the last decade, million tons

In 2018, the Yamalo-Nenets Autonomous District produced 11.427 million tons, which is 10% more than in 2017. This raw material is extracted by 17 enterprises in 24 fields. The leading positions are occupied by subsidiaries of Gazprom (7.424 million tons, or 65% of all condensate produced in the district). NOVATEK OJSC ranks second (2.811 million tons, or 24.6%), Rospan International CJSC (678.903 thousand tons, 5.9%) and Rosneft OJSC (256.439 thousand tons, 2.2%) rank third and fourth, respectively. In 2012, 12.3 million tons gas condensate were produced in the region. The growth of this indicator is associated with the development of the South Tambey gas condensate field and construction of a gas liquefaction plant.

2. Research

During field development and separation, unstable gas condensate is released from the gas mixture [2.5]. To deliver it, it is stabilized by distillation or kept at an atmospheric pressure and high temperatures to remove volatile fractions and obtain stable condensate.

The processes of preparing liquid hydrocarbons for transportation have been studied thoroughly. A number of technological units designed to prepare raw materials with a wide variety of physicochemical properties were developed. In preparing oil and gas condensate for long-distance transportation, it is necessary to isolate the dissolved gases, dehydrate and desalinate, and remove mechanical impurities, i.e. to provide the requirements for salable oil and gas condensate.

Condensate degassing is carried out at head facilities and compressor stations of the main gas pipelines. Condensation is usually degassed in several stages. The use of one-stage degassing of condensate, when the latter is sent directly to the atmospheric tank, is associated with maximum gas losses and significantly reduces the yield of commodity condensate.

Single-stage degassing is mainly used at compressor stations of gas pipelines. The two-stage degassing method is used: the condensate flows sequentially to the degassing ladder and the reservoir, reduces the loss of degassing gas and increases the condensate yield by 10-15%. An increase in the number of degassing stages over four slightly reduces the output of low-pressure gases and does not increase the yield of condensate. An increase in the number of steps increases capital investment and complicates the task of utilizing degassing gases [1].

As a result, it is necessary to develop more efficient degassing methods. Some researchers showed that the solution to this problem can be associated with the use of a vortex flow [3,4]. The swirling motion of a gas-saturated liquid forms a field of centrifugal forces, where components with different densities are separated from each other. In addition, intense mass transfer of liquid volumes with different gas saturation occurs, which inevitably leads to the formation of a gas-liquid interface with a turbulent structure, in which the gas evolution process significantly increases. Thus, ceteris paribus, the presence of a vortex should significantly accelerate degassing and increase the efficiency of the process.

The purpose of the work is to increase the efficiency of methods of preparing and separating multiphase multicomponent hydrocarbon systems by intensifying gas evolution processes.

The main objectives of the research are as follows:

1. To identify the problems of methods for condensate degassing and solving them in order to reduce or completely eliminate gas losses of degassing and environmental protection;
2. To select optimization criteria for the degassing of multiphase hydrocarbon systems;
3. To choose the optimal method for increasing the intensity of gas evolution processes and reducing the cost of commercial gas;
4. To substantiate the choice of dependences of the hydro-gas-dynamic and structural characteristics by experimental studies of the outflow of heterogeneous gas-saturated liquids in vortex chambers;
5. To develop a physical and mathematical model for the degassing of unstable condensate in vortex chambers, as well as to test the model in full-scale operating conditions.

The novelty of the research:

1. A system of criteria has been developed to optimize the degassing of multiphase hydrocarbon systems to select the most suitable method with technological and economic parameters;
2. Analytical and empirical dependencies of the main hydro-gas-dynamic parameters and geometric characteristics of the vortex chambers as condensate degassers have been established and justified;
3. A physical and mathematical model for the degassing of unstable condensate in vortex chambers has been developed. It is used to calculate parameters of separation devices depending on operating conditions.

The obtained analytical dependencies allow the calculation of parameters of the degassing stages when installing vortex chambers.

The process of degassing a saturated liquid in a vortex chamber was studied in the experimental unit. The diagram is shown in Figure 2.

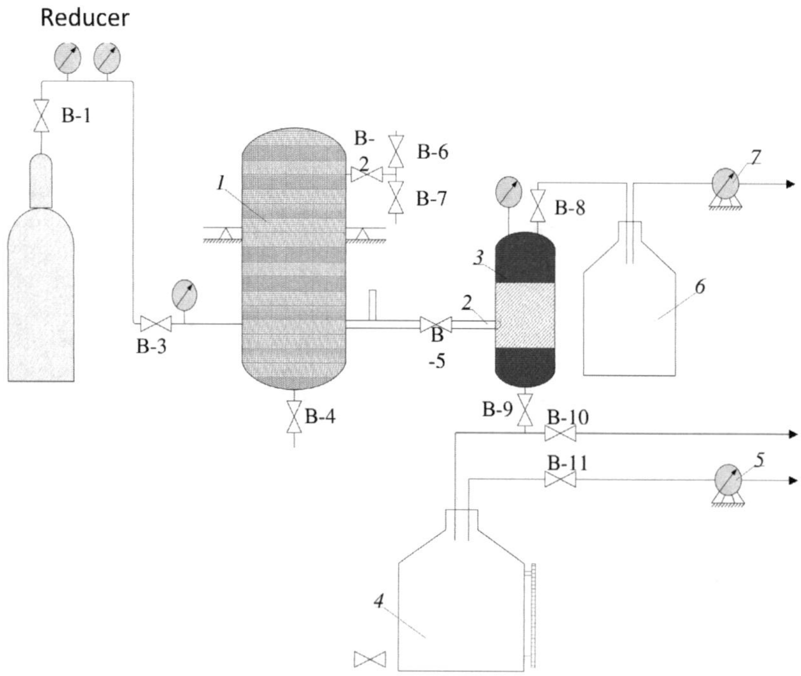

Figure 2. The experimental unit for studying the process of degassing a liquid in the vortex chamber.

The mixing tank 1 is designed for an operating pressure of 5 MPa and has a volume of about 60 liters. A sufficiently large volume of the mixing tank allowed us to obtain 6–10 experimental points in the experiments on the outflow of saturated and undersaturated liquids from nozzles using a single charge.

The container was fixed in a special stand, allowing it to rotate around the transverse axis in order to mix the liquid during sample preparation. The need for intensive mixing of liquid and gas is caused by the fact that gas dissolution by the method of barbation, which is most often used in laboratory plants, requires a more complicated installation scheme. To circulate through tank 1, it is necessary to install either a compressor or some other way to loop the gas supply line to tank 1. The method of dissolving gas in liquid by mixing the sample makes it possible to simplify the installation scheme, because the dissolution process is much faster. The time for preparing a saturated liquid sample at a given pressure during which the container rotates around the transverse axis was determined during the experiments and was evaluated by stopping the dissolution of carbon dioxide in water. The maximum equilibrium dissolution of carbon dioxide in water is achieved after 15 minutes. In all the experiments, the rotation time of the container during sample preparation was 20 min.

The process of degassing a saturated liquid in the vortex chamber is associated with kinetics and geometrical characteristics of the chamber and mass speed of the fluid entering the chamber through the nozzle. To select the main geometric characteristics of the vortex chamber, the results of [3] were analyzed.

IOP Conf. Series: Materials Science and Engineering **952** (2020) 012007 doi:10.1088/1757-899X/952/1/012007

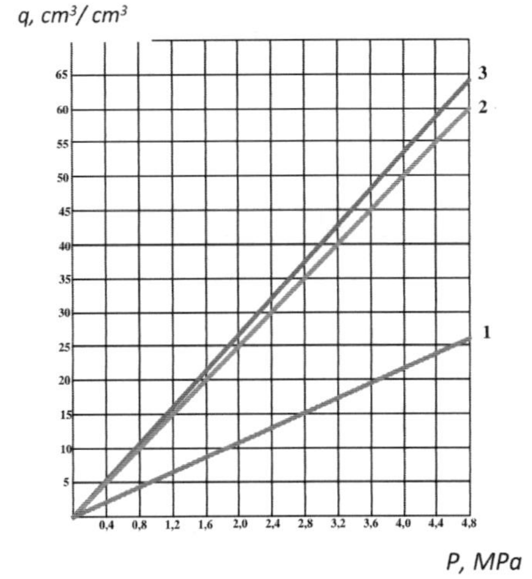

Figure 3. The solubility of carbon dioxide in a liquid: 1 - carbon dioxide in water; 2 - carbon dioxide in diesel fuel; 3 - natural gas in condensate

3. Results

The degree of liquid degassing which characterizes the completeness of this process is the main criterion for the efficiency of the vortex chamber. The degree of degassing ε is the ratio of the equilibrium gas saturation of the liquid at p and t in the vortex chamber to its actual gas saturation in the lower part of the chamber: $\varepsilon = q_p/q_к$

$$\varepsilon = 1 /[(q_{in}/q_p - 1)\, e^{(-n\tau)} + 1] \quad (1)$$

From equation (1), an expression for calculating the coefficient n (the coefficient of intensity of gas evolution) can be obtained, since intensity of the degassing process depends on its value.

$$n = \frac{1}{\tau} \ln\left(\frac{q_{нач} / q_k - 1}{1 - \varepsilon} + 1 \right) \quad (2)$$

where q_{in} – initial gas saturation, m^3/m^3;

$q_к$ - final gas saturation, m^3/m^3

When calculating n in the vortex chamber or other degasser, time τ is understood as the residence time of the liquid in the working cavity of the degasser.

The expression for determining the total amount of gas released is as follows:

$$Q = v(q_{in} - q_p)(e^{-n\tau} - 1) \quad (3)$$

The total height of the chamber consists of three components: heights of rotating gas and liquid phases and liquid column in the bunker space. Only a column of the rotating liquid phase is actively involved in the degassing process. The column of liquid located below in the bunker space does not participate in the degassing process. Therefore, the main geometric characteristic of the vortex chamber is the ratio of the diameter of the chamber to the height of the working space D / h_f. The optimal value is in the range of 0.9 ... 1.1

$$n = u_m\, [a_1\, (D / h_ж)2 + b_1\, D / h_ж + c_1] \quad (5)$$

The constants in equation (5) are determined empirically from the experimental data. Numerical values of the given constants: t = 2.9; A_1 = - 0.242 * 10-3; b_1 = 0.455 * 10-3; c_1 = - 0.129 * 10-3. The error does not exceed 7%.

The equation for the expiration of gas-saturated liquid (mass velocity):

$$u = [(1 — 0{,}8\ k)\ e^{(— 0{,}06\ \Delta q^* 1/d)} + 0{,}8k]\ \mu\ 2\rho\ \sqrt{(p_1 — p_2)} \quad (6)$$

where $k = (k / k_B)0.2$ — function of the ratio of the solubility coefficient of gas in the liquid to the solubility coefficient of carbon dioxide in water;

$\Delta q = q/q_P$ — degree of saturation of liquid with gas);

l/d — the ratio of the length of the cylindrical nozzle to its inner diameter;

μ — nozzle flow rate;
ρ —density of degassed liquid, kg/m

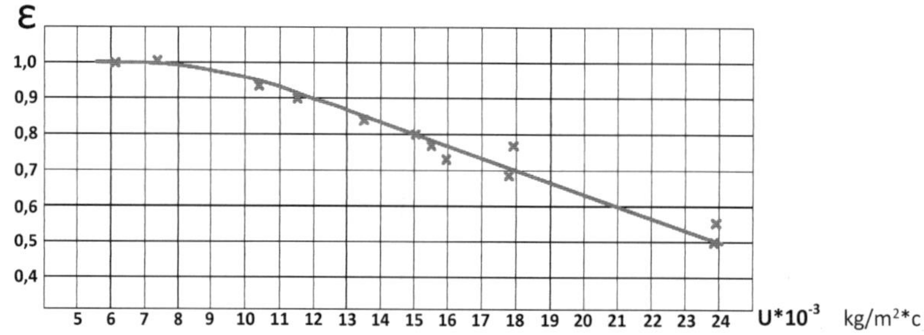

Figure 4. The dependence of degassing degree ε on mass speed of the outflow of raw materials through the nozzle at D/h_f = 0,5…1,0

The best degassing value is achieved at mass flow rates of less than 14.0 x 10^3 kg / (m^2 s).

The experimental studies of the vortex effect in gas-saturated liquids revealed possibilities for using vortex chambers for degassing oil and gas condensate and primary processing of hydrocarbon feedstocks. To identify all the possibilities of the vortex effect in liquids, much more effort is required.

Thus, the analysis allows us to identify the following advantages of the vortex chamber compared to other separation equipment:

1) simplicity of design, manufacture, installation and maintenance;
2) the absence of moving parts and mechanisms, the use of flow energy;
3) high unit power (productivity of one unit);
4) the ability to optimize the process of equipment operation under various operating conditions;
5) more complete use of the inner cavity of the vortex chamber and intensification of the degassing process (reducing the residence time for the gas-saturated liquid in the vortex chamber compared to other separators);
6) environmental friendliness, efficiency and manufacturability of equipment, making it possible to utilize gas-saturated mixtures, while eliminating the loss of light fractions of hydrocarbons, as well as the possibility of separation of light fractions by components by applying the step-by-step degassing method;
7) smaller dimensions for locating this equipment on offshore platforms and bulk sites in wetlands of Western Siberia;
8) lower metal consumption (mass of the vortex unit, comparable in its technological parameters with separators like NGS is 10 ... 13 times less).

As a result of the experimental study of the degassing of a saturated liquid in the vortex chamber, the influence of various factors on the efficiency of this process was specified.

4. Conclusion

1. The analysis of structural and operational characteristics of the vortex chambers confirmed their advantages over vertical and horizontal separators, as well as hydrocyclones.
2. An algorithm for calculating basic geometric, technological and technical and economic parameters of the vortex chambers was developed as condensate degassers depending on operating conditions.
3. As a result of the theoretical testing of the design calculation algorithm, the relative error in measuring the flow rate of saturated liquid, gas saturation and mass flow rate was determined, confirming the possibility of practical application of the algorithm.

References

[1] Trivus N A and Seidali-zade B M 1974 Optimum modes of condensate degassing step and its effectiveness, B Sat .: Development and exploitation of gas and gas condensate fields *VNIIEGazprom* **6** 17-22

[2] Dudin S M, Zemenkova M Yu and Shabarov A B 2005 Modeling of hydrodynamic processes in the pipeline transportation of hydrocarbons *Interstroimekh –2005: Proceedings of the International Scientific and Technical Conference* **1** 79-82
[3] Sinai E G, Lapin E A and Zaitsev Yu V 2002 *Separation of multiphase multicomponent systems* (Mocow: Nedra)
[4] Bakhmat G V 1981 Use a vortex effect for degassing of crude condensate *Oil and Gas Problems Tyumen* **49** 15-18
[5] Zemenkov Yu D, Markova L M, Prokhorov A D and Dudin S M 2009 *Collection and preparation of oil and gas* (Moscow: Academy)
[6] Kulikov A M, Petryakov V A and Seroshtanov I V 2015 Optimization of separation processes of multicomponent multiphase systems *Business magazine Neftegaz.RU* **11-12** 36-40
[7] Kulikov A M and Komarova E A 2010 To the question of preparation of unstable gas condensate *Problems of functioning of transport, International scientific-practical conference* pp 188-189
[8] Smulsky I I 1992 Aerodynamics and processes in the vortex kamerah (Novosibirsk: Nauka)
[9] Suhovich E P 1987 *Turbulence models for describing the anisotropy of turbulent transport in swirling flows. Methods of vortex flows with heterogeneous chemical. responds. sredoy* (Moscow: Nauka)
[10] Dudin S M, Zemenkov Yu D, Maier A V and Shabarov A B 2016 Research and Design of Thermophysical Gas-Liquid Mixture Parameters in Product Pipelines *IOP Conference Series: Materials Science and Engineering* DOI: 10.1088/1757-899X/154/1/012021

IOP Conf. Series: Materials Science and Engineering **952** (2020) 012008 doi:10.1088/1757-899X/952/1/012008

Thermal impact of buried reservoirs with industrial waste of the fuel and energy complex on frozen soils

M Yu Zemenkova[1], P Yu Mikhailov[2], A B Shabarov[1,2] and U Yu Shastunova[2]

[1] Tyumen Industrial University, 38, Volodarskogo Str., Tyumen, 625000, Russia
[2] Tyumen State University, 6, Volodarskogo St., Tyumen, 625003, Russia

E-mail: muzemenkova@mail.ru

Abstract. The oil and gas industry has been intensively developing in the far north of the Russian Federation. Underground tanks are repositories of waste from processed petroleum products, liquefied gas, condensate, process water, and other waste from the fuel and energy complex. High-quality underground disposal of industrial waste can eliminate leakage of chemicals into the soil and reduce the thermal effect on frozen soil. The article presents an algorithm for calculating the thermal effect of industrial "hot" waste of the fuel and energy industry on frozen soil. Heat fluxes from a 20,000 m3 tank are estimated, an area of soil thawing is calculated. An engineering solution was developed to reduce / eliminate this effect.

1. Introduction

One of the most important tasks of the economic and social development of territories is the use of the underground space of the permafrost zone for various economic needs of the fuel and energy industry. Therefore, the environmental issue of the efficient use of permafrost soils is very acute [5].

Underground tanks are repositories of waste from processed petroleum products, liquefied gas, condensate, process water, agricultural products, radioactive waste, etc. The disposal of industrial waste or the storage of other petroleum products is regulated by laws.

During the operation of underground tanks, there is a thermal effect of the product on frozen soils. The continuity of the moss cover is disturbed in the permafrost zone, which leads to the destruction of the chemical composition of soils, erosion and disruption of the natural balance of the ecosystem. Unfortunately, many manufacturers are forced to put up with the lack of effective methods of operating the fuel and energy industry in special climatic conditions. They pay substantial fines.

High-quality underground disposal of industrial waste will reduce the thermal effect on frozen soil, and reduce unforeseen production costs (fines). The negative temperature of soils will have a frosting effect on the waste, which can reduce their activity [5].

2. Problem statement

To study the thermal effect of the "hot" reservoir on frozen ground, we took a model of a vertical cylindrical concrete underground reservoir with a capacity of 20,000 m3, with known characteristics [6, 7] (table 1).

Table 1. Characteristics of the vertical cylindrical concrete underground tank and frozen soil

$V = 20\ 000\ m^3$ – tank volume $H = 9$ m - height $D = 54$ m - diameter $\Delta_w = 0{,}2m$, $\delta_b = 0{,}3m$, $\delta_c = 0{,}2m$ – thickness of the wall, bottom, cover, respectively. Tank material - concrete $\rho = 2300\ kg/m^3$ – concrete density $\lambda_{concrete} = 1{,}35$ W/(m·°C) – concrete heat capacity $C_0 = 840$ J/(kg·°C) – concrete heat capacity	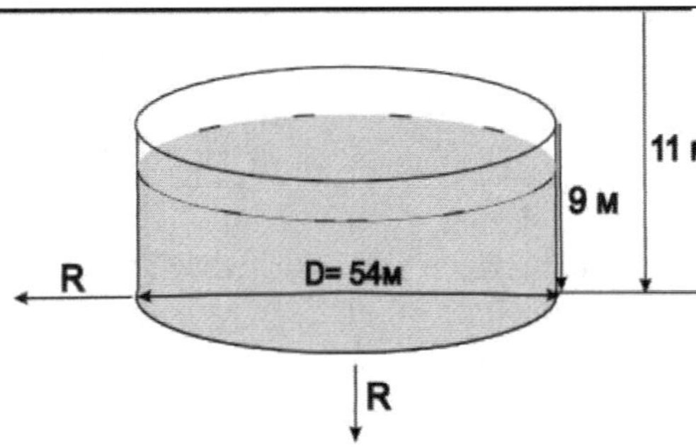

The tank was buried in permafrost soil. The height from the tank roof to the surface is 1.5 m. The soil is a homogeneous stationary medium. Density, heat capacity and thermal diffusivity depend on temperature changes [2].

The thermophysical characteristics of the soil in which the tank was installed are presented in table 2.

Table 2. Thermophysical soil characteristics

$t_{гр} = -7$ °C – soil temperature
$\lambda_{гр} = 3$ W/(m·°C) – soil thermal conductivity
$C_{гр} = 920$ J/(kg·°C) – soil heat capacity
$\rho_{гр} = 2083\ kg/m^3$ – soil density

Water with dissolved mechanical impurities was taken under the buried product. The initial temperature is 45 °C.

3. Calculation method

Since the tank, as a real heat source, has finite dimensions and acts for a finite period of time, when solving the problem, the fundamental solution to the heat equation can be used [3]. The tank was considered as a point source; and the time was divided into many infinite small intervals. From the heat source (reservoir) into the surrounding soil, the heat diverged simultaneously in three different directions: x, y, z. The process of cooling was modeled in the C ++ programming language.

To determine an area of soil thawing, it was decided to use the following dependence [2]:

$$T(r,t) = \frac{Q}{4\pi\lambda r} erfc\left(\frac{r}{2\sqrt{a\tau}}\right) \quad (1)$$

where Q – variable heat flux, which is formed from storage products in the tank, [W];
λ - soil thermal conductivity coefficient, [W / (m·°C)];
a – soil thermal diffusivity coefficient [m^2/s]; r - the distance from the heat source (tank) to the observation point, [m]; τ is time, [s].

The thermal diffusivity was determined by formula:

$$a = \frac{\lambda_g}{Cg\rho_g} \quad (2)$$

The distance from the tank to the heating point was determined by formula:

$$r^2 = x^2 + y^2 + z^2 \quad (3)$$

Since the heat from waste will be transferred to the soil, the soil will be cooled. The temperature of the buried product will decrease. Changes in the temperature of the tank can be determined by formula [1]:

$$T = T_0 + (T_H - T_0)e^{-\frac{\tau}{k_t'}} \quad (4)$$

where T_0 - ambient temperature, [K];

T_I - initial temperature of the buried product, [K];

τ - standing time, [sec];

k_t'- dimensionless heat transfer coefficient.

Expressing τ from this formula, we can determine the time during which the coolant is cooled to its ambient temperature.

Dimensionless heat transfer coefficientk_t' нcan be calculated by formula [1, 5, 8]:

$$k_t' = \frac{V_0 \cdot \rho \cdot C + m_0 \cdot C_0}{k_t \cdot S} \quad (5)$$

where V_0 - volume of the buried product, [m^3];

ρ - density of the buried product, [k /m^3];

C – heat capacity of the buried coolant product, [J / (kg · °C)];

m_0 - mass of the empty concrete tank, [kg];

C_0 - heat capacity of concrete, [J / (kg · ° C)];

k_t – heat transfer coefficient from the surface of the tank to the surrounding soil, [W / (m · ° C)];

S – total surface area of the tank, [m^2].

The dimensionless heat transfer coefficient (5) is an expression similar to the well-known formula of V.G. Shukhov for cooling the tank [8].

In formula (5), the heat transfer coefficient k_t is determined by complex equations that take into account heat transfer between the individual parts of the tank (bottom, roof, walls) and the liquid in the tank [8].

Under production conditions, it is desirable to have simpler equations or nomograms which can calculate the probable temperature drop with a variable flow rate of the disposal product and subsequent cooling of the tank [8].

The total heat transfer coefficient is determined by formula [1]:

$$k_t = \frac{k_1 \cdot S_1 + k_2 \cdot S_2 + k_3 \cdot S_3}{S} \quad (6)$$

where k_1, k_3 - heat transfer coefficient through the bottom, the tank roof, respectively, W/m^2 · deg);

S_1, S_3 – tank bottom area, m^2;

k_2 – heat transfer coefficient through the side wall of the tank, W/m^2 · deg;

S_2 – tank wall area, m^2;

S – total surface area of the tank, m^2.

To calculate the coefficient of heat transfer through the bottom, we use the formula [1, 5, 8]

$$k_1 = \frac{1}{\frac{1}{\alpha_{1d}} + \sum_{i=1}^{n} \frac{\delta_i}{\lambda_i} + \frac{\pi \cdot \delta_{concrete}}{8\lambda_g}}, \quad (7)$$

where δ_i, λ_i - thickness and thermal conductivity of the concrete wall of the tank bottom, insulation, etc .;

α_b- coefficient of heat transfer from the buried product to the inner wall of the tank, W/m^2 · deg;

λ_s - soil thermal conductivity coefficient, W / m2 · deg;

δ_c - thickness of the concrete wall of the tank, m.

The heat transfer coefficient α_{1b} is determined by the similarity method through dimensionless quantities. The mechanism of heat transfer is free convection [3, 5, 8]:

$$\alpha_{1b} = \frac{1}{2} \cdot \frac{\lambda_{waste}}{D_{res}} \cdot (Gr \cdot Pr)^{0,25}_{waste} \cdot \left(\frac{Pr_{waste}}{Pr_{bottom\,walls}}\right)^{0,25} \quad (8)$$

The dimensionless Grashof's numbers *Gr* and Prandtl's numbers *Pr* are calculated according to the generally accepted dependencies. The temperature correction coefficient $\left(\frac{Pr_{waste}}{Pr_{cbottom\,walls}}\right)^{0,25}$ is at the average temperature of the buried product Pr_{waste}, and the temperature of the tank bottom wall $Pr_{bottom\,wall}$. The temperature of the inner wall is based on physical considerations. At the end of the calculation, an adequacy of the temperature of the inner wall is checked. The calculation error is 3%.

The heat transfer coefficient through the wall k_2 can be determined by formula

$$k_2 = \frac{1}{\frac{1}{\alpha_{1st}} + \frac{\delta_{CT}}{concrete}}, \quad (9)$$

where α_{1w} - heat transfer coefficient from the buried product to the inner vertical wall of the tank, $W/m^2 \cdot deg$.

This coefficient can be found by formula in (8). The mechanism of heat transfer is also free convection [1].

$$\alpha_{1b} = \frac{1}{2} \cdot \frac{\lambda_{waste}}{H} \cdot (Gr \cdot Pr)^{0,25}_{waste} \cdot \left(\frac{Pr_{waste}}{Pr_{bottom\,walls}}\right)^{0,25} \quad (10)$$

The coefficient of heat transfer from the coolant to frozen soil through the roof k_3 was calculated by formula:

$$k_3 = \frac{1}{\frac{1}{\alpha_{1r}} + \frac{h_{gs}}{\lambda_{eq}} + \frac{\delta_{roof}}{\lambda_{concrete}}}, \quad (11)$$

where α_{1roof} – heat transfer coefficient between the buried product and the free gas space to the tank roof, $W/m^2 \cdot deg$;

h_{gs} –height of the gas space, m;

λ_{eq} – equivalent gas space thermal conductivity, $W/m^2 \cdot deg$.

Formula (12) is used to determine the equivalent coefficient of thermal conductivity of the gas-air mixture $\lambda_{экв} = \lambda_c \cdot \varepsilon_к$, (12)

where $\varepsilon_к$ - convection coefficient, λ_c - thermal conductivity of the gas-air mixture which are calculated by formulas (13), (14).

$$\varepsilon_к = 0{,}18 \cdot (Gr \cdot \mathrm{Pr})^{0.25} \quad (13)$$

If $\varepsilon_к < 10^3$, $\varepsilon_к = 1$.

$$\lambda_c = \lambda_{возд} \cdot (1 - C_1) + \lambda_{ср.} \cdot C_1 \quad (14)$$

The coefficient of thermal conductivity of the gas-air mixture with air humidity C_1 is determined in the gas space of the tank.

4. Results

By calculating the change in temperature of the buried product according to the above method, the dependence of temperature on the disposal time was obtained (see Figure 1).

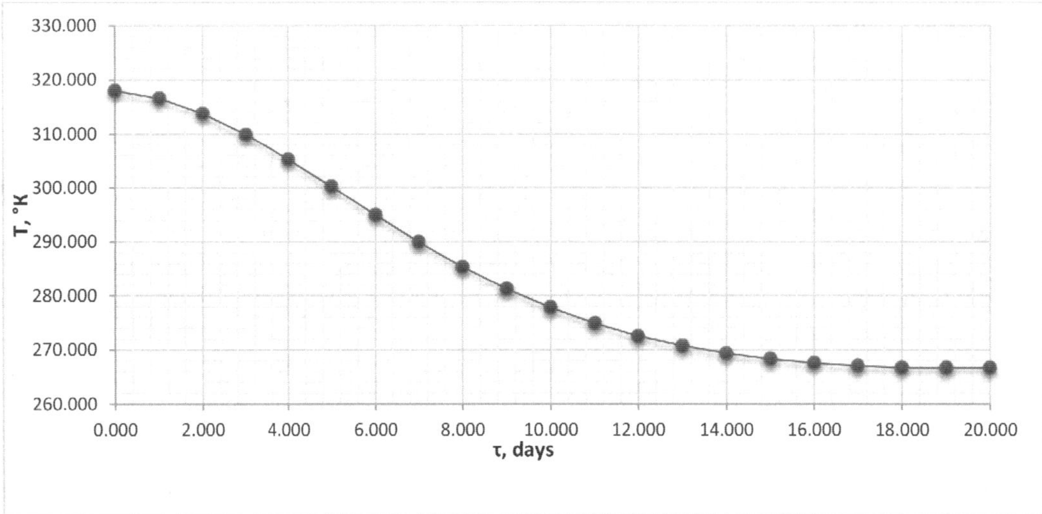

Figure 1. The dependence of temperature of the buried product on disposal time

The graph shows that 20 days after, the temperature of the "hot" waste becomes equal to the temperature of the surrounding soil.

Figure 2 shows a graph of the temperature taken at various fixed distances from the tank center (the tank is presented as a point source). It can be seen that on the first day the soil warmed up, which is indicated by the peaks of the curves, and then the soil temperature gradually began to decrease. The closer the observation point to the reservoir, the higher the temperature of the soil is.

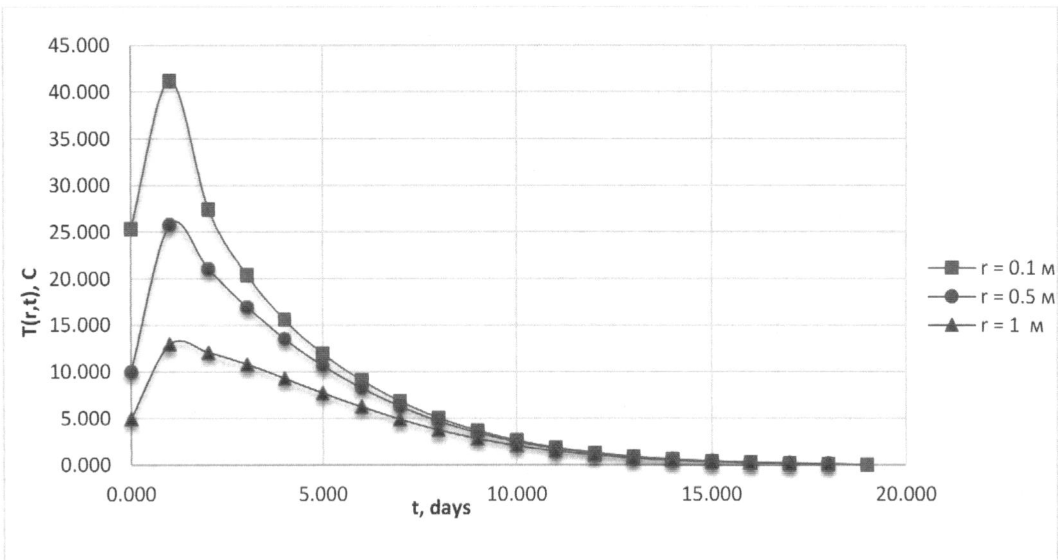

Figure 2. The dependence of soil temperature on disposal time at a fixed distance

Figure 3 shows that the temperature is maximum in the center of the tank and decreases monotonically with increasing distance r. The closer the moment of time t to the moment of the beginning of soil warming, the higher and narrower the peak of the curve T (r) is. Over time, heat gradually spreads in all directions, but if at any fixed point one can integrate (1) over the entire space, a value equal to Q can be obtained [3].

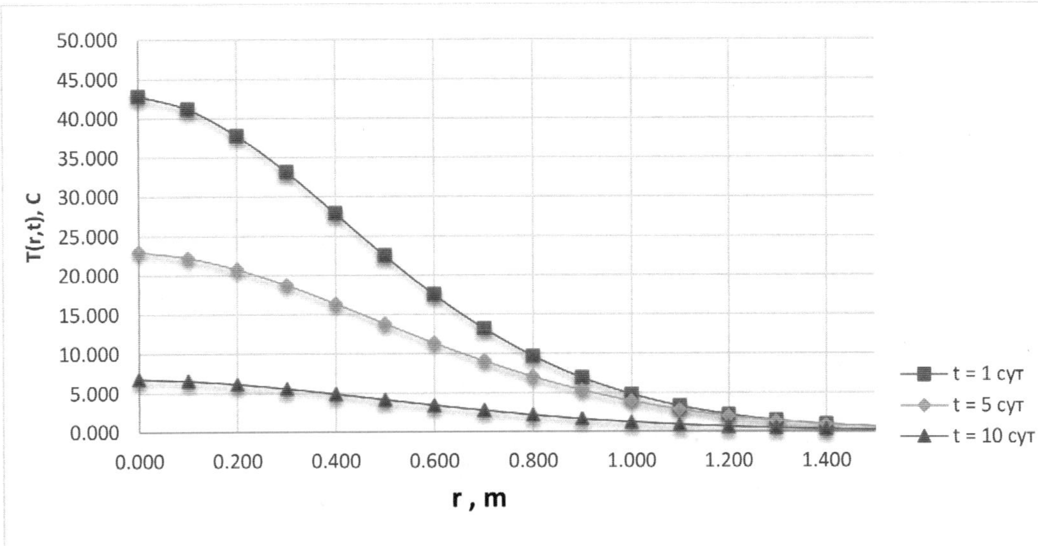

Figure 3. The dependence of soil temperature on the distance of the tank to any soil point at a fixed time

5. Engineering solution

To reduce the thermal effect of the tank with "hot" buried products on the soil, it was proposed to apply an insulation layer of calculated thickness that would reduce the thawing area [5].

For comparison, 4 insulation materials can be used:

- Mineral wool
- Foam glass
- Perlite
- Extruded polystyrene foam

Table 3 presents characteristics of the materials which affect the choice of insulation of the concrete tank in order to reduce the soil thawing effect.

Table 3. Comparative table of insulating materials

Characteristics	Mineral wool	Foam glass	Perlite	Extruded polystyrene foam
Heat conductivity coefficient, λ, W / (m·°C)	0,035	0,075	0,05	0,04
Insulation layer thickness, δiz, m	0,05	0,1	0,07	0,06
Cost, rub / m ^{2}	50	2000	105	160
Cost of complete isolation, t.rub	310,59	121423,6	652,239	993,888

Based on the results of the comparative analysis, one can conclude that the most advantageous insulating material is a 6 cm thick cushioning layer of extruded polystyrene. The standing thickness was selected so that the heat flux from the buried product causes minimal soil thawing.

Calculations of soil temperature in the presence of an additional layer between the soil and the wall of the concrete tank were made to justify the choice of an insulation material. The calculation results are presented in Figures 4 and 5.

Having analyzed the graphs, we can conclude that the selected material reduced the maximum temperature of soil heating by 6 times.

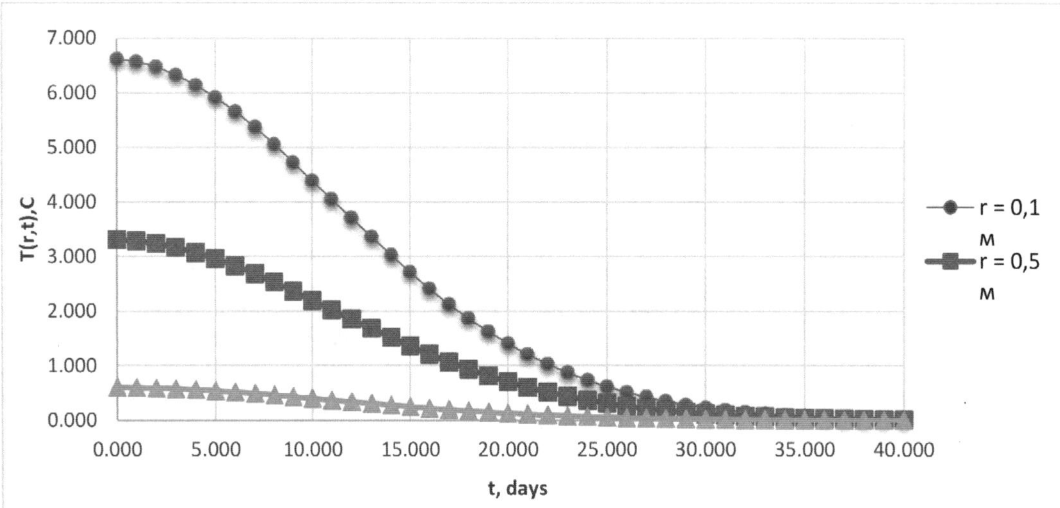

Figure 4. The dependence of soil temperature on the disposal time at a fixed point with insulation

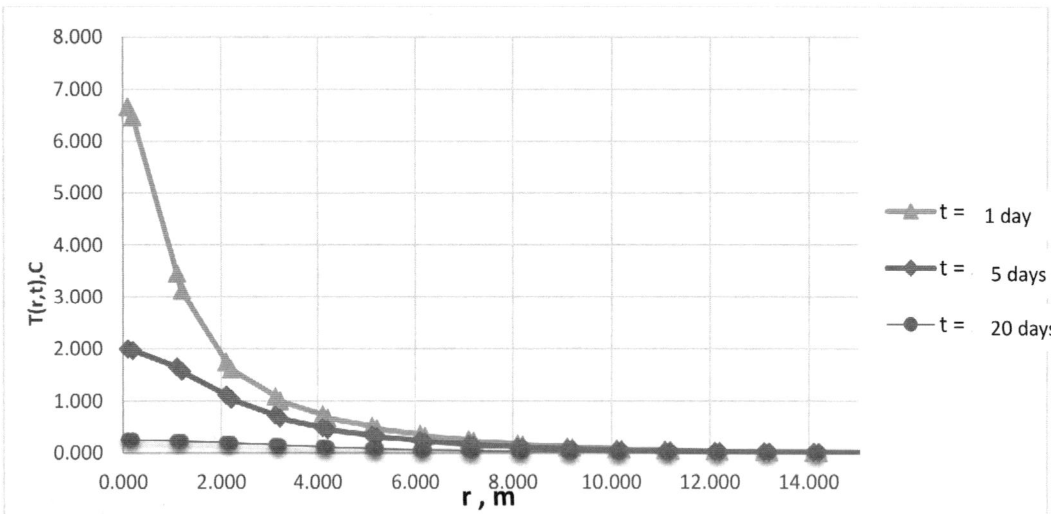

Figure 5. The dependence of soil temperature on the distance to the observation point at a fixed time with insulation

6. Conclusion

The thermal effect of the content of underground concrete tanks on frozen soil was studied. The method calculating the unsteady temperature field of the soil using the temperature of the cooled hot buried product was described. According to the calculations of the 20,000 m^3 tank, the product cools down during 20 days, while frozen soil is thawed at a distance of 1.4 meters from the tank center. In areas of humid soils, the thermal influence can be dangerous, it can lead to the bulge of the tank to the surface, which can destroy the tank. The application of thermal insulation between soils and the tank wall can be an alternative way to reduce heat losses.

References

[1] Bakhmat G V, Kislitsyn A A and Shastunova U Yu 2011 Methodology for calculating the thermal regime of a reservoir of the RVS type in winter during operation *Materials of the International scientific-practical conference on permafrost engineering dedicated to the 20th anniversary of the creation of NPO Fundamentstroyarkos LLC* (Tyumen: City Press) 301-306

[2] Gavriliev R I 2004 *Thermophysical properties of the components of the natural environment in the permafrost zone. Reference manual* (Novosibirsk: Novosibirsk Publishing House of the Siberian Branch of the RAS)

[3] Kislitsyn A A 2002 *Fundamentals of Thermophysics: lectures and seminars* (Tyumen: Publishing House of the Tyumen State University)

[4] Tugunov P I, Novoselov V E, Korshan A A and Shammazov A M 2008 *Typical calculations in the design and operation of oil depots and oil pipelines* (UFA: DesignpolygraphService)

[5] Sozinova D N 2016 *The thermal effect of underground tanks with industrial waste of the fuel and energy complex on frozen ground* (Tyumen State University, Institute of Physics and Technology. – Tyumen)

[6] 2004 STO 0030-2004 *Vertical cylindrical steel tanks for oil and oil products. Rules for technical diagnosis, repair and reconstruction* (Moscow: CJSC TsNIIPSK n.a. Melnikova)

[7] 2008 STO 02494680-0044-2008 *Vertical cylindrical steel tanks for storing liquid products. Rules for testing strength, stability and tightness (*Moscow: CJSC TsNIIPSK n.a. Melnikova)

[8] Shastunova U Yu 2018 *Heat and mass transfer in the system "hot reservoir - base - frozen soil,* dissertation of the candidate of physical and mathematical sciences, Tyumen

[9] Zemenkova M, Zemenkov Yu, Gladenko A and Podorozhnikov S 2016 Estimation of Emissions During Monitoring of Pipelines in the Dynamic Mode of Operation *MATEC Web Conf.* https://doi.org/10.1051/matecconf/20168604053

[10] Dudin S, Bahmat G, Zemenkov Yu, Voronin K and Shipovalov A 2017 Strategy for monitoring and ensuring safe operation of Russian gas transportation systems *MATEC Web Conf.* DOI: 10.1051/matecconf/201710606004

[11] Dudin S M, Zemenkov Y D and Shabarov A B 2016 Modeling the Phase Composition of Gas Condensate in Pipelines *IOP Conference Series: Materials Science and Engineering* **154(1)** DOI: 10.1088/1757-899X/154/1/012010

Justification of the choice of pump drive for pumping stations of main pipelines, taking into account the principles of energy and economic efficiency

A G Zakirzakov and Yu D Zemenkov

Tyumen Industrial University, 38, Volodarskiy str., Tyumen, 625000, Russia

E-mail: zakirzakovag@tyuiu.ru

Abstract. One of the key conditions for the dynamic development of oil trunk transportation is to reduce the cost of pumping whose important component is energy costs. Energy transportation costs can be reduced by choosing an optimal type of the drive for pumping stations during construction and reconstruction of existing oil pipelines. An alternative to the electric drive can be a gas turbine. In foreign countries, oil is transported through the gas turbine pump drive. In the conditions of changes in energy prices, it is necessary to develop a method for comparing types of drives of the pump unit, each of which has its own advantages and disadvantages, in order to compare factors such as technical characteristics of the drive, distance from power sources, and the availability of a gas supply system, etc. Development of the pump drive selection methodology that takes into account international and Russian experience is relevant and will allow us to justify the choice of aggregate drives when designing pump stations.

1. Introduction

The basis for increasing the economic efficiency of each country is the rational use of production resources, material and energy resources.

A series of world energy crises of the 1970s, as well as the embargo on oil from the Arab countries, which caused an increase in energy prices, prompted the world community to reduce the dependence of the economy on energy consumption. As a result, the energy conservation policy has become one of the solutions to the problem of economic efficiency. The states began to stimulate projects on energy conservation, including the efficiency of consumption of fuel and energy resources.

Huge material costs associated with the construction and operation of trunk pipelines, as well as the energy costs of oil and gas transportation necessitate a careful approach to the reliability of the trunk oil and gas pipeline systems.

2. Results

The calculations of the ratio of reduced costs of energy when using a gas turbine and an electric drive made by the Russian State University of Oil and Gas (Moscow) helped to determine the ratios of prices for fuel gas and electricity (Figure 1), as well as to build diagrams that determine the area of rational use of these drives, taking into account the ratio of energy prices and efficiency of a gas turbine engine. [3-7]

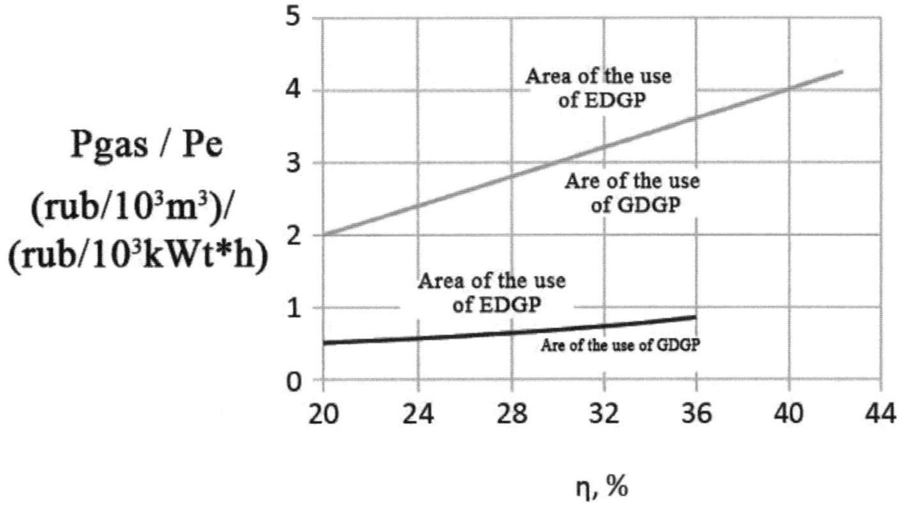

Figure 1. Areas of the rational use of gas turbines and electric drives

The data obtained show a critical ratio of electricity and fuel gas prices under the rational use of gas turbine and electric drives depending on their efficiency. It can be concluded that the gas turbine drive is more preferable at compressor stations of gas mains.

To transport oil, electric motors of main pumping units are widely used. The electric drive used on main pumps has many advantages, however its use entails additional costs associated with the use of electricity. In addition to direct additional financial costs to compensate for the increase in process for 1 kW * hour of electricity, it is necessary to take into account the cost of its supply. The construction of power lines for oil pipelines causes a number of difficulties. Firstly, the construction of power lines requires significant capital expenditures, because they lie in inaccessible sparsely populated areas; secondly, given the length of the conductive lines, which can reach several thousand kilometers, there are significant losses of electricity.

At the initial stage of development of oil pipeline transport, the choice of the type of drives for pumps was not difficult. There was only one source of energy - combustion energy of pumped products. However, in future, for geographically distant regions and regions with difficult climatic and geographical conditions, the electric drive began to be used. This was largely due to the electrification of the country. Over time, the electric drive of the main pumps has become virtually uncontested. The modern design system leaves no choice for a design engineer. Not all the factors that determine the choice may not be relevant. Technical solutions are different in different regions.

An alternative to the electric drive is a gas turbine. Oil is transported through the gas turbine pump drive abroad. A good example is the Transalyaskin trunk oil pipeline. CPC uses Siemens gas turbines (in regions with the undeveloped energy infrastructure).

Given the cost of construction of power lines and energy losses, the cost of transmission of 150 MW per 400 km is half the cost of building a thermal power plant of the same capacity.

The use of gas turbines is determined by the fact that a significant part of the main oil pipelines passes through the underdeveloped areas and areas located at considerable distances from the energy systems.

Single-shaft and twin-shaft aircraft, marine and industrial gas turbine engines with intermediate cooling of cyclic air, regeneration and utilization of heat of exhaust gases are used to drive pumps of the main oil pipelines. For these engines, it is possible to use various types of fuel, including prepared crude oil. To increase the efficiency, the fuel system and gas turbine engine automation are sometimes designed taking into account the use of two types of fuel - liquid and gaseous ones. The heat of exhaust gases can be used for household needs. Therefore, the efficiency of such gas turbine units reaches 31–35%, while the efficiency of aircraft and ship power plants rarely exceeds 23–25%.

When choosing the drive, it is necessary to take into account conditions of the construction area. It should be borne in mind that in order to improve the structure of the energy balance, it is desirable to use the drive which provides higher efficiency and runs on electricity generated by the least scarce and qualified fuel.

By comparing the efficiency of electric drive and gas turbine drives, it is possible to determine the most appropriate drive for the pump unit.

The efficiency of the electric drive is expressed as follows:

$$\eta_{pp} = \eta_{эс} \times \eta_{tr} \times \eta_{ee} \times \eta_{\text{р}}; \quad (1)$$

where η_{pp} – Efficiency generating power plants in the power system; η_{tr} – Power transmission efficiency; η_{ee} – engine efficiency; $\eta_{\text{р}}$ – Gearbox efficiency.

Drive applications are justified if the conditions are met

$$\eta_{pp} > \eta_{gt} \text{ or } \eta_{pp} \times \eta_{tr} \times \eta_{ee} \times \eta_{\text{р}} > \eta_{gt} \quad (2)$$

where η_{gt} – Gas turbine drive efficiency.

In order to determine the appropriate use of an electric drive, one can obtain expressions for comparing the efficiency of a power plant and a gas turbine:

$$\eta_{pp} = \frac{\eta_{gt}}{\eta_{tr}\eta_{ee}\eta_{\text{р}}}; \quad (3)$$

If

$$\eta_{tr} = 0.85;\ \eta_{ee} = 0.978;\ \eta_{\text{р}} = 0.985;$$

$$\eta_{pp} > \frac{\eta_{\text{гт}}}{0.85 \times 0.978 \times 0.985} = \frac{\eta_{\text{гт}}}{0.819} = 1.22\,\eta_{gt} \quad (4)$$

For the GTK-25, the efficiency is 0.29.

Therefore, if $\eta_{pp} > 1.22 \cdot 0.29 = 0.354$, the use of an electric drive is most appropriate.

The efficiency of gas turbine plants can be significantly increased by using heat of the exhaust gases (by-products).

3. Discussion

When using gas turbine units, the issue of fuel used is crucial. It is most advisable to use a pumped product as fuel for engines. Petroleum products such as kerosene, gasoline, diesel fuel correspond to the requirements for gas turbine fuel. However, in oil pipelines, the use of kerosene or diesel fuel for gas turbine engines is not economically feasible due to the high cost and difficulty of delivery. In cases where a gas pipeline is installed parallel to the pipeline, it is very efficient to use natural gas as fuel. When using the pumped crude oil as fuel, vanadium corrosion occurs on the parts of the combustion chamber and turbine. At high temperatures, corrosion develops rapidly, which leads to a sharp reduction in engine life. The use of special additives, washing oil with special aqueous solutions, limiting the initial temperature of gases to 650 ° C helps to slow down the corrosion process and clog the engine's gas-air path.

Oil should be prepared by separation or distillation methods or using additives. An example of oil preparation is the use of block distillation units of a column type. After oil rectification, 25-30% of light fractions are obtained, and the remainder (fuel oil) is pumped into the oil pipeline. The use of such a column is economically viable in areas which are remote from power sources and oil refineries, since the cost of the resulting fuel for gas turbine plants is half that imported.

The dependencies between the main operating parameters of the electric motor and pump are widely studied. In order to determine the relationship between the main operating parameters of gas turbines, the characteristics of various gas turbine engines were studied.

The rotational speed of a free turbine is independent of its throughput, which allows us to maintain pressure, outlet temperature, and air flow. Since the gas turbine works in one mode, the fuel consumption remains unchanged. There is a dependence of the main parameters of the turbine on the rotational speed of a free turbine. It is due to changes in the efficiency of a free turbine at different rotational speeds of the blades and constant adiabatic operation of the turbine. In such cases, there is a maximum relationship between power of the gas turbine of a free turbine and speed.

To determine the dependencies of the main parameters, we analyzed characteristics of both Russian and foreign gas turbine engines and experimental data:

$$\bar{N} = \frac{N_p}{N_{nom}}; \; \bar{G_{\text{т}}} = \frac{G_{\text{т}}}{G_{\text{т}\, nom}}; \; \bar{n} = \frac{n_p}{n_{p\, nom}};$$

An analysis of performance of the gas turbine engine allowed for the conclusion that for each value of turbine power, there are optimal values of the number of its revolutions. After analyzing the turbine speed at the peak of power, the optimal mode curve was created (Figure 2).

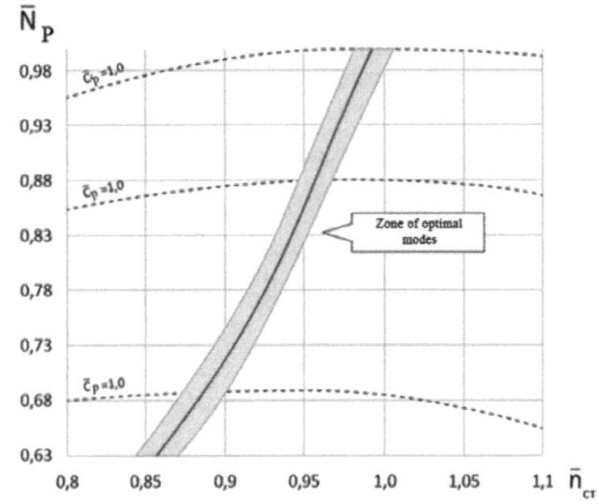

Figure 2. The zone of optimal modes of the power turbine

4. Conclusion

Given the efficiency of the turbine engine, the zone of optimal use of the gas turbine engine was calculated.

Having analyzed the existing pipeline transport systems and equipment of the main pumping stations and compared various drives, we can conclude that the electric drive used in 98% of cases is not alternative. The choice between different types of pump drives can be determined depending on the territorial location of the oil pumping station. In those regions where electricity is not expensive, an electric drive can be more economical than a gas turbine one. If it is necessary to change the operating mode of the pumping unit, thanks to the more economical method, the main pump driven by a gas turbine will be preferable.

In conditions of uneven oil supply, the use of a power turbine as a drive of main pumps can compete with an electric motor, and the optimum mode will save fuel gas.

References

[1] Antropova A B and Zakirzakov A G 2013 The use of associated petroleum gas *Oil and Gas of Western Siberia", materials of the international scientific and technical conference dedicated to the 50th anniversary of the Tyumen Industrial Institute*,Tyumen pp 64-66

[2] Gumerov A G, Gumerov R S and Akberdin A M 2001 Operation of equipment of oil pumping stations (Moscow: Nedra)

[3] Zakirzakov A G and Egorov A L 2015 Analysis of the state of the network of main oil pipelines of the Tyumen region based on statistical data *Modern problems of science and education* **1**
[4] Zemenkov Yu D, Shpileva V A, Podorozhnikov S Yu and Zakirzakov A G 2014 Energy-technological complexes in the design and operation of objects of transport and storage of hydrocarbon materials (Tyumen: Vector Book)
[5] Tyrylgin I V, Shpileva V A and Zemenkov Yu D 2012 Energy conservation and energy efficiency of the economy, production, transport of oil and gas in Russia *News of higher educational institutions. Oil and gas* **6** 29-33
[6] Shpileva V A and Zakirzakov A G 2016 Comparison of electric and gas-turbine types of drive of pumping stations of main oil pipelines *Successes in Modern Natural Sciences* **6** 202-206
[7] Shpilevoy V, Zakirzakov A and Shabarov A 2016 Comparison of electric and gas turbine drive types at pump stations *MATEC Web of Conferences* 01011
[8] Zemenkova M, Zemenkov Yu, Gladenko A and Podorozhnikov S 2016 Estimation of Emissions During Monitoring of Pipelines in the Dynamic Mode of Operation *MATEC Web Conf.* https://doi.org/10.1051/matecconf/20168604053
[9] Dudin S, Bahmat G, Zemenkov Yu, Voronin K and Shipovalov A 2017 Strategy for monitoring and ensuring safe operation of Russian gas transportation systems *MATEC Web Conf.* https://doi.org/10.1051/matecconf/201710606004
[10] Dudin S M, Zemenkov Y D and Shabarov A B 2016 Modeling the Phase Composition of Gas Condensate in Pipelines *IOP Conference Series: Materials Science and Engineering* **154(1)** DOI: 10.1088/1757-899X/154/1/012010

IOP Conf. Series: Materials Science and Engineering **952** (2020) 012010 doi:10.1088/1757-899X/952/1/012010

Directions for improving the motivational mechanics of modern oil and gas production companies

E L Chizhevskaya, J D Zemenkov and M J Zemenkova

Tyumen Industrial University, 38, Volodarskogo str., Tyumen, 625000, Russia

E-mail: chizel76@yandex.ru

Abstract. In the modern difficult financial situation when the need for highly qualified staff is increasing, the development of effective staff motivation systems is a crucial task. For any enterprise, regardless of its size and scope, it is necessary to update the payment and labor incentive systems as part of formation of a flexible motivational mechanism that allows it to gain maximum profit at minimum costs. For oil and gas production companies, the solution to this problem is complicated by the large number of personnel and a wide range of personnel qualifications, which increases the relevance of the task

1. Introduction

When the need for highly qualified staff is increasing, the development of effective staff motivation systems is a crucial task. Any operating company, regardless of its size and business sphere, has to update the payment and labor incentive systems in order to create a flexible motivational mechanism to gain maximum profit at minimum costs [1].

2. Materials and methods

According to the survey results, it was possible to identify the dependence of prevailing needs of workers on several criteria: category, education and age. Figure 1 shows the structure of needs of all company workers. The general pattern observed in all groups of respondents is the presence of material needs (45%), but security guarantees (15%) are no less important. The need for social contacts was identified in a tenth of the workers. Equally important are the needs for creativity and recognition (15%) (Figure 1).

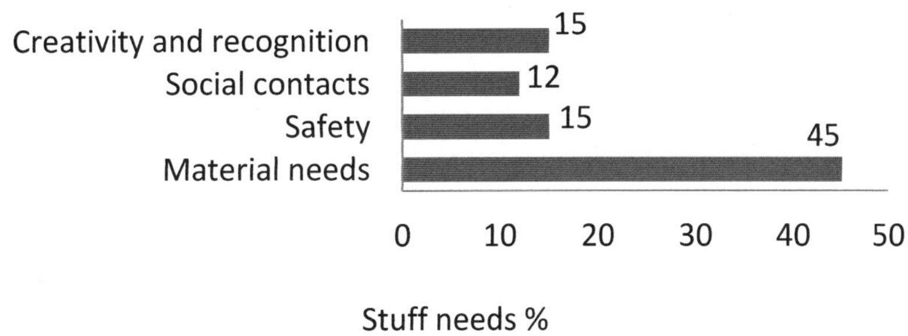

Figure 1. Distribution of needs of oil and gas company staff

For the production personnel, material needs are more important than for employees. These needs are of priority (45% - workers; 33% - specialists).

Safety is also important for the workers (22% and 31%, respectively). This might be due to the "practice of employers using the threat of dismissal as an" anti-stimulus "[5].

Social contacts as a need were emphasized by 15% of production personnel, and 18% of specialists. Recognition and creativity were indicated by the equal number of managers and employees (17% and 18%). According to I. Vardanyan, numerous studies of motivation factors indicate that people who have needs for creativity recognition feel motivated to work, because creativity contributes to the need for recognition. They try to generate useful ideas, provide them with a sufficient degree of recognition [2]. In the production staff, recognition and creativity occupy the penultimate position (2-3%).

Material needs are very important (about 50%) (Figure 2). The need for safety does not depend on the level of education. The desire to be competent is expressed by workers with higher professional education (57%), and secondary professional education (12%). It should be noted that only workers with higher education experience a pronounced need for creativity.

An analysis of the survey results showed that at the age of 25 years, material needs are more important, at the age of 26-69, they rank third. Safety is important for the age group of 46 to 60 years. The desire to be recognized is one of the leading needs for the group of 25 year-old workers and younger s and workers aged 26-45. In the latter age group, the need for self-expression takes the same place. The need for social contacts is the most pronounced need for employees of the pre-retirement age; for the age category "up to 25 years", this need is less evident.

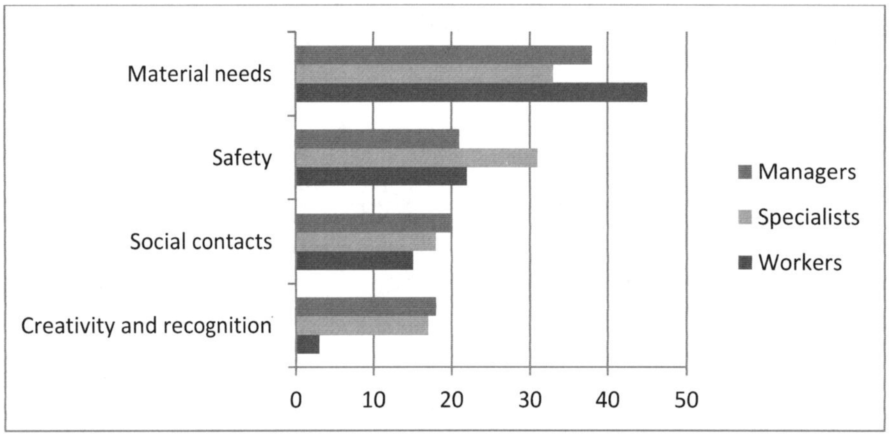

Figure 2. Distribution of staff needs by category

In general, according to the results of the study, we can conclude that in addition to material needs, the employees have other priorities. The least intangible needs were indicated by the production personnel, and workers with higher education aged 26-60 [7].

Safety guarantees are equally important for all categories, ages and jbs (from 13% to 31%). But for the respondents under 25 years with secondary vocational education, these needs are less improtant.

The need for social contacts is indicated by 15% -18% of the respondents aged 26-60. It is important for 55-60 year-old respondents (39%). 45% of managers and specialists aged 26-45 years with higher education have a need for recognition. 22% of employees of the same age and 36% of managers and specialists need self-expression (Figure 3).

The study revealed labor motives. The employee should be provided with external incentives. The questionnaire survey proved the importance of intangible factors; therefore, material incentives cannot serve as the only form of remuneration [6].

IOP Conf. Series: Materials Science and Engineering **952** (2020) 012010 doi:10.1088/1757-899X/952/1/012010

In order to analyze how the incentive system satisfies the needs of employees, one more questionnaire survey was conducted to assess the moral and psychological climate, work organization and conditions for self-expression.

3. Results and Discussion

The vast majority of respondents indicated that they were satisfied with relations with colleagues (34%), working hours (44%), and technical equipment (48%). However, all the employees said that they were completely dissatisfied with the diversity and independence of work, and career chances. Half the respondents were dissatisfied with relations with the management.

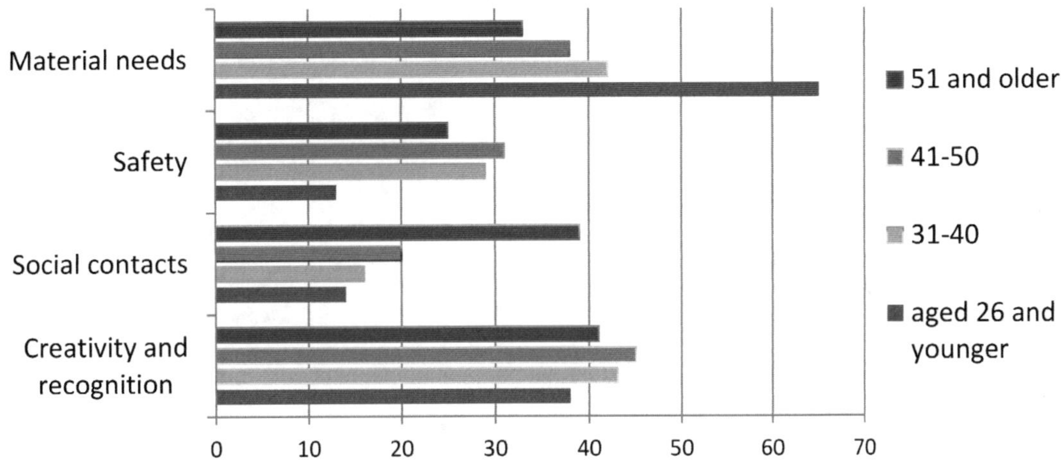

Figure 3. Distribution of staff needs by age groups

The next step in assessing the applied motivational mechanism was to study the opinion of managers (General Director, deputies, heads of departments and workshops, HR specialists).

According to the results, approximately 87% of managers understand staff incentives unambiguously. Some managers (69%) believe that "stimulation is money. Other respondents (28%) defined motivation as "creating comfortable conditions and n optimal climate for employees". The company's results are the same: staff incentives are a good monetary reward.

Only some managers combined goals and needs of employees and organizations and named all the factors that influence motivation. They realize the importance of an integrated incentive system [4,5].

Three of the four executives believe that the workers are satisfied with the existing labor incentive system. In addition, they believe that it is impossible to identify criteria for evaluating the work of workers, although, it is necessary to observe the following conditions:

- a shared goal;
- an analysis of jobs (their benefits), job descriptions;
- an analysis of technological processes and separation of functions between departments and employees;
- a clear organizational structure [3].

The company has implemented a package of social benefits. Many managers do not consider benefits as a motivation tool. Indeed, if benefits are not associated with specific results, they will not be incentives.

When determining the needs of personnel, managers put material needs in the first place, but the survey results refuted this belief of senior managers. Of 100% of possible incentives, 30% are used by managers: the opportunity to study, payment of expenses for food, one-time payments [8].

However, there is a group of incentives that are desirable for the staff, but not used by management: incentives for new ideas and their implementation, criticism, development of our own social infrastructure (medical care, sports), feedback, specification of existing criteria for remuneration, an opportunity for an employee to choose a remuneration form, development of "Career Planning and Development" program. Another drawback of the incentive system is the lack of feedback (employees' assessment of the motivation system).

As one of the options for improving the existing motivational mechanism and integrated personnel assessment, the competency-based approach can be used. It is based on the following evaluation criteria:

1. Labor productivity:
- volume of work performed;
- quality of work performed;
- timeliness.
2. Professional competencies:
- training;
- professional competence;
- ability to organize work.
3. Corporate competencies:
- initiative, focus on the active search for proposals;
- emotional endurance and psychological stamina;
- responsibility;
- contact;
- participation in contests, awards, thanks, titles.

4. Conclusion

Thus, the results of diagnostics of the motivation system implemented in the oil and gas company of the Russian Federation showed that the existing motivational mechanism does not make it possible to realize the labor potential. The primary task is to bring the goals and objectives of the company in line with the goals and tasks of each employee. The latter can be achieved by adjusting existing provisions on staff assessment, remuneration and labor incentives. Regular adjustment of the conditions and principles of assessment procedures will ensure the fullest realization of the potential of employees, motivate them to professional and creative development, and allow the company to form a loyal, stable and promising work team.

References

[1] Kuzmina Yu A and Rodina V N 2014 Improving the efficiency of the motivational mechanism of an oil and gas enterprise based on epy analysis of the level of satisfaction with staff work *Current problems and prospects of regional and sectoral development: Collection of scientific papers* pp 63-67

[2] Kuzmina Yu A, Chuba E V and Rodina V N 2013 Problems of non-monetary motivation at the enterprises of the oil and gas sector *Oil and gas of Western Siberia: Materials of the International scientific and technical conference dedicated to the 50th anniversary of the Tyumen Industrial Institute* pp 84-89

[3] Plenkina V V, Lenkova O V and Osinovskaya I V 2013 *Evaluation management in the personnel sphere: Textbook* (Tyumen: Tsogu)

[4] Plenkina V V, Chizhevskaya E L and Osinovskaya I V 2010 *Fundamentals of personnel management: Textbook* (Tyumen: Tsogu)

[5] Grigorieva P V, Voronin K S and Cherentsov D A 2018 Development of layout algorithm for pipeline, considering topographic features *IOP Conference Series: Materials Science and Engineering* 012022

[6] Voronin K S, Grigorieva P V and Cherentsov D A 2018 Estimating the cost of constructing and

operating a section of a pipeline in the search for its optimal route I*OP Conference Series: Materials Science and Engineering* 0120

[7] Chizhevskaya E L, M Y Zemenkova and Mayss N A 2020 Directions to Improve Effectiveness of Management of Management of the Industry Company Personnel *IOP Conf. Series: Earth and Environmental Science* **459** 052093 doi:10.1088/1755-1315/459/5/052093

[8] Zemenkova M Yu, Dubrovin D D, Shagbanov I F and Zemenkov Yu D 2019 I-res intelligent system for multi-factor risk control of accidents and the technical condition of the main gas pipeline *IOP Conference Series: Materials Science and Engineering* **663(1)**

Simulation of the stress-strain state of the working body of the system for the development of the pipeline trench

S P Pirogov[1,2], D A Cherentsov[1], K S Voronin[1] and A U Yakupov[1]

[1]Tyumen Industrial University, 38, Volodarskogo Str., Tyumen, Russia
[2]Northern Transural State Agrarian University, 7, Republiki Str., Tyumen, Russia

E-mail: piro-gow@yandex.ru

Abstract. The use of mechanical pressure gauges as pressure measuring devices is regulated by standard documentation. Also, manometric tubular springs – the sensitive elements of pressure gauges are used in various technology fields. Thus, the use of MTS in agricultural machinery allows reducing the traction resistance of tillage machines due to the effect of vibration when interacting with the soil, as well as improving the quality of the soil treatment process by adjusting the stiffness of the rack. This paper presents the results of calculating the stresses and deformations of the ripper shank structure, made in the form of a flexible tubular element when loaded with internal pressure and external force. The active development of the oil and gas industry is currently directed towards the remote territories of the Yamal peninsula and the territories of the Far East, which are characterized by complex natural and climatic conditions of construction. In this regard, it is necessary to develop new approaches and methods of construction in permafrost conditions. The paper presents the results of calculating the stresses and deformations of the structure of the ripper shank, made in the form of a flexible tubular element under loading by internal pressure and external force. The finite element method implemented in the "ANSYS" program was used for research.

1. Introduction

The development of the oil and gas industry is currently aimed at remote territories of the Yamal Peninsula and the territories of the Far East with promising deposits of hydrocarbon raw materials. The economic efficiency of the development of such deposits was influenced by the distance of fields from existing infrastructure, as well as by difficult natural and climatic conditions of construction.

In this regard, it is necessary to develop new approaches and methods of construction in the conditions of permafrost soil. According to the flow chart for pipeline trench development in permafrost conditions, the following process operations are included into the series of trunk line grid works:

- surface layout along the whole route using a bulldozer;
- geodesic survey of a trench;
- frozen soil loosening with a ripper-equipped bulldozer (Figure 1);
- excavation in a trench by a rotary bucket excavator;
- trench reworking by a single-bucket excavator up to project levels.

IOP Conf. Series: Materials Science and Engineering **952** (2020) 012011 doi:10.1088/1757-899X/952/1/012011

Figure 1. Ripper-equipped bulldozer for frozen soil loosening

The operation of frozen soil loosening can be optimized. The works [1, 2] showed that the vibration impact of a working element of tillage and sowing machines on cultivated soils ensures optimal agricultural requirements and reduces energy consumption. Thus, as an alternative to a stationary ripper it is proposed to use a structure where the working element is a flexible tubular element – a manometric tubular spring with a ripping tooth on the end (Figure 2).

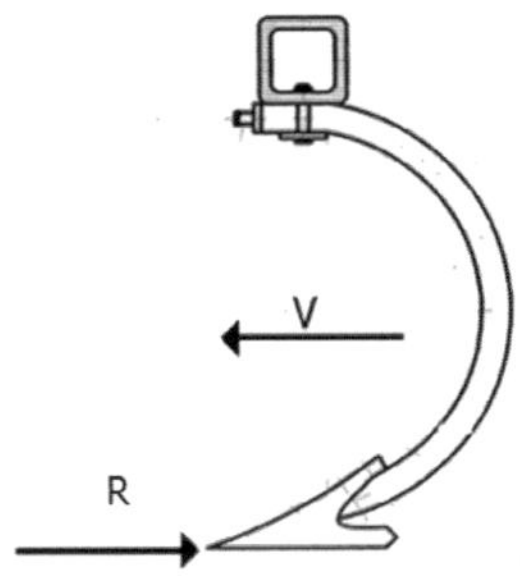

Figure 2. Working element of a trenching unit

The pressure change in the internal cavity of the manometric tubular spring makes cross sections to deform and the free end with a ripping tooth to make reciprocal movements. Variable pressure change will lead to oscillatory movements with certain amplitude and frequency, which depend on applied pressure parameters.

2. Materials and methods

The works of many researchers are devoted to the study of manometric tubular springs (MTS) and their vibration parameters [1, 2]. When the proposed design interacts with permafrost soils, it is necessary to limit the bulldozer movement speed since exceeding the speed limit violates the MTS integrity and, as a result, leads to the failure of the entire structure. The calculation of MTS pressure and deformation shall be carried out using the finite-element method in ANSYS software system.

3. Results

The simulation results obtained in ANSYS depend on the quality of the grid model. Earlier, in [2], a grid model was obtained, which increases the stability of simulation results for a sample tube with the following geometric characteristics: central angle – 180°, radius of curvature – 500 mm, major semiaxis of cross section – 25 mm, minor semiaxis of cross section – 12.5 mm, wall thickness – 2.5 mm, material – 36NHTU steel.

The grid model characteristics ensuring the required accuracy are as follows: grid method – Sweep, element size – 5 mm.

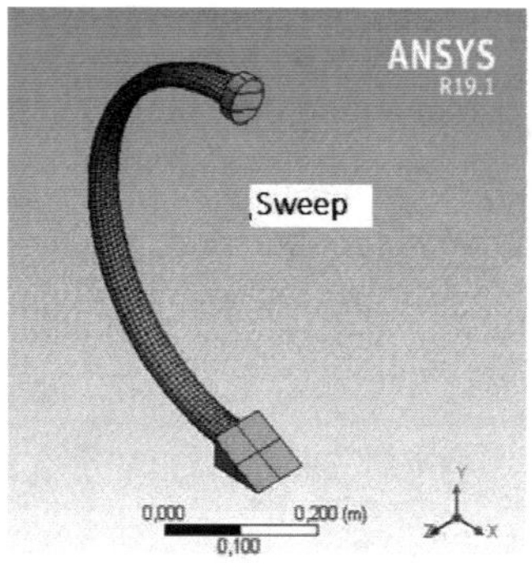

Figure 3. Grid model

Simulation experiments to assess the impact of internal pressure on the movement of the trenching unit working element and the maximum stresses in the considered structure were carried out. The results are shown in Figures 4 and 5.

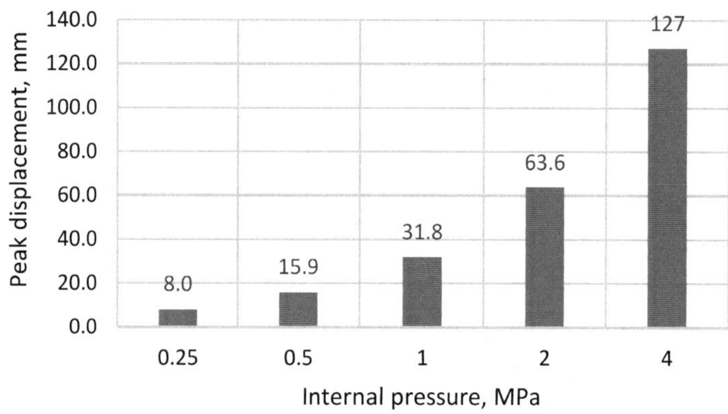

Figure 4. Impact of internal pressure on the end tube displacement

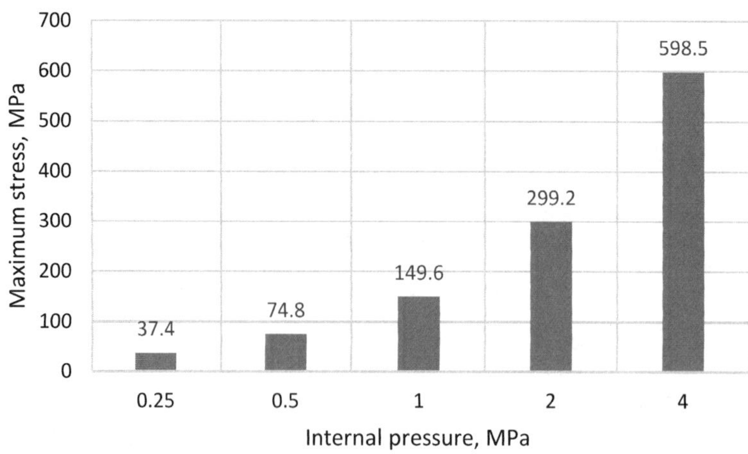

Figure 5. Impact of internal pressure on the maximum voltage

The analysis of the dependencies of movements and stresses on the internal pressure shows that at the initial loading – at 4 MPa – there is a linear increase of stresses and movements, while high pressures cause the deviation from linear dependence.

In order to determine the maximum speed of the bulldozer during the interaction of the proposed structure with permafrost soils, the dependence of the influence of the speed of movement of the bulldozer on the maximum horizontal force R was defined.

The calculation of the maximum horizontal force, at which the loss of static stability is observed and as a result the failure of the working element, was carried out in work [2] for different MTS cross sections and parameters. Thus, for plane-oval tubes the maximum possible R is 44.2 kN, for elliptical – 43, 8 kN, for eight-shaped – 45.9 kN. The maximum stresses, as well as the violation of MTS integrity, are observed at the anchorage base.

Permafrost soil characteristics: density – 1020 kg/m^3; freezing-thawing start temperature – minus 0.3°C, total humidity – 5.19 unit fractions, quantity of non-frozen water in frozen soil at permafrost temperature – 2.47 unit fractions, elastic modulus – 14 MPa.

The impact of bulldozer speed on the maximum horizontal force was determined in toolbox – Eigenvalue Buckling. The results of calculations are given in Figure 6.

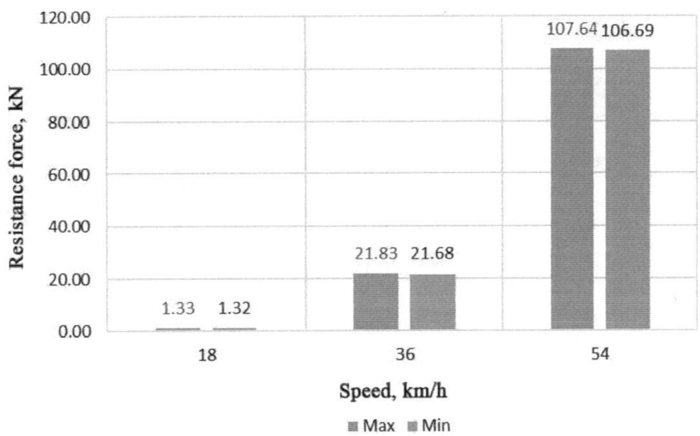

Figure 6. Impact of the displacement speed on the resistance force

The stress-strain analysis under the influence of the external force (1,000 N) showed that the end of the tube would move 30 mm. The necessary horizontal force for the buckling failure shall make at least 12,800 N. This confirms the sufficient strength of the considered structure.

4. Conclusion

The analysis of the dependencies of movements and stresses on the internal pressure shows that at the initial loading – at 4 MPa – there is a linear increase of stresses and movements, while high pressures cause the deviation from linear dependence.

The simulation of the working element in the form of MTS of the trenching unit showed that for the proposed MTS design the loss of stability will be observed at the horizontal force of 12.8 kN. In order to prevent buckling, it is necessary to limit the maximum speed of the bulldozer – not more than 25 km/h.

References

[1] Kokoshin S, Ustinov N and Kirgincev B 2016 Use of flexible tubular elements of the overhaul and tunnels reconstruction *Procedia engineering* **165** 817-828

IOP Conf. Series: Materials Science and Engineering **952** (2020) 012011 doi:10.1088/1757-899X/952/1/012011

[2] Pirogov S P, Ustinov N N and Smolin N I 2018 Mathematical Model of Stress-Strain State of Curved Tube of Non-Circular Cross-Section with Account of Technological Wall Thickness Variation *IOP Conference Series: Materials Science and Engineering* **357**

[3] Kokoshin S N 2018 Ensuring the depth of tillage by cultivators by changing the rigidity of the rack *Agricultural Bulletin of Stavropol Region* **1 (29)** 14-17

[4] Blednyh V V and Svechnikov P G 2014 Economical depth of main tillage *Tractors and agricultural machines* [In Russian - Traktory i sel'hozmashiny] **10** 34-35

[5] Fedorov S E 2016 *Improving the quality of surface tillage by regulating the stiffness of the elastic stand of the cultivator*, Diss. of candidate of technical Sciences, N.P.Ogarev Mordovia State University, Saransk

[6] Dmitriev S Yu and Dmitriev Yu P 2012 Automatic stiffness regulator of the elastic stand of the cultivator *Proceedings of GOSNITI* [In Russian - Trudy GOSNITI] **110 (1)** 85-87

IOP Conf. Series: Materials Science and Engineering **952** (2020) 012012 doi:10.1088/1757-899X/952/1/012012

The deployment of the collapsible piping through a large water obstacle

D I Melnikov, L V Seoev, M Yu Zemenkova and S Yu Podorozhnikov

Tyumen Industrial University, 38 Volodarskogo Str., Tyumen, 625000, Russia

E-mail: muzemenkova@mail.ru

Abstract. Portable pipelines have been used in the Armed Forces of our country since the early 1960s to transfer light oil products. They are used to pump light oil products in large volumes for considerable distances in any climatic and geographic conditions. Portable pipelines were successfully used to eliminate natural and technogenic emergencies. The paper presents an overview of the deployment of portable pipelines across water barriers. It shows the organization and technology of works during the deployment of a pipeline across the 2700 m barrier of the Volga River.

1. Introduction

Portable pipelines of the Armed Forces of the Russian Federation have gained prominence as a mobile and high-performance mode to transport low-viscosity fluids (light oil products, water). They are used to pump light oil products in large volumes for considerable distances in any climatic and geographic conditions. The efficiency of portable pipelines is most clearly confirmed by their large-scale use to supply fuel to the Limited Contingent of the Soviet Troops in Afghanistan. During 1980-1989, more than 5 million tons of light oil products, or 80% of the total volume of deliveries, were transferred to the territory of Afghanistan through portable pipelines with a total length of more than 1200 km [1].

The successful use of portable pipelines in emergency response is also well known, e.g. in Chernobyl [2], after the earthquake in Armenia [3], in the suppression of mass forest and peatland fires [4], to supply water to the population of the Republic of Crimea [5].

The advantage of using them is the ease of assembly. The pipes are connected without welding. This allows constructing a line with high productivity. Due to such capabilities, the portable pipelines are used in hard-to-reach locations. One of them is water barriers, including large ones.

2. Materials and methods

In order to develop skills to cross large water barriers using portable pipeline, the task was set to deploy a line to pump light oil products across the Volga River (Figure 1).

In order to select the method of crossing the water barrier using portable pipelines, the water area and coastal areas were explored along with the study of the river fleet documents.

According to survey results, the main characteristics of the water barrier were determined as follows: the width of the Volga River in the area of the construction of the underwater crossing of a portable pipeline amounted to 2,700 m, while at the distance of 500 m from the left bank there was a boggy island with shrubs and canes with a width of 350 m; the bed profile had a very uneven relief with sharp depth drops of up to 27 m and a flow rate of 1.4 m/s; intensive navigation.

Figure 1. Area of crossing the Volga river with portable pipelines

During the preparatory period, the accumulated experience of pipeline deployment across water barriers was studied.

The first known case – field collector pipeline laying with a diameter of 75 mm in 1937. The total width of the crossing area was 940 m, including across the Saifun River – 120 m, Smerinka duct – 125 m, the rest – across the boggy floodplain [3].

Further accumulation of experience in pipeline laying across water barriers took place during the Great Patriotic War. The examples of such works include the following: across the Volkhov River (280 m) with a carrier wire in spring 1942; across the Oka River in spring 1942; two lines on rafts across the Onega River in 1942; across the Don River in 1943; across the Dnieper River in the Dnepropetrovsk area (total 2720 m, including directly across the river 1070 m on rafts from empty metal barrels) in 1943; two lines with a length of 1120 m across the Visla River in 1945, first on ice, and with the onset of spring – with pipeline suspension to the pillars of the destroyed bridge [3].

A special place belongs to the pipeline laid during the Great Patriotic War across the Shlisselburg bay of the Ladoga Lake to supply fuel to besieged Leningrad. The pipeline represented a welded assembly. Its construction lasted 50 days, which at those times was quite short. During that time, a line of 21 km under water and 8 km on the shores of the lake was laid. During the period of operation, more than 40 thousand tons of light oil products were pumped through the pipeline [3].

The experience of laying pipelines across water barriers has not remained unused. Already in 1953 the pipeline crossings through the navigable Klazma River (width 135 m) and six non-navigational rivers (width up to 35 m) were laid during the maneuvers. The main method of laying is to pull pre-assembled and plugged lines afloat with their subsequent flooding. A year later, the Dnieper River and the Volga River, as well as the Minsk Reservoir, were crossed in the same during the maneuvers. Thus, the simplest way to cross water barriers, i.e. floating with subsequent flooding was initially used in practice [3].

It is worth mentioning another method. In 1959, during the maneuvers, a suspension cable was pulled across the Daugava River, to which a line was attached on safety hooks [3].

In 1960, when crossing the Volga River (about 1000 m wide), the floats from barrels on anchors were installed along the intended line of laying, which prevented the demolition of a pipeline by flow and breakage of pipes. After the floats were installed, the upstream pipeline was pulled afloat, then pressed against the floats by its current, and flooded with water from the shore. The works were successful, but this method turned out to be too labor-intensive, and the installation of anchors much increased the time of laying. Besides, it was necessary to temporarily stop navigation [3].

However, the attempts to use it on water barriers at a flow rate of more than 0.5 m/s were not successful – the pipeline was demolished by high-speed flow of water from the supposed laying line to the bottom of the river and broke along its body or in the joint. It was necessary to find a new technical solution. An example of this is crossing the Elba River in the German Democratic Republic in 1960. The width of the river is 300 m, the current speed is up to 10 m/s. A new method has been proposed and implemented – pulling a line, previously assembled on the bank, along the bottom of the river with its simultaneous flooding. In 1961, this method, which was improved by a special crown, was successfully used in pipeline laying across the Dnieper River. A cable was attached to the crown, delivered to the opposite shore and attached to the tractor, which dragged the line across the river [3]. This method became the main in practice and was repeatedly applied during the maneuvers, as well as in crossing the Amu Darya River with a width of 860 m with a portable pipeline, when a Limited Contingent of the Soviet Troops entered Afghanistan. The assembly of the cable-connected line on the pontoon bridge was successfully tested here with its subsequent lowering into the river and flooding [1].

In the 1990s of the 20th century during the use of a portable pipeline to pump oil from the remote areas of Siberia such methods of crossing the water barriers, generally small water barriers, as the placement of a line in a bearing structure – a truss, bigger diameter pipes, etc. were used [3].

Unsuccessful attempts to use the method of afloat crossing include the maneuvers at the Volga Reservoir in 1979. The length of the crossing was 9.5 km. The flow speed was only 0.2 m/s, but the strong wind (13-16 m/s) and waves up to 1.8 m high presented serious difficulties. There was almost no flow in the nearby backwater, but the waves and wind were the same as on the main river line. With the use of a harbor pusher tug (power – 300 hp, afloat traction – 3 tf) it was possible to extend afloat a line of 5 km long in the backwater, but the pipes broke under the influence of wind and wave impact. The attempts to pull a line of up to 2 km through the channel were not successful either [6-8].

The review of the accumulated experience highlighted the following proven ways of routing a field pipeline through water barriers:

- afloat with subsequent flooding (with line support by floats);
- pulling along the river bottom with simultaneous flooding;
- using load-bearing structure (pillars of a broken bridge, a truss, bigger diameter pipes, suspension cable, etc.);
- on rafts (floats).

Each of them has certain advantages and disadvantages and is applicable in a purely specific environment.

The analysis of the above methods of crossing the water barriers taking into account local peculiarities showed that the most acceptable method is the method of laying a portable pipeline across the Volga River along the bottom with its simultaneous flooding.

Considering the longer length of the subsea part of a pipeline, a separate "preventer" was rigidly attached along the entire length of the pipeline in order to prevent possible rupture of pipe joints during pulling. The second cable was used as a traction cable to pull the pipeline from the left bank to the right bank using a heavyweight tractor. A float of empty barrels was attached to the head portion of the line to prevent the pipeline crown from catching underwater in the bottom relief roughness.

The new technical solution was the simultaneous connection of pumping and compressor stations to the drag line of the pipeline. This was ensured by filling the line with water to flood it (pumping station) or emptying from water with air to surface the line when pulled, when the situation on the navigation channel (compressor station) would allow.

The required traction effort was calculated taking into account the following factors:

- the greatest traction force will be required when the whole line has crossed the river and each bank has 100 m of pipeline;
- the total length of the laying line is assumed equal to 2900 m;
- the length of the line located at the river bottom by the moment of laying completion is 2350 m;
- the length of the line on land (including an island 350 m wide) is 550 m.

Based on the above factors, the required traction force F (kgf) is determined by the formula [6]:

$$F = 10(l_д + 2*l_c),$$

where $l_д$ and l_c – length of lines at the bottom and on land respectively, m.

The design value of the maximum required traction force at the moment of laying completion was 34,500 kgf or 338,100 N. A metal cable as per GOST 7688-80 with the 33 mm diameter was used to pull the pipeline, which withstands breaking strength equal to 588,000 N. safety factor made 1.74. The traction cable was connected both to the crown and to the safety cable of the portable pipeline (diameter – 22 mm, breaking strength – 26,370 kgf or 258,500 N).

The preventer was designed to reduce the longitudinal load from the pipes of the laying line, which reduced the possibility of rupture of their joints. The cable was attached to pipes by couplings (Figure 2).

The calculations also took into account the permissible breaking strength not only of the cable, but also of the pipes made of 16GS steel with the allowable stress [σ] = 490 MPa. Taking into account the thickness of the pipe walls and the tensile strength of the 16GS steel, the maximum breaking strength of the pipe by metal made 196,710 N. Thus, the total breaking resistance of the cable with the diameter of 22 mm and the pipe metal equals 455,210 N, which is 1.35 times the maximum required strength for pipeline laying thus guaranteeing its protection against breakage.

Figure 2. Attachment of the preventer cable to the pipeline

Thirty people were laying the pipeline through the water barrier. Mobile pumping units MPU-100/200M and fuel-transfer stations FTS-160 were used as pumping facilities. The floating transporter PTS-2, four push boats BMK-MT and pontoons were used to carry out the work on the water. The track plotter БАТ-2was used as a tractor.

Pipes and pipeline equipment loaded into cars were transported through the Volga River by ferry. The delivered equipment was placed on special sites where it was prepared for deployment, including the completion of pipe joints with removable parts.

At the first stage of the pipeline crossing the water barrier, the works were carried out to prepare the traction cable with a total length of 3000 m for pulling to the opposite bank of the river. The cable sections with the length of 500 m using the crane-manipulator of the maintenance workshop were unwound from bays and laid on shore, after which they were fixed to each other.

A complex barrier to pulling the cable was the boggy island, which required the use of a floating transporter and a boat. The cable of the original length of 1000 m was laid on two pontoon sections and transported by boat to the island with simultaneous laying on the bottom of the river. The boggy section, which in some places was filled with canes up to 3 m high, did not allow getting into the island neither a floating conveyor, nor a boat, so the cable was pulled by hand, and where it seemed possible a conveyor hoist was used. The end of the cable was taken to the opposite shore of the island (Figure 3).

Figure 3. Dragging of a traction cable through the boggy island

Eight pontoon sections were required to lay and transport the remaining 2000 m long cable (Figure 4). Using four boats it was towed to the opposite bank of the island and after fastening the sections of the cable to each other the pontoons were transferred to the right bank of the river with simultaneous laying of the cable to the bottom.

At the same time, the works were carried out on the shore to deploy the pump station. The minor depth of the river at the shoreline required the installation of FTS-160 (for hydraulic pipeline testing) on a three-section pontoon (Figure 5).

Figure 4. Laying of a traction cable on pontoon

Figure 5. Fuel transfer station FTS-160 on a pontoon

The process diagram of the fuel transfer station consisting of one MPU-100/200M and FTS-160 pumping station ensured stable operation of the pipeline, as well as provided the possibility of its pneumatic and hydraulic unloading.

Pipe sections were installed on the shore in advance. The area allowed to mount 8 sections up to 120 m long (20 pipes in each). To ensure pipeline tightness, all sections were hydraulically tested using FTS-160 (Figure 6).

Figure 6. Hydraulic test of assembled pipeline sections

To avoid burying the pipeline crown into the bottom ground while it was being pulled through the water barrier, an original float consisting of four empty 200-litre barrels with a bottom-welded metal sheet that served as a kind of sleds was applied. Considering the weight of the entire structure, the total design load capacity of the float was 535 kg. In order to determine the location of the crown taking into account the maximum depth of the channel, two buoys were tied to the float with 20 m long cables (Figure 7).

Figure 7. Pipeline float

Another new technical solution was the use of an insert-trap to fill the pipeline with water, which was mounted after the first two hermetically sealed pipes. The tightness of the two pipes ensured their positive buoyancy (38 kg). Thus, the total load capacity of the float was 573 kg, which allowed the crown of the pipeline to be kept in the floated state [9-13].

The BAT-2 track plotter (Figure 8) was used as a tractor to pull the pipeline. Taking into account the connection of a hoist, its characteristics met the design requirements.

On the pontoons, the track plotter was transferred to the opposite bank of the river. The construction of the crossing across the Volga River using portable pipelines began after connecting the hoist of the tractor with the traction cable and its tension.

The pipeline entered the water along sections for the length of the unwound cable of the tractor hoist. After the end of the next section of drawn pipes reached the water line, the next section was connected to it, which was successively attached to the safety cable.

Figure 8. Track plotter BAT-2

In order to reduce the weight of the pipeline and reduce the coefficient of friction against the soil during the period of pulling through the water barrier, water was displaced from its inner cavity by the compressor station, as a result of which the whole line was afloat (Figure 9). This significantly reduced the load on the tractor hoist.

Figure 9. Pulling the portable pipeline afloat

After passing 500 m of clean water, the crown of the pipeline reached the boggy island. The sunken tree roots and the beds of canes created some difficulties in overcoming the island (Figure 10), which caused the need to increase the traction force on the tractor to 50 tons by connecting the pulley block tackle to the hoist.

Figure 10. Pipeline crown inlet to boggy island

A compressor station was also used during piping through a 1850 m wide section of large water. When the water was replaced with air, the crown of the pipeline floated thus making it easier to pull the entire section (Figure 11).

The organization of works when crossing of the navigable channel required particular attention, which was caused by the need to fit within short intervals between ship journeys.

After the crown reached the right bank of the river, the entire undersea part of the 2,700 m pipeline was subjected to a hydraulic test.

During the pipeline closing period, we faced considerable difficulties in removing it from the water, so the operations were carried out in two stages [14,15].

Figure 11. Displacement of water from pipeline by air

At the first stage, a short section of 500 m was pulled. Near the water line on the nearby shore of the island, one connection was dismantled, and water was displaced from the inner cavity of the pipeline by means of a compressor station, as a result of which the afloat pipeline was easily pulled to the land.

Due to a combination of negative factors, the second stage proved to be more difficult. The navigation intensity on the river prevented the method of pulling the entire pipeline afloat. The length of the pipeline section was 2,200 m, of which 350 m laid on the island was almost completely tightened into the boggy soil, which affected the increase in the total pipeline weight, and the limited terrain prevented the tractor from pulling the sections of more than 50 m from the water. The sections of pipes pulled from the water were dismantled, and to secure the traction cable the crown was re-connected to the next section.

After dismantling, the pipes were loaded on pipe trucks and taken to their storage site.

3. Conclusion

Complex work related to the deployment of a portable pipeline across a large water barrier with a width of 2700 m was performed successfully. A number of new technical solutions, which had never been previously used in portable pipeline practice, contributed to this success:

- connection of a pulled line along its entire length to the safety cable;
- use of original float fixed on a pipeline crown;
- periodic flooding and lifting of the line to the water surface depending on the time of ship journeys along the navigable channel.

During the maneuvers, due to the introduction of new technical solutions, for the first time a portable pipeline was successfully laid across the water barrier with a width of 2700 m. Undoubtedly, the experience gained will serve to further improve the methods of crossing large water barriers with portable pipelines.

References

[1] Danilchenko I G, Vinogradov V A, Kartashov A G and Sirenko A M 2009 Fuel service in Afghan War (M.: Gralia)

[2] Sereda V V, Danilchenko I G and Melnikov D I 2015 Use of field main pipelines to increase fire safety of the Russian Federation *Fundamental research* **10** 309-313

[3] Shein K G, Sereda V V and Danilchenko I G 2005 *50th anniversary of pipeline troops. Historical sketch. 2nd edition* (Moscow: Gralia)

[4] Sereda V V, Melnikov D I and Danilchenko I G 2014 Use of field main pipelines in emergency response *Military thought* **9** 29-36

[5] Melnikov D I, Kopnyshov S L and Sedykh N I 2015 Peculiarities of providing drinking water to the settlements of Crimea *Fires and Emergency Situations* **1** 87-91
[6] 1982 *Field main pipelines of increased capacity. Operating Manual* (M.: Voenizdat)
[7] Sereda V V and Ulit'ko A V 2014 Role of Chemmotology at the Present Stage *Chem Technol Fuels Oils* **50** 363-368 DOI: 10.1007/s10553-014-0534-0.
[8] Ostrovskaya V M, Sereda V V, Prokopenko O A et al. 2013 Indicator Tubes for Determining Anti-Knock Additives in Automotive Gasolines *Chem Technol Fuels Oils* **49** 444–450 DOI: 10.1007/s10553-013-0468-y.
[9] Ziatdinov A M, Sviridova A G, Sereda V V and Skrylnik P G 2009 In situ conduction ESR and theoretical studies of graphite intercalation by antimony pentafluoride *Applied Magnetic Resonance* **18**
[10] Ziatdinov A M, Sviridova A G, Sereda V V et al 2009 In situ conduction ESR and theoretical studies of graphite intercalation by antimony pentafluoride *Appl Magn Reson* **35** 221 DOI: 10.1007/s00723-008-0156-y.
[11] 2017 Decree of the President of the Russian Federation No. 176 *Strategy of ecological safety of the Russian Federation for the period up to 2025*
[12] Concept of Improvement of the State Policy in the Field of Industrial Safety until 2020, available at: http://docs.cntd.ru/document/902354089.
[13] Zemenkova M Yu 2019 *Methods of reducing technological and environmental risks in the transportation and storage of hydrocarbons* (Tyumen: TIU)
[14] Zemenkova M, Zemenkov Yu, Gladenko A and Podorozhnikov S 2016 Estimation of Emissions During Monitoring of Pipelines in the Dynamic Mode of Operation *MATEC Web Conf.* DOI: 10.1051/matecconf/20168604053
[15] Dudin S M, Zemenkov Yu D, Maier A V and Shabarov A B 2016 Research and Design of Thermophysical Gas-Liquid Mixture Parameters in Product Pipelines *IOP Conference Series: Materials Science and Engineering* DOI: 10.1088/1757-899X/154/1/012021

Calculation of the speed descrioptions of the upward swirling gas flow in conditions of lateral wind action

A G Obukhov

Tyumen Industrial University, 38, Volodarskogo Str., Tyumen, 625000, Russia

E-mail: agobukhov@inbox.ru

Abstract. The paper presents the results of numerical simulation of lateral wind impact on three-dimensional non-stationary air flows in ascending vortex flow of artificially created tornado in the stationary operating mode. The mathematical model represents a complete system of Navier-Stokes equations taking into account the viscosity and thermal conductivity of the moving gas, as well as the effects of gravity and Coriolis. Using the explicit difference scheme and the corresponding choice of initial and boundary conditions, the solutions of the complete system of Navier-Stokes equations in the computational domain representing a rectangular parallelepiped are numerically obtained. All components of gas flow speed are calculated at fixed time points and instantaneous current lines are built taking into account the constant horizontal wind speed. The calculations showed that the result of the wind action on the ascending vortex flow is an unsymmetrical change in the circumferential velocity, its uneven deformation at opposite sections, and the total displacement of the vortex in the wind direction. Besides, the displacement and the curvature of the vertical vortex part in wind direction are established, the "separation" of some instantaneous current lines from vertical rotating part and the area in the vortex center increasing in diameter, free from current lines is fixed. In the studied time change range, the wind resistance of the air vortex and stable operation of the used computational scheme are observed.

1. Introduction

The relevance of numerical modeling of complex three-dimensional non-stationary flows of viscous compressible heat-conducting gas in natural atmospheric vortices and gas pipelines is caused by the lack of convincing theory and numerical algorithms that adequately describe such flows. The creation of such a theory and calculation schemes is a serious task both in terms of mathematical justification and from a computational point of view. The results of the studies based on adequate models can be used as the basis for minimizing the negative destructive effects of atmospheric vortices, as well as for improving the efficiency of gas pipeline systems and reducing energy costs, as they are more appropriate to their actual operating conditions. At the same time, the analytical studies allow establishing the mathematical solutions to the considered problems. Numerical modelling provides specific numerical characteristics of the studied flows, including for their comparison with operational data and with experimental observations.

The pattern of occurrence and stable functioning of the ascending vortex air flow [1] implies the presence of a sufficiently long-term ascending air flow and the accompanying radial air flow, as well as the effect of the Coriolis force imparting substantial circumferential velocity to air particles at the bottom.

The fact of vortex occurrence and its direction is strictly mathematically proved by the corresponding theorems [2, 3] and confirmed experimentally [4, 5].

The observations of natural ascending vortex flows [6] made it possible to hypothesize the existence of a boundary separating the external quiet air from the air moving in the ascending vortex flow. In gas dynamics such boundary is called a contact surface [7, 8]. Therefore, in order to create a stable upward vortex flow under laboratory conditions, an air-impermeable vertical cylindrical pipe with an exhaust fan guiding air through the pipe from bottom to top is used as the contact surface [3, 9].

Based on experimental results, all gas-dynamic parameters [3, 10], including speed and energy characteristics of the three-dimensional non-stationary ascending vortex gas flow of a specific scale caused by vertical blowdown, were numerically modeled and calculated. The numerical calculations of this gas flow made it possible to give specific proposals and recommendations for potential large-scale experiment to twist large air masses – artificial tornado.

The previously obtained numerical results [10-12] form the basis for further study in the gas-dynamic theory of destructive atmospheric vortices. In particular, the numerical simulation of the phenomenon of the curvature of the vertical part of the ascending vortex flow – an "arm" of visually observed natural tornadoes – seems quite interesting [6]. This effect clearly indicates significant stability of the atmospheric vortex and does not yet have quite adequate scientific explanation. The purpose of this study is numerical simulation of artificial tornado and calculation of its speed characteristics under lateral wind impact.

2. Mathematical model

A complete system of Navier-Stokes equations is used as a mathematical model describing complex three-dimensional non-stationary flows of compressible viscous heat-conducting gas (atmospheric air), which in dimensionless variables taking into account the action of gravity forces and the Coriolis force in the vector form can be presented as follows [3]:

$$\begin{cases} \rho_t + \mathbf{V}\cdot\nabla\rho + \rho\,\mathrm{div}\,\mathbf{V} = 0, \\ \mathbf{V}_t + (\mathbf{V}\cdot\nabla)\mathbf{V} + \dfrac{T}{\gamma\rho}\nabla\rho + \dfrac{1}{\gamma}\nabla T = \mathbf{g} - 2\mathbf{\Omega}\times\mathbf{V} + \dfrac{\mu_0}{\rho}\left[\dfrac{1}{4}\nabla(\mathrm{div}\,\mathbf{V}) + \dfrac{3}{4}\Delta\mathbf{V}\right], \\ T_t + \mathbf{V}\cdot\nabla T + (\gamma-1)T\,\mathrm{div}\,\mathbf{V} = \dfrac{\kappa_0}{\rho}\Delta T + \dfrac{\mu_0\gamma(\gamma-1)}{2\rho}\Big\{\Big[(u_x - v_y)^2 + \\ + (u_x - w_z)^2 + (v_y - w_z)^2\Big] + \dfrac{3}{2}\Big[(u_y + v_x)^2 + (u_z + w_x)^2 + (v_z + w_y)^2\Big]\Big\}. \end{cases}$$

In this system:
t – time;
x, y, z – Cartesian ordinates;
ρ – gas density;
$\mathbf{V} = (u, v, w)$ – gas speed vector with projections on the corresponding Cartesian axes;
T – gas temperature;
$\mathbf{g} = (0, 0, -g)$ – gravity acceleration vector;
$\gamma = 1.4$ – polytropic exponent for air;
$-2\mathbf{\Omega}\times\mathbf{V} = (av - bw, -au, bu)$ – Coriolis acceleration vector;
$a = 2\Omega\sin\psi,\ b = 2\Omega\cos\psi,\ \Omega = |\mathbf{\Omega}|;$ $\mathbf{\Omega}$ – angular speed vector of the Earth rotation;
ψ – latitude of the point ψ – beginning of the Cartesian axes $xyzO$ rotating with the Earth;
$\mu_0 = 0.001$, $\kappa_0 \approx 1.46\mu_0$ – constant values of dimensionless viscosity and thermal conductivity coefficients.

The computational domain (Figure 1) represents a rectangular parallelepiped with the lateral length $x^0 = 1$, $y^0 = 1$ and $z^0 = 0.04$ along the axes Ox, Oy and Oz, respectively, that is filled with a three-dimensional grid of intersection nodes of the three plane families $x = x_i$, $y = y_j$, $z = z_k$, where $x_i = i \cdot \Delta x$, $y_j = j \cdot \Delta y$, $z_k = k \cdot \Delta z$, $0 \le i \le L$, $0 \le j \le M$, $0 \le k \le N$. The differential steps on three spatial variables are as follows: $\Delta x = x^0 / L$, $\Delta y = y^0 / M$, $\Delta z = z^0 / N$, $L = 200$, $M = 200$, $N = 20$.

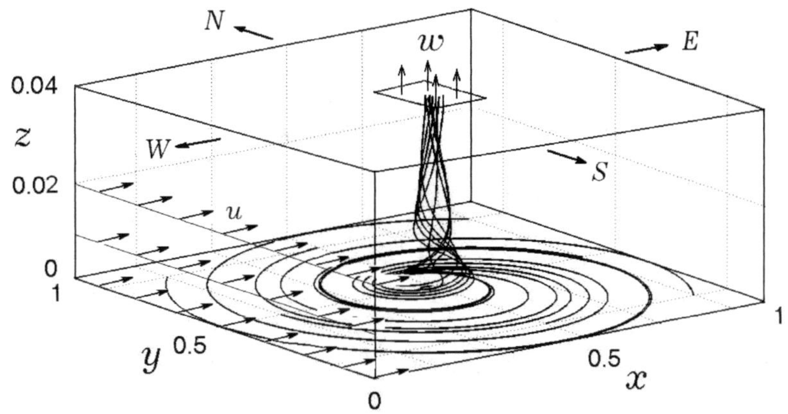

Figure 1. Computational domain

The values of the desired five functions characterizing the air flow in the artificially created tornado at the moment $t = t_s$ of its setting to the stationary operating mode are taken for initial conditions in all internal nodes of the computational domain [11, 12]:

$$u = u(x,y,z,t_s),\ v = v(x,y,z,t_s),\ w = w(x,y,z,t_s),\ T = T(x,y,z,t_s),\ \rho = \rho(x,y,z,t_s).$$

The boundary conditions on the faces of the computational domain are set as follows:

The density on four side faces of the computational parallelepiped $x = 0$, $x = x^0$, $y = 0$, $y = y^0$ is taken equal to the values from the stationary distribution of atmospheric air density [3]

$$\rho\big|_{x=0,\,x=x^0} = \rho_0(x,y,z), \qquad \rho\big|_{y=0,\,y=y^0} = \rho_0(x,y,z),$$

$$\rho_0(z) = (1-kz)^{\nu-1},\quad k = \frac{lx_{00}}{T_{00}},\quad l = 0.0065\ \text{К/м},\quad x_{00} = 50\ \text{м},\quad T_{00} = 288^o\text{K},\quad \nu = \frac{\gamma g}{k}.$$

The density on the lower $z = 0$ and upper face $z = z^0$ is subject to flow continuity. This means that the density per region boundary belong to linear extrapolation normal to the given boundary surface from the inside of the computational domain [3].

For temperature on all lateral faces the values from the stationary distribution of free air temperature are specified

$$T\big|_{x=0,\,x=x^0} = T_0(x,y,z), \qquad T\big|_{y=0,\,y=y^0} = T_0(x,y,z), \qquad T_0(z) = 1-kz.$$

The temperature at the lower $z = 0$ and upper face $z = z^0$ corresponds to the symmetry condition. At the same time, the temperature values are calculated from the condition of equality to zero of their derivative by normal to this face.

Impermeability conditions are set at the lower and upper faces. The third velocity component equals zero $w\big|_{z=0,\,z=z^0} = 0$, and the first and second velocity vector components are determined from the symmetry condition, i.e. they are considered to be zero of their derivative normal to the face. Besides, through a square hole with size 0.1×0.1 in the center of the upper face, a vertical speed $w = 0.03$ is set, which simulates the vertical purge with air at a speed of 10 m/s through a pipe with a diameter of 5 m.

On the side faces of the computational domain, the continuity conditions are set for all velocity vector components normal to the faces, and two tangential velocity components are calculated from the symmetry condition. Additionally, a normal velocity component equal to wind speed $u = 0.03$ is set on the left (west) face $x = 0$. Thus, a horizontally directed wind action on the ascending vortex flow of artificially created tornado is simulated.

The values of functions from two consecutive time layers are used to approximate the derivatives over time, and the central differences of function values are used to approximate the derivatives over spatial variables.

The three-dimensional non-stationary flow is calculated according to a clear difference scheme of transition from one n time layer to the next $n+1$ time layer with constant specified step Δt. The values of all required functions at all internal points of a rectangular parallelepiped are calculated. The required function values are then defined at all internal points of each of the six faces. The values of all the required functions at the inner points of all twelve edges of the rectangular parallelepiped are found as the arithmetic mean of the two intermediate values determined by linear interpolation according to the values of functions at two points on the normals to the edge at each face.

The calculations were performed with the following input parameters: scaling dimensional values of density, speed, distance and time are equal accordingly to $\rho_{00} = 1.2928$ kg/m^3, $u_{00} = 333$ m/s, $x_{00} = 50$ m, $t_{00} = x_{00} / u_{00} = 0.15$ s. The differential steps – as three spatial variables $\Delta x = \Delta y = 0.005$, $\Delta z = 0.002$, and the time step – as $\Delta t = 0.001$.

3. Results

The calculation analysis of speed characteristics of air flow in artificial tornado at the presence of the wind load on the lower half of its bottom part is given. Figures 2-7 show the diagrams of three velocity components for height $z = 0.2$ m and for two fixed moments of time $t_1 = 1$ s, $t_2 = 10$ s. The numbers of the computational grid nodes are placed along axes Ox and Oy, along the axis Oz – gas velocity in dimensionless values.

Figures 2-3 show the calculated surfaces of the first gas velocity component as a graphical representation of functions of two variables $u(x, y)$ for the above height and fixed times.

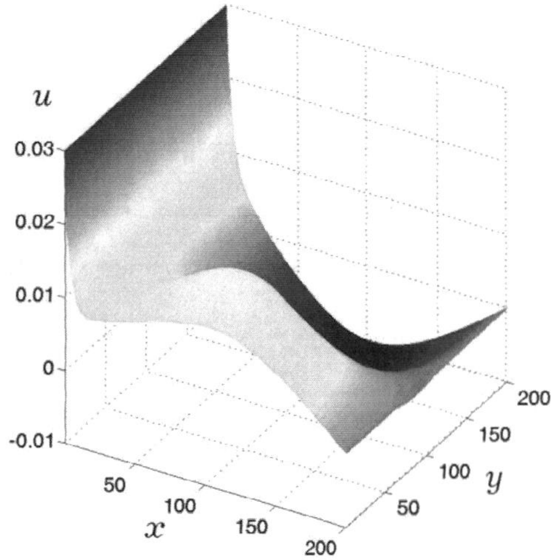

Figure 2. Speed u for $t_1 = 1$ s

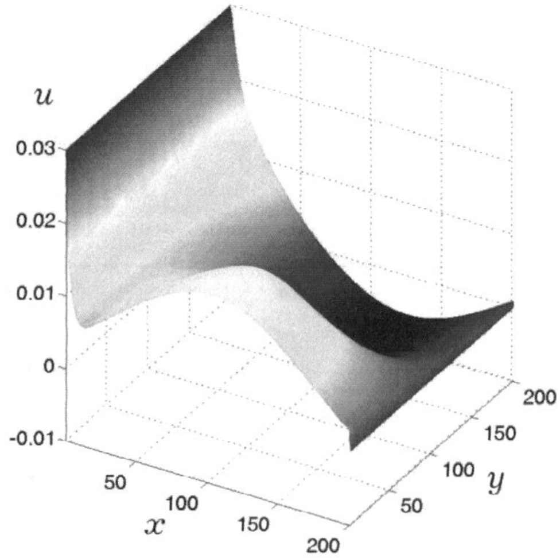

Figure 3. Speed u for $t_2 = 10$ s

In the figures given, the diagrams of the first speed component u reflect the presence of the specified wind speed 0.03 (dimension value 10 m/s) on the west (left) face $x = 0$ of the computational domain. The horizontal axially directed movement of air loses its speed quite quickly to the values 0.008 (dimensional value 2.7 m/s) when interacting with an axis flow (Figure 2). The oncoming movement of air in the wind flow and the northern part of the vortex results in their interaction, as a result of which the wind speed is subtracted from the circumferential velocity of the vortex, which as a result takes a negative value −0.01 (dimensional value −3.33 m/s), and the vortex is completely displaced eastwards (Figure 3).

At the same time, the air movement in the wind flow coincides with the air movement in the southern part of the vortex. As a result, the wind speed is added with the circumferential velocity of the vortex in the southern part, reaching the total velocity of 0.02 (dimensional value 6.66 m/s).

Thus, the result of the wind action on the ascending vortex flow is its uneven deformation in the northeast direction, an unsymmetrical change in the circumferential velocity, and the total displacement of the vortex in the wind direction.

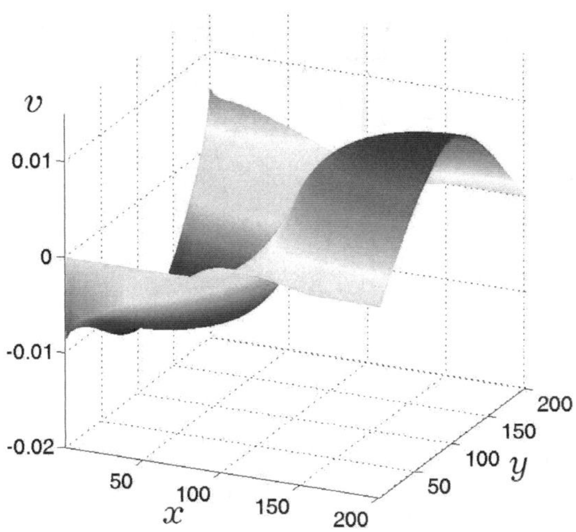

Figure 4. Speed v for $t_1 = 1$ s

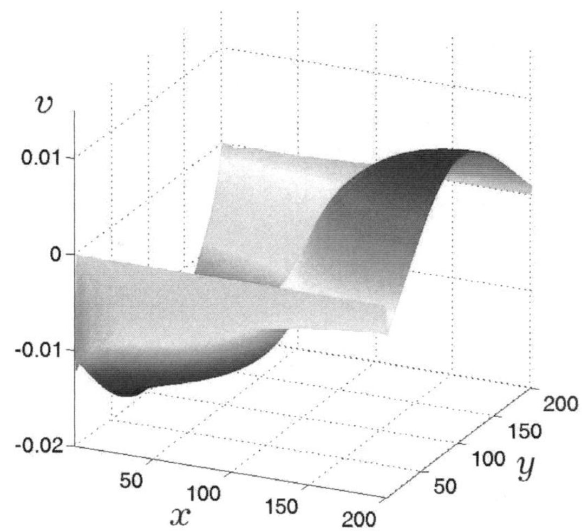

Figure 5. Speed v for $t_2 = 10$ s

Figures 4-5 show the calculated surfaces of the second gas velocity component as a graphical representation of the functions of the two variables $v(x, y)$ for the above-mentioned height and fixed times. The figures illustrate the behavior of the second component v of the gas flow rate.

During the first five seconds of the estimated time, the absolute value of this velocity component directed to the south of the computational domain towards the face $y = 0$ (left side in the figures) increases within $0.01 \div 0.015$ (dimensional values $3.33 \div 5.0$ m/s). This is caused by the fact that during this period of time there is an intensive addition of two perpendicular directed flows – the flow of air in the vortex directed to the south and the specified wind flow directed to the east (Figure 4). The second half of the estimated time is characterized by the growth rate reduction of this velocity component in the specified part of the computational domain with its simultaneous displacement towards the eastern part and an increase in velocity in this part of the vortex directed to the north (Figure 5). The above-described process of changing the second velocity component suggests deformation and displacement of the vortex as a whole in the northeast direction at initially given wind velocity.

Figures 6-7 show the diagrams of the third velocity component w for the considered time moments. It is typical to have a local increase of this velocity component to a value of 0.0025 (dimensional 0.83 m/s) in the center of the computational domain corresponding to the square air flow in the upper face of the computational domain. Besides, the figures show an area of increased speed values w to values 0.003 (dimensional 1 m/s) near the left (west) face of the computational domain, which at the distance of the order of 10 calculation nodes (2.5 m) decreases to zero. There is also a slight increase in vertical velocity values in the vicinity of the north-west vertical rib of the computational domain over time, the presence of which can be interpreted as a result of the interaction of opposite air flows.

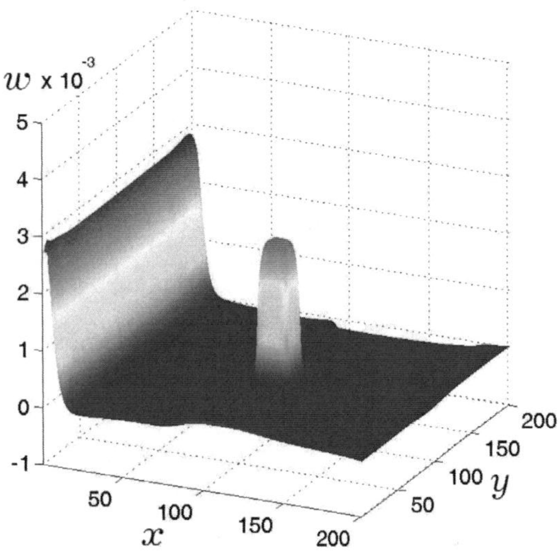

Figure 6. Speed w for $t_1 = 1$ s

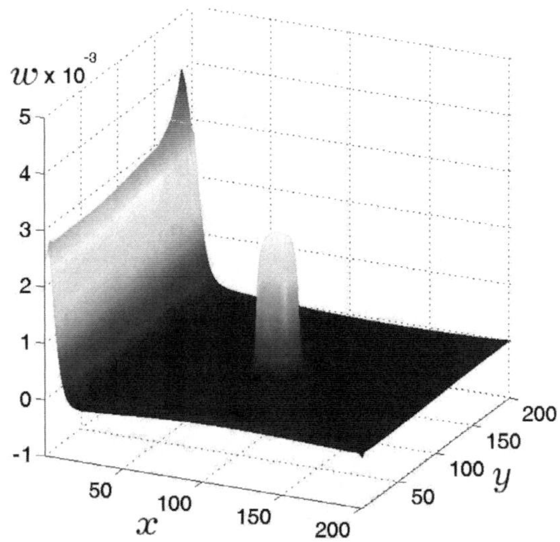

Figure 7. Speed w for $t_2 = 10$ s

Numerical calculations of three velocity components make it possible to build instantaneous current lines of complex three-dimensional non-stationary flows in the considered ascending vortex flows of the artificial tornado when exposed to external air flow. Figures 8-9 show instantaneous current lines in lateral projection perpendicular to the abscissa axis. In the figures, the number of instantaneous current lines is the same and is taken from the horizontal surface adjacent to the plane $z = 0$.

The figures with instantaneous current lines correspond to the same moments of time $t_1 = 1$ s, $t_2 = 10$ s. the distances in dimensionless values are set along the axes of coordinates.

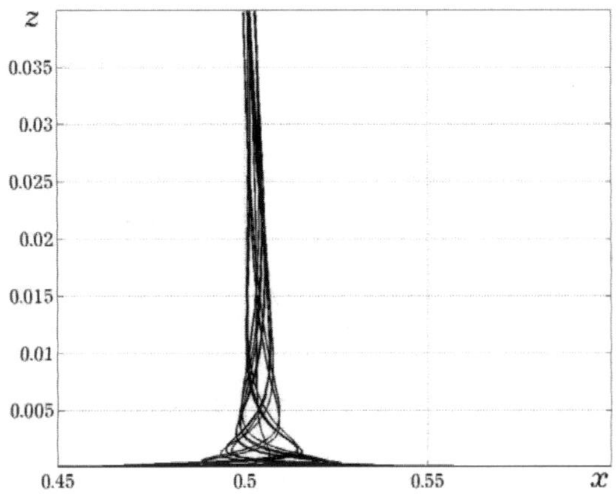

Figure 8. Current lines for $t_1 = 1$ s

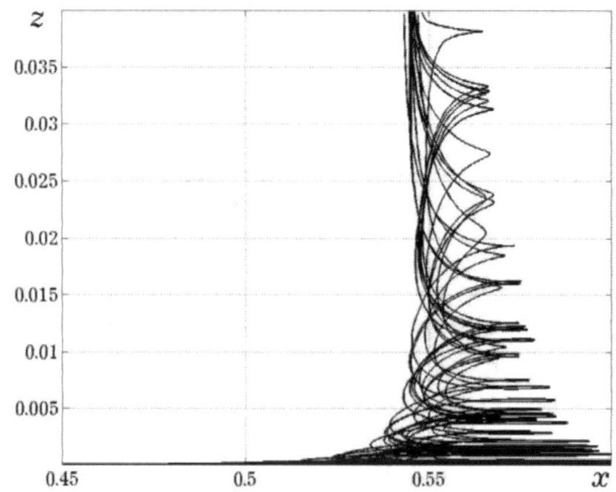

Figure 9. Current lines for $t_2 = 10$ s

As can be seen from the calculations, during the calculated period of time the vortex center in the bottom of the lower part shifted in the east direction (to the right) to a distance of 0.08 (dimensional value 4 m), and the vortex center at the top of the vertical part in the same direction to a distance of 0.04 (dimensional value 2 m). At the same time, the whole vertical part of the vortex – its "arm" – is subject to bending, convexity in the east direction (to the right). Besides, the figures showing instantaneous current lines clearly show an increase in vortex diameter starting at the bottom and moving to the vertical part thereof.

Figures 10-11 show a plan view of the same instantaneous current lines, from which it can be seen how the vortex is deformed and displaced in the north-east direction under the influence of a given wind flow. The result of such deformation can also be considered the appearance of an increasing area free of current lines in the center of the vortex (Figure 11). Finally, the built instantaneous current lines allow observing how the vortex gradually bends from the south side through the flow of the external wind flow.

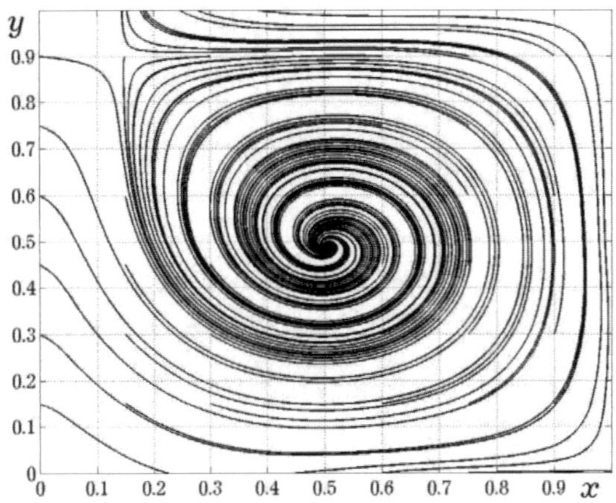

Figure 10. Current lines for $t_1 = 1$ s

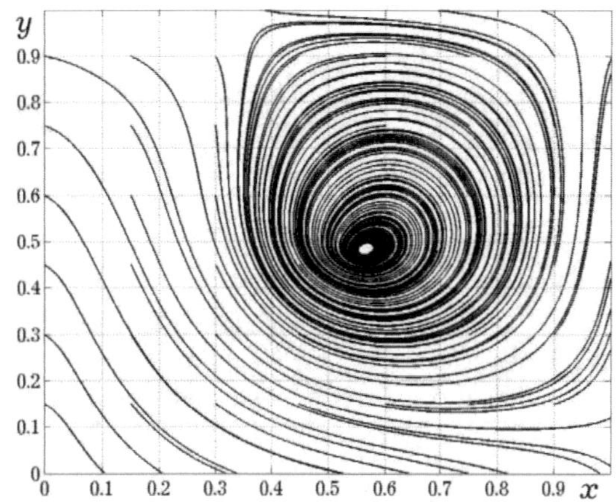

Figure 11. Current lines for $t_2 = 10$ s

Figures 12-13 represent instantaneous current lines at corresponding times at some angle. They give the overall picture of wind action on the vortex and highlight another feature.

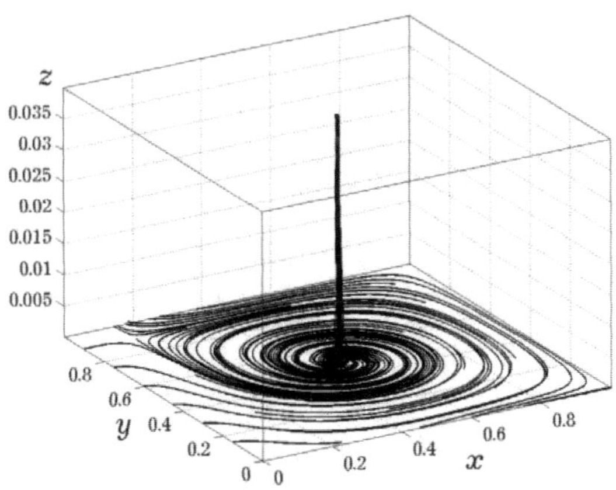

Figure 12. Current lines for $t_1 = 1$ s

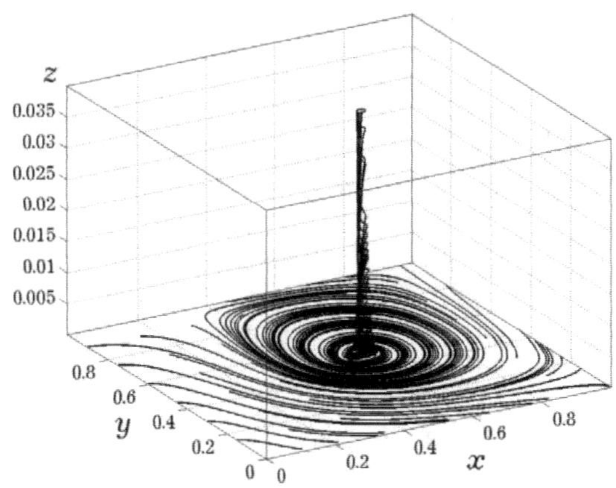

Figure 13. Current lines for $t_2 = 10$ s

Figure 13 shows that some current lines begin to "break off" from the vertical part. The effect can also be seen in Figure 9. This can be explained by both the substantial deformation of the vortex flow in the northeast direction and the different velocity of the circumferential gas flow in different parts of the vortex.

4. Conclusion

The study presents the numerical simulation of lateral wind action on three-dimensional non-stationary air flows in the ascending vortex flow of artificially created tornado in the stationary operating mode. Using the explicit difference scheme and the corresponding choice of initial and boundary conditions, the solutions of the complete system of Navier-Stokes equations in the computational domain representing a rectangular parallelepiped are numerically obtained. All flow speed characteristics and instantaneous current lines under the gravity conditions and the Coriolis force, as well as taking into account constant horizontal wind speed are calculated at fixed moments of time.

Displacement, curvature of the vertical part of vortex in wind direction, increase of vortex diameter in the lower part passing into the vertical part is established. There is a “break” of some current line from the vertical part and the appearance of an increasing area free of current lines in the center of the vortex.

In the studied time change range, wind resistance of air vortex and stable operation of the used computational scheme are observed.

References

[1] Bautin S P 2008 *Tornado and Coriolis force* (Novosibirsk: Nauka)
[2] Bautin S P 2009 *Characteristic Cauchy problem and its application in gas dynamics* (Novosibirsk: Nauka)
[3] Bautin S P, Deryabin S L, Krutova I Yu and Obukhov A G 2017 *Destructive atmospheric vortices and rotation of the Earth around its axis* (Yekaterinburg: USURT)
[4] Bautin K V, Bautin S P and Makarov V N 2013 Experimental confirmation of the possibility of creating an air flow twisted by the Coriolis force *USURT Bulletin* **2 (18)** 27-33
[5] Bautin S P and Makarov V V 2016 Creation of air flow twisted by the Coriolis force using a two-meter diameter pipe *USURT Bulletin* **4 (32)** 39-45
[6] Nalivkin D V 1984 *Tornados* (Moscow: Nauka)
[7] Ovsyannikov L V 2003 *Lectures on the basics of gas dynamics* (Moscow: Izhevsk, Institute of computer research)
[8] Loytsyansky L G 1970 *Fluid and gas mechanics* (Moscow: Nauka)
[9] Bautin S P and Obukhov A G 2013 Mathematical simulation of the near-bottom section of an ascending twisting flow *High Temperature* **51 (4)** 509-512
[10] Abdubakova L V and Obukhov A G 2014 Numerical calculation of speed characteristics of three-dimensional ascending twisted gas flow *News of higher educational institutions. Oil and gas* **3** 88-94
[11] Volkov R E, Obukhov A G and Terehova N V 2018 Numerical calculations of energy characteristics in modeling the entry into stationary mode of artificial tornado *News of higher educational institutions. Oil and Gas* **1** 84-92
[12] Volkov R E and Obukhov A G 2018 Three-dimensional instantaneous current lines of gas flows during simulation of setting to stationary mode of laboratory tornado *News of higher educational institutions. Oil and Gas* **2** 71-76

IOP Conf. Series: Materials Science and Engineering **952** (2020) 012014 doi:10.1088/1757-899X/952/1/012014

Complex thermophysical modeling of processes in the foundation soil of oil pipelines in the Arctic and offshore conditions

V V Golik, Yu D Zemenkov and I F Shagbanov

Tyumen Industrial University, 38 Volodarskogo str., Tyumen, 625000, Russia

E-mail: darkpaint@mail.ru

Abstract. The paper presents the results describing the mathematical model in operation and testing methodology aimed at heat engineering calculations needed to do for multilayer oil pipelines. A section of the pipeline running under difficult geological conditions is under simulation. The methodology aimed at calculating the thermal processes and its results are described. The simulation was carried out by using some insulating materials with a constant internal diameter of the pipeline and a pumping mode. The simulation was carried out with the universal software system of end-element modeling – ANSYS and with a module developed by the authors for TPS thermotechnical calculations.

1. Introduction

In the modern world, where the production rate is increasing, the growing demands are made upon the construction and operation of oil and oil products transportation facilities. For this, the oil production zones are expanding, more and more new fields are being developed; most of them are located in the northern part of Western Siberia and in the Arctic shelf.

Due to the extreme temperatures of these regions, some special requirements are made for the purposed technologies, constructions, necessary technical tools, as well as the production technological schemes, preparation methods, including the methods on recovery, storage and transportation of the hydrocarbon products.

For trouble-free operation of all of the above, it is needed to make a large number of calculations even at the development stage of all facilities involving in extraction and transportation of the oil products. The particular attention should be paid to pipeline networks laid in permafrost soil.

The pipelines laid in the permafrost soils have a number of serious problems that appear only after a sufficiently long period of time. All of them are associated with soil body thawing, which used to be in a frozen state. When thawing, soil characteristics begin to change, which entails a change in a pipeline design position, which leads to a change in a pipe geometry (deformation), a change in pumping conditions, and, as a consequence, to some accidents. It is also necessary to consider that all of the above also affect buildings and constructions located in the vicinity of the underground pipelines.

In view of the above, it seems relevant to develop a refined and effective methodology aimed at the thermal calculation of oil pipelines that will take into account the influence of heat transfer of oil and oil products on the environment. This method will enable to examine the effect of a pipeline on the permafrost soils of the Arctic zone and will allow to do more correct selection of material for laying [1-3].

2. Materials and methods

The thermal effect of the underground pipelines on the dynamics of seasonal soils freezing is determined by combination of many factors. The first group of the factors is: average annual, maximum and minimum temperatures of the transported product. The second natural factors are represented by some seasonal variations in air temperature. In other words, the set up task can be solved within the framework of a joint study of heat transfer processes inside a pipeline and in freezing soil with the conjugation of temperature and heat flux in a pipe-ground system.

The most complete mathematical model reflecting the processes of freezing and thawing in the moist soils in the continuous medium is the Stefan problem. The moving interface between zones with different aggregate states of pore moisture in the soil, which is not known in advance, puts the Stefan problem on a par with the most complex nonlinear problems solved by mathematical physics [4, 5].

At the first stage of the study, taking into account the above conditions, a calculation model of the heat distribution process around the underground oil pipeline was developed (Figure 1)

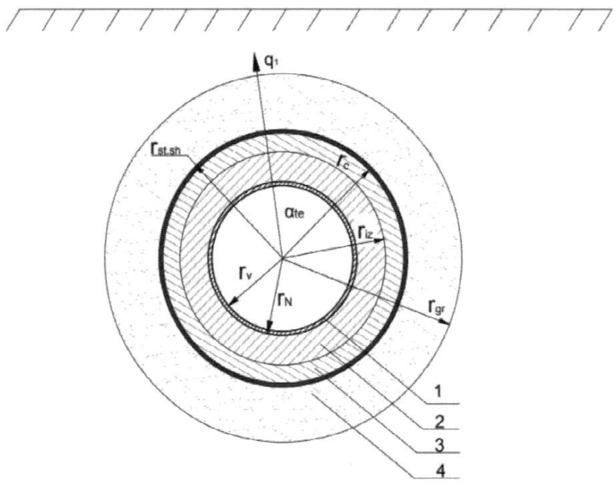

Figure 1. Calculation model of thermal interaction happening to underground oil pipeline in permafrost soils:
1. steel pipe; 2. thermal insulation of polyurethane foam; 3. nanomodified concrete coating; 4 permafrost soil.

The mathematical model of the thermal interaction of a pipeline with radius r buried to a depth H in the soil describes a heat transfer in the soil [6-8]:

$$c_{ef}\frac{\partial t}{\partial \tau}=\frac{\partial}{\partial x}\left(\lambda\frac{\partial}{\partial x}\right)+\frac{\partial}{\partial y}\left(\lambda\frac{\partial}{\partial y}\right), \tag{1}$$

where t is a temperature of soil;

τ – estimate time.

Assuming that the phase transition of pore soil moisture occurs at a certain temperature range Δt_i, effective heat capacity C_{ef} [9, 10]:

$$c_{ef}=\begin{cases} c_{T}, \quad at\ t>t_{ph}+\Delta t; \\ \rho_x c-\chi\gamma\dfrac{dW(t)}{d\tau}, at\ t\in\left[t_{ph}-\Delta t, t_{ph}+\Delta t,\right], \\ c_{M}, \quad at\ t>t_{ph}-\Delta t, \end{cases} \tag{2}$$

where t_{ph} is the temperature of the phase transition of moisture;

χ – latent heat of the phase transition of moisture;

W – content of unfrozen moisture;
γ – the specific gravity of the soil skeleton;
ρ_x – the coefficient taking into account the density of the soil during subsidence of a pipeline.
The indices T and M correspond to the thawed and frozen state of the soil:

The second stage required verification of all previously described models in a program format to confirm the adequacy of the model. After working out the 3D model and designating all the necessary parameters and boundary conditions, a program was needed to choose. The choice focused on the program in which the simulation of the thermal processes as close as possible to the real conditions. It was the universal software system of end-element modeling – ANSYS [11].

The calculation of the parameters was carried out in the Transient Thermal package where a 2D model of the pipeline located in the soil body was loaded. In this calculation, the 2D model was used to speed up the calculations and minimize the computer-based load, and also this model allowed calculating a halo for a long time in the shortest possible time. To build up a more accurate calculation of the halo, the method of developing a triangular grid (Tetrahedrons) was used, and the cell size was adjusted (0.08 m), which will optimize the load in the calculation and obtain the correct data (Figure 2).

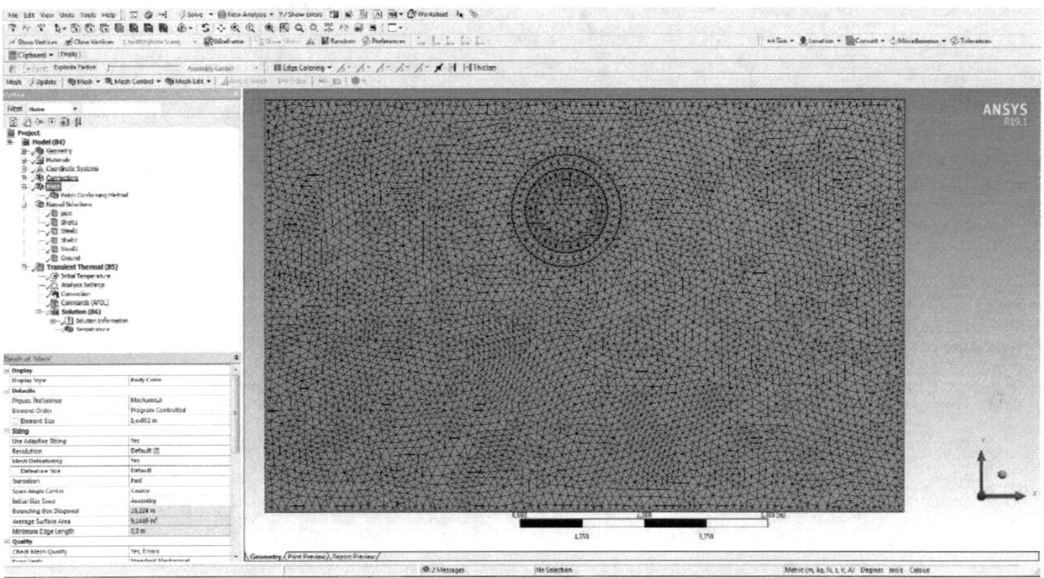

Figure 2. 2D soil model with the developed mathematical grid

The model also sets the parameters of the magnitude, type and properties of the materials and soils (thermal conductivity, density, specific heat) and the function of changing the temperature on the soil surface during a year was approximated (Figure 3) [12-14].

The last step before starting the calculation is to configure the solver of our thermal analysis model and select the boundary conditions in the program. In the Analysis Settings menu, the section Step-Controls, we set the time step for stationary thermal calculation. In the stationary calculation, time is used only to track the load history. The choice of a suitable time step value is very important because it affects the accuracy and stability of the solution:

• If the step is too small, oscillations can occur in the solution, which can lead to physically inadequate temperature values;

• If the step is too large, the temperature gradients may not be adequately resolved.

As an output, a small initial time step can be set up and the automatic selection function to increase the step size as far as possible can be used. To accurately determine the acceptable time step for our

IOP Conf. Series: Materials Science and Engineering **952** (2020) 012014 doi:10.1088/1757-899X/952/1/012014

problem, we use the Fourier number - this is a dimensionless time (Δt/t) that quantitatively determines the ratio of thermal conductivity to stored heat for an element with a width Δx: where r and c are the average density and specific heat, respectively.

$$F_o = \frac{4\,K\,\Delta t}{\rho C\,(\Delta x)^2} \tag{3}$$

Next, the model is calculated by using all the boundary conditions and the solvers available in the program for this type of calculation. After completing the calculation in this software package, it is possible to display not only a graphic image of the temperature halo, but also graphs and various dependences (Figure 4) [15].

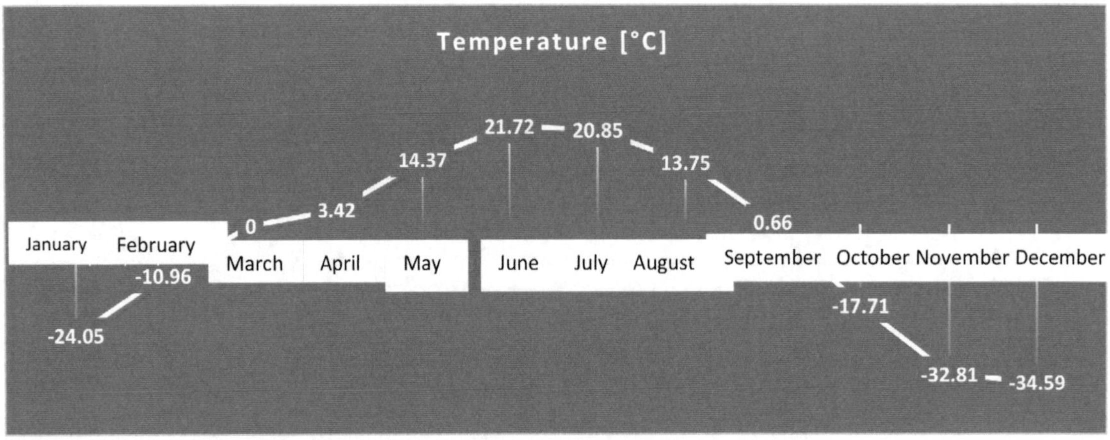

Figure 3. Schedule of temperature changes on soil surface during a year

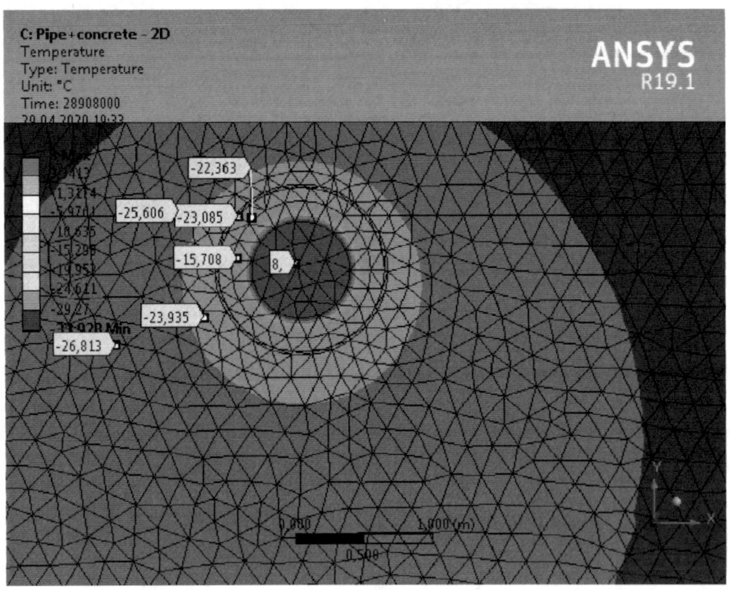

Figure 4. Graphic image of thawing ground halo

3. Conclusion

The values obtained as a result of modeling were compared with the data obtained by the authors in the developed TPS software block based on Visual Basic for Applications (VBA). As a result, the data discrepancy was 18%. At the moment, the modeling result is satisfactory. For a more accurate calculation, it is needed:

- to conduct additional soil studies in the Arctic zone;
- to clarify the grid on the model in the most important places;
- to verify the data of the author's mathematical model.

Also, due to the mathematical modeling according to the developed methodology, we suggest that it is advisable to use multilayer concrete pipes in the most emergency sections in the Arctic region.

The developed mathematical model contains a large number of variables, including coefficients developed by the team of the authors under the guidance of Professor B.V. Moiseyev. These coefficients allow describing more accurately the processes taking place in the "pipe-soil" system over a sufficiently long period of time.

References

[1] Eds Grigoriev V A and ZorinaV M 1982 *Heat and mass transfer. Thermotechnical experiment. Guidelines* (Moscow: Energy Publishing House)

[2] Zemenkov Yu D, Moiseev B V, Bogatenkov Yu V, Nalobin N V and Dudin S M 2015 *Energy technological complexes in design and operation of transport and storage facilities for hydrocarbon raw materials* (Tyumen: TSOGU)

[3] Moiseev B V, Kushakova N P and Nalobin N V 2022 Mathematical modeling of unsteady temperature interaction of heat pipes and seasonally freezing soils *Composite building materials: Proceedings of the international scientific-practical conference, Penza: PenzGASA* pp 261-264

[4] Zemenkova M Y, Shastunova U, Shabarov A, Kislitsyn A and Shuvaev A 2018 Physical and mathematical modeling of process of frozen ground thawing under hot tank *IOP Conference Series: Materials Science and Engineering* **357**

[5] Zemenkova M 2018 Methodology of Monitoring of Hydrocarbon Transportation Hydraulic Reliability *Advances in Intelligent Systems and Computing* **692** Springer Cham https://doi.org/10.1007/978-3-319-70987-1_72SP

[6] SP 25.13330.2012 *Foundations on permafrost soils.*

[7] Golik V V, Moiseev B V, Gulkova S G and Zemenkov Yu D 2018 Mathematic simulation of the effect of a buried oil pipeline on permafrost soils *IOP Conference Series: Materials Science and Engineering* **445**

[8] GOST 25100-2011 *Soils. Classification* (Amended)

[9] GOST 26263-84 *Soils. Laboratory method for determining the thermal conductivity of frozen soils*

[10] RD-75.180.00-KTN-198-09 2009 *Unified technological calculations of the main oil pipelines and oil product pipelines facilities*

[11] Fedorova O B and Chizhevskaya E L 2014 Adapting intellectual property evaluation methods to the region brand evaluation *Middle - East Journal of Scientific Research* DOI: 10.5829/idosi.mejsr.2014.19.1.12477

[12] Balikaeva M B, Chizhevskaya E L, Grevtseva Ya G, Kotlyarova I O, Volkova M A 2018 Innovative technologies as a means of the development of future engineers' professional mobility abroad *IOP Conference Series: Materials Science and Engineering* doi: 10.1088/1757-899X/441/1/012007

[13] Chizhevskaya E L, Fedorova O B and Kot A D 2013 Regional branding as factor of achievement of competitiveness of the territory *World Applied Sciences Journal* **28 (2)** 185-188

[14] Stepanov O, Moiseev B, Chekardovskiy M, Aksenov B and Shapoval A 2017 Physical and mathematical conditions of non-stationary thermal conditions of the underground air channels *International Journal of Applied Engineering Research* **12** 10110-10113

[15] Moiseev B, Zemenkov Y, Nalobin N and Dudin S 2017 Thermal calculations of underground oil pipelines *MATEC Web of Conferences*

Use of collapsible pipelines for joint work with main oil product pipelines

D I Melnikov M Yu Zemenkova and S Yu Podorozhnikov

Tyumen Industrial University, 38, Volodarskogo str., Tyumen, 625000, Russia

E-mail: muzemenkova@mail.ru

Abstract. The paper describes development stages for dismountable pipelines to work together with the major oil pipeline. Dismountable pipelines are used for transporting refined oil volumes over great distances in any climatic and geographical conditions. A dismountable construction provides technological pipeline schemes for rapid transportation and distribution of refined oil products to storage tanks. Subsequently, these products from containers can be utilized for various purposes.

1. Introduction

Dismountable pipelines were developed for the country's armed forces with a view to transporting fuel for military equipment.

Integrated experience of using dismountable pipelines in various situations over the years has allowed for a definite system of views to be shaped towards their possible applications in various fields.

2. Materials and methods

The scope of dismountable pipelines can be very wide. They have long been used for transporting water and oil for the needs of the national economy [1-4].

The main indicators of dismountable pipelines encompass a deployment rate and pumping capacity. The deployment rate depends on the type of pipeline, availability of forces and means, proficiency of specialists, adopted task execution plan, climatic conditions and the terrain. Table 1 presents the main characteristics of dismountable pipelines.

The deployment rate is defined in kilometers of the pipeline installed and filled within a day and may amount to:

– DFP-150 using mechanical devices – up to 60 km/day (without mechanical devices – 25-30 km/day);

– DFP-100 using mechanical devices – 70-80 km/day (without mechanical devices – 35-40 km/day).

The idea behind the special exercises was to master the procedures of stocking oil products through dismountable pipelines.

The main objectives of the exercises were:

- working out possible ways to connect dismountable pipelines to MOP;
- organizing documentation procedure to ensure product supplies from MOP to a dismountable pipeline for further transportation;

- elaborating the procedure for determining the quantity and quality control of products accepted from MOP;
- mastering skills required to ensure the deployment, operation and reeling-up of dismountable pipelines;
- shipping oil through a pipeline from a transfer terminal to a MOP (OTT–MOP) to a stocking site with a capacity of 600 m^3;
- organizing documentation procedure to ensure product transfers from a dismountable pipeline to a stocking site;
- product transfers to road transport through a mass delivery section;
- organizing documentation procedure to ensure product transfers from the stocking site to road transport.

To ensure compliance with the plan, special exercises were deployed:

- 10 km section of a dismountable pipeline;
- stocking site with a capacity of 600 m^3;
- mass delivery section for road transport.

Table 1. Main characteristics of dismountable pipelines

Indicator	DFP-100	DFP-150
Pumping capacity, t/day	1200	3000
Working pressure, MPa (kgf/cm^2)	6.0 (60)	
Test pressure, MPa (kgf/cm^2)	7.5 (75)	
Pipeline arrangements	Mechanic using MST-100 and manual	Mechanic using PLC-150V and manual
Pipe connection	"Bell"	
Conventional pipe length, m	6.0	
Conventional pipe diameter, mm	100	150
Pipe weight, kg	36,2	80,9
Pipe material	Steel 16GS	
Angular mobility in joints, degrees	1.5 – 2.0	
Longitudinal mobility, mm	up to 5	

An upstream end of the dismountable pipeline was a MOP junction point. The dismountable pipeline was connected to the MOP via an oil transfer terminal [5-7]. OTT–MOP was performed in two options: through a "hot tapping" method and through an access manhole (AM) of the dismountable pipeline (DP) to MOP (Figure 1).

Figure 1. Connection of a dismountable pipeline to OTT–MOPL

OTT–MOP was developed and manufactured by Transneft PJSC together with the specialists of the Armed Forces fuel service. The main specifications and characteristics of OTT–MOP are given in Table 2 [8-11].

Table 2. OTT–MOP specifications and characteristics

Indicator	Specifications
Type of the pipeline connected to OTT–MOP	DFP-150 high-capacity field pipeline (DFP-100)
Diameter, mm	150 (100)
Supply, m^3/h, at least	90
Delivery rate, m^3/day, at least	1800
Working pressure, MPa (kgf/cm^2)	до 6.3 (63)
Test pressure, MPa (kgf/cm^2)	1,25 P_{work}
Type of junction to the pipeline	Bell joint
Type of junction to MOP	Flanged or welded joint
Working substance	Oil products
Filtration fineness, mm, at least	4
Relative error of mass measurements in the flow range from 10% of the nominal flow to the nominal,% (of the volume), at least	±0.25
Measurement equipment error,% (from the upper limit of measurements)	1.5
Time to put OTT–MOPL, mounted on the pipeline as AM, into operation (to the delivery of oil products), h	1.0
Time to put OTT–MOPL, mounted on the pipeline via hot tapping devices shortly before the delivery of oil products, into operation (to the delivery of oil products), h	24.0

The dismountable pipeline was connected to the OTT–MOP using an MPT-15.4 – Bell adapter that came with a junction point and was stored in an AM. The adapter was directly connected to a Flange – MPT-15.4 quick-disconnect joint installed directly in the AM.

The dismountable pipeline was installed based on linear and auxiliary equipment that came with the DFP-100 and DFP-150 kits, and on linear prototypes for a metal composite DP-MC-100 pipeline. The dismountable pipeline was assembled in one line (Figure 2).

Figure 2. Section of DP-MC-100 pipeline

The dismountable pipeline was used during special exercises within four stages:
- deployment preparation;
- pipeline deployment;
- pipeline operation;
- pipeline reeling-up.

Deployment preparation involved a series of activities to select and verify the availability of linear and auxiliary equipment, shutoff and control valves enclosed in the DFP-100 and DFP-150 kits, prototypes for the DP-MC-100 metal composite pipeline, mobile pumping units (MPU), air compressor station (ACS). The said components were also delivered to the region to host special exercises.

Pipeline deployment encompassed the following activities:
- loading and removal of linear and auxiliary equipment, shutoff and control valves enclosed in the DFP-100 and DFP-150 kits and DP-MC-100 kits to the pipeline deployment site;
- transporting MPU, ACS to the pipeline deployment site;
- assembling a linear part of the pipeline;
- assembling a process piping of the pumping station, consisting of two MPUs and technological units on the pipeline;
- filling the pipeline with a test working medium (water);
- hydraulic testing of the pipeline;
- emptying the pipeline from the test working medium.

Pipeline operation involved the following activities:
- filling the pipeline with a product;
- shipping the product through the pipeline;
- simulating an emergency situation on the pipeline and taking measures to handle it by the patrol and emergency team (Figure 3).
-

Figure 3. Emergency response on dismountable pipeline

Pipeline reeling-up involved the following activities:
- emptying the pipeline from the product;
- dismantling the pipeline;
- dismantling the process piping of the MPU pumping station;
- loading and delivery of linear and auxiliary equipment, shutoff and control valves enclosed in the DFP-100 and DFP-150 kits, linear prototypes for the DP-MC-100, MPU, ACS to permanent storage.

The linear prototypes for the DP-MC-100, equipment enclosed in the DFP-100 and DFP-150 kits were delivered by the following means of transport:

- from storage to the load station – by road;
- from the load station to the delivery station – by rail;
- from the delivery station to the pipeline deployment site – by road.

KamAZ-6350 cars were used as motor vehicles.

DFPL was assembled using an installation tool coming bundled with the DFP-100 and DFP-150 kits.

Transferring the product at the upstream and delivery ends of the pipeline required universal samplers to be mounted from the DFP-150 kit [12].

At the upstream end of the pipeline there were:
- product sampling terminal;
- ACS connection unit;
- process piping of the pumping station, consisting of two MPUs;
- separator trigger.

Table 3. Technical specifications of MPU-100/200K

Parameter	MPU-100/200K
Overall dimensions, mm:	
- length	5750
- width	1890
- travel height	2275
Travel mass, kg	3860
Carrier – trailer	2-PN-2
Engine: type	Four-stroke eight-cylinder diesel
- brand	YaMZ-238G
Operational shaft rotation frequency, rpm:	
- engine	1700
- pump	4330
Engine operating power, kW	125
Engine: type	Centrifugal two-stage
- brand	H6x2M4
- feed, m^3/h	130
Pressure, MPa:	
- outlet	Up to 6.0
- inlet	0.1-3.0
Capacity, l:	
- fuel tank	140
- engine lubrication	33.5
- engine cooling systems	40
Fuel consumption, kg/h	27
Oil consumption, kg/h	0.3
Crew, person	1
Deployment time from traveling to working (from working to traveling), min	60 (20)

The sampling terminal is designed for sampling and measuring the density of the shipped product.

The ACS connection unit is designed to connect an air compressor station to a DFPL section and supply compressed air in order to clean it from foreign objects, as well as to displace from it the test medium and product.

The separator trigger is designed for performing technological operations to start a separator during filling, sequential pumping of the product and emptying the dismountable pipeline.

The process piping of the pumping station is designed to perform technological operations for filling and shipping the product through the dismountable pipeline.

At the delivery end of the pipeline there were:

- separator receiving unit;
- sampling unit;
- block valve stations.

The separator receiving unit is designed to receive the separator in the process of emptying the dismountable pipeline from the test medium and product.

The block valve station is designed for tight separation of adjacent sections of the dismountable pipeline.

The product was shipped through the pipeline using mobile pumping units MPU-100/200K.

Technical specifications of MPU-100/200K are given in Table 3.

In order to ensure stable shipping of the product, the pumping station consisting of two MPU-100/200K, main and reserve (Figure 4), was located at the upstream end of the pipeline.

Figure 4. Pumping station at the upstream end of the pipeline

At the delivery end of the pipeline, a stocking site was constructed with a capacity of 600 m^3 (Figure 5).

The stocking site is designed to keep oil stocks, as well as receiving and delivering them to various means of transport. The main technical characteristics of the site are given in Table 4.

Table 4. Main technical characteristics of the site for oil stocks

Characteristic	Indices
Tank module capacity, m^3, at least	600
Tank capacity, m^3	50
Operating temperature range, C	from minus 60 to plus 60
Tank Material	thermoplastic polyurethane
Deployment time, h	15
Crew, person	8
Cargo turnover, m^3/day	1,000

The site included:

- manifold block with fittings and sleeves (Figure 6);
- elastic tanks (Figure 7);
- pumping and transfer module (container);

- quality control laboratory (container);
- warehouse of materials and spare parts (container);
- reactive lightning protection;
- lighting kit;
- camouflage nets;
- diesel generator (stand-alone);
- loader;
- a site for mass distribution of the product to vehicles (Figure 8);
- fire kit.

Figure 5. Development of stocking site

Figure 6. Manifold block with fittings and sleeves

Figure 7. Elastic tanks

Figure 8. Mass delivery section for road transport

From the terminal, the product was shipped through the mass delivery section to road transport. The total volume of product shipped through the pipeline during the exercise was 600 m^3 [13].

3. Conclusion

The joint special exercises resulted in:

- ways to connect a dismountable pipeline to MOP;
- documentation procedure to ensure product supplies from Transneft PJSC;
- procedure for determining the quantity and quality control of products accepted from MOP;

- sampling procedure at the upstream and delivery ends of the pipeline;
- activities towards the deployment, operation and reeling-up of dismountable pipelines;
- actions of pipefitters when the product is shipped through the pipeline from MOP to the stocking site;
- documentation procedure to ensure product transfers from the dismountable pipeline to the stocking site;
- actions of the crew responsible for the delivery of the product to road transport;
- documentation procedure to ensure product transfers from the stocking site to road transport.

References

[1] Petrov V G 1982 *High-capacity field pipelines. Manual* (M.: Voenizdat)
[2] Shein K G, Sereda V V and Danilchenko I G 2005 *The pipeline troops are 50 years old Historical background* (Moscow: "Gralija")
[3] Sereda V V, Melnikov D I and Danilchenko I G 2014 The use of field pipelines in response to emergency situations *Military Thought* **9** 29-36.
[4] Melnikov D I, Kopnyshov S L and Sedykh N I 2015 Features of the operation to provide drinking water to settlements of the Crimea *Fires and emergencies* (Mosocw: Academy of State Fire Service of the Ministry of Emergencies of Russia) **1** 87-91
[5] Sereda V V, Danilchenko I G and Melnikov D I 2015 The use of field pipelines to improve the fire safety of the Russian Federation Basic research (Mosocw) **10** 309-313
[6] Sereda V V, El'kin A V and El'kin A V 2016 Third-Generation Dismountable Assembled Pipelines from Composite-Made Pipes *Chem Technol Fuels Oils* **51** 618–622 DOI: 10.1007/s10553-016-0648-7
[7] Sereda V V and Ulit'ko A V 2014 Role of Chemmotology at the Present Stage *Chem Technol Fuels Oils* **50** 363–368 DOI: 10.1007/s10553-014-0534-0
[8] Ostrovskaya V M, Sereda V V, Prokopenko O A et al. 2013 Indicator Tubes for Determining Anti-Knock Additives in Automotive Gasolines *Chem Technol Fuels Oils* **49** 444–450 DOI: 10.1007/s10553-013-0468-y
[9] Ziatdinov A M, Sviridova A G, Sereda V V et al. 2009 In situ conduction ESR and theoretical studies of graphite intercalation by antimony pentafluoride. *Appl Magn Reson* **35** 221 DOI: 10.1007/s00723-008-0156-y
[10] Ziatdinov A M, Sviridova A G, Sereda V V, Skrylnik P G 2009 In situ conduction ESR and theoretical studies of graphite intercalation by antimony pentafluoride *Applied Magnetic Resonance*
[11] Zemenkova M, Zemenkov Yu, Gladenko A and Podorozhnikov S 2016 Estimation of Emissions During Monitoring of Pipelines in the Dynamic Mode of Operation *MATEC Web Conf.* DOI: 10.1051/matecconf/20168604053
[12] Dudin S, Bahmat G, Zemenkov Yu, Voronin K and Shipovalov A 2017 Strategy for monitoring and ensuring safe operation of Russian gas transportation systems *MATEC Web Conf.* https://doi.org/10.1051/matecconf/201710606004
[13] Dudin S M, Zemenkov Yu D, Maier A V and Shabarov A B 2016 Research and Design of Thermophysical Gas-Liquid Mixture Parameters in Product Pipelines *IOP Conference Series: Materials Science and Engineering* DOI: 10.1088/1757-899X/154/1/012021

IOP Conf. Series: Materials Science and Engineering **952** (2020) 012016 doi:10.1088/1757-899X/952/1/012016

Studying quality of diagnosing rolling stock at service enterprises of automobile transport

V N Krasovsky, N I Krasovskaya and A V Sycheva

Tyumen Industrial University, 38, Volodarskogo str., Tyumen, 625000, Russia

E-mail: krass51@mail.ru

Abstract. The article focuses on one of the car service market components being a car diagnostic system and, in particular, on the effectiveness of the activities of diagnostic operators. The objective of this study was an experimental assessment of the effectiveness of the proposed methods for improving the training of diagnostic operators. The results of testing hypotheses on the correct diagnosis by a group of specialists on the technical condition of passenger car engines are presented.

1. Introduction

Under the conditions of modern market relations, automobile service is developing rather progressively. The market of car services is sale of services over maintaining vehicles as well as restoring capability in the event of its loss over the entire period of use (fig. 1). This includes not only monitoring and diagnostics services but also warranty services, seasonal maintenance and repair of both a car and its components, including the body.

After analyzing the list of services [4, 5, 6, 7] provided by car service centers, it should be noted that car diagnostics is an integral part of the market for these services and currently such services are more and more in demand.

Diagnosing faults can become quite simple if it is done logically. The first step in diagnosing any malfunctions should be to determine the cause of malfunction as accurately as possible. By accidental search, it is also possible to diagnose a malfunction. However, this may require a large investment of time and replacement of serviceable parts, which is disadvantageous.

The engine may not start for many reasons, and you will save time and effort by turning from one test to another, which increases the chances of finding a malfunction.

There are many malfunctions that interfere with the normal car operation, engine starting. However, there are much more malfunctions causing its poor operation. Although experienced diesel mechanics can often find the cause of a malfunction immediately, specialists who do not have such an experience are advised not try to guess the reason. Using a methodical and systematic approach, it is possible to save time and money and increase the likelihood of detecting a malfunction.

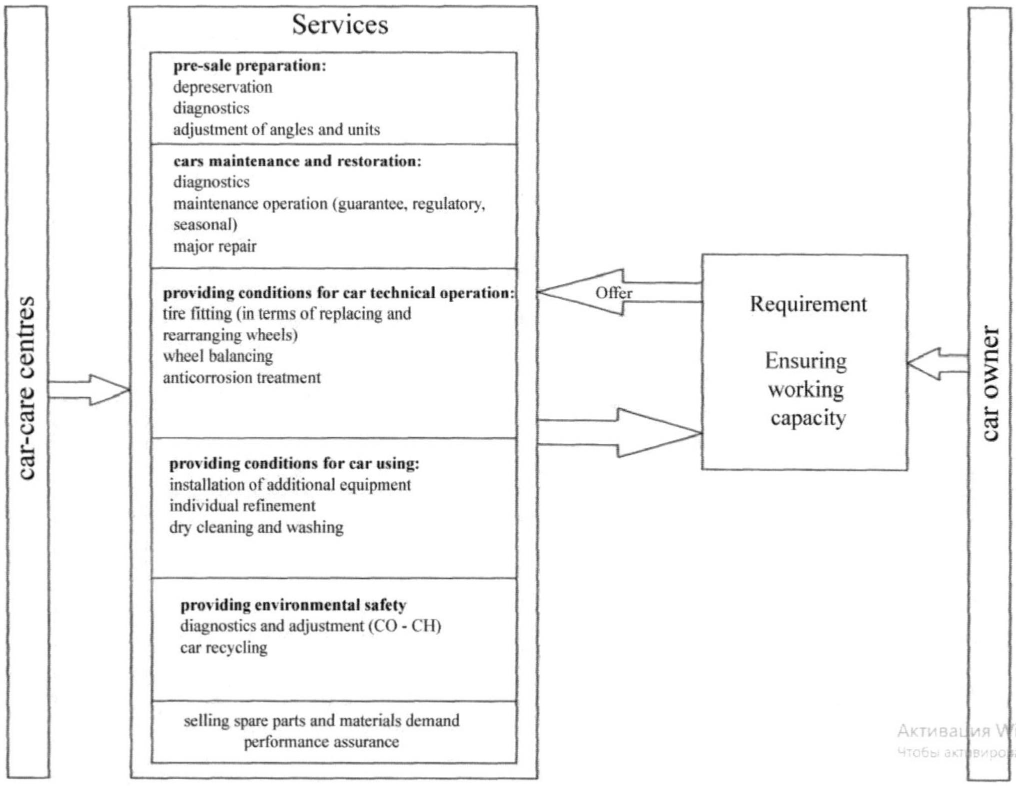

Figure 1. Car service market structure

To determine the cost of maintenance and repair of automobiles and their units, it is necessary to have sufficiently complete information on all facilities and planned volumes of work. Lack of such information as well as its incorrect presentation causes great difficulties in coordinating interaction between engineering, technical, financial and economic services. In the end, the decision is made not on the basis of objective criteria but subjectively, based on the prevailing power balance at wheeled transport enterprise [5, 6].

Currently, the number of cars equipped with electronic fuel injection systems has sharply increased, which has caused a significant increase in demand for services related to the diagnosis of technical condition and repair of such cars. Engine management systems need more thorough and effective diagnostics. Herewith, such a factor as a vehicle technical condition concerning engine management systems becomes very important since it affects the traction and speed indicators of a car as a whole. The problem of diagnosing a technical condition and repair of an engine management system can be solved by using new special technical means and measuring instruments.

Adaptation algorithms frequently incorporated into control systems hide malfunctions that gradually arise during operation. Therefore, the same engine satisfactory operation has to be judged by the following indirect signs that not every diagnostic operator can detect: increased fuel consumption, deterioration of dynamic characteristics, etc. It should be noted that some devices like sensors and actuators give errors that exceed their maximum permissible values. And this fact cannot be ignored, since diagnosis in this case becomes inferior, and, therefore, unreliable. [8, 11, 16].

This is quite enough to state that the diagnosis of complex automotive equipment, which would be reliable and, therefore, effective needs new methods, new algorithms, and new technical means. At the same time, it is necessary for both developers and manufacturers of automotive equipment to think not only about those who work on the assembly line, but also about those who will operate vehicles, diagnose and repair them.

Diagnostics is the basis of high-quality maintenance and repair. An important link in this chain is a person (operator-diagnostician) and their professional qualifications, knowledge, skills and experience, while the errors in the diagnosis and determination of repair volumes lead to repeated work, which means significant financial losses not only to customers, but also entire enterprises, etc. [7, 9, 14].

2. Materials and methods

The purpose of this study was an experimental effectiveness evaluation of the proposed methods aimed to improve diagnostic operators training.

According to the task to use the mathematical apparatus of operations research [1, 2], two hypotheses were suggested. They are as follows: zero H_0 being the proportion of specialists in the initial group of experts distinguishing the proposed values of the diagnostic parameter, that is, diagnosing accurately, are the same for the studied groups of experts with various qualifications; alternative H_1 being the proportion of specialists in the initial set of experts who distinguish the proposed values of the diagnostic parameter, that is, diagnosing precisely, are not the same for the various groups of experts studied [1].

To do this, an evaluation experiment was conducted. It implyied the correct diagnosis of the technical condition of a car engine (the brand was not taken into account) by a group of specialists in the amount of 60 people. In the course of the experiment, the following restrictions were observed: the subject of the study was diagnosing at Tyumen maintenance stations; the object of study was a car engine; the measured diagnostic parameter was the change in compressed air leaks through the cylinders of a warmed-up engine, a device used for measurements was the pneumotester K-272; the age of diagnostic operators was not taken into account; confidence coefficient value accepted in the study was as follows: a = 0.95; deviations of the measured parameter values in a positive or negative direction from the real one were not taken into account. Only downswings or upswings of the actual value of the diagnostic parameter (first-order error αd_p being false malfunction, second-order error β_{dp} being skip malfunction, respectively) was recorded.

Table 1. Initial conditions

Diagnostic Value	Qualification of a diagnostic operator						Row totals	
	Average		Average technical		Highest			
	Value	Notation	Value	Notation	Value	Notation	Value	Notation
Smaller	10	A	15	B	4	C	29	A+B+C
Actual	7	D	9	E	4	F	20	D+E+F
Larger	3	G	6	H	2	I	11	G+H+I
Column totals	20	A+D+G	30	B+E+H	10	C+F+I	60	---

To assess the hypotheses put forward in this paper, we used the Pearson χ^2 criterion [1]. The classical formula for calculating χ^2 has the following form [2]:

$$\chi^2 = \sum_{i=1}^{r}\sum_{j=1}^{c}\frac{(f_0 - f_e)^2}{f_e} \quad (1)$$

where: r is the number of columns in the source table r = {i, where i = 1 ÷ r};

c is the number of rows in the source table c = {j, where j = 1 ÷ c};

f_o is observed (empirical) frequency of expert falling within a given interval (cell of the original table);

f_e is expected (estimated) frequency of experts falling within a given interval.

Table 2. Data for calculating the criterion χ^2

Cell	f_o	f_e	f_o-f_e	$(f_o-f_e)^2$
A	10	9.66	0.34	0.12
B	15	14.5	0.50	0.25
C	4	4.83	-0.83	0.69
D	7	6.66	0.34	0.12
E	9	10.00	-1.00	1.00
F	4	3.33	0.67	0.45
G	3	3.66	-0.66	0.44
H	6	5.50	0.50	0.25
1	2	1.83	0.17	0.03

Calculations of the expected frequency f_e were performed for each cell of the original table. To do this, the final frequencies of the column and the row in which a cell was located were multiplied, and then the resulting product was divided by N.

Table 3. Calculation results

Cell	f_e	$(f_o - f_e)^2$	$\frac{(f_o-f_e)^2}{f_e}$
A	9.66	0.12	0.01
B	14.50	0.25	0.02
C	4.83	0.69	0.14
D	6.66	0.12	0.18
E	10.00	1.00	0.10
F	3.33	0.45	0.14
G	3.66	0.44	0.12
H	5.50	0.25	0.05
I	1.83	0.03	0.02

Having performed the corresponding calculations by the described method, we obtained the value $\chi^2 = 0.78$. A critical value χ^2 was selected from the tables for a given α level. If the calculated value χ^2 exceeds the critical value or is equal to it, the hypothesis H_0 can be rejected.

To select the critical value of χ^2, it is necessary to determine the number of degrees of freedom:

$$S = (r-1)(c-1) = (3-1)(3-1) = 4 \tag{2}$$

r is the number of columns in the source table;

c is the number of rows in the source table.

The critical value of χ^2 at $\alpha = 0.95$ and $S = 4$ was 0.711. In this case, we rejected the H_0 hypothesis and adopted the alternative H_1 hypothesis.

3. Conclusion

1. As a result, an almost complete quantitative ratio of true and error diagnoses was revealed.
2. It is obvious that using a one-factor model in the presented example, it is not possible to make a qualitative analysis of the results.
3. To ensure the possibility of a better analysis, it is necessary to introduce an additional factor, for example, age limit, time of day, etc.

References

[1] Ventzel E S 1972 *Operations Research* (Moscow: "Soviet Radio")
[2] Gmurman V E 2004 *Probability Theory and Mathematical Statistics* (Moscow)
[3] Dyakov I F 2011 Probabilistic and statistical assessment of the car uptime *Automotive Transport Magazine* **11** 8-11
[4] Karagodin V I, Krasovsky V N and Skrypnikov S A 1999 The use of information technology in the centralized repair of automobile engines according to their technical condition *Problems of creating information technology* (Minsk: MAIT) pp 81-90
[5] Krasovsky V N 2007 Centralized repair of vehicle assemblies by technical condition by manufacturers (Tyumen)
[6] Kuznetsov E S 2008 *Management of technical operation of cars* (Moscow: Transport)
[7] Miroshnikov L V, Boldin A P and Pal V I 2008 *Diagnosing the technical condition of cars at motor transport enterprises* (Mosocw: Transport)
[8] Mushik E and Muller P 1990 Methods of making technical decisions (Moscow: Mir)
[9] Starikov A *Application of neural networks for classification problems*, available at: http://www.basegroup.ru/library/analysis/neural/classification/
[10] Shibanov G P 1983 Quantitative assessment of human activity in human-technology systems (Moscow: Mechanical Engineering)
[11] Patan K Witczak M and Korbicz J 2008 Towards Robustness in Neural Network Based Fault Diagnosis *Int. J. Appl. Math. Comput. Sci.* **18 (4)** 443–454
[12] Cornak S 2007 Selected Methods of Vehicles Maintenance in Operation Stage *Machines, Technologies, Materials* **2-3** 14-16
[13] Yasnitsky L N 2008 Introduction to Artificial Intelligence (Moscow: Publishing Center "Academy")

IOP Conf. Series: Materials Science and Engineering **952** (2020) 012017 doi:10.1088/1757-899X/952/1/012017

A comprehensive analysis of the properties of reservoir rocks to determine the productive oil horizons

R V Klyuev[1,2], I I Bosikov[1], V Ch Revazov[1], K S Krysanov[2] and B V Dzeranov[3,4]

[1] North-Caucasian institute of mining and metallurgy (state technological university), 44, Nikolaeva Str., Vladikavkaz, 362021, Russia
[2] Moscow Polytechnic University, 38, B. Semenovskaya Str., Moscow, 107023, Russia
[3] Geophysical Center RAS, 3, Molodezhnaya St., Moscow, 119296 Russia
[4] Geophysical Institute, Vladikavkaz Scientific Center RAS, 93a, Markov Street, Vladikavkaz, 362002, Russia

E-mail: kluev-roman@rambler.ru

Abstract. The purpose of the research is to study the reservoir and filtration properties of reservoir rocks of the Aptian horizon of formation No. 1 within the Kovylnoye field. The initial data were reservoir-filtration and geological-geophysical parameters within the exposed rocks of the reservoirs of the Aptian horizon of formation No. 1, for wells No. 1 and No. 2, which were drilled at the search and evaluation stage of work: porosity coefficient, permeability coefficient, rock resistivity, oil-saturated thickness. High values of effective porosity, permeability, and oil and gas saturation are found in biomorphic, organogenic, and clastic carbonate rocks where the void space has not been subjected to secondary changes (salt deposits), resulting in reservoirs characterized by low capacitance and filtration properties. Analysis of carbonate reservoirs gives values of permeability up to 0.3-1 mm^2 and porosity of 20-35 %. Rocks are lumpy, loose, weakly cemented with cement content up to 10%. Their initial water saturation in the oil deposit does not exceed 5-20 %. The research has shown the effectiveness of conducting a comprehensive analysis of reservoir rock properties to determine the oil and gas content of the site.

1. Introduction

In the Prikumsky district, oil and gas content was detected in deposits from the Permian-Triassic to the Neogene ages. Several oil and gas complexes of the following age have been identified within this section interval: Permian-Triassic, Jurassic-Cretaceous, and Paleogene-Neogene.

Permian deposits are the most ancient, from which oil inflows are obtained.

In recent years, commercial oil flows have been produced from the fractured-cavernous carbonate reservoirs of the Lower Triassic age related to deposits of Neftekumsk Suite (Vostochnaya, Vostochno-Bezvodnenskaya, Pushkarskaya, Russkii Khutor Severnyi, etc.).

Mixed oil and gas and pure gas condensates were produced from sand layers of the Middle Triassic age in the Sukhokum, Vaidzhanov, and other areas.

A slight inflow of oil is obtained from the Upper Triassic limestones of Zakumskaya and Novokolodeznaya areas.

Jurassic and Cretaceous deposits are regionally oil and gas-bearing within the entire Eastern Stavropol territory.

Industrial inflows of oil were produced from stratum VII of Jurassic age in the Maksimokumsk area, from stratum V of Jurassic age in Polevaya, Molodezhnaya, Prignanichnaya, Russkii Khutor Yuzhnyi, Russkii Khutor Severnyi, and other areas.

Oil was obtained from the upper Jurassic in the Ozek-Suatskaya and Uvarskaya areas, and a gas-condensate Deposit was discovered in the third layer of the Jurassic.

Sandy-aleurolite-carbonate strata of the Lower Cretaceous deposits are connected on the squares Border, Youth, Ozek-Suat a, Feather, Russian farm Northern, non-oil inflows received in the Volynsk area on the Manych area in well No. 1 of the Ilmen VIII of the Lower Cretaceous reservoir received inflow water with 25% oil.

The upper Cretaceous carbonate reservoirs are associated with oil inflows in areas: East Bezbednosti, Winter Rate.

A slight show of oil in Paleogene deposits was identified at the Ozek-Suat field, and oil occurrences were also noted at the Vostochno-Bezvodnenskoye field.

Testing of wells to study the oil content of the Kovylnoye field was performed in a significant stratigraphic range from Triassic to Maikop deposits.

Layer N0. 1.

Oil-saturated sandstones of the No. 1 formation represent the lower part of a single Deposit in the strata. The water-oil contact (WOC) was opened at the absolute level of -3100 m timed to the structural uplift. The lift was opened by well 1.

An independent Deposit was opened by well 1 within the Kovylnoye area. The boundaries of the Deposit were previously determined by the position of the WOC at the absolute level of -3100 m. The boundaries of the Deposit, of course, are somewhat expanded in comparison with the operational calculation of reserves. Taking into account that the ratio of oil-saturated and water-saturated sandstones exposed by perforation is 2/3, the predominance of water in the product testing is quite understandable.

The Deposit of the Kovylnoye area has a rounded appearance. The height of the Deposit in the VIII layer is 10 m. The reservoir is Plast, vaulted.

The prospects for deposits of the formation are mainly related to the Kovylnoye area, where the area of the Deposit in layer VIII is probably larger than expected.

2. Materials and methods

In this work, a capacitive-filtration and geological-geophysical parameters of reservoir rocks of the Aptian horizon Grass field was evaluated in comparison with similar rocks East Bezbednosno field. To perform this task, the variability graphs of reservoir-filtration and geological-geophysical parameters were constructed and their correlation analysis was performed [1, 2].

For this purpose, data from the geo-information system (GIS) and data from laboratory studies of reservoir cores [3-6] obtained during drilling of wells No. 1 and No. 2 of the Kovylnoye field were used. The productive Aptian horizon of formation 1 is well lit with core material. For all samples under laboratory conditions, the following parameters were determined: porosity coefficient, permeability coefficient, resistivity, and effective thicknesses.

3. Characteristics of the lithological composition of reservoirs of the Aptian horizon

The Aptian productive horizon at the Kovylnoye Deposit is lithologically represented by a stratum of interlayers of light gray and greenish-gray sandstones, fine-and medium-grained, quartz-glauconite, oil-saturated, with traces of fauna, with dark gray mudstones, weakly silty, layered with traces of fauna and flora of poor preservation and dark gray siltstones, clay, mica massive, with rare thin layers of Sandstone or Dark gray mudstone.

Vertical cracks made by calcite are observed in the core samples. There are also layers of Sandstone, mixed-grained, mica, with rare inclusions of glauconite and carbonaceous material.

Limestones are mostly Micrograin structure with an uneven shell fracture. There are inclusions of spherical organogenic formations, the number of which varies in different areas - from 10-15% to 25-70%. Rare secondary leaching voids are observed, unevenly located in the rock, not interconnected, oval-rounded shape, size 0.02-0.3 mm, formed by the dissolution of calcite.

Allothigenic components of Sandstone are represented by quartz, feldspars, micaceous minerals, and accessory minerals. Operational objects of neighboring fields of similar age have a similar material composition.

4. Comparative analysis of reservoir properties

In addition to core studies, the results of geophysical studies [7-10] and hydrodynamic studies of wells in the Kovylnoye field were used to evaluate reservoir properties.

A comparative analysis of the results of research of the Kovylnoye field for wells No. 1 and No. 2 is presented in figures 1-4.

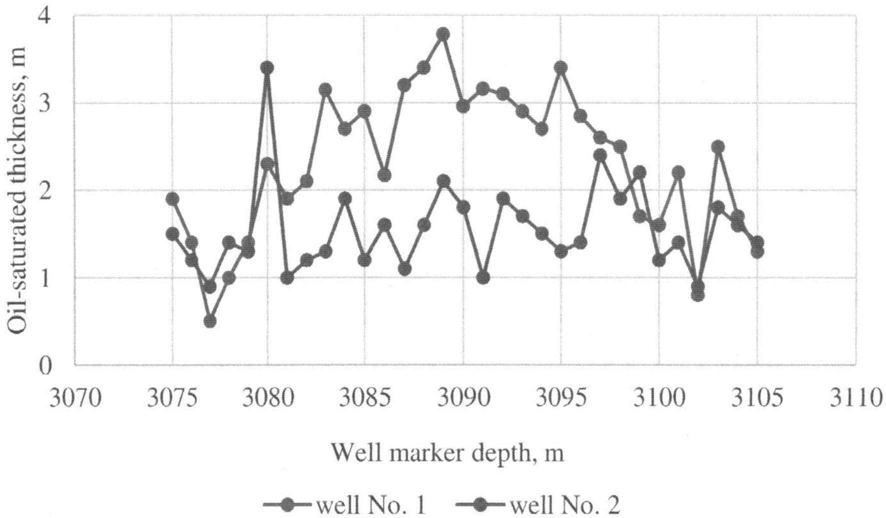

Figure 1. A graph of the variability of oil-saturated thickness with changes in the depth of wells No. 1, 2, Kovylnoye.

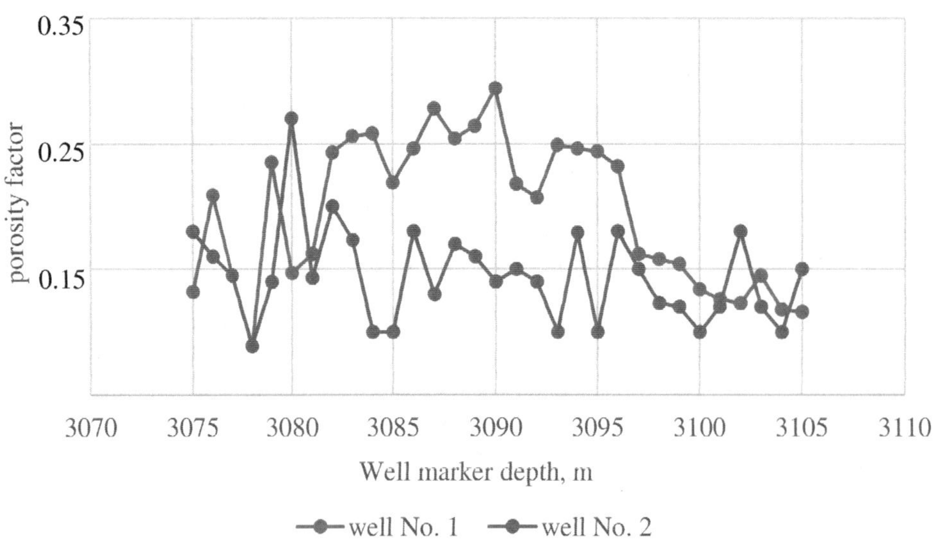

Figure 2. Graph of the variability of the porosity coefficient of rocks with changes in depth for wells No. 1, 2 Kovylnoye.

The ratio of the porosity of rocks (Figure 2) is an increase in the values in interval Upstage skyline reservoir number 1 (from 3080 3095 to – field No. 1; from to 3099 3105 – field No. 2). The maximum value is observed in the range from 3085 to 3091-for field No. 1 and at a depth of 3100 m-for field No. 2.

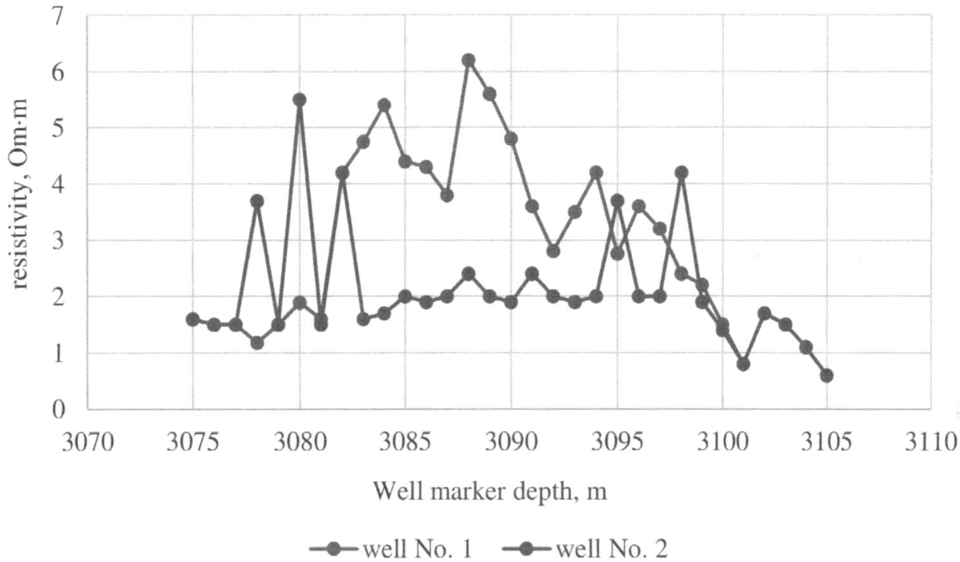

Figure 3. The variation graph of rocks resistivity with depth for wells No. 1, 2, Kovylnoye.

On the graph of rock resistivity (Figure 3), there is an increase in the value in the range of the Aptian horizon of formation No. 1 from 3084 to 3097 – for field No. 1 and from 3097 to 3104– for field No. 2. The maximum value was observed at a depth of 3.090 m in field 1 and 3.100 m in field 2.

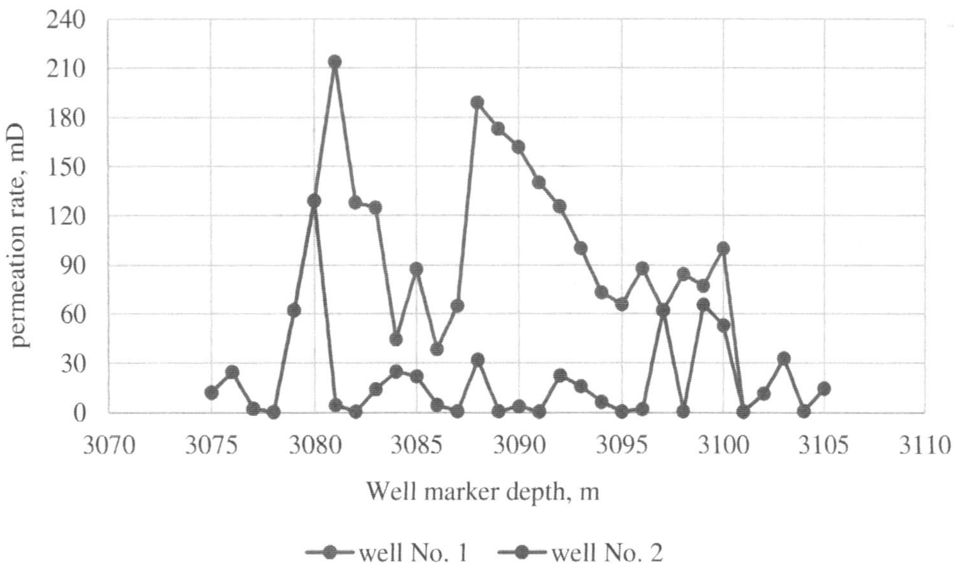

Figure 4. Chart of the variability of the permeability coefficient with changes in depth for wells No. 1, 2 Kovylnoye.

The graph of rock permeability coefficient variability (Figure 4) shows an increase in the value in the range of the Aptian horizon of formation No. 1 from 3087 to 3101 – for field No. 1 and from 3099 to 3103– for field No. 2. The maximum value was observed at a depth of 3,099 m - for field No. 1 and depths of 3,100 and 3,103 m-for field No. 2.

Based on the analysis findings of the distribution graphs of indicators in wells No. 1 and 2 Kovylnoye, it can be concluded that the formation is not uniform, there are layers of mudstones with other reservoir-filtration and geophysical characteristics. The intervals (3080-3090 – for well No. 1) and (3096-3106 – for well N0. 2) are oil-bearing horizons [11, 12], which is indicated by increased relative resistance values.

Correlation coefficients [13, 14] between the oil-saturated thickness (osth), porosity coefficient (por), resistivity (r) and permeability coefficient (perm) obtained during the correlation analysis of research results for wells No. 1 and 2 of the Kovylnoye field:

- for field No. 1:

$k_{osth,por} = 0.71$; $k_{osth,r} = 0.74$; $k_{osth,perm} = 0.59$; $k_{por,r} = 0.83$; $k_{por,perm} = 0.52$; $k_{r,perm} = 0.57$;

- for field No. 2:

$k_{osth,por} = 0.31$; $k_{osth,r} = 0.43$; $k_{osth,perm} = 0.67$; $k_{por,r} = 0.35$; $k_{por,perm} = 0.34$; $k_{r,perm} = 0.27$.

The following pairs of parameters have a significant positive correlation: porosity coefficient and oil-saturated thickness of rocks, rock resistivity and porosity coefficient, resistivity, and oil-saturated thickness [1].

The dependence of porosity on resistance is observed, which indicates the presence of hydrocarbons in the reservoir rocks, i.e., with increasing porosity, the resistance increases, which is also observed in the graphs [15, 16].

5. Conclusion

Having studied the distribution of indicators at the Kovylnoye field for two wells (No. 1 and No. 2) on the same horizon, we can conclude that these indicators differ very slightly, which indicates the uniformity of the reservoir properties. The quality indicators remain unchanged. Thus, on the territory of the Grass, you can expect similar rates within the entire area prospect reservoir number 1.

The analysis showed the dependence of porosity on resistance, which indicates the presence of hydrocarbons in the reservoir rocks, i.e., with increasing porosity, the resistance increases.

The use of a comprehensive analysis of reservoir rock properties makes it possible to determine the effective areas of the oil and gas potential of the Kovylnoye territory.

References

[1] Juang C, Zhang J, Shen M and Hu J 2019 *Engineering Geology* **249** 148-161
[2] Afzal P, Heidari S, Ghaderi M and Yasrebi A 2017 *Ore Geology Reviews* **91** 278-295
[3] Pradhan B 2015 *Geoscience Frontiers* **6** 791-792
[4] Haldar S 2018 *Mineral Exploration (Second Edition)* 69-84
[5] Asadzadeh S ans Filho C 2016 *International Journal of Applied Earth Observation and Geoinformation* **47** 69-90
[6] Kuzin A, Grishchenkova E and Mustafin M 2017 *Procedia Engineering* **189** 744-751
[7] Kortiev L I, Klyuev R V, Kulumbegov R P, Kortiev A L, Bosikov I I, Gavrina O A and M Z Madaeva 2019 *IOP Conf. Series: Materials Science and Engineering*(Tyumen) **663**
[8] Klyuev R V, Fomenko O A, Gavrina O A, Sokolov A A, Sokolova O A, Plieva M T, Kabisov A A and Ikoeva E Yu 2019 *IOP Conf. Series: Materials Science and Engineering*(Tyumen) **663**
[9] Plieva M T, Gavrina O A and Kabisov A A 2019 *Int. Multi-Conf. on Industrial Engineering and Modern Technologies (FarEastCon)* (Vladivostok) **Inspec Accession Number** 19229305
[10] Tsidaev B S, Logachev A V, Golik V I 2019 *Sustainable Development of Mountain Territories*

IOP Conf. Series: Materials Science and Engineering **952** (2020) 012017 doi:10.1088/1757-899X/952/1/012017

1 98-104
[11] Brekhuntsov A, Nesterov I, Nechiporuk L 2017 *Russian Geology and Geophysics* **58** 362-370
[12] Spiro D 2014 *Journal of Economic Dynamics and Control* **48** 159-175
[13] Hajirezaie S, Pajouhandeh A, Hemmati-Sarapardeh A, Pournik M and Dabir B 2017 *Journal of Molecular Liquids* **229** 89-97
[14] Wang F, Zhao J 2019 *Journal of Petroleum Science and Engineering* **178** 376-388
[15] Schimmel M, Liu W, Worrell E 2019 *Earth-Science Reviews* **194** 455-471
[16] LI Y, WU Sh, HOU J and LIU J 2017 *Petroleum Exploration and Development* **44** 603-614

IOP Conf. Series: Materials Science and Engineering **952** (2020) 012018 doi:10.1088/1757-899X/952/1/012018

Development of an algorithm for determining the optimal start time of oil wells to reduce the peak load

R V Klyuev[1,3], I I Bosikov[1], V Kh Tavasiev[2] and O A Gavrina[1]

[1] North-Caucasian institute of mining and metallurgy (state technological university), 44, Nikolaeva Str., Vladikavkaz, 362021, Russia
[2] North Ossetian state university named after K L Hetagurov, Vatutina Str., 46, Vladikavkaz, 362025, Russia
[3] Moscow Polytechnic University, 38, B. Semenovskaya Str., Moscow, 107023, Russia

E-mail: kluev-roman@rambler.ru

Abstract. The subject of this work is the development of an algorithm for determining the start time of wells in the automatic re-activation mode to save power during peak hours. This measure is relevant for the consumer and for the energy-supplying organizations, because the redistribution of load on peak hours or night zones will provide the most economically balanced operation of the enterprise, and will allow the most efficient use of sources of electrical energy supply organization. The algorithm development will provide operational control and management of the operating mode of the considered equipment in automatic mode. Besides, this will reduce the influence of the human factor on the operation of these consumers and minimize deviations in the hourly planning of electricity consumption, as well as reduce the cost of paying for consumed power. The short-term prospect of implementing this development will ensure optimal power consumption of the periodic well stock during peak load hours, and, consequently, will provide savings in the material resources of the enterprise.

1. Introduction

According to the dynamics of oil production, there are 4 stages of oil field development. The first stage is the development of the facility, the second is permanent production, the third is a decrease in oil production, and the fourth is the final stage.

In the oil and gas production department considered in this paper, there are already fields that are at the 3rd and 4th stages of development, which are characterized by a high degree of waterlogging of wells. Some of the wells have insufficient flow rate (inflow) due to the physical properties of the reservoir. In this regard, measures are being developed to switch wells to the automatic re-activation mode (hereinafter-ARM) to maintain the dynamic liquid level and save electrical energy.

Since 2014, every year there has been an increase in wells operating in the ARM mode (figure 1). From the analysis of this well stock, it was revealed that today there are 627 electric submersible installations (hereinafter referred to as ESI) operating in a periodic mode.

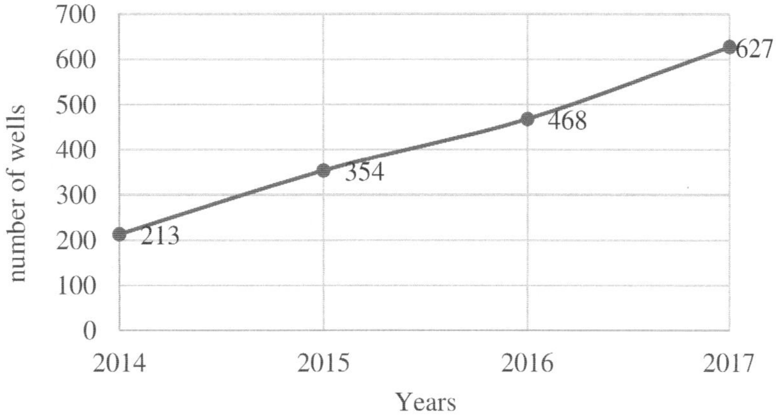

Figure 1. Analysis of changes in the stock of wells operating in the ARM

The company pays for the actual electrical energy consumed and the purchased (consumed) power.

The volume of purchased (consumed) power is defined as the average monthly consumption of the enterprise during peak hours, during which the maximum aggregate consumption was observed for the subject of the Russian Federation in which the enterprise is located.

Peak load hours are the time interval of the day when the load on the power supply network and generating equipment are the highest [1-3].

However, the start and the duration of time intervals for planned peak hours may change each month depending on the length of daylight, total load, and other factors. Statistics show that the time interval of the maximum hours of the same month does not change significantly each year.

Power is fixed in the power system during the billing hours, which are determined only on business days of the month at the time of peak electricity consumption in the region of the Russian Federation. At the same time, the consumer does not have information about what exact hour will be calculated, so it is not known when it is necessary to reduce consumption.

Every year, statistics are made on the time intervals during which the load in the power system is maximum. The maximum consumption range in winter is from 3 to 5 hours a day, in summer-from 6 to 12. That is why it is necessary to reduce the power not at a certain hour that falls during peak hours, but throughout the entire time interval that falls during peak hours.

Reducing electricity consumption during the billing hour will reduce power costs.

2. Materials and methods

The algorithm presented in this paper was implemented in a program created in the Microsoft Office Excel office Suite.

The analysis of the operating mode and consumption of electricity and capacity was carried out based on the data obtained from the automated power management system, the software tool «unified system for working with the well Fund», the information system «portal technologist», the oilfield complex «Alfa» and other sources.

3. Analysis of well operation mode

When analyzing the consumption system of an oil company, consumers were identified who have a re-short-term operation mode, namely wells with a set periodic operation mode (ARM) [4-8]. In this case, wells are operated with periodic stops of ESI operation for liquid accumulation.

According to electric power engineering, these wells are individual receivers of electric energy with a cyclic nature of the load schedule (figure 2), which is described by the formula:

$$t_{cycle} = t_{idle} + t_{work} = const, \quad (1)$$

where t_{idle} – idle time; t_{work} – work time; t_{cycle} – cycle time.

To date, the time when wells are put into operation is not regulated, and the well is started if there is a request to change the mode of operation of the well.

According to optimizing electricity and capacity consumption in the context of the daily schedule, it is necessary to optimize the start-up time of installations so that the idle hours of wells fall on peak load hours.

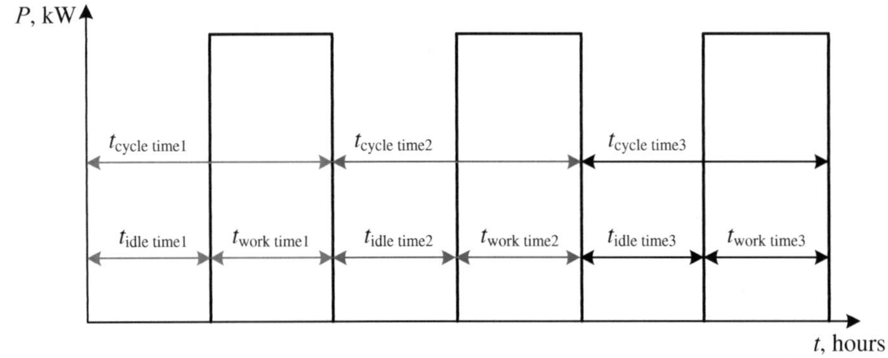

Figure 2. The mode of operation

If it is not possible to completely match the idle hours with the maximum load hours, the task must be reduced to a minimum match between the working hours and the maximum consumption period.

4. Optimization of the operating mode based on the algorithm

To solve this problem, an algorithm was developed to determine the optimal start-up time of the well [9-10]. The principle of operation of the proposed algorithm is based on the search for the optimal start-up time depending on the time of receipt of the request to change the operation mode of the well. To implement the launch time determination, you must perform the following operations:

1. Select the main parameters for calculation (operating time, downtime, and acceptable well stop time).

2. Determine the current date.

2.1. If the current date is the last working day of the month or a weekend or holiday, which may be followed by an interval of non-working days and not followed by working days of the current month, then calculation should be made for the first working day of the following month.

2.2. Determine the max hours and their duration.

3. Select the ESI involved in the calculation:

3.1. If the work cycle of wells is less than or equal to 1 (short-term operation), they are not used in the calculation.

3.2. If the daily cycle of the ESI is not a multiple of 24 hours (+/- 2 hours), then the possibility of adjusting the mode is considered, provided that the daily volume of oil production is maintained; otherwise, the installations are not used in the calculation.

3.3. The wells are divided into 2 groups:

– the first group of wells with the downtime from 4.5 to 10 hours a day;

– the second group – wells with the downtime of more than 10 hours per day.

4. The start of cycle:

4.1. If the initial startup time (the time of a request receipt for changing the operation mode) is before the interval of max hours, then the reference point is assumed to be equal to the interval end of max hours, we plot from it the maximum number of periods of the well (*N*). We obtain the time of start. Otherwise, the reference point must be assumed to be equal to the end of the max interval of the next day (i.e. +24 hours). The calculation is made from the next day to the current day.

4.2. If the downtime period of the well exceeds the interval of max hours, start the well at any time, provided that the well will not operate during max hours.

5. Determine the optimal well start time and display it on the screen.

6. Provide the opportunity to start the well at the optimal time.

7. Shift to the next month:

7.1. Five days before the end of the month, an alert is generated, which offers the possibility of recalculating the start time of wells for the next month. The responsible specialist confirms/does not confirm the restart of wells. If there is a confirmation of the restart, the system sends a command to complete the well cycle on the last day of the month (the first working day of the next month), and the start time is calculated. A command is given to automatically turn on the well at the specified time. If the restart is not confirmed, the well does not change its operating mode.

7.2. The calculation are made relative to the ESI of clause 3.3:

– for the first group of wells in the calculation period, if there is one forecast interval of maximum hours with a total duration of no more than 6 hours per day (applicable for 7 months of the year);

– for the second group of wells, applicable during the year.

7.3. The possibility of switching should be checked every month based on the conditions of section 7.2.

Well start-up is performed remotely in the software tool "remote control and operational control of objects". To assess the practical applicability of the proposed algorithm, it was implemented in a program created in the Microsoft Office Excel office Suite. The main window of the developed program is shown in figure 3 and is a simple interface with data entry cells, where the user needs to independently set the operating time, downtime, and select peak hours.

On the panel, there is a button «Determine the launch time», pressing which activates the algorithm, and the optimal start time of the well appears in the start time output cell. Clicking the «Clear» button deletes the data entered by the user. The prototype showed that the algorithm was correct.

Since the main task of this work is to develop an algorithm and not software, the following assumptions were introduced to simplify the calculation at the design stage:

1) we consider options for working wells only with integer values of hours of operation and downtime;

2) the start time of the well is rounded to an integer (minutes are not taken into account).

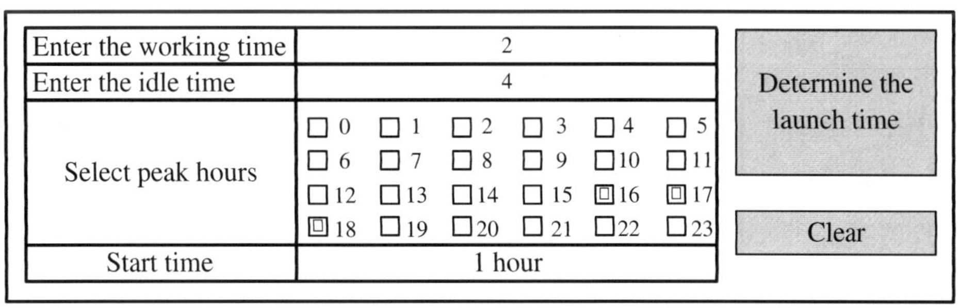

Figure 3. Main application window

5. Calculation of economic efficiency

The cost-effectiveness calculation was performed for November, since the range of planned maximum load hours in this month is the most averaged. After downloading the data from the "unified system for working with the well Fund" software, it was found that 627 wells in the ARM mode were involved in the work for this period.

However, the launch time was determined for only 456 of them. The rest of the wells (171 units) were not included in the calculation for the following reasons:

1) 72 wells have a working period of no more than 1 hour. It is not advisable to consider this option since these wells will work every hour;

2) 82 wells have non-integer values of operating hours and downtime;

3) 17 wells have a period value that is not an integer divisor of the number 24;

The quantitative components of the selected wells are clearly shown in the diagram shown in figure 4.

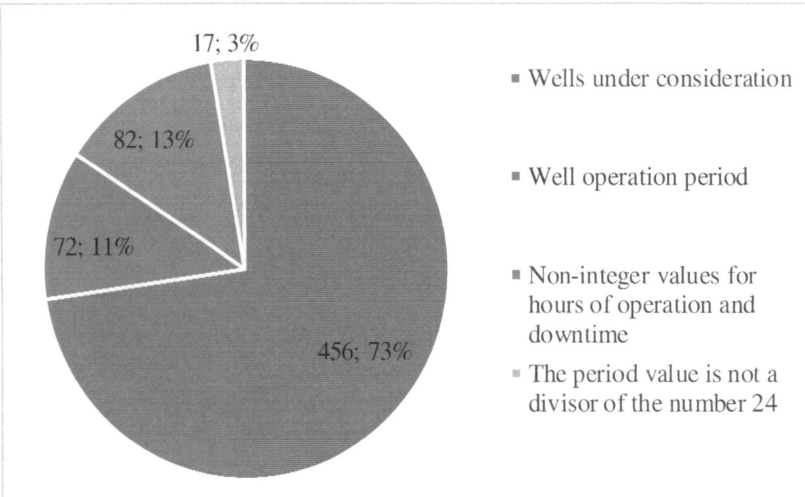

Figure 4. Quantitative components of selected wells

After determining the optimal time for putting wells into operation, a well start-up schedule is created as an example. As mentioned above, the start time of the wells is rounded off. Scheduled peak hours in November are 4 pm, 5 pm, and 6 pm.

Having constructed daily graphs of power consumption (figure 5) at the start time of wells before the adjustment and at the found optimal start-up time, the power consumption during peak loads were analyzed [11-15].

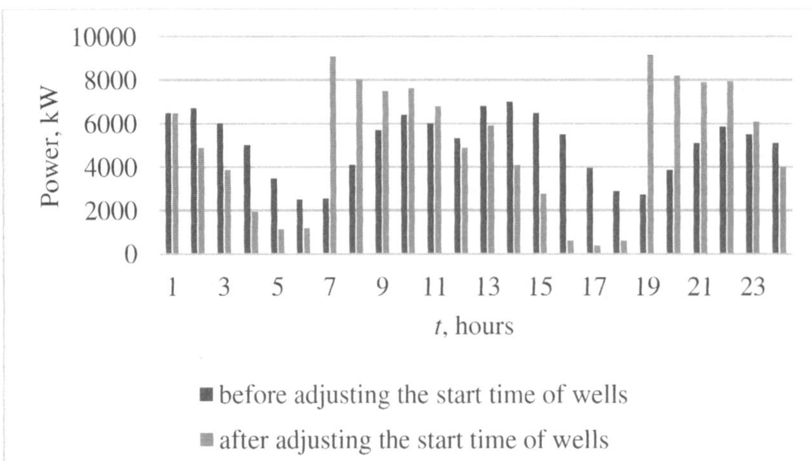

Figure 5. The daily schedule of power consumption before and after adjusting the start time of wells

IOP Conf. Series: Materials Science and Engineering **952** (2020) 012018 doi:10.1088/1757-899X/952/1/012018

At the same time, the graphs in figure 5 show that the maximum power consumption in the power system is significantly reduced. Based on the above, it was found that the implementation of this measure can bring economic benefits by reducing the cost of active capacity. The purchasing capacity is determined during the hours when the maximum power is consumed not by the enterprise, but by the entire region. To calculate the amount of power consumed, you need to know the estimated hours when the power consumption is recorded. During the calculations, the annual economic efficiency of the enterprise will be 26 million rubles.

6. Conclusion

The increase in the number of wells being converted to ARM mode requires solving the problem of optimizing their start-up to reduce power costs by distributing the load on the hours of semi-peak or night zones.

To solve this problem, an algorithm was developed for determining the optimal start-up time of the well. The practical applicability of the algorithm was evaluated in a program created in the Microsoft Office Excel office Suite (as a prototype).

It is planned to implement regulation of the operation mode of this group of electric receivers and launch an automatic generation of reporting forms, which will be used to control wells and thus minimize deviations in hourly planning of electricity consumption and reduce the cost of paying for power consumption.

Thus, by modernizing the existing system of consumption, the company can save a large amount of its budget. It should be noted that this algorithm can be applied to any electric receiver that has a periodic operation mode and meets the above requirements.

References

[1] Praveen M, Sivakrishna Rao G 2020 *Procedia Computer Science* **167** 2599-2605

[2] Lakshmi Sh, Ganguly S 2019 *Journal of Energy Storage* **22** 208-218

[3] Mehrjerdi H 2019 *Energy* **181** 916-926

[4] Lang J, Zhao J 2016 *Chinese Journal of Chemical Engineering* **24** 1423-1430

[5] Li J, Cao L, Guo B and Zhang X 2018 *Journal of Petroleum Science and Engineering* **160** 510-518

[6] Xie K, Nian Y, Cheng W 2018 *Energy* **163** 1006-1016

[7] Inchauspe J, Li J, Park J 2020 *Journal of Policy Modeling* **42** 536-556

[8] Husaini D, Puah Ch, Lean H 2019 *Energy* **171** 1000-1008

[9] Kortiev L I, Klyuev R V, Kulumbegov R P, Kortiev A L, Bosikov I I, Gavrina O A and M Z Madaeva 2019 *IOP Conf. Series: Materials Science and Engineering* (Tyumen) **663**

[10] Klyuev R V, Fomenko O A, Gavrina O A, Sokolov A A, Sokolova O A, Plieva M T, Kabisov A A and Ikoeva E Yu 2019 *IOP Conf. Series: Materials Science and Engineering* (Tyumen) **663**

[11] Ming B, Liu P, Cheng L, Zhou Y and Wang X 2018 *Energy Conversion and Management* **171** 528-540

[12] Lee W, Kang B, Jung J 2018 *Energy* **161** 963-973

[13] Burdzieva O, Zaalishvili V, Kanukov A and Maisuradze M 2016 *International Journal of GEOMATE* **10** 1693-1697

[14] Buryanina N S, Korolyuk Yu F, Maleeva E I and Lesnykh E V 2018 *Sustainable Development of Mountain Territories* **3** 404-410

[15] Tsidaev B S, Logachev A V, Golik V I 2019 *Sustainable Development of Mountain Territories* **1** 98-104

Main aspects of the oil industry within the sustainable development concept

N Fomenko and Al Mashkhadani Mohammed Jasim Hamad

Moscow State University of Civil Engineering Moscow, Russian Federation

E-mail: fnata77@mail.ru

Abstract. The main aspects of the industry development within the concept of sustainable development are considered in the article by the example of the oil industry. Sustainable development means the collaboration of three principal components, including ecology, society, economy. According to the carried-out analysis, the field of application and the stages of development of this concept are pointed out. The authors defined the economic, social, technology, ecological and political factors, influencing the development of the oil and gas industry. Thus, the application of the sustainable development concept in the oil industry will allow supporting the balance between the economy and the environment without causing damage to the environment, keeping resource base and providing stable improvement of quality of the population life.

1. Introduction

The first half of the 21st century is characterized by global changes in the field of science, equipment and technologies. The Internet, artificial intelligence, robotics and many other things became the realities of today. It is the century of boundless technological capabilities. However, the current state of the society is the subject to many crisis phenomena, which are realted, first of all, with the technicalization of all the spheres of human activity. The important consequences of these phenomena include the emergence of contradictions in the relations of human with the nature, of person with the society, the increase in number of technogenic and natural disasters. On September 15, 2015 the members of the UN offered the concept and some purposes in the field of sustainable development for the solution of these problems [1]. These purposes are aimed at the poverty eradication, preserving of resources of the planet and ensuring wellbeing for everyone. Within the adopted concept seventeen purposes are distinguish. Each of the purposes contains its indicators. The term of achievement is defined as the interval up to 15 years of this concept implementation.

The sustainable development concept is the target model of the civilization development, which proceeds from the need to provide the world balance between the solution of social-and-economic problems for the recent generations and preserving the environment for them [2]. The principle of the balanced economic, social and eco-friendly development is the basis for the target model of the civilization development [3].

The significant contribution to forming of the methodology of the sustainable development concept was made by such specialists as: Akimova T. Bobyliov A., Verstina N., Vinnokurov And., Danilova-Danilyan V., Tolstykh of T. Shimov O., etc. [4-8].

Sustainable development supposes long existence and development throughout unlimited time, therefore the creation of the concept within the triad: ecology-society-economy is a labour-consuming task. Thereof, the solution of great social, ecological and economic tasks has to be considered at the world and national level. The practical realization of the sustainable development concept, accepted by the UN, demands active participation of all the society sectors: public authorities, business structures, and society in general. It is an operated process, which assumes the development of economy and society, taking into account the careful use of natural resources.

The sustainable development concept really assumes certain restrictions in the field of the use of natural resources, but these restrictions are not absolute, they are relative and connected with the modern level of the equipment as well as with the social organization and with the capability of the biosphere to cope with the consequences of human activity. But technical aspects and aspects of the social organization can be taken under control and improved. That will open the way to a new era of economic growth [9]. The civilization development cannot be separable from the environment. The international community joined in the solution of these problems; sustainability strategies are developed in the majority of the countries, the relevant ministries and departments appeared. The governments of the leading states of the world have developed and accepted their sustainability strategies of the countries since the beginning of the 1990s. But the accepted strategies were not implemented, they were purely declarative in many respects.

2. Materials

The sustainable development happens in three main areas: economic growth, preserving of natural resources and environment, social development.

It is possible to allocate three main stages of the development of the Sustainable Development Concept application in the international community:

1) sustainable development is considered as the high level of individual income, i.e. its realization supposes steady rise. It in turn results in the capability of the national economy to reach an annual gain of GDP, providing its excess over the rate of the population growth;

2) implementation of the sustainable development concept is considered as the basic component of the social indicators improvement (liquidation of poverty, bad income distribution, unemployment, etc.); that is possible due to systematic increase in the economic growth rates;

3) the introductin of the sustainable development concept provides "the general promotion of the society as whole" through satisfaction of basic needs of the person in addition to the achievement of his or her subjectivity and human feelings and granting the liberty of choice. At this stage there is understanding that poverty is one of the factors, creating the threat of lag for the developing countries. At this stage "International Committee on the Development and Environment" made the report, in which the new sustainable development concept, meeting the needs of modern generation without exhausting the future resources, is created. It is necessary to find the balance among the exhaustion of the available resources, such as oil, the population size and the requirements of development, without consequences for the level of future generations living.

It is obvious that the idea of the sustainable development changes, the needs of people are specified, the technologies are reoriented, the mentality of the states changes, etc. It means that the implementation of its principles has to be considered step by step and has to be based on harmonization of interaction of the international community with the nature. So, some structural and territorial shifts in the world economy are necessary.

3. Results and Discussion

The special role in the implementation of the obstacles, set by the Strategy belongs to the industrial enterprises. Industrial enterprises bear a great share of the responsibility in the development of the country and society in general [10]. The enterprises face many the directions, capable to provide the increase in the activity efficiency within the sustainable development concept. The main directions include: the development of a technique of the sustainable development strategy implementation in the industry in general and at certain enterprises in particular and the improving of the competitiveness of industrial enterprise.

The leading role in the industry is assigned to the oil industry which is one of strategic components of the world economy growth and development of world power, the guarantor of revenues of the state budget, ensuring activity of all the industries in the national economy as well as of economic, political and social stability in the society.

World oil market has its own features and undergoes significant changes during the last decades. During 2011-2016 the world oil extraction increased by 9.3%. Saudi Arabia, Russia and the USA, which make up the share of 38.4% of all world oil extraction in 2016, are the traditional market leaders according to the production indicators. A bit different situation on oil reserves. So as of the end of 2019 the leader is Venesula (300.878 mlr. barrel), Saudi Arabia (266.4 mlr. barrels) is on the second place, the USA (35.2 mlr. barrels) is on the tenth place. The rating of the countries on the oil inventory level as of the end of 2019 is presented in Figure 1.

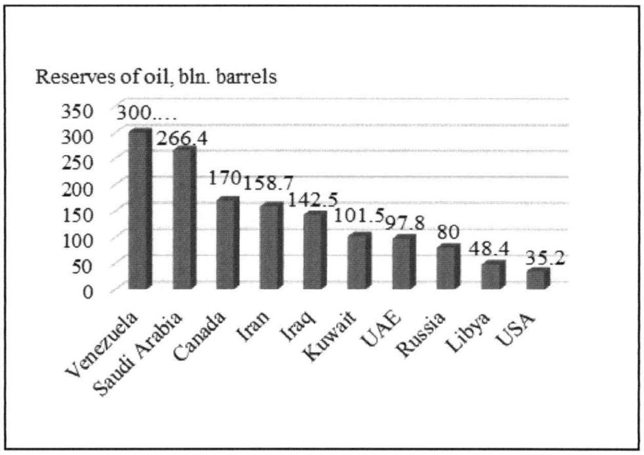

Figure 1. Top-10 of the countries according to the oil inventory level in 2019 [2]

Crude oil Trade is the main component of the international trade and any change in prices has a huge impact on the international trade balance. So according to the Research financial institution of the Ministry of Finance of the Russian Federation, in January, 2019 practically all the OPEC countries OPEC have the decrease in oil production in comparison with last month (Table 1). That is within the agreement on "freezing" of oil extraction that could not avoid affecting on the revenues of these countries' budgets. However, even at the same time the receipts from oil extraction make up the most part of revenues of the budgets of the oil exporting countries.

It is possible to distinguish the following major factors influencing the development of the oil and gas industry:

1) economic factors (the national economy growth rates and GDP; volatility of the oil prices; production cycles of the economy development; investment injections in the industry);

2) social factors (rates of the population growth; globalization and urbanization);

3) technological factors (innovation development in the field of oil production and alternative power engineering; the use of clever technologies; application of the innovation forms and methods of production organization and work organization);

4) ecological factors (application of the sustainable development concept, offered by the UN; world regulations and rules of the environmental protection);

5) political factors (maintenance of the investment climate; improvement of the right regulation in the field of pricing and rules of oil extraction).

In 2016 the largest exporters of crude oil in the world were Saudi Arabia – 7.463; Russia – 5.08; Iraq – 3.804; Canada – 2.742; The UAE – 2.408 in mln barrels a day [11]. However, according to the Lukoil forecasts, Iran and Iraq will be the main sources of growth of traditional oil extraction among the members of Organization of the Petroleum Exporting Countries (OPEC) till 2030 [12].

Table 1. Dynamics of oil production in opec countries

Countries	Target level of oil production		Oil production		Absolute deviation from the target value	
	according to the old conditions	according to the new conditions	December, 2018	January, 2019	December, 2018	January, 2019
Algeria	1.039	1.025	1.05	1.041	0.011	0.016
Angola	1.673	1.481	1.491	1.416	-0.182	-0.065
Gabon	0.193	0.181	0.194	0.193	0.001	0.012
Iraq	4.351	4.512	4.712	4.669	0.361	0.157
Congo	-	0.315	0.331	0.31	-	-
Kuwait	2.707	2.724	2.8	2.71	0.093	-0.014
Nigeria	1.8	1.685	1.74	1.792	-0.06	0.107
UAE	2.874	3.072	3.223	3.078	0.349	0.006
Saudi Arabia	10.06	10.311	10.562	10.213	0.504	-0.098
Ecuador	0.522	0.508	0.521	0.515	-0.002	0.007
Equatorial Guinea	0.178	0.123	0.11	0.117	-0.068	-0.006
11 OPEC countries	-	25.937	26.733	26.051	-	0.114
Iran *	3.797	-	2.758	2.754	-1.039	-
Libya *	1	-	0.947	0.895	-0.053	-
Venezuela*	1.972	-	1.165	1.106	-0.807	-
14 OPEC countries	32,16 **	-	31,603	30,806	-0,562	32,16 **

* According to the new conditions Iran, Libya, Venezuela received release from the transaction since January 2019.

** For which According to former terms of transaction the target level of production of oil was not established except Congo

As for the field of sustainable development, the activity of the oil companies is performed in collaboration with other industries. In such conditions the target orientation to the simultaneous and interconnected increase in the resource potential production, the increase in financial results, providing social and ecological safety amplifies. Thus, the implementation of the sustainable development Concept in the oil industry is characterized by the following aspects of sustainable development [13]:

1. Ecological aspect. The process of the sustainable development of the industry requires the rational and balanced management of the environment. The sustainable development philosophy is based on the fact that exhaustion of natural resources has a harmful impact on the development and economy in general. Therefore, it is necessary:

- to develop a peculiar balance of the economic and ecological system without exhaustion of natural resources, at the same time to consider and observe ecological safety;
- to eliminate poverty and to revise the structure of income distribution in the sake of the interests of the people below the poverty line. The solution of these matters is important not only in the terms of

social justice, but also in the terms of environment protection and also for the achievement of "ecological balance".

2. Human aspect. This aspect should be considered from the position of the increase in growth rates of the oil industry, when preserving stability of dynamics of the population growth rates without having a harmful impact on natural resources. Therefore, the value of sustainable development is the basis of the distribution of resources of oil production and refining for ensuring satisfaction of the basic needs of people (education, medicine, providing with clear drinking water and ecological products, solution of housing problems, etc.).

3. International aspect. The scientists revealed close interrelation between the nature of the global economic system and the exhaustion of natural reserves. The majority of the countries tries to increase the rates of extraction of raw materials (including oil), thereby providing economic development and growth. In such conditions protection of natural resources goes by the wayside or it is performed within complete isolation from the outside world. That affects the ecological component of the environment extremely harmful.

Thus it is possible to prove that the oil industry is the perspective direction within the implementation of the sustainable development:

1. It is of special interest as a financing source of the capital, consumer income and services stimulating economic development of the state, industry and the separate enterprises, it also allows to attract and perform additional investments at the expense of the high level of financial return from the oil export.

2. It is the raw resource which can be used for the industrial base creation and can be one of the production elements for other industries.

3. It has an opportunity to act as the employer for citizens of the oil exporting countries. It allows one to improve the situation in the labour market, to provide the growth in incomes of the population, to reduce the unemployment level and to reduce the quantity of the needy population.

3. Ecological aspect. The development of the oil industry does not always contradict the ecological postulates. Rationalization of the production processes and the energy use can be the source of the increase in the production efficiency, and at the same time it meets the nature protection purposes, saving non-renewable resources. However, it is impossible to disprove the fact of the existence of some environmental impacts as well as the problems, related with oil production and refining. These are the problems of water purification, disposal of gas, leak of oil and gas from pipes or tanks and so on. The ecological aspects allowing one to prevent environmental disasters are considered by the development programmes of the oil industry for the elimination of such consequences.

The environment and the oil industry are continuously interconnected. The development of the industry is not possible due to degradation of the environment and exhaustion of natural resources. At the same time, it is impossible to protect and preserve the environment in case of limiting the economic growth. Maintenance of such a balance allows one to raise the level of the population living by the achievement of sustainable development. It is obvious that the majority of oil-producing countries consider the environmental standards and follow global technological developments in this area.

4. Conclusion

Thus, it is possible to conclude that the main task of the sustainable development concept in the oil industry, as one of the main sources of income of importing countries, is the maintaining a certain balance between the economic development of the country, industry, its aspiration to the maximize profit on the one hand, and the long-term interests, which consist in preserving of resource base and permanent improvement of quality of life, human potential, on the other hand.

References

[1] UN *Purposes in the field of sustainable development*, available at: https://www.un.org/sustainabledevelopment/ru/sustainable-development-goals/

[2] Evteev S A and Perelet R A 1989 *Our common future: Report of the International Commission on Environment and Development* (Mosocw: Progress)
[3] Shaab 2019 *Sustainable development mechanism of the company on the basis of inter-organizational integration* (Moscow)
[4] Akimova Yu and Moseykin N 2009 Sustainable development economy (Moscow: Ekonomika Publishing House)
[5] Bobylev S N and Zaharov V M 2011 *Modernization of economy and sustainable development* (Moscow: Ekonomika Publishing House)
[6] N. Verstina N 2017 Formation of tools of resource providing management at the enterprise of the industry of construction materials in modern conditions *MATEC Web of Conferences* **106** 08091
[7] Vinnokurov A A 2007 Concept and criteria of sustainable development in the course of development of strategy of vertically integrated oil companies *Problems of regional ecology* **3**
[8] Granberg A G et al 2002 *Strategy and problems of sustainable development of Russia in the 21st century* (Moscow: Ekonomika Publishing House)
[9] Tolstykh T et al 2018 Digital innovative production on the basis of formation of an ecosystem of services and resources *Russian Journal of Industrial economics* **11 (2)** 159-168
[10] Shimova O S 2010 *Sustainable development* (Minsk: BGEU)
[11] 2017 *OPEC Annual Statistical Bulletin,* available at: https://www.opec.org/opec_web/static_files_project/media/downloads/publications/ASB2017_13062017.pdf
[12] Dev*elopment of the world market till 2035,* available at: http://www.lukoil.ru/Business/Futuremarkettrends
[13] UNDP 2010 *Human Development Report 2014: Making Humanity Progress: Reducing Vulnerability and Improving Sustainability*

IOP Conf. Series: Materials Science and Engineering **952** (2020) 012020 doi:10.1088/1757-899X/952/1/012020

Determining the location of the well's production collecting site at an oil field

K S Voronin, D A Cherentsov and A U Yakupov

Industrial University of Tyumen, 38, Volodarskogo str., Tyumen, 625000, Russia

E-mail: voronin_tsogu@mail.ru

Abstract. The paper considers a possible solution to the problem of optimizing the location of a well's production collecting site at an oil field. Depending on the location of the collecting site, the length of the pipelines going from the well clusters changes. Minimizing the total length of the oil collecting pipelines network will lead to lower capital costs for the construction of a well's production collection system.

1. Introduction

When designing a well's production collection system at an oil field, the question of choosing the optimal location for the well's production collecting site arises.

Well clusters are connected by pipelines to the collecting site. Depending on the location of this site, the length of pipelines going from well clusters will vary. Pipelines can also have different diameters, depending on the capacity of the well. Reducing the total length of these pipelines leads to lower capital costs for their construction.

Determining the location of the well's production collecting site at which the cost of constructing a pipeline network will be minimal taking into account the different diameters of the sections is an urgent optimization problem.

2. Materials and methods

Consider the case of determining the location of the well's production collecting site from N wells on flat terrain. Let the well clusters be located at points with coordinates (x_i, y_i), the desired optimal coordinates of the collecting site location will be denoted by (x, y). All well clusters are connected to the production collecting site by pipelines (Figure 1).

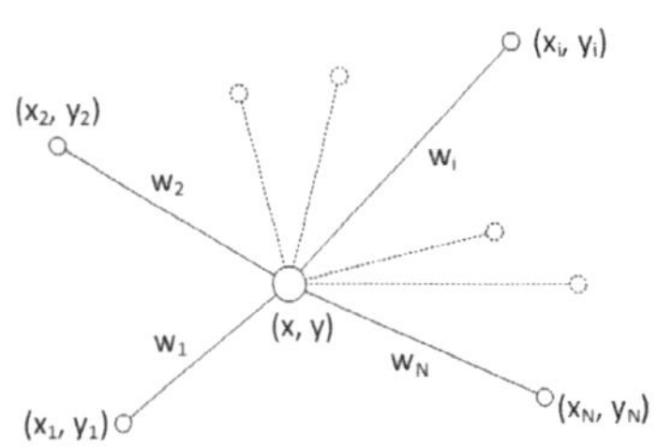

Figure 1. Schematic diagram of the pipeline network from well clusters to the production collecting site

IOP Conf. Series: Materials Science and Engineering **952** (2020) 012020 doi:10.1088/1757-899X/952/1/012020

Since the diameters of these pipelines depend on the capacity of the wells and can be different, we assign to each pipeline a weight (w_i) that characterizes the relative unit cost of the pipeline section construction:

$$w_i = \frac{C_i}{C_0} \tag{1}$$

where C_i – unit cost of the i-th section of the pipeline construction;

C_0 – unit cost of the reference section of the pipeline construction.

We take the cost of constructing a network of pipelines from well clusters to the production collecting site as the optimization function, which can be determined by the formula [7]:

$$C = C_0 \sum_{i=1}^{N} w_i \sqrt{(x - x_i)^2 + (y - y_i)^2} \tag{2}$$

where x_i, y_i – i-th well cluster coordinates;

x, y – production collecting site coordinates;

w_i – weight of the pipeline from the i-th well cluster.

We'll find the minimum of this function using the Newton method [1-4]. To do this, we need to set arbitrary initial coordinates (x_k, y_k) from which we will start the search. For example, it can be taken as the arithmetic mean value of the well clusters coordinates:

$$x_k = \frac{\sum_{i=1}^{N} x_i}{N}; \qquad y_k = \frac{\sum_{i=1}^{N} y_i}{N} \tag{3}$$

Find the gradient of the cost function:

$$\nabla C = \begin{vmatrix} \dfrac{\partial C}{\partial x} \\ \dfrac{\partial C}{\partial y} \end{vmatrix} = \begin{vmatrix} C_0 \sum\limits_{i=1}^{N} w_i \dfrac{x - x_i}{\sqrt{(x - x_i)^2 + (y - y_i)^2}} \\ C_0 \sum\limits_{i=1}^{N} w_i \dfrac{y - y_i}{\sqrt{(x - x_i)^2 + (y - y_i)^2}} \end{vmatrix} \tag{4}$$

Calculate the value of the cost function gradient at the point (x_k, y_k):

$$\nabla C(x_k, y_k) = \begin{vmatrix} C_0 \sum\limits_{i=1}^{N} w_i \dfrac{x_k - x_i}{\sqrt{(x_k - x_i)^2 + (y_k - y_i)^2}} \\ C_0 \sum\limits_{i=1}^{N} w_i \dfrac{y_k - y_i}{\sqrt{(x_k - x_i)^2 + (y_k - y_i)^2}} \end{vmatrix} \tag{5}$$

Find the absolute value of the gradient at the point (x_k, y_k):

$$|\nabla C(x_k, y_k)| = C_0 \sqrt{\left(\sum_{i=1}^{N} w_i \frac{x_k - x_i}{\sqrt{(x_k - x_i)^2 + (y_k - y_i)^2}}\right)^2 + \left(\sum_{i=1}^{N} w_i \frac{y_k - y_i}{\sqrt{(x_k - x_i)^2 + (y_k - y_i)^2}}\right)^2} \tag{6}$$

Next, check the condition for finding the minimum of the cost function with a given accuracy:

$$|\nabla C(x_k, y_k)| < \delta \tag{7}$$

where δ – required accuracy for finding the minimum of the function.

If condition (7) is satisfied, then we can assume that the optimal location coordinates of the well's production collecting site are found with a given accuracy δ [5-7]:

$$x = x_k, \qquad y = y_k \tag{8}$$

If condition (7) is not satisfied, then we need to find the inverted Hessian matrix at the point (x_k, y_k):

$$H = \begin{vmatrix} \frac{\partial^2 C(x_k, y_k)}{\partial x^2} & \frac{\partial^2 C(x_k, y_k)}{\partial x \partial y} \\ \frac{\partial^2 C(x_k, y_k)}{\partial y \partial x} & \frac{\partial^2 C(x_k, y_k)}{\partial y^2} \end{vmatrix} \tag{9}$$

where

$$\frac{\partial^2 C(x_k, y_k)}{\partial x^2} = C_0 \sum_{i=1}^{N} w_i \left(\frac{-(y_k - y_i)^2}{\sqrt{((x_k - x_i)^2 + (y_k - y_i)^2)^3}} + \frac{1}{\sqrt{(x_k - x_i)^2 + (y_k - y_i)^2}} \right)$$

$$\frac{\partial^2 C(x_k, y_k)}{\partial y^2} = C_0 \sum_{i=1}^{N} w_i \left(\frac{-(x_k - x_i)^2}{\sqrt{((x_k - x_i)^2 + (y_k - y_i)^2)^3}} + \frac{1}{\sqrt{(x_k - x_i)^2 + (y_k - y_i)^2}} \right) \tag{10}$$

$$\frac{\partial^2 C(x_k, y_k)}{\partial x \partial y} = C_0 \sum_{i=1}^{N} w_i \frac{-(x_k - x_i)(y_k - y_i)}{\sqrt{((x_k - x_i)^2 + (y_k - y_i)^2)^3}}$$

Inverted Hessian matrix:

$$H^{-1}(x_k, y_k) = \frac{1}{\frac{\partial^2 C(x_k, y_k)}{\partial x^2} \cdot \frac{\partial^2 C(x_k, y_k)}{\partial y^2} - \left(\frac{\partial^2 C(x_k, y_k)}{\partial x \partial y} \right)^2} \cdot \begin{vmatrix} \frac{\partial^2 C(x_k, y_k)}{\partial y^2} & -\frac{\partial^2 C(x_k, y_k)}{\partial x \partial y} \\ -\frac{\partial^2 C(x_k, y_k)}{\partial y \partial x} & \frac{\partial^2 C(x_k, y_k)}{\partial x^2} \end{vmatrix} \tag{11}$$

Using the recurrence formula, we calculate the coordinates of the desired point in the next step:

$$\begin{vmatrix} x_{k+1} \\ y_{k+1} \end{vmatrix} = \begin{vmatrix} x_k \\ y_k \end{vmatrix} - H^{-1}(x_k, y_k) \cdot \nabla C(x_k, y_k) \tag{12}$$

Substitute (4), (10) into (11):

$$\begin{vmatrix} x_{k+1} \\ y_{k+1} \end{vmatrix} = \begin{vmatrix} x_k \\ y_k \end{vmatrix} - \frac{1}{\frac{\partial^2 C}{\partial x^2} \cdot \frac{\partial^2 C}{\partial y^2} - \left(\frac{\partial^2 C}{\partial x \partial y} \right)^2} \cdot$$

$$\cdot \begin{vmatrix} \frac{\partial^2 C}{\partial y^2} & -\frac{\partial^2 C}{\partial x \partial y} \\ -\frac{\partial^2 C}{\partial y \partial x} & \frac{\partial^2 C}{\partial x^2} \end{vmatrix} \cdot \begin{vmatrix} C_0 \sum_{i=1}^{N} w_i \frac{x_k - x_i}{\sqrt{(x_k - x_i)^2 + (y_k - y_i)^2}} \\ C_0 \sum_{i=1}^{N} w_i \frac{y_k - y_i}{\sqrt{(x_k - x_i)^2 + (y_k - y_i)^2}} \end{vmatrix} \tag{13}$$

Calculations according to formulas (5) – (13) are repeated until the condition (7) is satisfied. The above calculations can be represented as an algorithm (Figure 2) [8-11].

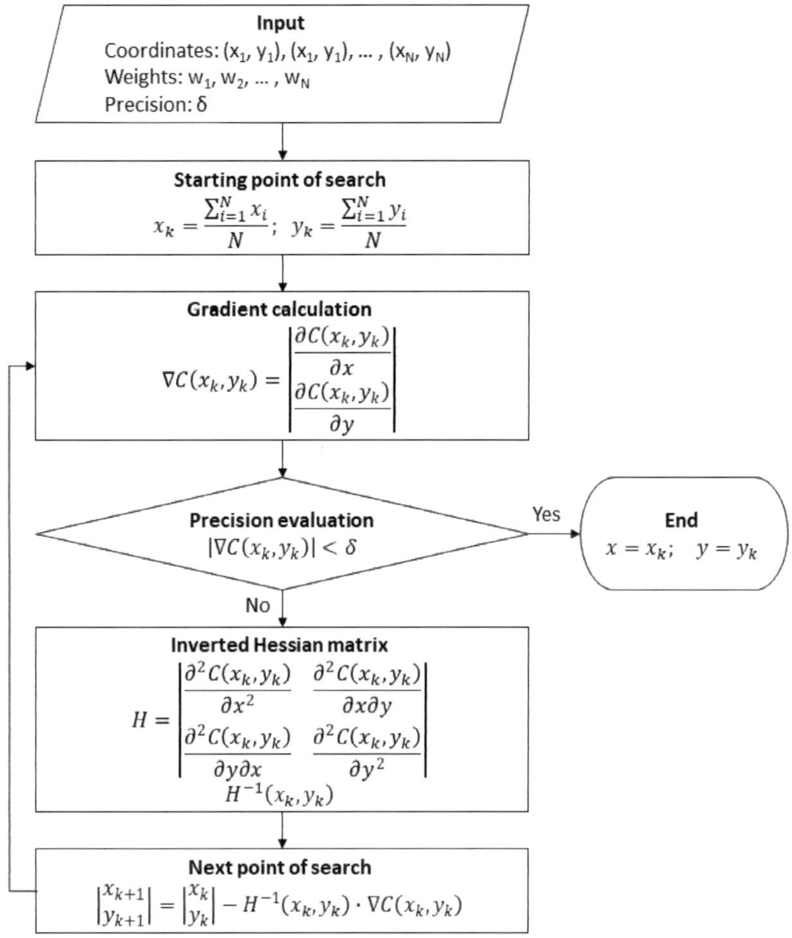

Figure 2. Algorithm of finding the well's production collecting site optimal location.

3. Conclusion

The cost of a pipeline laying depends not only on its length and diameter. There are other factors that are not taken into account in this method. For example, the technique of laying the pipeline, the amount of equipment involved, etc., depends on the terrain and the geological conditions. As a result, this also affects the unit cost of the pipeline laying.

Thus, the presented method allows us to find the optimal location of the well's production collecting site at the oil field with a given accuracy in the conditions of a flat terrain within the considered area.

References

[1] Amosov A A, Dubinskiy Yu A and Kopchenova N P 1994 *Computational methods for engineers* (Moscow: High school)
[2] Babin L A et al. 1970 *Selection of main pipeline routes* (Leningrad: Nedra)
[3] Borodavkin P P, Berezin V L and Ruderman S Yu 1974 *Choice of optimal routes of main pipelines* (Moscow: Nedra)
[4] Dolgirev O M 2015 Simple comparative analysis when choosing the best trace option for pipeline construction *Chemical technology* **9** 33

[5] Egorov Ya, Sharonin K A, Fursov I S and Nemtinov K V 2010 Information and logical model of tracing of technological pipeliness *Bulletin of the Tambov State Technical University* T. **16** **(4)** 826-830
[6] Grigorieva P V, Voronin K S and Cherentsov D A 2018 Development of layout algorithm for pipeline, considering topographic features *IOP Conference Series: Materials Science and Engineering* **012022**
[7] Voronin K S, Grigorieva P V and Cherentsov D A 2018 Estimating the cost of constructing and operating a section of a pipeline in the search for its optimal route *IOP Conference Series: Materials Science and Engineering* **012002**
[8] Shalay V, Zemenkova M, Zemenkov Y et al 2016 Modeling Parameters of Reliability of Technological Processes of Hydrocarbon Pipeline Transportation *MATEC Web of Conferences* **73** 01029 DOI: 10.1051/matecconf/20167301029
[9] Spasskaya K A 2014 Criteria for choosing a variant of the direction of the railway route, taking into account the development of landslide processes *Internet-journal "Naukovedenie"* **3** May - June 2014
[10] Zemenkova M, Shalay V, Zemenkov Y and Kurushina E 2016 Improving the Efficiency of Administrative Decision-Making when Monitoring Reliability and Safety of Oil and Gas Equipment *MATEC Web of Conferences* **73** DOI: 10.1051/matecconf/20167307001
[11] Chizhevskaya E L, Zemenkova M Y and Mayss N A 2020 Directions to Improve Effectiveness of Management of Management of the Industry Company Personnel *IOP Conf. Series: Earth and Environmental Science* **459** 052093 doi:10.1088/1755-1315/459/5/052093

Issues of ensuring the design durability of reinforced concrete supports of the main pipelines of oil refineries

P A Fedorov, M M Abdullin, A M Fatykhova and M M Fattakhov

Ufa State Petroleum Technological University, 1, Cosmonauts Str., Ufa, 450062, Russia

E-mail: Fpa_idpo@mail.ru

Abstract. The aggressive environment of oil refineries contributes to increased corrosion of reinforced concrete supports of trunk pipelines. This is primarily due to the increased concentration of acid gases. The article presents results of the analysis of the gas-air environment of an oil refinery and its impact on primary support measures. It was established that when using typical reinforced concrete supports made from B15 W2 concrete, the protective layer with a thickness of 15 mm becomes neutralized 3-6 years after the operation. To ensure continuous 25-year operation, it is necessary to increase thickness of the protective layer to 35 mm. To increase the durability of the structure, it is proposed to replace steel reinforcement with "metal-fiberglass" or hybrid fiberglass one.

1. Introduction

Pipeline transport is one of the most popular in the oil, refining and petrochemical sectors of the national economy. Its main advantages are efficiency and quick transportation of liquids and gases. However, there are a number of disadvantages: low life cycle of pipelines and building structures; danger of damage to the environment due to technological straits; great variability of construction methods, depending on the construction area. These shortcomings cause pipeline failures contributing to the premature decommissioning of building structures and formation of foci of technological disasters. The most common causes of failures [1] are the use of low-quality building materials and products, errors in the design of structural solutions, violation of building codes, performance of poor quality construction and installation works, etc.

One of the common methods for building pipelines in industrial zones is overhead construction on special overpasses. In the oil refining plants, overpasses with reinforced concrete columns are most widely used. This is due to their increased fire resistance compared to the metal ones. A typical solution for reinforced concrete supports of overpasses involves the use of concrete with a compressive strength class B15 or B20, with a waterproofing grade of W2 and W4. The thickness of the protective layer is 15 - 20 mm.

The life cycle of these supports is small. In [2], it is indicated that when examining the supports of the pipeline, which is an element of the technological scheme for processing natural gas, 15 years after the operation, vertical and transverse cracks, and violations of the continuity of the concrete protective layer (flakes and chips) were revealed. The total number of columns with defects was 64%.

It is characteristic of reinforced concrete used in pipeline transport facilities that in most accidents it has insufficient primary protection measures (thickness of the protective layer, cement consumption, etc.) and lacks secondary protection measures. Another cause of accidents is a change in the operating environment, for example, in the transition from a dry mode to a wet one.

The main ways to increase the durability of reinforced concrete include: the use of dense concrete and plaster compositions (waterproof grade W8 and more); rational thickness of the protective layer; additional coating of steel reinforcement with special protective compounds, etc.

It is necessary to assess the resistance of reinforced concrete in an aggressive environment of the oil refining complex, increase its durability and develop effective protection measures. Therefore, the aim of this study is to determine the influence of an aggressive environment on the reinforced concrete supports of overpasses located on the territory of an oil refinery.

2. Analysis of the aggressive environment of the refinery

The formation of defects and damages in the pipeline supports is associated with unfavorable temperatures and humidity, as well as aggressiveness of the environment.

Many technological processes occur at high pressures and temperatures of up to 500 °C [3]. Technological devices are located in open areas; therefore, these processes affect the temperature-humidity mode of operation. For example, during the winter period, the air temperature in the apparatus zone can range from + 2 ° C to + 35 °C at a temperature of five days in the region equal to -35 °C. Relative humidity varies from 20 to 90%. In summer, the temperature reaches + 40 ... + 50 ° C.

In processing hydrocarbon feedstocks and its products, acidic gases such as hydrogen sulfide, sulfur dioxide, nitrogen oxides, carbon monoxide, hydrocarbons, carbon dioxide and others are released into the atmosphere. The main reasons are as follows: technological needs (any complex has special burners for continuous or periodic gas combustion); uncontrolled emission due to tightness damage (the main cause of metal corrosion); emergency reset, etc. In [4], it is indicated that the main technological installations that contribute to the emission of acid gases are: combustion plants emitting sulfur dioxide, nitrogen oxides and solid particles, and others. When burning gases, combustion products in the atmosphere are distributed from the flare unit at a distance of 3 ... 4 km [5].

Concentration measurements in the areas of installation of overpass columns using the "HT2000" digital gas analyzer showed that the concentration of carbon dioxide is about 0.10 ... 0.15% in the winter period. Measurements were taken on the outside of the Ufa refinery in 2018. The concentration of CO_2 in a radius of 50-100 m from the flare unit reaches 0.40% when winter is poor. Measurements were taken on the outside of the Salavatnefteorgsintez plant.

An analysis of the operating environment by the degree of impact on reinforced concrete of the overpass columns, taking into account the indexation provided by GOST 31384-2008 "Protection of concrete and reinforced concrete structures from corrosion", showed (Figure 1) that the existing exposure conditions have the following indices: XC4 - corrosion of reinforced concrete as a result carbonation of concrete in zones of alternate wetting and drying under moderate aggressive effects of gas; XD2 and XD3 - reinforced concrete is exposed to chlorides, including salts and industrial waters containing chlorides (overpass supports are often performed through roads both for the plant use and streets). In addition, reinforced concrete of the aboveground part of the overpass support is affected by alternate freezing and thawing without combiners for vertical XF1 and horizontal XF3 faces, the effect of erosion due to the presence of large aerosol particles in the air of the production zone, and the effect of acid rain, which contribute to the corrosion of concrete in liquid environments. In special cases, there is a moderate impact of an aggressive environment on the foundations due to technological strains XA2.

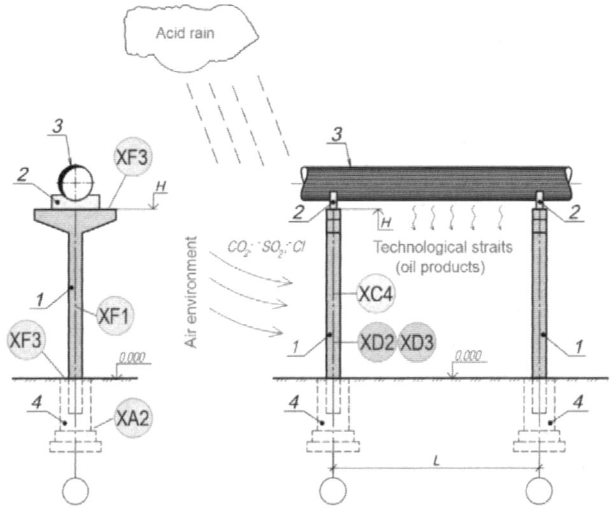

a

b

Figure 1. Indexing of aggressive media on reinforced concrete overpasses of a process pipeline: a - schematic diagram of the impact of an aggressive environment; b - general view of the reinforced concrete support with exposed reinforcement: 1 - reinforced concrete support; 2 - supporting structure of the pipe "lezhka-opora"; 3 - trunk pipeline; 4 - column foundation under the support

The impact of CO_2 on the protective layer of the reinforced concrete structure carbonizing concrete lowers the pH of the pore fluid from 11.8 to 9 in the area where the steel reinforcement is located. If aggressive sources (chlorides, water, oxygen) get in the environment, the steel corrodes and cracks the protective layer. Corrosion processes in reinforced concrete are more intensive with an increased concentration of CO_2, temperature and humidity of 50-60%.

3. Determination of the rational thickness of the protective layer by the criterion of CO_2 exposure

Currently, the use of concretes of low and especially low permeability is a common method in monolithic construction. When using these types, the class of concrete becomes B25 or higher. This is due to a lower water-cement ratio, the use of formulations including special additives, fillers. It should be noted that the cost of structures made of these concrete is increasing, which affects technical and economic indicators of the construction of an elevated trunk pipeline. A more rational option is to use these types of concrete to repair the existing supports.

A more optimal option is to determine the rational thickness of the protective layer of concrete. Under the action of CO_2, the purpose of thickness of the protective layer will depend on the kinetics of concrete carbonization. Currently, there is no single approach to the analytical calculation of this indicator. Many researchers describe this process by Fick's First Law in the form of a differential mass transfer equation. An analytical model describing the kinetics of concrete carbonization is as follows [6]:

$$x = \sqrt{\frac{2 \cdot D_{eff} \cdot C}{m_0} \cdot t}, \qquad (1)$$

$$m_0 = 0{,}4 \cdot Cem \cdot p \cdot f,$$

where D_{eff} – effective coefficient of diffusion of CO^2 in concrete, cm^2 / s; C –CO_2 concentration in relative units; m_0 – cement stone reaction capacity; t – impact time, c; Cem – cement consumption, kg / m ^{3}; p - the amount of basic oxides in cement in terms of CaO in relative quantities by weight; f –

the degree of neutralization of concrete, equal to the ratio of the number of basic oxides that have interacted with carbon dioxide, to their total amount in cement.

Based on the results of accelerated testing methods for fine-grained concrete of different densities at a CO_2 concentration of 5% and 75% humidity, a mathematical model taking into account the transition to the actual gas concentration was proposed in [7].

$$x = \left(0{,}9 \cdot \frac{W}{Cem} - 0{,}34\right) \sqrt[n]{t} \cdot \frac{C}{C_0}, \quad (2)$$

where $\frac{W}{Cem}$ – water cement ratio; C – actual concentration of CO_2, %; C_0 – reference concentration of CO_2 equal to 0.03 % n – the root index, depending on the sample density varying within 1.9…2.45; t – impact time, year.

To analyze the kinetics of carbonation of concrete of the overpass supports according to models (1) and (2), the following dependences were obtained (Figure 2).

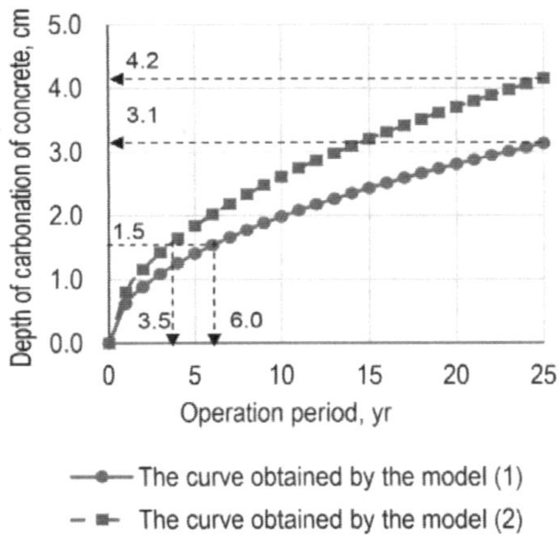

Figure 2. Dependence of the carbonation depth of concrete W2 on the flyover operation period

The following parameters were taken as initial data. The design service life is 25 years (7.9 × 108 s), adopted for structures operating in aggressive environments of oil refineries in accordance with GOST 27751-2014 "Reliability of building structures and foundations. Key Points. " The water-cement ratio of 0.65 is characteristic for B15 concrete [8]. The effective coefficient of diffusion of CO_2 in concrete for a water-cement ratio of 0.65 is 2.4 × 10-4 cm^2/s was adopted on the basis of experimental data obtained using the method of electrical conductivity of concrete. For the manufacture of reinforced concrete supports, cement "CEM II / V-Sh 32.5N" with a flow rate of 250 kg / m3 is used in the amount of CaO 64.61%. The average value of the degree of neutralization of concrete is 0.6. The average concentration of CO_2 in the refinery zone is 0.1%. The root index, which depends on the density of the sample for a water-cement ratio of 0.65 is 1.98.

Figure 2 shows two forecasts of concrete carbonization - optimistic and pissmestic. Both forecasts showed that the design thickness of the protective layer of 15 mm will be achieved 3 ... 6 years after the operation. These data are subject to the results of an inspection of the support structures 5 years after the repair [2]. Analyzing the data obtained, we can conclude that the most rational thickness of the protective layer of concrete should be at least 35 mm with concrete density of W2.

4. Improving the durability of reinforced concrete supports due to the use of metal-fiberglass reinforcement

An innovative way to increase the durability of reinforced concrete structures is to replace steel reinforcement with corrosion-resistant polymer materials with a high tensile strength. One of the most promising materials is non-metallic fiberglass reinforcement. Their significant disadvantages are: low modulus of elasticity; brittle fracture, inability to use a welded frame - usually connected by knitting (using steel wire, plastic clamps) or bonding with a polymer composition such as epoxy adhesive, etc.

To improve the mechanical properties, it was proposed to add a steel core [9–11] (Figure 3). This armature is called "metal-fiberglass" or "hybrid" fiberglass.

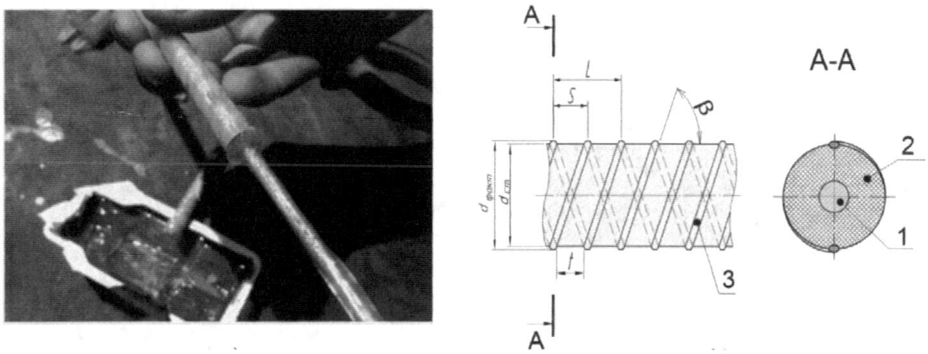

a b

Figure 3. Metal-fiberglass (hybrid) rod: a - the process of installing a steel core in a sample; b - is a diagram of a metal-fiberglass rod; 1 - steel core; 2 - a bundle of fiberglass threads; 3 - multiple winding

According to [9], fiberglass reinforcement with a diameter of 10 mm was used as the main rod, the steel core was made of fittings of a smooth profile "BP-I" with a diameter of 3 mm. These samples were obtained by excavating the middle part of fiberglass reinforcement with a diameter of 3.5 mm. Epoxy adhesive was initiated into the hole and the steel rod was inserted. After hardening, the samples were subjected to tensile and thermal stress tests. According to the results of tensile tests, an increase in the modulus of elasticity of metal-fiberglass reinforcement was obtained. The stress-strain state of rods was established in [10], they exhibit bilinear behavior, i.e., there is a certain amount of plasticity compared to the brittle fracture of a conventional fiberglass rod, the elastic modulus reaches 145 MPa, depending on the ratio of the outer and core diameters. The studies described in [11] showed that for a hybrid carbon fiber core with a total diameter of 12 mm and a core diameter of 8 mm, the elastic modulus was 224 GPa. The production of such rods should be carried out according to the traditional technology by assembling a bundle of filaments and a core followed by their subsequent treatment with epoxy copound. Focusing on the above data, we can assume that this type of reinforcement can reduce material consumption and increase the corrosion resistance of the structure.

5. Conclusion

Studies have shown that when using class B15 concrete with a waterproofing grade W2, the thickness of the protective layer of reinforced concrete supports of 15 mm will be neutralized 3 ... 6 years after the operation in the oil refinery. To ensure a design life of 25 years, the rational thickness is 35 mm. Otherwise, it is necessary to reduce the permeability of concrete, or use secondary protection measures. One of the effective ways is to replace steel reinforcement with "metal-fiberglass" or hybrid one.

References

[1] Dushin V A and SHammazov A M 2008 *Overhaul of Linear Parts of Main Oil Pipelines* (Ufa: OOO «Monografiya»)
[2] Kasimov V G 2017 *The russian scientific-methodical conference "University complex as a regional center of education, science and culture"* 746-751
[3] Suhanov V P 1979 *Oil Refining* (Moscow: Vysshaya shkola)

IOP Conf. Series: Materials Science and Engineering **952** (2020) 012021 doi:10.1088/1757-899X/952/1/012021

[4] Speight J G 2020 Petroleum Refining and Environmental Control and Environmental Effects *Fossil Energy* ed. Malhotra R (New York: Springer New York) pp 101–131
[5] Laden F, Schwartz J, Speizer F E and Dockery D W 2006 *American J. of Resp. and Crit.Care Medicine* 2006 **173(6)** 667-672
[6] Alekseev S N and Rosenthal N K 1976 *Corrosion Resistance of Reinforced Concrete Structures in an Aggressive Industrial Environment* (Moscow: Stroyizdat)
[7] Fedorov P A, Anvarov A R, Lutsyk E V, Latypov V M and Latypova T V 2016 *Int. J. of App. Eng. Research* **11(11)** 7439-7445
[8] Podmazova S A 2004 *Stroitel'nye materialy* **6** 8-9
[9] Abdullin M M, Fattahov M M, Fedorov P A and Andriyanova S O 2012 RU, Patent No. 120981 *Specification*
[10] Ju M, Lee S and Park C 2017 *Int. J. of Conc. Str. and Materials* **11(4)** 677-686
[11] Zhou Y, Zheng Y, Sui L, Hu B and Huang X 2020 *Materials* **13** 234

Current state and perspectives of the gas industry of Russia

L Sh Makhmudova[1], Kh Kh Akhmadova[1], A A Pimenov[2], Zh T Khadisova[1] and Z A Abdulmezhidova[1]

[1] Grozny State Oil Technical University named after academician M.D. Millionschikov, 14/53, Avtorkhanov st., Grozny, Chechen Republic, 364902, Russia
[2]Samara State Technical University, 244, Mologvardeiskaya st., Samara, 443100, Russia

E-mail: janna_h@list.ru

Abstract. The dynamics of energy consumption in the world market is analyzed. The natural gas demand growth forecast is provided. Gas will be the most crucial energy source in the next 30 years which can be considered to be the "era of gas". It is shown that natural gas is a promising source of energy and raw materials of petrochemical chemistry. Unlike oil and coal, gas will have strengthened its position by increasing its share in the total balance of primary energy carriers from 22 to 24-26% by 2035. By the middle of the 21st century, its share will have been 25 -27% in the world and 56% in Russia. The use of natural and associated petroleum gas in Russia is compared with global trends. In Russia, the use of useful gas components does not meet potential capabilities of the raw material base and does not comply with the world practice. The rational use of hydrocarbons can be organized through the cooperation of gas processing. The backlog of Russian gas processing is described. Currently, global gas processing is in the GTL era. In order to increase the profitability of producing synthetic liquid hydrocarbons, a combination of existing and GTL enterprises is promising. Developments of Russian scientists and world experience allow us to create GTL-production in almost all developed regions of Russia. The conclusion about the possibility and necessity to develop GTL-production is drawn.

1. Introduction

Natural gas plays an important role in the global energy balance due to its environmental friendliness, economy and technological effectiveness. Compared to other mass energy resources (coal, oil), gas produces less harmful emissions, its reserves are significant, production and transportation are relatively cheap, and its processing into chemical products requires is less expensive.

However, Russia, which is a powerful gas power with huge reserves of natural gases and gas condensates, does not use all the potential resources of the raw material base, and useful gas components are processed worse than in developed countries. There is an urgent need to formulate priority areas for increasing the efficiency of production and use of natural gas. It is relevant to carry out a scientific analysis of the gas industry by comparing it with global trends. The purpose of this study is to assess the current state of the gas industry in Russia and determine effective directions of processing natural and associated petroleum gases to produce valuable chemical products.

2. The role of natural gas in the fuel and energy industry

In the Russian economy, the fuel and energy complex (FEC) and exports of its products occupy a unique position, their dynamics affects the stability of the national economy. "Gas will be the most crucial energy source in the next 30 years which can be considered to be the "era of gas" [1].

The annual world energy consumption will grow by 1.3% in the next 20 years. Natural gas, unlike oil and coal, will strengthen its position by increasing its share in the total balance of primary energy carriers from 22 to 24-26% by 2035 [2]. By the middle of the 21st century, it could reach 25-27% in the world, and 56 % in Russia. Until 2035, world gas consumption will continue to grow by 1.8% per year [1].

Data on world gas production for 2011-2017 (billion m^3) are given in Table 1.

According to the estimates of [4], in 2017 in Russia, natural gas production grew by 8.1% and reached 690.9 billion m^3, which was high for the post-Soviet period. In 2018, it was 727.6 billion m^3 [5], in 2019 - 739 billion m^3 [6] and increased by 1.8% compared to the previous year, which was facilitated by an increase in global demand and domestic consumption. About 13% or 95 billion m^3 fell to the share of petroleum gas.

An analysis of the data shows that Russia ranks second after the United States, which produced more than 1000 billion m^3 of marketable gas in 2019 [7].

Table 1. Data on gas production in the leading gas producing countries for 2011-2017 (billion m^3) (according to [3])

Countries	Years				
	2011	2012	2013	2016	2017
USA	648,5	681	685,4	749,2	751,1
Russia	607,0	656,0	604,7	579,4	690,9
Canada	144,4	157,0	141,4	152,0	174,1
Iran	159,9	158,0	166,8	202,4	226,9
Qatar	145,3	160,0	177,6	181,2	182,8
Norway	101,4	115,0	108,7	116,6	120,2
China	109,0	107,0	122,2	138,4	136,6
Saudi Arabia	92,3	95,0	100,0	109,4	110,9
Other	1282,8	1271,0	1297,1	1323,0	1242,6
World gas production	3290,2	3400,0	3403,9	3551,6	3636,1

In Russia, current reserves of natural gas are 48.805 trillion m^3, and the volume of exploration of undeveloped raw materials is 24.6% [8]. The regional structure of gas production is presented in Table 2.

Table 2. The regional structure of gas production in Russia in 2018 (according to [9])

Regions	Volume of gas extraction	
	billion m ^{3}	%
Yamal-Nenets Autonomous District	591,2	81,2
Khanty-Mansi Autonomous District	34,9	4,8
Sakhalin region	32,4	4,5
Orenburg region	16,2	2,2
Astrakhan region	14,2	2,0
Krasnodr region	9,4	1,2
Other	30,1	4,1

The largest oil and gas company is Gazprom of Russia, which has the richest natural gas reserves (33.4 trillion m^3). Its share in the world gas reserves is 17%, in the Russian gas reserves - 72%. Gazprom accounts for 12% of global and 66% of Russian gas production. Currently, the company is

IOP Conf. Series: Materials Science and Engineering **952** (2020) 012022 doi:10.1088/1757-899X/952/1/012022

implementing large-scale projects for the development of gas resources of the Yamal Peninsula, the Arctic shelf, Eastern Siberia and the Far East, as well as a number of hydrocarbon exploration and production projects abroad [9]. Thus, the volumes of natural gas production in the world, including in Russia, tend to increase, which indicates the relevance of gas as a source of high-quality fuel and gas chemistry raw materials.

The situation with associated petroleum gas is somewhat more complicated. This is due to the fact that the resources of associated petroleum gas are more limited - they are determined by the scale of oil production and its low content in the oil produced. Accordingly, the gross volume of associated petroleum gas produced with oil is less by 10 times [10]. The energy costs of associated petroleum gas processing are significantly higher, since, unlike high-pressure natural gas, associated petroleum gas is released from oil at a low pressure. Associated petroleum gas is much more complicated than natural gas, which complicates the technology of its processing, but increases its value as a raw material.

3. Comparative analysis of the use of gas hydrocarbon feedstock for gas chemistry

In Russia, the use of useful gas components does not meet potential capabilities of the raw material base and comply with world trends. For example, over the past 25 years, gas condensate reserves have increased 100 times, and its extraction does not exceed 3-6 million tons per year [11].

The rational use of hydrocarbons can be organized through the cooperation of the gas processing and petrochemical industries [2, 5, 11, 12].

Modern large oil and gas companies have well-developed gas-chemical industries. The share of petrochemical products and profits of the largest oil and gas companies (Exxon Mobil, Shell, Total, etc.) are 8-10% of the total production [12].

The global petrochemical industries have different raw material orientations. Japan, South Korea, most European countries use liquid hydrocarbon feedstock, mainly straight-run gasoline fractions (naphtha). These countries cooperate in the oil refining and petrochemical industries. Other countries (USA, Canada, the Middle East, primarily Saudi Arabia, Venezuela) use gaseous raw materials (ethane, propane, butanes and other hydrocarbons). These countries are building cooperative ties between the gas processing and petrochemical industries. Some of these countries (e.g., the United States) produce ethane and other shale gas hydrocarbons. In the countries of the Middle East, gas condensate is used as a raw material [13, 14].

The problem of supply of Russian petrochemical and chemical industry enterprises with hydrocarbon raw materials can be solved by using natural and associated petroleum gases and gas condensates, which are not used rationally.

In the total volume of natural gas produced, methane is mainly used as furnace and household fuel. The share of ethane, propane, butanes and heavier hydrocarbons is much smaller [15].

Natural gas exported from Russia is processed into extract valuable hydrocarbons used to produce petrochemical products. The development of high-tech gas chemistry would make it possible to obtain high value-added products and increase export revenues, which is especially important in the conditions of instability of the world energy markets and relatively low prices for hydrocarbons.

For 2013, there is information on 2,000 gas processing plants and plants that process more than 4 billion m^3 of gas per day [2]. The largest number of gas processing plants are located in the USA and Canada –700 and 979, respectively (with a total capacity of 840,2 and 554.0 billion m^3 / year) [10], 36 gas processing enterprises are located in Russia, which is incommensurable with Russian reserves of natural and oil gas. Large gas processing facilities are located in Saudi Arabia, Mexico, Kuwait, Iran and other countries. In these countries, liquid products extracted from natural and associated gases are raw materials for the petrochemical industry [14].

Only 7.5% of natural gas is processed in Russia. IN 2005 in the USA, about 90% of natural gas was processed. Russia is losing at least 19–20 million tons of light hydrocarbons per year [16].

The uniting of oil and gas refining and petrochemical industries led to the creation of the largest centers for the production of petrochemical products (petrochemical clusters) in Texas and Louisiana.

In these states, there are 263 out of 700 American gas processing plants and 46% of gas processing capacities of the USA [15]. In the USA, small units of shallow gas processing are located at the fields preparing gas for transportation to large gas processing plants for subsequent deep processing [17].

In Russia, 0.4 million tons of ethane have been extracted from natural and associated petroleum gas, and in the United States, this volume is 9 million tons per year. In 1977, a 3,000-kilometer pipeline was built to transport ethane, ethylene, propane and butane from western Canada to the United States; ethane is used to produce ethylene, its production volume and consumption structure determine the level of development of the organic synthesis industry. About 65% of propane are produced from petroleum and natural gases, which are used as a pyrolysis feedstock. In addition to ethane and propane, n-butane, isobutane, n-pentane and isopentane are extracted from natural and petroleum gases, which are used in the production of many valuable products: synthetic rubber, communal butane and motor fuels. When cleaning and processing natural and associated gases, large quantities of cheap sulfur, helium and other inorganic products are produced. In the production of helium, the USA ranks first. Helium is used in space rocket technologies. Canada is the second largest sulfur producer in the world. For Russia, it is promising to extract helium from gas-helium raw materials of Eastern Siberia and Yakutia, which contain large amounts of helium - up to 1 percent or more [15].

Recently, the process of liquefying and transporting liquefied natural gas by water has been developing rapidly, which makes it accessible to almost any country in the world. It is expected that in the USA, an increase in imports of liquefied natural gas will increase 2-3 times [18-21].

The results of a comparative analysis of gas industries in different countries confirm that Russia lags in the gas processing.

Currently, global gas processing is at a stage called the “GTL era” (conversion of natural gas into higher hydrocarbons, fuels and chemical products).

The main products of the synthetic liquid fuel production are: high-quality environmentally friendly diesel fuel, which fully complies with Euro-5 requirements; synthetic light naphtha, which can be used as an environmentally friendly component of motor fuels (an analog of the straight-run gasoline fraction) and high-quality raw materials for petrochemical industries; synthetic oil (instead of synthetic motor fuels) [22, 23].

Currently, there are four large-tonnage GTL plants: Mossel Bay (South Africa); Bintulu (Malaysia); Oryx (Qatar); Pearl (Qatar). New GTL plants are going to be constructed in the future [24, 25].

Various foreign companies have created GTL pilot plants in the USA, Brazil, Great Britain, South Africa, and Japan. 55 projects of GTL technologies with a total capacity of about 2 million bar with a total gas consumption of 166 billion m^3 / year are being implemented [26]. Technological solutions for producing liquid hydrocarbons from natural gas directly from synthesis gas according to the Fischer-Tropsch reaction and through the methanol production stage have been developed and implemented.

There is no similar production in Russia, and leading developments are mainly based on the use of the second variant of synthesis through methanol. However, the accumulated world experience allows us to create GTL-production in almost all developed regions of Russia.

According to experts of the All-Russian Scientific Research Institute of Natural Gases and Gas Technologies, the use of natural gas from remote fields by converting it to synthetic liquid fuels is a more promising area [27].

The creation of GTL-production in Russia is relevant for the following five reasons:

- oil independence of energy;
- GTL product - environmentally friendly fuel;
- GTL is an alternative way of exporting natural gas and pipeline independence;
- possibility of utilizing associated gas;
- profit from the production of synthetic fuels. In Russia, Gazprom, Rosneft and Lukoil officially announced plans to develop this new technology [25].

Over the past decade, a number of Russian energy companies (Gazprom PJSC, Irkutsk Oil Company LLC, etc.) have evaluated the possibilities of creating large-capacity GTL plants in Russia using the technologies of Western companies (Shell, Japan GTL, Syntroleum) [24, 28]. However, due to the calculated risks of the GTL business, plans to build Russian plants have not been implemented.

Currently, there are a number of innovative organizations in Russia: LLC Gazohim Techno, LLC Energosyntop-Engineering, LLC Infratechnologies, Institute of Petrochemical Synthesis of the Russian Academy of Sciences, LLC CITIS, etc. GTL technologies are being developed, but most of the projects are at the stages of theoretical development or pilot implementation and require additional research [28-34].

In order to increase the profitability of producing synthetic liquid hydrocarbons, it is necessary to unite existing enterprises and new GTL ones.

On the basis of Novokuybyshevsk Oil Refinery of Rosneft, a pilot industrial unit for the production of synthetic oil is going to be launched for the first time. The plant capacity will be 300 tons of synthetic hydrocarbons per year. It is planned to use the technology for converting natural and associated petroleum gas into high-quality synthetic liquid and solid hydrocarbons by the Fischer-Tropsch technology.

Novomoskovsk Institute of Nitrogen Industry began to manufacture plants for the production of methanol from gas. In 2010, Russia's first pilot plant for producing methanol from natural gas with a capacity of 40 thousand tons per year was installed at the Yurkharovskoye gas condensate field in Western Siberia [32].

The GTL industry can be considered in two key aspects: large-capacity and small-capacity production. Given the capital intensity and risks of GTL large-scale production, as well as its competition with LNG projects and gas pipeline transport, small-scale production is a promising direction for the GTL industry. In this area, GTL micro- and minichannel technologies, which have been developed in recent years [35], are of interest.

Thus, in Russia there is no industrial GTL production, and developments of Russian scientists and accumulated world experience allow us to create GTL production industries in almost all developed regions of Russia.

4. Conclusion

It is important for Russia to choose industrial development areas where it has competitive advantages and potential and natural prerequisites for the innovative development. This potential exists in the gas industry, since Russia has the largest world reserves of natural gas. However, the use of useful components of natural and associated gas does not meet potential capabilities of the raw material base and does not comply with the world trends. At the same time, world experience shows the effectiveness of small installations for shallow gas processing directly at the fields, preparing it for transportation to large gas processing plants for subsequent processing.

Developed and implemented technological solutions for the production of liquid hydrocarbons from natural gas will allow the use of gas from remote fields through its conversion and solve the problem of reliable supply of petrochemical and chemical industries with hydrocarbon raw materials. To do this, it is advisable to develop a GTL-production industry in Russia and combine it with gas processing enterprises. These are an alternative way of exporting natural gas, pipeline independence and a possibility of utilizing associated gas from oil production.

Acknowledgement

This research was supported by the Ministry of Education and Science of the Russian Federation within the agreement 05.607.21.0311. The unique identifier for the project is RFMEFI60719X0311

References

[1] Ed. Makarova A A, Mitrova T A and Kulagin V A 2019 *Forecast of the development of energy in the world and Russia* (Institute for Energy Research of the Russian Academy of Sciences. Energy Center of the Moscow School of Management Skolkovo)

[2] Aksyutin O E 2017 Competitiveness of Russian gas in world markets *Gas industry* **4 (762)** 8-11

[3] *Statistical Review of World Energy 2017* (Center for Research in Economics, Energy and Politics, Heriot-Watt University) available at: http://www.bp.com/content/dam/bp/en/corporate/pdf/energy-economics/statistical-review-2017/br- statistical- review-of-world-energy-2017-full- report.pdf.

[4] 2017 Oil industry: 2017 results and short-term prospects *Energy Bulletin* **56** 10-11

[5] 2020 Green certificates: world experience and plans in Russia *Energy Bulletin* **80** 10-11

[6] Analytical center under the government of the Russian Federation. Ministry of Energy of Russia 2020 *Statistical Digest. Russian fuel and energy complex in 2019: production results* available at: https://news.myseldon.com/en/news/index/222942466.pdf.

[7] Analytical center under the government of the Russian Federation. Ministry of Energy of Russia 2018 *Statistical Digest. Russian Fuel and Energy Complex 2017. Natural gas* available at: https://ac.gov.ru/archive/files/publication/a/17267.pdf.

[8] Golubeva I A 2015 Gas processing enterprises of Russia. Gas processing enterprises of Gazprom Group *Oil Refining and Petrochemicals* **1** 18-26

[9] Eder L, Filimonova I, Nemov V, Komarova A and Shumilova S 2018 Gas industry in Russia: achievements and prospects *National Industry Journal. Oil and gas vertical* **1** 24-37

[10] Kurbankulov S R, Fakhrutdinov R Z, Ibragimov R K, Zinnurova O V, Ibragimova D A 2016 Problems and prospects of using associated petroleum gas in oil fields *Vestnik Technological University* **19 (12)** 55-59

[11] Braginsky O B 2014 The rational use of hydrocarbons in the oil and gas complex of Russia *Science and Practice of the Russian University of Economics n.a. G.V. Plekhanov* **1 (13)** 26 - 29

[12] Gorbaev A V and Gorlenko N V 2019 Assessment of environmental and economic damage during the combustion of associated petroleum gas at the Yarakta oil and gas condensate field *XXI century. Technosphere Safety* **4 (3)** 366–374 https://doi.org/10.21285/2500-1582-2019-3-366-374.

[13] Rajovic V, Kiss F, Maravic N and Sera O 2016 Environmental flows and life cycle assessment of associated petroleum gas utilization via combined heat and powerplants and heat boilers at oil fields *Energy Conversion and Management* **118 (15)** 96-104 https://doi.org/ 10.1016/j.enconman.2016.03.084

[14] Berlin M A 2013 Inconvenient companion *Sphere. Oil and gas* **1 (34)** 106-110

[15] Golubeva I A, Rakhmetov K Kh and Rodina E V 2017 Gas processing enterprises of Russia. Reconstruction and development prospects of gas processing enterprises *Oil Refining and Petrochemicals* **6** 30-36 https://doi.org/10.24411/2310-8266-2019-10201

[16] Barkan Sh M and Kornev A V 2017 Prospects for the Use of Associated Gas of Oil Development as Energy Product *International Journal of Energy Economics and Policy* **7 (2)** 374–383

[17] Braginsky O B 2019 The Development of Domestic Petrochemical Chemistry: Course Adjustment *Petrochemical Chemistry. Development strategy* **1** 5-10 https://doi.org/10.24411/2310-8266-2019-10101

[18] Mescherin I V 2016 Optimization of technologies for liquefying natural gas in order to increase the economic efficiency of the process *Territory "NEFTEGAZ"* **3** 146-152

[19] Kirillov N G, Yarygin Yu N, Drozdov Yu V and Belozerova T B 2010 Liquefied natural gas: analysis of the global and prospects of domestic production *Gas chemistry* **2** 23-29

[20] Tikhonov S 2018 LNG out of politics *Oil and gas vertical. Analytics and forecasts*

[21] *PRESS RELEASE JSC Gazprom,* available at: http://www.nrg.org.ru/publication/detail.php?ID = 1328
[22] GTL.URL Market Overview, available at: http: // www. metaprocess.ru/ fi les / GTL.pdf.
[23] Arutyunov V S and Lapidus A L 2005 The role of gas chemistre in world power engineering *Herald of the Russian Academy of Sciences* **75 (4)** 337-346
[24] Eliseev O L 2008 Technologies "gas to liquid" *Russian Chemical Journal* **LII (6)** 53-62 https://doi.org/10.1134/S107036320911395
[25] Khasin A A 2008 Overview of technologies for the production of coolants developed by Shell and Sasol *Gas chemistry* **6** 38-47
[26] Karabanova A N and Bessel V V 2017 The Effectiveness of GTL Technology For Energy Supply To Oil And Gas Facilities In The Remote Russian Areas *Territorija «NEFTEGAZ» - Oil and Gas Territory* **1–2** 82–87
[27] Kuznetsov A M, Savelyev V I and Bakhtizina N V 2012 GTL industry: status and prospects *Scientific and Technical Bulletin of Rosneft Oil Company* **2** 44-49
[28] Zagashvili Yu V, Efremov V N, Kuzmin A M and Lischiner A A 2017 The synthesis gas production complex for small-capacity methanol production *Neftegazokhimiya* **1** 24-36
[29] Soldatov A P, Tsodikov M V, Parenago O P and Teplyakov V V 2010 New nanosized catalytic membrane reactors for hydrogenation with stored hydrogen: Prerequisites and the experimental basis for their creation *Russian Journal of Physical Chemistry* **84 (12)** 2102-2109
[30] Khorsanov V L, Ivanov A P and Blokhin A I 2004 Methane Conversion Using Water Vapor of Fast Nuclear Reactors I*nternational Scienti fi c Journal for Alternative Energy and Ecology ISJAEE* **6 (14)** 5-7
[31] Razmanova S V and Machula I A 2015 Analysis of the competitive advantages of production of synthetic liquid fuels *Oil and Gas Geology. Theory and practice* **10(4)** 15-18 https://doi.org/10.17353/2070-5379/38_2015
[32] Arutyunov V S, Savchenko V I, Sedov I V, Makaryan I A, Shmelev V M and Aldoshin S M 2014 New concepts for the development of small-scale gas chemistry *Neftegazokhimiya* **4** 19-23
[33] Arutynov V, Troshin K, Nikitin A, Belyaev A, Arutyunov A, Kiryushin A and Strekova L 2020 Selective oxycracking of associated petroleum gas into energy fuel in the light of nev data on sele-ignition delays of methane-alkane compositions *Chemical Enginering Journal* **381** 122706 https://doi.org/10.1016/j.cej.2019.122706
[34] Braginsky O *Will the "era of ethane" come in Russia? New chemical technologies. Analytical portal of the chemical industry,* available at: http://www.newchemistry.ru/letter.php?n_id=3637.

IOP Conf. Series: Materials Science and Engineering **952** (2020) 012023 doi:10.1088/1757-899X/952/1/012023

Temporary and operational transport roads in construction of transfer linear structures and their scientific and technical support

L I Kortiev[1], R V Klyuev[1,3], F G Totchiev[1], S D Vaneev[1] and N I Babunska[2]

[1] North-Caucasian institute of mining and metallurgy (state technological university), 44, Nikolaeva Str., Vladikavkaz, 362021, Russia
[2] Todor Kableshkov University of Transport, Geo Milev Str., 158, Sofia, 1574, Bulgaria
[3] Moscow Polytechnic University, 38, B. Semenovskaya Str., Moscow, 107023, Russia

E-mail: kluev-roman@rambler.ru

Abstract. Due to the economic and technical progress of countries and regions in Europe, the necessary and possible transport communications, power lines (power lines), various pipelines and other linear structures were laid through the Alpine Mountains. The Caucasian Mountains in this part are strongly lagging behind. The railway crosses the Main Caucasian Ridge (MCR) only in the western part up to the Black Sea coast, and the automobile road is in the Central part through the Cross Pass along the Georgian Military Highway and the Roksky Pass along the Military Ossetian Road. The Gas pipelines and power lines were laid in the same areas. At the beginning of the 11th century, the Dzuarikau - Tskhinval pipeline was laid to provide gas to the population of South Ossetia. In the peak position this gas pipeline is the highest one on a global scale being of a particular organizational interest. Of the same interest is the construction of a road through the Roksky Pass, hereinafter referred to as Transkam. Organizationally, these two linear structures had to be laid on the automobile road as a temporary one and a technologically advanced soil and gravel road on the gas pipeline at an absolute height of more than 3,000 meters. These roads have important social, economic, agricultural and strategic importance which is considered in this article.

1. Introduction

When designing and building transhipment railways, highways, gas pipelines, power lines, it is necessary to lay temporary access roads to ensure transport links between the production sites and the coordinating center independently on the opposite slopes [1-4]. In common practice, for example, during the construction of the Transkam through the Roksky Pass, temporary «pioneer roads» were designed and built independently for the northern slope of 20 km and the southern one of 33 km.

During the active construction process, there was a need for a dirt road through the Main Caucasian Ridge (MCR) to transfer construction equipment from site to site. In the northern section the earthmoving equipment was idle but in the southern section it was not enough.

Previously a unit of tracked vehicles was diverted via Baku or Sochi: the northern section of Zaramag – Alagir – Beslan – (by car on a trailer) – Beslan – Baku – Tskhinval by rail and from

Tskhinval to the southern section – also by car which took about 2 weeks using the locksmith-automobile link. The experts in terrain conditions proposed to lay a 13 km dirt road through the MCR separating the organizational and administrative production road sections. The proposal was based on the economical and mathematical methods, taking into account the determination of energy costs for transporting road equipment by rail and the cost of constructing a dirt road using the formula:

$$E = \sum_{i=1}^{n} N_i \eta_i t_i$$

where E - energy consumption for the work performed;
N_i - the number of equipment units used to perform the work (in units);
η_i - engine power of the applied equipment (in HP);
t_i - equipment usage time (per hour).

According to calculations, it was determined during that Transkam construction, the cost of a dirt "service road" building was paid off within 1.5 years and it could be used for the entire period of the road of and the main tunnel construction, which took about 7 years. During these seven years, the problem of imbalance in earthmoving equipment between sites was removed. The bulldozers «Komatsu» and «Caterpillar» easily extended the Main road and after 5 hours began the work on the other slope.

For the same period, this road was prescribed to be used «for the needs of agriculture, for example, for driving livestock to pastures, using hay lands, as well as for tourism purposes».

Following the completion of the first stage of Transkam, the problem of laying power lines and looping the North Caucasian and Transcaucasian electric networks arose. The designers took advantage of the presence of the road and the power line was laid a direction through the MCR along the Magsky pass.

During the war in 2008 in South Ossetia, the transfer of the military units to Tskhinval in order to force the peace for the Georgian army troops was carried out through the Roksky tunnel. The service road under consideration could be an option for the transfer of such troops in the event of the Roksky tunnel portals destruction. Thus, the strategic significance of such roads should not have been ignored.

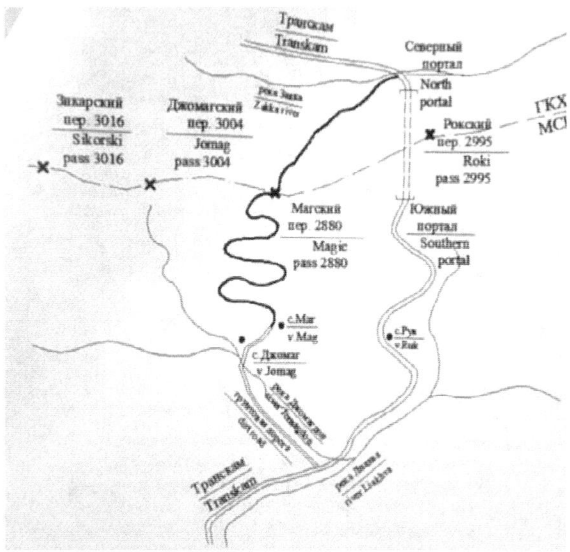

Figure 1. A schematic plan of Magsky tractor roads

They could have a significant role in the development of tourism. In Soviet times, there was an all-Union pedestrian tourist route along the MCR along the path through the Roksky pass, which was canceled due to dangerous movement conditions along a narrow path and the occurrence of a tourist's fall down the slope with a fatal outcome. This route may well be used via a service tractor road where high – traffic vehicles can be brought to any point of the pass section. The scheme of the considered and promising use Magsky tractor dirt road is shown in Figure 1.

2. Materials and methods

The paper uses data obtained during the study of the gas pipeline between North Ossetia and South Ossetia.

3. Analysis of the gas pipeline between North Ossetia and South Ossetia functioning

Comprehensive studies of the gas pipeline between North Ossetia and South Ossetia were conducted [5-10].

The Dzuarikau – Tskhinval gas pipeline was laid through the Alagir district of RNO-Alania and the Dzhava and Tskhinvali districts of the RNO – Alania region in the Central part of the MCR near the borders of North Ossetia and South Ossetia on the Kudar pass at an unprecedented height of 3148 m above sea level. The length of the pipeline route is 168.7 km laid in particularly difficult mountain conditions. The Dzuarikau – Tskhinval gas pipeline is not the first line through the MCR. In Soviet times, the Stavropol – Tbilisi – Yerevan gas pipeline was built along the historical Georgian Military road through the Krestovsky pass at an absolute height of 2495 m which is almost 0.5 km lower than the Kudar pass.

The altitude position of the Dzuarikau – Tskhinval gas pipeline is shown in Figure 2.

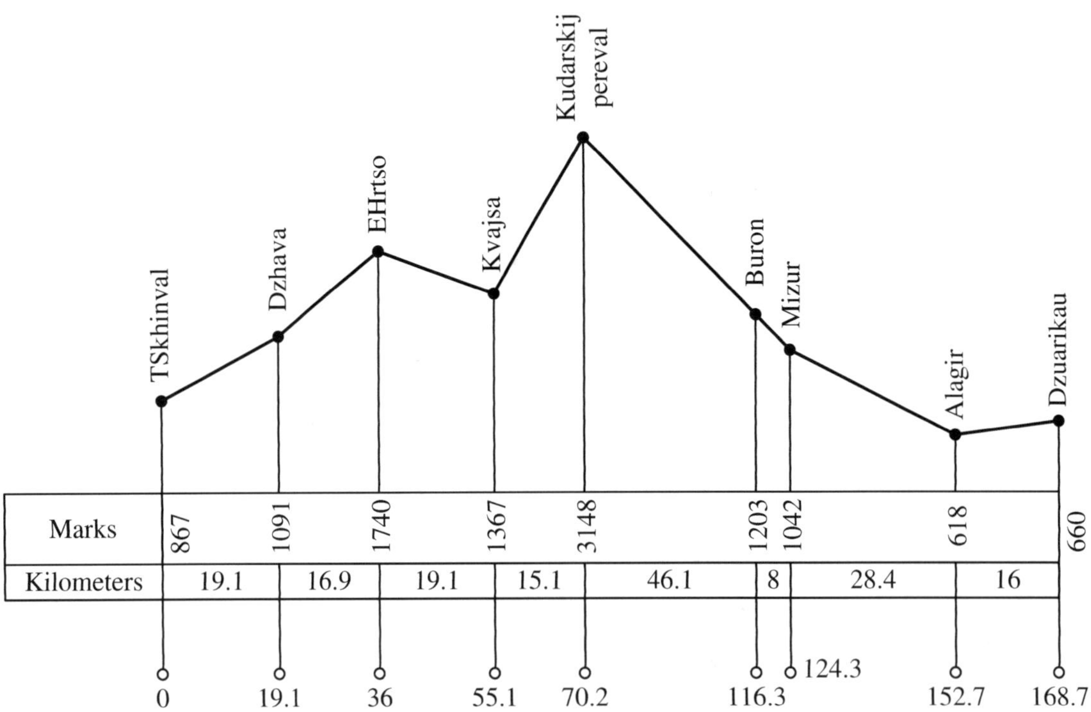

Figure 2. Longitudinal profile of the gas pipeline

The gas pipeline plan and gas distribution stations (GDS) is shown in Figure 3.

The gas pipeline under study is the highest in the world. Its end sections on both sides of about 18 km are laid in flat conditions, then the route of about 40 km in the South is laid in mountainous conditions, and in the North in the cramped conditions of the Ardon river and the middle part of the gas pipeline from Kobet to Chiba about 50 km passes in particularly difficult mountain conditions. The strip of the gas pipeline of Tskhinvali to Kvajsa and from Dzuarikau to the planned construction of a tourist complex «Mamison» is laid using relatively perfect and safe access roads.

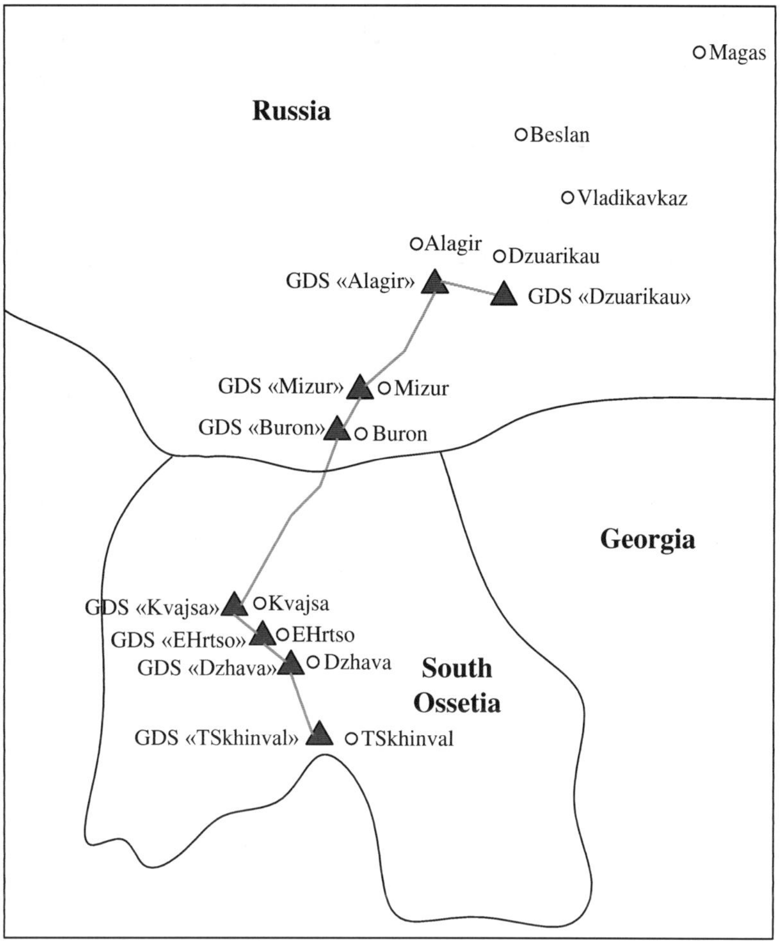

Figure 3. Dzuarikau – Tskhinval gas pipeline plan

On the remote mountain sections, the track works on a dirt and gravel road attract special attention. On the northern slope the future mountain tourist complex «Mamison» is located directly at the beginning of a complex section of the operating gas pipeline and after the introduction of this tourist complex. The necessity of a convenient exit to the South, towards Kvajsa and further to Sachkhere and to Kutaisi via Oni – Ambrolauri, will appear. The Western Georgia, Abkhazia and even the Turkish Black Sea coast will become points for these routes attraction.

Taking this in mind the future significance of a particularly remote mountain section, a dirt and gravel road and the gas pipeline itself should be laid on this section with as few slopes with a small radius on curves, which would be possible with scientific and technical support for the road safety project [11-15]. This would give more opportunities for the rapid implementation of transport links between the tourist and recreational complex and transport routes in the south-west direction.

Figure 4 shows fragments of the complex sections of the operating gas pipeline road that are technically, technologically and topographically available for reconstruction in order to transfer the road to a safe channel for the road traffic, taking into account its operation as a seasonal – summer one. And for year-round operation, laying a tunnel in the technical terms is not particularly difficult.

Figure 4. Fragments of road sections along the Dzuarikau – Tskhinval gas pipeline

4. Conclusion

Thus, the above-mentioned positive aspects in the use of the dirt service road through the Magsky pass in social, agricultural, tourist and strategic terms will certainly be implemented on the operated road of the Dzuarikau – Tskhinval gas pipeline. In addition, if it is brought to a safe state, it can become a route of attraction for all the countries of the Black Sea coast from Sukhum and Batumi via Kutaisi-Kvajsa to the future Mamison tourist complex.

5. Acknowledgments

The study was carried out with the financial support of the RFBR and the Ministry of Education and Science of the Republic of South Ossetia in the framework of scientific project No. 19-511-07001.

References

[1] O'Connor K and Fuellhart K 2020 *International Encyclopedia of Human Geography (Second Edition)* 421-428
[2] Liu H, Love P, Sing M, Niu B and Zhao J 2019 *Transportation Research Part D: Transport and Environment* **77** 615-626
[3] Wessel J 2019 *Transport Policy* **78** 42-57
[4] Kopiec A, Siguencia L, Szostak Z and Marzano G 2019 *Procedia Computer Science* **149** 508-514
[5] Plieva M T, Gavrina O A and Kabisov A A 2019 *Int. Multi-Conf. on Industrial Engineering and Modern Technologies (FarEastCon)* (Vladivostok) **Inspec Accession Number** 19229305
[6] Burdzieva O, Zaalishvili V, Kanukov A and Maisuradze M 2016 *International Journal of GEOMATE* **10** 1693-1697
[7] Buryanina N S, Korolyuk Yu F, Maleeva E I and Lesnykh E V 2018 *Sustainable Development of Mountain Territories* **3** 404-410
[8] Tsidaev B S, Logachev A V and Golik V I 2019 *Sustainable Development of Mountain Territories* **1** 98-104
[9] Kortiev L I, Klyuev R V, Kulumbegov R P, Kortiev A L, Bosikov I I, Gavrina O A and M Z Madaeva 2019 *IOP Conf. Series: Materials Science and Engineering* (Tyumen) **663**
[10] Klyuev R V, Bosikov I I, Gavrina O A and Atrushkevich V A 2020 *Bezopasnost' Truda v*

Promyshlennosti **7** 46-53
[11] Wang D 2019 *Transportation Research Part D: Transport and Environment* **77** 212-223
[12] Bystrov V and Kozak N 2018 *Transportation Research Procedia* **36** 103-107
[13] Malygin I, Komashinskiy V and Korolev O 2018 *Transportation Research Procedia* **36** 487-492
[14] Petrova T, Chistyakov E and Sorvacheva Y 2017 *Transportation Research Procedia* **20** 505-510
[15] Davidović J, Pešić D, Lipovac K and Antić B 2020 *Transportation Research Procedia* **45** 333-342

IOP Conf. Series: Materials Science and Engineering **952** (2020) 012024 doi:10.1088/1757-899X/952/1/012024

Modern technologies of project management and implementation in the construction industry

A A Ermakov, A N Korkishko, S A Smirnov and A S Shapovalov

Tyumen Industrial University, 38, Volodarskogo str., Tyumen, 625000, Russia

E-mail: anton_ermakov_94@mail.ru

Abstract. The article considers the technology of management and implementation of project activities, from the perspective of project management in the field of innovation and multifunctionality of work in the field of construction and production. The current existing approaches to project management in various manufacturing fields and also such special tools as the construction extension of project management standards have been examined. As follows from the analysis, the main features of the considered standards are highlighted and conclusions were reached on the necessity to use modern technologies in project management.

1. Introduction

Nowadays the problem of project management and implementation technologies in the construction industry, especially as part of infrastructure development, is a complex task. Information technologies and specialized software are increasingly used in the modern construction business.

The modern market of construction activities is characterized by active participation of foreign firms, investors, architects, designers, implementation of new design and construction technologies and equipment, new construction and finishing materials and expansion of activities. The issue of development, and most importantly, the introduction of new construction technologies in practice instead of the previous ones is becoming an increasingly urgent component.

2. Project management

Project management is a set of measures to coordinate human, financial, information and material resources throughout the project cycle, aimed at achieving the project objectives. The whole set of operations for the implementation of the project is interconnected in time and space. Ensuring such a linkage of operations is a very difficult task.

Project management in the field of design of construction facilities is a systematic application of management methods and tools in order to produce sets of project documentation with high quality of design considerations and design, in volumes and in terms determined by the relevant contracts.

There are two necessary components of project management.

Revision Management. Design and detailed documentation of the project constantly changes and requires coordinated and orderly work of all participants of the construction. In view of this the effectiveness of the management of this process is characterized, first of all, by a regulated procedure for making changes and well-timed informing related departments. Sometimes the best solution is to create a Project Office at the construction site for these purposes.

Innovative approach. It is the visualization of design solutions using modern CAD programs. This means it became possible to work in a team using all options of 3D design. Automation allows each member of the team responsible for his structural element to exchange 3-D models with each other. The quality of project documentation increases many times using this possibility.

3. Types and examples of project management standards

The methods of project management and implementation technologies are numerous and diverse.

The current project management standards can be divided into national and expanded geography of application [3]. National standards are the standards of the following countries: France (AFITEP), Great Britain (APM Body of Knowledge, BSI BS 6079, OSCEng), Germany (DIN 69901, V-Modell), USA (NASA Project Management), Switzerland (VZPM).

Examples of standards with extended geography of application include the following: A Guide to the Project Management Body of Knowledge (PMBOK Guide); ISEB Project Management Syllabus; ISO 10006: 2003, Quality management systems - Guidelines for quality management in projects; Microsoft Solutions Framework (MSF); Oracle Application Implementation Method (AIM); PRINCE2 (Puojects In a Controlled Environment) [3].

At the current stage of development special products, for example, software for project management become popular as a standard for project management and implementation. Services can be automated with the following products: Agresso; Augeo Software; BIM, CA Clarity; Epicor Software IRIS Software Group SharpOWL; Lawson; Maconomy Microsoft Project Professional. In Russia, Microsoft's Project Professional software products have become widespread.

However, it should be noted that primarily methods of project management are general purpose. Only a few of them, such as PMBOK and EUPS (Eurasian Project Management Standard) have specialized extensions for construction projects.

4. Modern project management standards and their features

The ANSI PMBDK GUIDE (APG) standard, which contains a fairly detailed description of generally accepted project management technologies, is meant to be used internationally. Project management processes are divided into five main groups: initiation, planning, execution, regulation and completion. These stages are similar to the PDCA methodology, but have a number of differences, in contrast to the Deming cycle, in other words, APG is primarily focused on the management of individual stages of the project, and not on the management of a set of projects that include integration processes. Therefore it is be noted that APG is dominated by the principles of effective control over the execution of the composed plan, GAP analysis of the actual progress from planned and timely reasonable adjustment targets. [4]

The approach to the development of tools for the competent setting of goals was QFD (QFD - Quality Function Deployment), which was proposed by the Japanese [4]. Firstly this method began to be used in the field of automotive engineering, and then began to spread in other different industries. The results of research and development of the American Institute of project management (PMI – Project Management Institute) have been disseminated to ensure a large increase in the level of regulation and management of resources. [5] The set of management tools offered by the Institute has gained worldwide distribution and recognition. [7]

However, not all classical approaches to project management are quite applicable to the implementation of projects related to the development of fundamentally new products or services. It is related to the fact that such projects are usually implemented in conditions of high uncertainty about their goals and content.

Changing the goals and content of the project has far-reaching consequences in the context of choosing effective management techniques. The classical project management tools for the case of innovative projects are losing their relevance.

Thus, in the conditions of rapid changes, continuous innovation activity on a global scale, there is an objective need to develop new, "non-classical" techniques that would better reflect the conditions of high uncertainty regarding the most important strategic issues. One of the answers to this objective necessity was the development of a new concept of project management Agile Project Management (APM).

APM is a "non-classical" approach to project management. This method of project management consists of many rapid iterative planning and development cycles, which are applicable to projects, allowing the project team to constantly evaluate the developing project and receive instant feedback from users and project participants. The team studies and improves project performance as well as the method of work in each successful cycle. After well-established planning, needs identification and solution outline, the stage is completed, with the project going through iterations with already more detailed planning processes, needs analysis and implementation, taking the form of waves.

In addition, the APM methodology allows us to approach the process from the point of view of situational moments, using diagnostic tools functions in the process. This approach allows for instant changes in the final product when new requirements are received. Firstly APM was positioned as an approach to project management in the field of information technology, but after a while it became clear that in the conditions of rapid changes in many areas of business, the scope of APM can be significantly expanded in areas such as: production and construction. [6] In the construction industry the APM methodology is used as a methodology for planning the stages of work from design to completion.

5. Features of project management and implementation for the construction industry

Project Management Institute Construction Extension to A Guide to the Project Management Body of Knowledge (PMBOK® guide)

Despite the growing interest in project management, the increase in the number of models and standards, the tendency to develop more accurate programs and standards for project management in the production sector, nowadays, only a few of them can offer extensions of special significance, directly for construction projects.

One such product is an Extension for the construction field of Project Management Institute Construction Extension. PMBOK is a framework. Many people call it the standard, but it's not quite true. Most people don't know what the PMBOK Guide is not a standard, in spite of the fact that PMI (Project Management Institute) stamp of the ANSI standard on cover, the standard is only Project Management Standard for Managing a Project, measuring just under 40 pages. The PMBOK manual itself is much more extensive – 616 pages (in the Russian edition 587).

After the publication of the third edition of the guide, there was a need to create separate extensions and bring them into full compliance with the standard itself. In this expansion, the basis is the norms and features generally accepted for construction projects, which are not common for the types of projects of different production areas. This extension also addresses specific practices and features found in construction projects. This Supplement applies the generally accepted PMBOK standards to the construction industry, and aims to improve the efficiency of project management in construction.

Construction projects involve a large number of people and often project teams speak different languages while in the same country. These facts complicate communications, and therefore slow down the work of the project.

Eurasian project management standard. Extension for the design of construction projects.

ESUP is an integrating standard developed by the professional community on the basis of world experience, values and goals of Eurasian civilization. Corporate project management standards are an important part of management systems in many successful companies. ESUP reflects the best practices in this area by developing a prototype of a corporate project management standard designed to be customized to a wide range of projects. As such this basic document contains the main provisions, which, if necessary, are detailed in the industry extensions and standards of organizations. [8]

The extension of the ECUP standard for the design of construction objects was developed with the aim of its use in design organizations and self-regulatory organizations for the design of construction objects in the development of their own corporate standards for similar purposes and documents of quality management systems, taking into account the individual characteristics of such organizations.

Corporate expansion of the Eurasian standard of project management in the field of design of construction objects hereafter referred to as Esup_K_projection of construction objects is developed taking into account the world achievements of project management in this applied area. [8]

The main feature of the application of Esup_K_projection of construction projects for this area is the principal multiplicity of simultaneously executed projects, the totality of which is a single portfolio with many projects.

6. Conclusion

As a result of the conducted research it was found out that in the existing standards and literature there is no uniform approach to construction of the General system of project management. National standards exist in about 50 countries. In Russia there is no universally recognized national (state) standard. The PMI standard is popularized in Russia privately or as a public initiative.

Because of the growing interest in project management, the number of models and standards in this area is constantly increasing. This reflects a natural tendency to adjust more precisely to regional, national, industry and corporate governance needs.

As a result of the study of different methodologies and standards for the management and implementation of General-purpose projects, that is, used in various areas of production activity, two specialized extensions of PMBOK and ESUP for the construction area were considered.

A distinctive feature of specialized extensions is the accepted practices and features found in construction projects.

As for PMBOK, this extension is a very important development for this specialization, as it ensures the consistency of approaches around the world. In other words, the application of this development leads the management of construction projects to a single standard, a common language.

As a rule, these projects are independent from each other, and only in rare cases they form project programs (for example, projects for the development of microdistricts or industrial sites). In this regard, the task of Expanding the ESUP standard for the design of construction projects is to distinguish the areas of application of project and portfolio management.

From all the above it follows that the peculiarity of modern project management technologies is that these technologies work in all areas of applications, allowing to effectively implement projects, to ensure mutual understanding and interaction of project participants, high reliability of achieving goals. Basically project management helps to decide to justify the feasibility of investments, to develop an optimal scheme for financing works, supplies of materials and equipment.

References

[1] Airoyan Z A and Korkishko A N 2016 Project management of oil and gas complex on the basis of information modeling technologies *BIM-technologies* **4 (43)**

[2] 2018 *Architecture Scientific and technical and urban planning and production journal. Co-temporal problems and trends in architecture*

[3] Brown D T 2010 *Guide to the PMI-NY project management knowledge*

[4] 2018 *Deloitte forensic Mining is the Foundation of a successful project. Project management in construction,* available at: https://www2.deloitte.com/content/dam/Deloitte/kz/Documents/finance/ConstructionProjectManagement/KZ_Deloitte%20Construction_Mining_EN.pdf

[5] 2010 *Eurasian project management standard (corporate version). Extension for design of construction projects Version 1.5* (Moscow)

[6] Mustafa Malla Osman Shadi 2015 Features of management of innovative projects in the field of construction, *Statistics and Economics* **3**

[7] Orekhov S A and Seleznev V A 2008 *Fundamentals of corporate governance* (Moscow: Market DS)

IOP Conf. Series: Materials Science and Engineering **952** (2020) 012024 doi:10.1088/1757-899X/952/1/012024

[8] Popov V L 2008 *Management of innovative projects* (Moscow: INF-RA-M)

IOP Conf. Series: Materials Science and Engineering **952** (2020) 012025 doi:10.1088/1757-899X/952/1/012025

Analysis of the development of global energy production and consumption by fuel type in various regions of the world

D S Shalaeva[1], O I Kukartseva[1], V S Tynchenko[1,2], V V Kukartsev[1,2], S V Aponasenko[1] and E V Stepanova[2,3]

[1] Reshetnev Siberian State University of Science and Technology, 31, Krasnoyarsky Rabochy Av., Krasnoyarsk, 660037, Russian Federation
[2]Siberian Federal University, 79, Svobodny pr., Krasnoyarsk, 660041, Russian Federation
[3]Krasnoyarsk State Agrarian University, 90, Mira Av., 660049, Krasnoyarsk, Russian Federation

E-mail: vlad_saa_2000@mail.ru

Abstract. This article is devoted to the analysis of the geographical structure of the world energy industry. The article analyzes trends in global energy production and consumption from 1990 to 2018. Statistical indicators of energy production and consumption are also considered. The study notes that the leading countries in GNP production are the largest energy consumers in the world. The authors consider the issues of distribution of production and consumption of energy resources across the world, changes in the relationship between the leaders in primary energy production and energy consumption. Electricity consumption and production by regions of the world are considered, and the share of global energy intensity of energy by fuel type is determined. In addition, the article analyzes the world consumption of natural gas and coal, as well as its extraction. Changes in world demand for petroleum products and their consumption are considered.

1. Introduction

Energy is the basis for the development of the economy of the modern world. Energy consumption in a country is determined by industrial and transport needs, as well as the needs of the population [1]. Russia was the second largest energy consumer after the United States in 1990 [2]. The growing economies of Asian countries have changed the global division in energy consumption. Economic activity in most industrialized countries increased in the 1990s, resulting in increased energy needs [1].

Russian energy efficiency is not exceptionally high. In Russia, energy consumption declined significantly in the 1990s, and by the turn of the Millennium it was in line with the European average, that is, higher than in southern Europe, but noticeably lowers than in Norway, Finland, and Sweden, and significantly lowers than in the United States and Canada [3]. A comparison of countries shows that the higher the energy consumption, the higher the standard of living, income, solvency, production levels, etc. Although in Russia the level of energy consumption is generally at the European level, the level of energy consumption by the population has significantly decreased than in developed industrial countries [4].

In recent years, many analysts have realized the danger of another wave of growth in global energy consumption. However, the largest developing countries – China and India – are currently increasing their per capita energy consumption at an increasingly rapid pace. Rapid population growth and high energy intensity of the economy, the energy needs of these countries are growing rapidly, given the continued economic growth of developing Asian countries. Energy consumption in Africa and Latin America is growing at a faster rate, and even per capita energy consumption has increased in the European Union.

2. Energy consumption and production by fuel type

Today, all the energy used by humans does not exceed 0.3% of the photosynthetic energy that supports life on Earth. However, in the future, what is important is not the level, but the growth rate of human energy consumption. In the last century, the dynamics of global energy consumption indicates its clearly exponential growth, which can disrupt the planet's thermal balance and lead to catastrophic climate changes [5].

Over the entire historical period of human development, the level of human use of anthropogenic energy increased at the end of the 19th century, reaching 1 billion tons of fuel. And over the past century, it has grown almost 17 times, reaching 16.5 billion tons of fuel in 2006. In other words, the average level of energy consumption has increased by several orders of magnitude since biblical times.

In 2018, global energy consumption increased significantly due to sustained economic growth and growing demand in China, which has been the world's largest energy consumer since 2009 (figure 1).

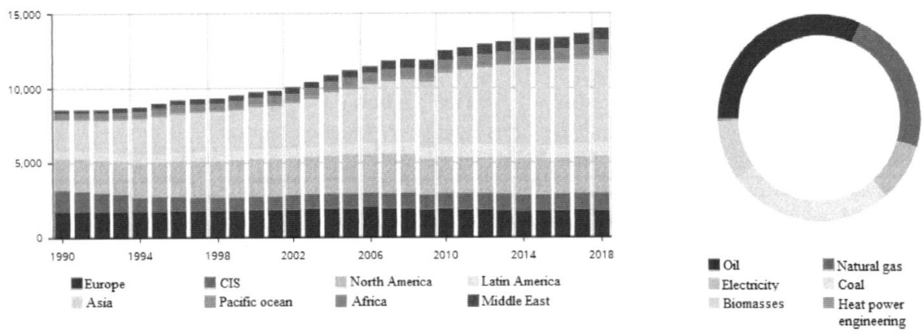

Figure 1. Diagram of global energy consumption by fuel type.

China's energy consumption showed the highest growth since 2012, mainly due to electricity production, strong industrial demand and increased fuel consumption in the transport sector, which was helped by a growing fleet. In the United States total energy consumption reached a record 2.3 Gtoe in 2018, which is 3.5% more than in 2017, partly because of weather stipulations. On the other hand, in the European Union, energy consumption fell by 1%, and in particular in Germany by 3.5%, partly due to falling electricity consumption, a mild winter, reduced consumption and improved energy efficiency.

Global energy production continued to grow in 2018 (2.8%), exceeding the historical trend. The United States and China have made a main contribution to increasing global energy production, with growth of 54% in 2018 [6]. In the European Union, energy production continued to decline due to a small reduction in nuclear power production, depletion of oil and gas resources, and climate policies. Despite an increase in hydroelectric power production after a dry year and a moderate increase in energy consumption, this decline is occurring.

Global energy intensity (total energy consumption per unit of GDP) it decreased by 1.3% in 2018, which is slightly below the historical trend (-1.6% per year on average from 2000 to 2017). The dynamics of energy intensity is shown in figure 2. Levels and trends of energy intensity vary significantly across the world, reflecting differences in the structure of the economy and advances in energy efficiency. Between 2000 and 2018, energy intensity in China increased by almost 40%, and last year – by 2.7% due to energy efficiency policies focused on energy-intensive industries. For some time, China has developed and applied energy intensity reduction targets in response to energy-intensive industries, which has led to high demand for energy efficiency services.

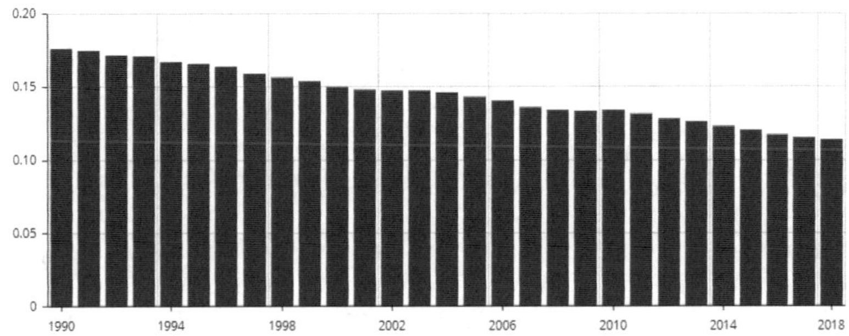

Figure 2. Diagram of energy intensity of all types of fuel.

In the US, energy intensity increased (+ 0.6%) in 2018 compared to a downward trend (-1.9% per year) in 1990-2017. Energy efficiency improvements continued in the European Union, the region with the lowest energy intensity in the world. However, this was facilitated by weather conditions (mild winter).

Since 2000, the energy intensity in the CIS region has been steadily decreasing (-2.7% per year), but remains the highest in the world (75% higher than the global average). The high energy intensity in the CIS, the Middle East, China and other developing countries in Asia is due to the dominance of energy-intensive industries, exporting countries and low energy prices, which do not contribute to energy efficiency.

3. Electricity consumption and production by region of the world

The energy sector of the former USSR developed as a single economic complex for a long period of time. The formation of independent States on the territory of the USSR led to radical changes in the development of the electric power sector.

At this time, changes in the political and economic situation in the country began to have a significant negative impact on the development and operation of electric power: power production and power consumption is reduced. Indicators of the quality of electrical energy deteriorated. The number of consumer restrictions and blackouts has increased, and electricity supplies to Eastern European countries have significantly decreased.

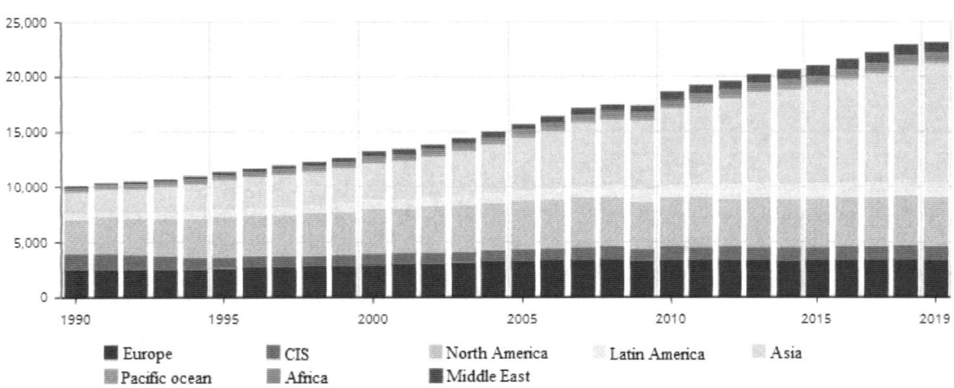

Figure 3. Diagram of global electricity consumption.

Most of the increase in global electricity consumption is accounted for Asia (almost 80%, almost 60% in China). Demand for electricity in China has accelerated on the back of sustained economic growth and industrial demand. Demand also increased in India, South Korea, Japan, and Indonesia (figure 3) [7].

In the US, electricity consumption decreased by 1% in 2017, and recovered in 2018 (+ 2.2%). Most of this growth was in the residential sector (+ 6.2%), mainly due to increased electricity consumption for household appliances and air conditioning. Economic growth and industrial demand also contributed to the increase in the consumption of electricity in Canada, Brazil and Russia. It also increased in Africa, especially in Egypt and the Near East, which was promoted by Iran.

As in 2017, electricity consumption in Europe remained stable in 2018: it declined in France and Germany, stabilized in other major countries (Great Britain, Italy, Spain), and grew in the Netherlands, Poland, and Turkey. The main growth in global electricity production occurred in Asia (+ 6.1%) in 2018: China accounted for almost 60% of global growth due to high demand and rapid capacity development [7].

Electricity production also increased in the USA (+ 3.6%), as weather conditions and economic growth stimulated electricity consumption, while in Canada it decreased slightly. Electricity production continued to grow in Russia (economic recovery), the Near East, and Africa. Latin America remained stable as growth in Brazil and Mexico was offset by a strong decline in Venezuela due to political tensions. In Europe, electricity production remained stable, despite growth in France and Turkey due to increased production of hydroelectric power and renewable energy. On the other hand, electricity production fell in Belgium, Germany, Italy and the UK (mild winter).

4. Global natural gas consumption

Global gas consumption accelerated in 2018 thanks to the efforts of the United States and China, which accounted for about two-thirds of the additional consumption. In the US, gas demand grew by 10% in 2018, the highest growth rate in the past 30 years, driven by electricity (+15 GW of new gas-fired power plants) and construction (figure 4).

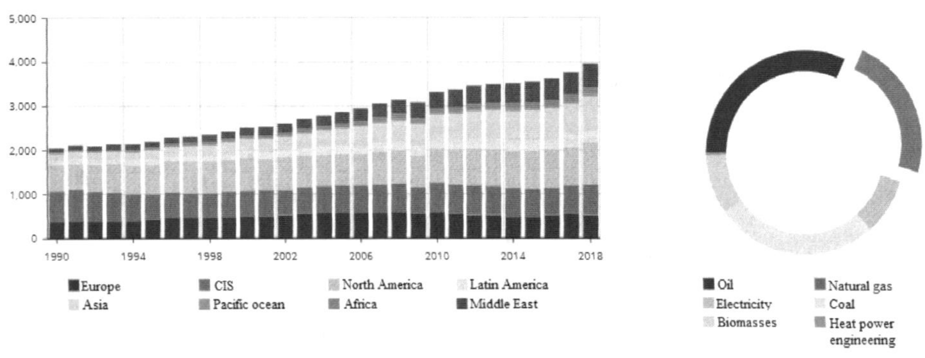

Figure 4. Diagram of global natural gas consumption.

Gas consumption also increased in China (+ 18%), according to the policy of replacing coal with gas in the electricity and heat supply sectors. Growth was also seen in India and South Korea due to sustained economic growth. However, in Japan, consumption declined as the restart of nuclear reactors reduced the need for gas-powered electricity.

Gas consumption also continued to grow steadily in Russia (although at a slower pace than in 2017) and accelerated in Canada, Iran, and Algeria. Despite economic growth, gas consumption has declined in Europe, especially in Turkey, France, Germany and Italy, due to rising temperatures, increased availability of nuclear and hydro power, and increased production of renewable energy.

In the United States, gas production increased (+11.5%), with the world's largest gas company accounting for 45% of the increase, triggered by recent events in the Permian basin and Haynesville, as well as an increase in domestic shale gas consumption. In the United States, shale gas currently accounts for about 70% of the country's gas production.

Growth in gas production occurred in Russia (+6.7% in 2018), which was facilitated by a significant increase in domestic demand, as well as growth recorded in Iran after the launch of new stages of projects in the fields from South Pars. In Australia, gas production continued to grow (+ 15%) due to the launch of new LNG lines in 2017 and 2018. In Egypt, gas production grew very rapidly (+20%) due to the start of new stages of the project in the West Nile Delta. In Europe, gas production continued to fall (-15%) as the Netherlands reduced domestic gas production.

5. Global coal consumption and production

Growth in coal consumption was driven by India and China, the two largest coal-consuming countries, with Turkey and Russia also contributing to demand growth in 2018 (figure 5).

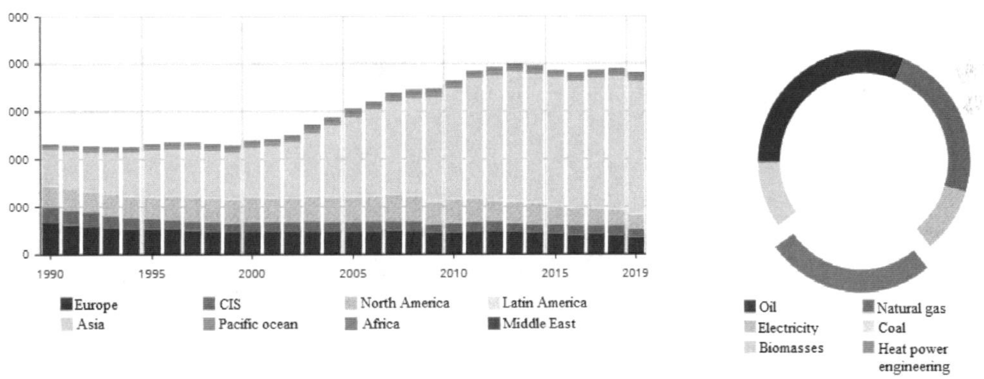

Figure 5. Diagram of global coal consumption.

In China, which accounts for almost half of global coal consumption, coal consumption has increased for the second year in a row, mainly due to electricity generation and certain industries such as steel, chemical and cement industries. In 2018, coal consumption increased again due to slowing economic growth and concerns about gas supplies for the transition from coal to gas heating. This contradicts previous efforts to green the economy while maintaining prosperity.

The United States has seen the largest drop in coal consumption (-4%), reaching its lowest level in 40 years as a result of decommissioning coal-fired power plants, stricter emissions standards, and the availability of cheaper natural gas for power generation. In Europe, coal consumption has fallen for the sixth year in a row due to climate policy, increased competition from renewable energy and gas sources, and the increased cost of CO2 emissions (tripling in 2018) in the European Union; in Turkey, by contrast, coal demand has increased by 11%.

China has strengthened its position as the world's largest producer of coal and lignite (45% of global production). In 2018, the country approved more than 45 billion yuan ($6.7 billion) for new coal mining projects. The recent gas shortage in the country has weakened the government's motivation to switch from coal to gas for indoor heating and has kept the demand for coal alive. China's coal and lignite production accounts for 70% of global growth.

Increased coal imports to China supported a strong international coal market, increasing production in Australia, Indonesia and Russia, the three main coal suppliers.

In India, production increased significantly (+5.3% in 2018) due to domestic demand and the government's desire to reduce its dependence on imports. In the US, coal production has fallen to its lowest level in 39 years, despite rising exports. In the European Union, coal production continued to decline.

6. Global demand and consumption of petroleum products

Global demand for petroleum products increased in 2018, increasing by 0.9%, due to expansion in the transport and petrochemical sectors. Increased demand for petroleum products led to increased consumption in 2018, driven by the development of petrochemicals, especially in the US, and increased fuel consumption in the transport sector (figure 6).

The largest increase in demand was recorded in Asia, mainly in China and India (growth of 6% and 2.7%, respectively) and, to a lesser extent, in Thailand and Indonesia (decline in Japan). Consumption of petroleum products also increased by 2.5% in the United States, mainly due to industrial demand. In Russia, consumption of petroleum products increased slightly (+0.5%) and decreased slightly in the European Union (-0.6%), as demand in industry and transport remained unchanged. Consumption of petroleum products in Latin America, which has been declining since 2014, reached a maximum decline in 2018 (-3.3%). This was due to lower oil production in Mexico, rising prices, and the introduction of international sanctions in Venezuela [8].

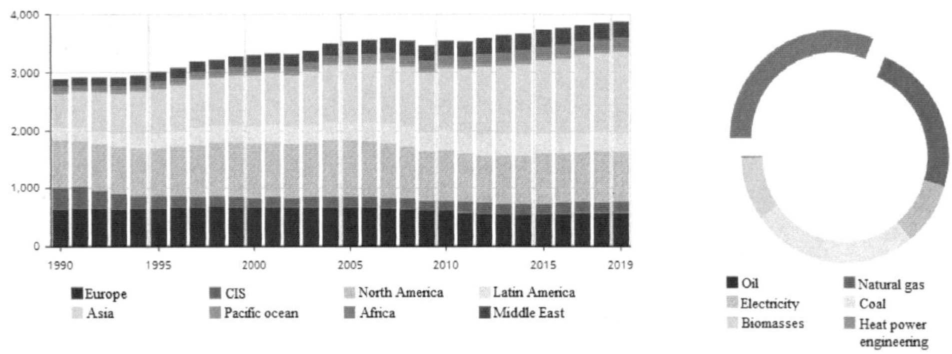

Figure 6. Diagram of global consumption of petroleum products.

Production of petroleum products increased in 2018 (+ 0.8%), although it is lower than the historical average (+ 1%). The bulk of global growth (58%) was in Asia, primarily in China and India. The Near East, especially Kuwait, saw a decline in production after last year's growth, as companies in the Near East sought growth opportunities in international markets, as well as in Latin America, where Venezuela and Mexico continued to reduce oil production [9].

Oil products in the European Union, especially in Germany, Great Britain and France, continued to fall. In the Near East, the recent closure of refineries and a growing desire to process crude oil locally has led to a decline in oil refining volumes in the EU [10].

7. Conclusion

Thus, energy production is the foundation of industrial civilization and the condition for the existence of modern material culture. In China, economic growth and growing demand have led to an increase in global energy consumption. Natural gas consumption increased in all regions of the world. But on the other hand, coal consumption in Europe has declined due to climate policy. With the development of petrochemicals, the demand for petroleum products increased, which led to an increase in consumption.

References

[1] Apergis N and Payne J E 2010 Renewable energy consumption and growth in Eurasia *Energy*

IOP Conf. Series: Materials Science and Engineering **952** (2020) 012025 doi:10.1088/1757-899X/952/1/012025

Economics **32(6)** 1392-7

[2] Mukhametshin R Z, Kryukova N I, Beloborodova A V, Grinenko A V and Popova O V 2019 Implementation of efficient energy policy in Russia: Energy consumption monitoring and problem analysis *International Journal of Energy Economics and Policy* **9(4)** 224

[3] Zhang Y J 2011 Interpreting the dynamic nexus between energy consumption and economic growth: Empirical evidence from Russia *Energy Policy* **39(5)** 2265-72

[4] Burakov D and Freidin M 2017 Financial development, economic growth and renewable energy consumption in Russia: A vector error correction approach *International Journal of Energy Economics and Policy* **7(6)** 39-47

[5] Halova G O 2017 Analysis of the changing role of renewable energy in global energy production and consumption *Economics of industries and regions. Innovation and investment* **1** 113-8

[6] *American Wind Energy Association (AWEA),* available at: https://www.awea.org/

[7] *Asia-Pacific Economic Cooperation Database (APEC),* available at: https://www.apec.org/

[8] Serzhantova M P, Tynchenko V S, Bukhtoyarov V V, Petrovsky E A, Kukartsev V V and Tynchenko V V 2019 Automation of mass fraction determination of water in petroleum in the laboratory *J. Phys.: Conf. Ser.* **1384** 012058

[9] Grigoriev L M 2013 Economic growth and energy demand *HSE Economic j.* **3** 390-406

[10] Serzhantova M P, Tynchenko V S, Bukhtoyarov V V, Bashmur K A, Kukartsev V V and Tynchenko V V 2019 Automation of laboratory studies in the determination of the mass of mechanical impurities in petroleum *J. Phys.: Conf. Ser.* **1384** 012059

IOP Conf. Series: Materials Science and Engineering **952** (2020) 012026 doi:10.1088/1757-899X/952/1/012026

Classification and features of projects in the oil and gas industry

A S Bovkun[1], O V Arkhipkin[2],I S Koroduk[3] and A A Krupchenko[4]

[1]Irkutsk National Research Technical University, 83, Lermontov str., Irkutsk, 664074, Russia
[2] Irkutsk State University, 1, Karl Marx St., Irkutsk, 664003, Russia
[3] Baikal State University, 11 Lenin St., Irkutsk, 664003, Russia
[4] Saint Petersburg State University, 7-9 Universitetskaya nab., Saint Petersburg, 199034, Russia

E-mail: Bovas87@yandex.ru

Abstract. Project management has been widely used in various fields and sectors of the economy over the past years. Projects implemented in the oil and gas industry have significant differences between them. To choose the optimal approach to managing an oil and gas project, firstly, the distinctive features of the project type should be identified in advance. The use of project classification makes it possible to optimize methods and tools for managing specific projects and increase the efficiency of their implementation. The article analyzes the classification of projects. The main classification features and types of projects most commonly used in scientific literature and practice are identified. In the course of the research, it was found that the presented approaches to the classification of project management do not sufficiently take into account the branch of economy in which they are implemented. Therefore, the classification of projects in a particular sector of economy may have its own classification features applicable only to the specific industry. The oil and gas industry is considered as a separate branch of economy. The article considers the most well-known oil and gas production projects implemented in recent years. The analysis of the projects has shown there is a number of features that are different from other projects. Therefore, for detailed classification of oil and gas projects, the author's classification features of projects significantly affecting the financial costs and timing of implementation of oil and gas projects are presented.

1. Introduction

In recent years, more and more literature on project management has been published.

The variety of projects being implemented is extremely large. Project classification allows to clearly rank promising and ongoing projects, and, as a result, set achievable goals and realistic deadlines for attaining targets, and attract optimal resources for their successful implementation.

Projects in the oil and gas industry have their own characteristics, including attraction of large investments, usage of capital-intensive technologies, execution of a large amount of work, and presence of special forms of partnerships (joint venture or production sharing agreement). Projects differ both in terms of scale and timing of implementation and in the type of activity: oil production and transportation, development of new fields and infrastructure construction. Moreover, they differ in such characteristics as the level of coverage/number of participants (national, international), ways to raise funds (borrowings, loans, project funding), by reason of occurrence (reengineering, reorganization, restructuring), and other [1]. There are many project classification systems in the literature. Thus, projects are divided according to class, type, scale, duration, complexity, and other factors [2].

2. Theory

The analysis of the classification of projects by various authors shows that a lot of researchers classify projects by similar criteria (table 1). The classification criteria shown in Table 1 are the most common.

Table 1. Project classifications

Author	Classification features	Project types
I. I. Mazur, V. D. Shapiro, V. I. Voropayev, I. P. Voicu, A. V. Polkovnikov, M. F. Dubovik, V. N. Burkov, D. A. Novikov	By project scale (size)	1. Small 2. Medium 3. Large (megaproject)
I. I. Mazur and V. D. Shapiro, V. I. Voropayev, I. P. Voicu, A. V. Polkovnikov, M. F. Dubovik, V. N. Burkov, D. A. Novikov	By the composition and structure of the project and its subject area	1. Monoproject 2. Multiproject 3. Megaproject
I. I. Mazur and V. D. Shapiro, V. I. Voropayev, I. P. Voicu, A. V. Polkovnikov, M. F. Dubovik, V. N. Burkov, D. A. Novikov	By main fields of activity	1. Social 2. Economic 3. Organizational 4. Technical 5. Mixed
I. I. Mazur, V. D. Shapiro, I. P. Voicu, A. V. Polkovnikov, M. F. Dubovik, V. N. Burkov, D. A. Novikov	By the nature of the subject area	1. Educational and training 2. Research 3. Investment 4. Innovative 5. Combined
I. I. Mazur and V. D. Shapiro, V. I. Voropayev, I. P. Voicu, A. V. Polkovnikov, M. F. Dubovik, V. N. Burkov, D. A. Novikov	By the duration of the implementation period	1. Short-term 2. Medium-term 3. Long-term
I. I. Mazur and V. D. Shapiro, V. I. Voropayev, I. P. Voicu, V. N. Burkov, D. A. Novikov	By the complexity level of project implementation	1. Simple 2. Complex (organizationally complex, technically complex, complex in terms of resources, complex in an integrated manner) 3. Highly complex
I. I. Mazur, V. D. Shapiro, I. P. Voicu	By the requirements for quality and ways to ensure it	1. Defect free 2. Modular 3. Standard
I. I. Mazur and V. D. Shapiro, V. I. Voropayev, I. P. Voicu, G. Diethelm	By status of participants	1. National 2. Foreign 3. Cross-national

Sources: compiled by the authors based on data analysis [3,4,5,6].

However, some authors implement new approaches to project classification.

For example, D. Locke [7] believes that all projects could be classified into 4 categories:

1. Projects in the field of engineering, construction, petrochemical, mining and quarrying operations (carried out across the territory affected by factors that are remote from the contractor's head office).
2. Industrial projects for the development of new products or for the production of equipment parts (projects are implemented at the factory or across the main territory of the organization).
3. Management projects (imply management involvement or coordination of activities to achieve an end result that cannot be defined as a piece of equipment or construction object).
4. Research projects (aimed at research with a high risk of failure).

Based on Locke's classification, oil and gas projects could be classified as category 1. However, the majority of projects in the oil and gas industry are implemented through scientific research so they should be classified as category 4.

A. V. Vlasov [8] supplements the classification of projects in the oil and gas industry with such classification features as the climatic conditions at the project site and the landscape area of the project implementation which significantly affect the financial costs, terms of implementation, and also have an increased level of risk. It is difficult to disagree with A. V. Vlasov because the peculiarity of the climate in the Russian Federation is determined by its large geographical extent, size and length. The project implementation budget will depend directly on climate conditions.

Vlasov divides the classification of the project landscape territory into 2 types: land and water territory. However, in our opinion, the landscape territory should be divided into water, land and mixed territories. An example is Sakhalin 2, which is being implemented on Sakhalin island. The project is developing the Piltun–Astokhsky and Lunsky oil and gas fields located in the sea of Okhotsk. The extracted oil and gas flow through the Trans-Sakhalin pipeline system through the united onshore technological complex to the ‘Prigorodnoye’ production complex which includes a liquefied natural gas plant and an oil shipment terminal.

It should be noted that projects implemented in the oil and gas producing region have significant differences between them.

For this, the implemented large oil and gas production projects such as Yamal LNG, Sakhalin-2, Snohvit Project, Power of Siberia have been analyzed [9,10,11,12,].

The analysis of these projects has a number of features that are different from each other:

1. Each project has a specific course of action;
2. Projects are implemented in harsh climatic conditions;
3. The fields are located mainly in undeveloped areas significantly remote from the centres of industrial and energy development of large regions.
4. Observed could be the dependence of oilfield objects of industrial and civil construction on the volume of recoverable reserves of oil (gas), the size of the oil (gas) reservoir, quality characteristics of the hydrocarbon product.
5. There is no proper infrastructure across the territory of the project implementation;
6. There are high energy costs, complex logistics of oil and gas pumping;
7. The project execution is accompanied by the construction of small settlements in remote areas which is tied to natural reserves;
8. Projects are implemented across water area with a significant distance from land which places additional difficulties on the project implementation;
9. The project implementation is due to significant financial costs;
10. The timing and overall duration of the project are not clearly defined and are largely probabilistic.
11. The implementation of projects involves the introduction of a huge number of innovative technologies.
12. Projects are implemented with the direct support and participation of the state.

IOP Conf. Series: Materials Science and Engineering **952** (2020) 012026 doi:10.1088/1757-899X/952/1/012026

3. Experiment

For detailed classification of oil and gas production projects, it is proposed to highlight the following classification features (Table 2).

The classification presented by the author should be considered in more detail. It should be noted that the main characteristic of an oil and gas project is the field of implementation, i.e. the activity for which the project is being implemented. This paper highlights areas of activity that are distinctive and specific to the oil and gas sector, such as: exploration, transportation, oil and gas processing, and other. This division will help to identify the characteristics and specific risks in oil and gas projects.

Table 2. Classification of projects in the oil and gas industry (developed by the author)

№	Feature	Project type
1	Project focus	Geological exploration Transportation Oil and gas processing Sales of petroleum products Liquefied natural gas projects Modernization of the oil refinery Plant capacity increase
2	Scale	By costs: small, medium, large By structure: monoprojects, multiprojects, megaprojects By time: short-term, medium-term, long-term By coverage: regional, state, international
3	Status of participants	Being implemented: By company itself By a group of companies With the support of the state As a joint venture Internationally
4	Extent of project funding	State Private (own, borrowed, or attracted funds) Combined
5	Distance from markets for product sales	Transportation costs to reach product markets: Low Moderate High
6	Distance from settlements	A settlement is located close to the project site There is no settlement

One of the important differences between projects in the oil and gas sector is the scale, i.e. the size. Small projects include those that require capital investment of 10 to 15 million dollars [1]. Such projects are quite simple to implement, including: reconstruction of fields with small reserves, introduction of new equipment and technologies. Medium-sized projects are characterized by a large amount of investment and the presence of more complex business processes, for example, projects to develop a new field or conduct global events to maintain production levels. In turn, large projects include projects that require huge investment of $500 million or more.

Moreover, projects are divided into mono-, multi- and megaprojects by structure. A monoproject is a specific project with a specific goal and limited resources allocated to it. A multiproject consists of several monoprojects. A megaproject is a targeted programme that includes multi- and mono- projects united by a common goal and resources [13]. Main characteristics of a megaproject are duration of 5-7 years, capital intensity (the project is financed by both borrowed funds and project funds), impact on the economic and social environment of the region or country.

Generally, large megaprojects are implemented in conjunction with major oil and gas partners along with the support of the state.

A fairly high percentage of megaprojects is being implemented in the oil and gas sector. The volume of investment in the industry is steadily growing as it is necessary to increase production capacity to meet the demand from emerging markets and search for new fields.

Thus, the Yamal LNG investment project [11], is being implemented on the Yamal Peninsula, beyond the Arctic circle, based in the Yuzhno-Tambeyskoye field. The project is being implemented on the basis of a joint venture with the Russian oil and gas company Novatek and several major international partners that have a certain share in the equity capital of Yamal LNG: Novatek 50.1%, total 20.0%, China National Petroleum Corporation (CNPC) 20.0% and the Chinese state Silk road fund 9.9%.

The Yamal LNG scale of implementation could be attributed to megaprojects: the duration is 35 years, it affects the economic and social environment of the region and the country, and it is financed by borrowed and own funds.

The main goal of the investment project is to create a production of liquefied natural gas with a capacity of up to 16.5 million tons per year based in the Yuzhno-Tambeyskoye field on the Yamal Peninsula. Despite the fact that the project is private, the state is interested in its implementation and provides support from its side. This is due to the fact that one of the main goals of the state is to increase Russia's presence in the world energy market. In the international LNG market, the main exporters are the following countries: Qatar, Australia, Malaysia, Algeria, Indonesia and other [14]. In 2014, 19 countries were exporting liquefied natural gas. Russia ranks 8th in the world in terms of LNG exports, with a market share of 4.5%. By 2035, the share of Russian LNG exports is planned to increase to 12% due to the expansion of production [15].

It should be noted that the goal and strategy of the project correspond to the energy strategy of Russia until 2030 as well as the strategy of social and economic development of the Yamalo-Nenets Autonomous district. Thus, Yamal LNG is a strategic project for Russia.

Notably, due to the harsh cold climate conditions of the Arctic, less specific energy is required for gas liquefaction compared to the amount necessary for similar projects located in the southern zones, which allows for higher production volumes.

Furthermore, state resources and services were provided for the project implementation on preferential terms. Thus, the state allocates a large amount of investment to the development of the region's infrastructure, namely, the construction of a seaport, an icebreaker and tanker fleet, and the Sabetta airport .

The next step in support of Yamal LNG is the adoption of a law on the liberalization of LNG exports. Until 2013, Gazprom has had a monopoly on the export of liquefied gas in Russia while other companies must have entered into agency agreements with the monopoly. However, in 2013, the Russian State Duma passed a law according to which companies could directly supply liquefied gas abroad if their license provides for the construction of an LNG plant or allocation of the extracted gas for liquefaction to such a plant. Thus, owing to the introduction of the new law, Novatek would be able to supply LNG to other countries without intermediaries after the launch of the plant.

It should be noted that LNG production in Russia is a new direction for the development of the industrial sector compared to other countries such as Qatar or Indonesia. According to a study conducted by the Norwegian analytical centre 'Sigra Group', the Yamal LNG project will not bring economic benefits to the state. However, despite the economic inefficiency, the implementation of Arctic projects leads to new competencies for Russian companies and an increase in their international competitiveness [16].

To minimize risks, companies sign a joint venture agreement. On the one hand, this interaction contributes to risk differentiation but, on the other hand, it adds complexity to project management.

When implementing such megaprojects, there is a high risk of violating the obligations of one of the stakeholders, and thus the project implementation is put at risk. With the number growing, project management becomes more complex, requires more control and establishment of an effective communication system, also there might be an increase in the risk of evasion of obligations or violation of conditions by one of the project participants.

At the stage of project planning, it is important to determine the markets for product sales, to provide for the total costs incurred in connection with the maintenance and operation of main gas pipelines as well as all structures intended for storage and transport of oil and gas.

The costs of gas transportation largely depend on the properties of the gas, the transportation distance and the diameter of the main gas pipeline.

The cost of pumping oil and petroleum products through separate main pipelines is not equal. It is influenced by the diameter of oil and oil products pipelines, the viscosity of products, the pipeline load level, natural and climatic conditions.

It should be taken into consideration that when calculating prices for gas, oil and petroleum products for consumers, the costs of their transportation, storage and marketing are taken into account and along with covering the costs the profit is also considered.

In our opinion, for a more detailed classification of oil and gas projects, such classification feature as the place of the project realization to the markets is to be highlighted and, in this feature, transportation costs of delivering product to consumer are characterized as: low, moderate, high.

Also, when classifying projects, the distance from settlements should be taken into account. The presence of a settlement close to the project site could lead to saving of costs for building its own infrastructure, and the settlement would be provided with an opportunity to develop the socio-economic situation.

4. Conclusions

Thus, the classification features of oil and gas projects presented by the authors make it possible to determine the specifics of the project implementation in the oil and gas industry and apply appropriate approaches to the project budget.

In the theoretical aspect, the additions to the classification of oil and gas projects presented by the authors ensure the logic and consistency of further scientific research in the implementation of project management in the oil and gas industry. In practical terms, the presented additions to the classification of oil and gas projects enable optimization of methods and tools for managing specific projects and make it possible to increase the efficiency of their implementation.

References

[1] Barkhatov V D, Gabrielov A O and Litvin Yu V 2011 *Report of niigazekonomika LLC on the requirements for economic and mathematical models for evaluating the effectiveness and risks of oil and gas projects* (Moscow)
[2] *Europe in figures — Eurostat Yearbook 2011* (Luxembourg: Publications Office of the European Union)
[3] Burkov V N and Novikov D A 1997 *Project management: Scientific and practical publication.* (Moscow: SINTEG-GEO)
[4] Voicu I P 20102 *Project Management: lecture Notes* (Pskov: Pskov state University)
[5] Mazur I I and Shapiro V D etc. 2010 *Project Management : textbook. aid for students enrolled in the specialty "organization Management"* (Moscow: Publishing House "Omega-L")
[6] Polkovnikov A V and Dubovik M F 2015 *Project Management* (Moscow: Eksmo)
[7] Locke D 2004 *Fundamentals of project Management* (Moscow: "HIPPO")
[8] Vlasov A V 2015 Classification of projects in the oil and gas industry *Vestnik MGIMO* **5 (44)**

198-203
[9] Sakhalin energy 2014 *Report on sustainable development*, available at: http://www.sakhalinenergy.ru/ru/media/sd_report/
[10] David Wood & DWA Energy Limited 2015 *Barents Sea Gas Liquefaction Approaches Reality at Snohvit – Innovation at a Price*, available at: http://www.dwasolutions.com/images/BarentsSeaLNGDWJan07.pdf
[11] Yamal LNG 2015 *Disclosure of YAMAL LNG project information,* available at: http://yamallng.ru/progress/disclosure-of-information/
[12] Plyaskina N I, Kharitonova V N and Vizhina I A 2015 "the Power of Siberia" - a driver or threat to gas chemical clusters in the East of Russia? *ECO* **4**
[13] Panyagina A E 2012 Hikes to understanding and classification of risks *Modern economy: problems, trends, prospects* **6**
[14] Immaneni A and Mastro C 2004 A Structured Approach to Building Predictive Key Risk Indicators *Operational Risk: A Special Edition of The RMA Journal* 32-47
[15] Kossov V V 2008 Megaproject-stop the depopulation of Russia *World of Russia* **1** 89-101
[16] Negreeva V V and Abarkina D V 20106 Yamal LNG: new risks and opportunities of the Russian fuel and energy sector in the Arctic *Economics and environmental management* **4**

Use of methods for water suppression in the Severo-Orekhovskoe field

V F Dyagilev[1], T V Dyagileva [2]
[1] Tyumen Industrial University, Nizhnevartovsk branch, 628616, Tyumen Region, Khanty-Mansiisk autonomous district - Yugra, Nizhnevartovsk, Western industrial hub, Panel 20, Lenina St., 2/P, Bld 9, Office 107
[2] Tyumen Industrial University, 38, Volodarsky St., Tyumen, 625000, Russia

E-mail: glibazval.dyagilev@yandex.ru

Abstract. The subject of the study is the use of methods for water suppression in the productive stratum of the Severo-Orekhovskoe field. Based on the results' analysis of the use of water suppression methods in the Western Siberia fields, recommendations were developed for the application of these methods in the Severo-Orekhovskoe field. It is shown that only technologies based on gel-and sediment-forming compositions having selective operation in layer-by-layer inhomogeneous stratum meet the conditions of productive strata of the Severo-Orekhovskoe field for all boundary parameters of application. During the study of the specifics of the field-geologic conditions of the Severo-Orekhovskoe field, effective water-shutoff compounds and technologies based on them were proposed: for strata A2-1, A1-3 - gel-forming compounds ACC-1, TSS-1, and "Geopan". The reagents proposed for use for water suppression operations have a selective effect in layer-by-layer inhomogeneous permeability stratum and have a good filtration in low-permeability reservoirs. The efficiency of using the water suppression technology based on the thermotropic reagent "RV-3P-1" for high-temperature strata BV, JV. It is established that relatively low temperature conditions expand, on the one hand, the range of water suppression technologies. For technologies based on acrylamide polymers (injection of cross-linked polymer compounds, visco-elastic compounds, concentrated polymer compounds, Temposkrin polymer-gel system, etc.), the only limitation is the low permeability of productive strata AB1-2, AB1-3.

1. Introduction
Based on the generalization and analysis of scientific and technical literature on the water suppression technology, we can classify isolation methods into non-selective and selective. This separation is determined by the technological, physicochemical, and mechanical properties of the material.

2. Literature Review
Identification of effective methods of using hydrocarbon resources is the research task of the modern scientific community (Etminan et. al., 2017) [1], which is reflected in the discussions of domestic (Vakhrushev et. al., 2016 [2], Surguchev and Sharbatova, 1988 [3]) and foreign scientists (Lei Wang et. al., 2017 [4], Xian'e et. al., 2017 [5], Wang and Sheng, 2017 [6]), focused on the search and development of methods for predicting the technological efficiency of production and assessment of oil reserves. The study relevance of the existing methodology for the development of natural resources of oil and gas complexes (Seigman et. al., 2010 [7], Report, 1999-2014, 2015 [8]) is undeniable. The

availability of modern methods and principles for assessing the current state of oil reserves in the Northern territories of the Russian Federation demonstrates the need to review and re-evaluate existing methods for predicting oil reserves production. This focuses on updating existing methods for predicting the technological efficiency of production and assessment of oil reserves.

3. Methods

Non-selective isolation methods (NSIM) – are methods using materials that, regardless of the medium's saturation with oil, water, and gas, form a screen that does not break down over time in reservoir conditions. The main requirements for NSIM are the exact identification of the treated watered interval and exclusion of reducing the permeability of the productive oil-saturated part of the stratum. For this purpose, cement, foam cements, polymer cements, technical devices such as drillable packers and overlapping devices are mainly used [9].

Selective isolation methods (SIM) are those methods where materials are used that are injected into the entire perforated part of the stratum. In this case, the resulting sediment, gel, or hardening substance increases the filtration resistance only in the water-saturated part of the stratum, and there is no plugging of the stratum's oil part. The selective effect of chemicals is based on the difference of physico-chemical qualities of stratum fluids (oil and water) and physico-geological features of the formation's structure, determining hydrodynamic conditions of the collector [9].

4. Results and discussion

For productive strata characterized by high layer-by-layer inhomogeneity, it is preferable to use selective technologies for water suppression with the formation of temporary screens – "soft technologies".

Selective materials can be used in the form of water-swelling polymers, emulsions, organosilicon compounds, sediment-forming compositions, and anhydrous grouting mortar. Water-swelling polymers are pumped as a suspension in an inert fluid, at a time of water contact, they swell, increasing in volume by 100-300 times, thereby displacing water and increasing the resistance to its movement. Emulsions are used in the wells development in the oil-saturated interval, as a result of their application, the viscosity of the blocking screen decreases. Organosilicon compounds interact with water, resulting in a durable gel. Sediment-forming compositions mixed with formation water form a sediment (10–50% volume), plugging the watered interval. Finally, anhydrous hydrocarbon-based grouting mortars, at a time of water contact in the stratum, form a high-strength impermeable stone, while no stone is formed in the oil zone.

By chemical nature, reagents for water isolation and elimination of behind-the-casing flows are divided into three groups. The first group consists of compositions based on organic materials: viscoelastic mud and cross-linked polymer compositions based on polyacrylimide; water-swelling polymers; "Hpan" and "Givpan" polymers; oxyethyl- and carboxymethylcellulose; compositions based on organic formaldehyde resins.

The second group includes compositions based on organoelemental compounds: organosilicon (AKOR, AKOR-2, AKOR-4, AKOR-BN, product 119-204, Water-repelling organosilicon liquid-11); organoaluminum; organotitanium, etc.

The third group is represented by reagents based on inorganic materials and includes cements, silicates (liquid glass — hydrochloric acid, silicate-alkaline solutions, silinom) and aluminosilicates (nepheline, zeolite-containing component, aluminum salts (chloride, aluminum sulfate, potassium alumen with alkalis), reagents "Galka", "Thermogel", VIS-1.

Depending on the formation mechanism of water-shutoff masses and the physicochemical principles of influence on the host medium, five groups of selective chemical reagents can be distinguished (figure 2):

- curable - chemical reagents forming, after entering the stratum, a water-shutoff mass, soluble in oil and insoluble in an aqueous medium;

- gel-forming - chemical reagents leading to the formation of spatial gel-like systems with an inorganic or organic solid phase of high degree of dispersion with aqueous or non-aqueous dispersion medium;
- sediment-forming - chemical reagents under formation conditions leading to the setting of insoluble sediment in water-saturated zones;
- hydrophobizators - reagents, the use of which is based on the water-proofing of the rocks surface in the bottom-hole zone, which leads to a decrease in the phase rocks permeability to water and, consequently, to an increase in the filtration resistance for its movement;
- foam systems are formed as a result of the chemical reagents interaction, which allows blocking the path of water movement as a result of gas bubbles sticking to the surface of water supply channels and the filming of colloido-dispersed compounds.

After entering the reservoir, the curing reagents form a water-shutoff mass that is soluble in oil and insoluble in water. Today, synthetic curing reagents are most in demand their consumption per well is minimal, and the use does not require special equipment [10].

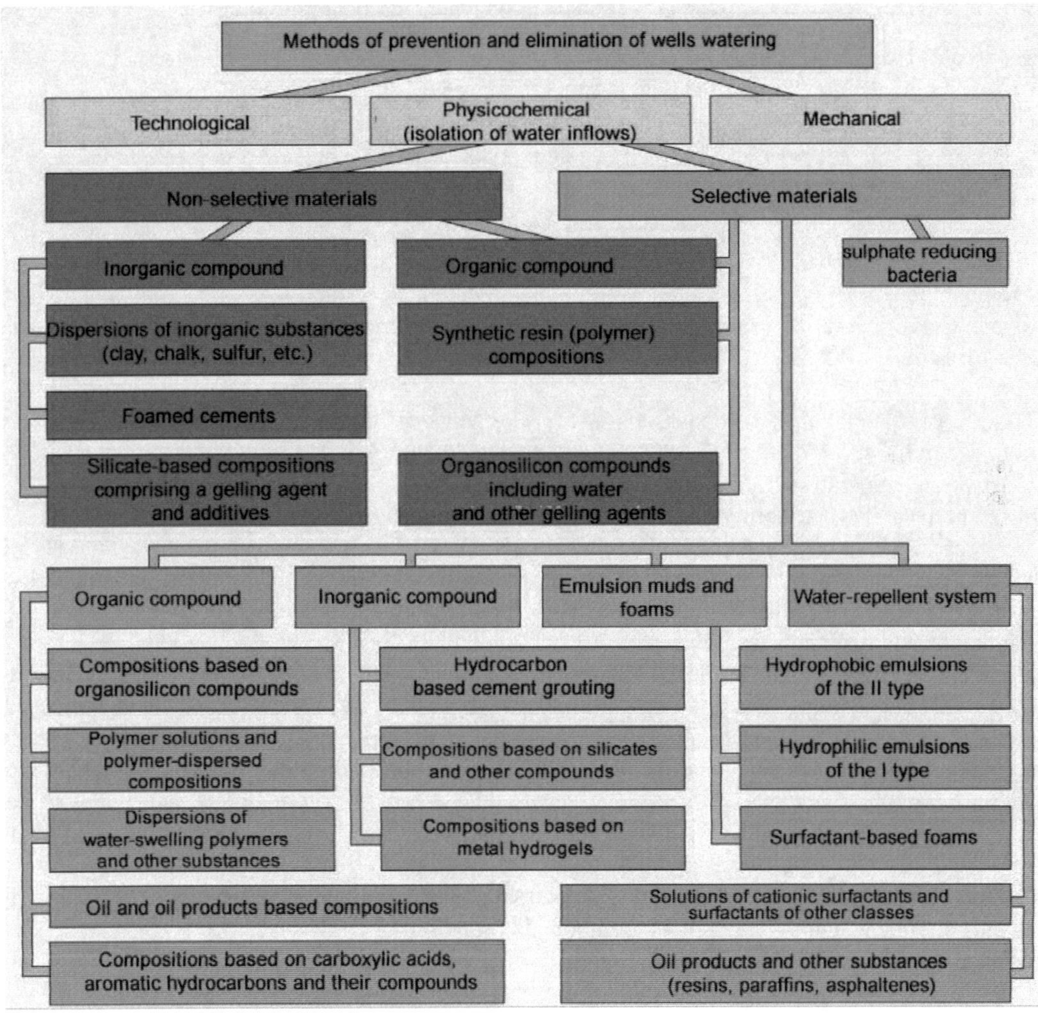

Figure 1. Classification of water suppression methods [10]

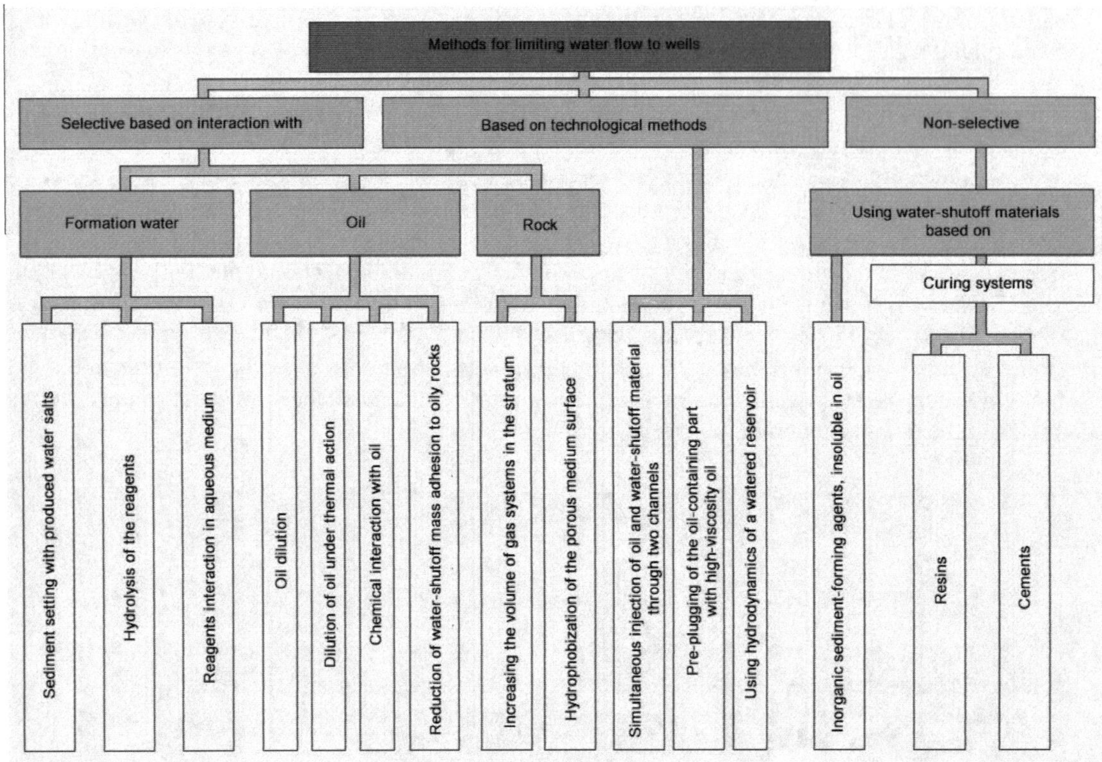

Figure 2. Classification of water suppression methods according to the mechanism of water-shutoff masses formation

The use of water-shutoff compositions in the sediments of the Severo-Orekhovskoe field is very limited, due to the specifics of the geological and physical conditions of these objects:

- formation temperature (54 °C or more);
- high permeability variability of strata (layer-by-layer heterogeneity);
- low reservoir properties of certain benches of AB, JV 1 strata (the stratum is represented by VI-V collectors (permeability from 0.002 to 0.01 μm^2), less often IV (permeability up to 0.1 μm^2) classes according to Hanin. As shown by field practice, such parameters exclude or severely restrict the use of PE methods, and cause their low efficiency [9].

For the conditions of the productive strata of the Severo-Orekhovskoe field, only technologies based on gel-and sediment-forming compositions having selective operation in layer-by-layer inhomogeneous stratum correspond to all boundary parameters of application. Relatively low temperature conditions expand, on the one hand, the range of water suppression technologies. For technologies based on acrylamide polymers (injection of cross-linked polymer compounds, visco-elastic compounds, concentrated polymer compounds, Temposkrin polymer-gel system, etc.), the only limitation is the low permeability of productive strata AB1-2, AB1-3 (in some cases).

For technologies based on water-soluble polianionites (injection of Givpan polymer with crosslinking agent), inorganic and organic silicon compounds (injection of AKOR, KROS, ETS-40 solutions of water glass — silicate-alkaline effect and other ethylsilicates), the limiting factor for their application is the low average permeability of the rocks of AB 1-3 strata [11].

Suspension or disperse systems have low filtration capacity in collectors with permeability less than (0.02-0.03 μm^2).

field-geologic conditions of the Severo-Orekhovskoe field, effective water-shutoff compounds and technologies based on them were proposed: for strata $A_{2\text{-}1}$, $A_{1\text{-}3}$ - gel-forming compounds ACC-1,

TSS-1, and "Geopan". The reagents proposed for use for water suppression operations have a selective effect in layer-by-layer inhomogeneous permeability stratum and have a good filtration in low-permeability reservoirs.

Technologies based on the proposed reagents are aimed at increasing the oil recovery of terrigenous strata and are intended for:

- restrictions on water inflow to producing wells;
- reducing non-productive injection in the formation-pressure maintenance system;
- alignment of the injection and selection profile;
- elimination of interstratal flows between productive strata in the bore-hole annulus and the contour discharge of injected water into the aquiferous part of strata [12].

Technologies based on thermogel-forming compositions (GFC "RV-3P-1", GFC "GALKA" GFC "METKA" are applicable only in the conditions of highly permeable strata BV6,8 medium permeable strata JV1 (stratum temperature of layers over 80^{0}C). For technologies based on thermogel-forming reagents, the high temperature of the strata is a necessary factor in the formation of an ecologically friendly aluminum hydroxide gel. As field practice shows, cross-linked or gel-forming polymer systems based on known polyacrylimide are not recommended for use in high-temperature strata (they are destructured at a temperature above 75-80 °C); emulsion systems (they are thermally stable for a long time only up to a temperature of 85 °C) [12].

Field-statistical criteria for optimal application of the "Geopan " reagent based technology:

- eff. oil-filled thickness more than 2 m.;
- collector-terrigenous, porous-fractured;
- permeability of productive strata more than 0.01 μm^2;
- acceleration not less than 50 m^3/day;
- during well treatment oil flow rate is more than 1.5 t/day;
- during complex treatment (injection + reactive producing) the total oil flow rate by element (block) should be more than 3.5 t/day.
- abrupt change (20% or more) in the water cutting of the produced liquid at the source of impact;
- for producing wells, the reservoir pressure is not less than 0.7 hydrostatic pressure;
- availability of residual recoverable reserves.

Criteria for applying the technology based on the "ACC-1" reagent:

- terrigenous heterogeneous reservoir (number of permeable intervals >2);
- carbonate content is not more than 10%;
- permeability from 0.01 μm^2 and more.
- the technology works in waters of different mineralization;
- Criteria for successful application of "RV-3P-1" reagent technology:
- Stratum temperature - not less than 70 ° C
- Collector - terrigenous.
- Permeability coefficient-not less than 0.05 μm^2.
- Number of permeable intervals - more than 2.
- Formation water mineralization - without restrictions.
- Water cutting by the source of impact is 50–90 %.

A wide range of process conditions:

- Injected water is used in working solutions;
- high processability (liquid commodity form), use of standard equipment;
- well treatment at temperatures up to -35 °C;
- incorrectly installed gel barrier is eliminated with 0.5 % hydrochloric acid solution;
- reagents are non-toxic, do not violate the ecology.

IOP Conf. Series: Materials Science and Engineering **952** (2020) 012027 doi:10.1088/1757-899X/952/1/012027

5. Conclusion

For the field-geologic conditions of the Severo-Orekhovskoe field, effective water-shutoff compounds and technologies based on them are proposed: for strata A_{2-1}, A_{1-3} - gel-forming compounds ACC-1, TSS-1, and "Geopan". The reagents proposed for use for water suppression operations have a selective effect in layer-by-layer inhomogeneous permeability stratum and have a good filtration in low-permeability reservoirs. The water suppression technology based on the thermotropic reagent "RV-3P-1" are recommended for high-temperature strata BV, JV.

References

[1] S. Reza Etminan, Brij B. Maini, Zhangxin Chen 2017 Determination of mass transfer parameters in solvent-based oil recovery techniques using a non-equilibrium boundary condition at the interface *Fuel* **120** 218–232

[2] Vakhrushev V V, Melnikov V N and Moskvitin S A 2016 Generalization of experience of development of the target area YUV1 of fields of LLC "LUKOIL - Western Siberia" to increase the validity of the forecast of technological parameters *Neftyanoe hozyajstvo* **8** 44–47

[3] Surguchev M L and Sharbatova I N 1988 *Cyclical effects on heterogeneous oil reservoirs* (Moscow: Subsoil)

[4] Lei Wang, Ye Tian, Xiangyu Yu, Cong Wang, Bowen Yao, Shihao Wang, Philip H. Winterfeld, Xu Wang, Zhenzhou Yang, Yonghong Wang, Jingyuan Cui, Yu-Shu Wu 2017 Advances in improved/enhanced oil recovery technologies for tight and shale reservoirs *Fuel* **210** 425–445.

[5] Xian'e, Li, Zhi, Xu, Hongyao, Yin, Yujun, Feng, and Hongping, Quan 2017 Comparative Studies on Enhanced Oil Recovery: Thermoviscosifying Polymer Versus Polyacrylamide *Energy fuels* 31 (3) 2107–3328

[6] Xiukun Wang, James J. Sheng 2017 Effect of low-velocity non-Darcy flow on well production performance in shale and tight oil reservoirs *Fuel* **190** 41–46

[7] Seigman Yu V, Gumerov O A, Karimov R M and Shamaev G A 2010 Hydrodynamic methods of regulation of oil field development *Current issues of oil and gas field development at late stages. Technologies. Equipment. Security. Ecology: materials of the scientific and practical conference* p. 258

[8] 2015 *Report on the main technical and economic indicators of exploitation of deposits of CCI "Langepasneftegaz" for the period 1999-2014*

[9] Kaplan L S and Kaplan A L 2004 *Reference book of the oilman. Part 2* (Ufa: USPTU)

[10] Efimov N K 2011 Water suppression technology in oil wells and ways to improve the efficiency of the squeeze job *PTOGJ "Engineering practice"* **7** 2-17

[11] Starkovsky A V 2011 An integrated use of physical-chemical impact technologies for the enhanced oil recovery *STJ "Oil industry"* **5** 92-95

[12] 2013 Analysis of the achievement of project targets for 2012 and development of measures for the implementation of project solutions for 2013 for OJSC "SN-MNG" **14**(Nizhnevartovsk)

IOP Conf. Series: Materials Science and Engineering **952** (2020) 012028 doi:10.1088/1757-899X/952/1/012028

Preliminary zoning of negative impact of the projected oil waste polygon at the Samotlor field

I S Aitov and A T Khudaiberdiev

Tyumen Industrial University, Nizhnevartovsk branch, Bld 9, 2/P, Lenina St., Nizhnevartovsk, 628600, Russia

E-mail: aziz.5199@mail.ru

Abstract. This paper provides a comprehensive analysis of the building expediency of a polygon for processing oil wastes in KhMAD-Yugra. Preliminary research has shown the environmentally dangerous impact of its project location relative to Nizhnevartovsk city, village Izluchinsk and village Bolshetarkhovo, as well as other important objects. The method of location estimation based on the main rules of life safety is proposed. Its application will reduce the environmentally dangerous impact and make it possible to choose the most optimal location of the polygon. An environmental pollution assessment scale has been developed, presented in tabular form for convenience. The area of zones of different dangerous influence degrees is calculated. The article describes the increase in the probability of earthquakes, wildland fires and floods in the area of the Samotlor field polygon due to global warming. The article also considers the prospect of using GIS programs in modeling environmental phenomena that affect oil and gas production facilities.

1. Introduction

The object of the study is the negative impact of the oil waste polygon.

Subject-the optimal location of the oil waste polygon relative to localities, strategic locations (structures) in case of natural and man-made accidents for minimizing negative consequences.

The purpose is to develop evaluation criteria for choosing the optimal location of the oil waste polygon.

Objectives:

1) consider the dynamics of natural hazards (fires, floods, earthquakes) under the influence of global warming;

2) analysis of negative consequences and zoning of the territory at the project location of the oil waste polygon;

3) development of evaluation criteria for choosing the optimal location of the oil waste polygon.

Emergency oil spills, oil-contaminated land, oil sludge waste – the main and serious problem for the environment, the health of Yugra population, and life safety. In 2018, 3624 accidents occurred in the oil industry of Yugra. About 4.7 million tons of waste are generated annually [1]. More than 4700 hectares of district land, in total, taking into account the previous years' spills, are polluted and require reclamation. In addition, today, considering 2017, there are more than 1,600 sludge pits on the territory of the fields – places where oil waste is drained. In the Khanty-Mansi Autonomous Okrug-

Yugra, more than 80 % of pollutants, entering water objects, are contained in the wastewater of chemical and petrochemical enterprises [2].

The relevance of the polygon for processing oil wastes in our district is not in doubt. Since it is necessary to solve the problem of processing of oil wastes in KhMAD-Yugra. This problem is really enormous and is caused by a significant amount of accumulated and annually generated waste, their negative impact on the environment.[3] It may be necessary to make changes to environmental legislation affecting the conduct of tenders with the selection of contractors on the principle of the presence or absence of modern non-polluting technologies. Today, oil companies and subsurface users conclude agreements with external agencies that plan to eliminate these oil-contaminated wastes and sludges. But in fact, only a small part of them are able to comprehensively solve this problem. Many of them often do not have their own bases and facilities to place, store, and recycle waste. They are often inearthed or thrown into inaccessible places. In the course of the research we studied various works on the environmental hazard assessment of oil and gas industry facilities [4-8].

The described polygon was planned to be built in 2012 on the territory of the Samotlor field in the area of the dismantled CS-16. The construction was supposed to be carried out in accordance with the Federal target program "Reducing risks and mitigating the consequences of natural and man-made emergencies in the Russian Federation until 2015". The program was approved by Decree of the Government of the Russian Federation № 555 of 07.07.2011. The first section of this program indicates that the number of natural hazards and major man-made disasters on the territory of the Russian Federation has been growing annually over the past decade. The document mentioned natural disasters related to natural hazards and fires, water accidents. As well as man-made accidents and terrorist acts are the main sources of emergency situations and pose a significant threat to the citizens' security, the national economy and, as a result, to the sustainable development and national security of the Russian Federation [9]. The polygon was supposed to work on the basis of innovative technologies and equipment, reduce the number and area of oil-contaminated sites and accumulations of oil wastes. Experts assured that as a result of the project implementation, the threat of an emergency will be reduced, and it will even be possible to return previously infected lands and water objects to circulation. Theoretically, everything looked fine and safe. But in fact, it turned out the opposite since the polygon became a source of increased danger.

2. Materials and methods

3 public hearings were held on this project, initiated by the public, citizens and environmental specialists, who were against its construction and considered that the polygon poses a serious environmental threat [10]. But the directions indicated by us (radius of negative impact, earthquakes, fires and floods) were not considered.

In the course of our work, we conducted research in these directions. We believe that the polygon is necessary to solve the problem of processing oil-wastes, since there is no unified scheme for processing or recycling waste. They are often buried in the ground or dumped along rivers and lakes. But it is required to choose another place for its construction. The former project location actually included Nizhnevartovsk, Izluchinsk some other localities in the zone of negative environmental impacts, as well as a strategic location – Nizhnevartovsk conventional thermal power station (SDPS), located in the Izluchinsk village.

It is doubtful that choosing the location of this polygon, the natural hazards that arise periodically in the area, such as seismic hazards, floods and fires, have been taken into account, which in the perspective of global warming may increase in scale.

The trend of global warming is no longer a myth, but a scientifically based fact. We live in an unstable period in terms of climate fluctuations. It lasts about 500-700 years, compared to others that last millions of years, this is a rather short period [11].

All these changes, which seem anomalous, are just characteristic of the period in which we live. The coming warming is similar to the Holocene climate optimum — the Atlantic period (about 5-6 thousand years ago).

2.1. Analysis of negative consequences from the polygon in the case of project location.
Seismic hazards continue to pose a significant threat to the population [12].

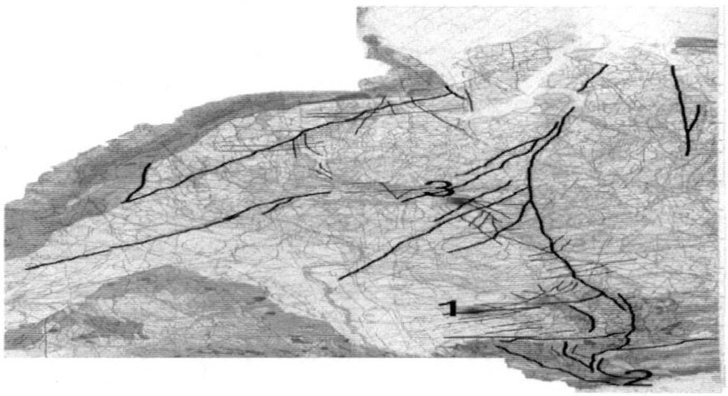

Figure 1. A schematic map showing the connection of the geological structure with the earthquake's manifestations along the tectonic faults of the foundation of the Western Siberian young epipaleozoic plate.
Explanation of figure 3. 1 – earthquake of 1966 in the area of Kamen'-na-Obi; 2 – the district of the Altai earthquake in 2003 with the epicenter within the Kuray mountain range; 3 – earthquake of 2003 in the area of the lake Samotlor and Nizhnevartovsk.

According to figure 1, the project location of the polygon is located in the zone of intersections of longitudinal and transverse breaks (lineaments) of the crystalline basement. If the phases and amplitudes of seismic waves coincide at the intersection of longitudinal and transverse lineaments, then the tremors are intensified, otherwise, the tremors are reduced [13]. In the case of an earthquake hazard, accidents may occur at the polygon, as a result of which the depressurization of sludge pits and containers with oil wastes and the inevitable leakage of waste into the environment is possible, as well as their further ignition. As a result, the pollution zone will expand, the area of which includes Nizhnevartovsk, and Izluchinsk village.

Since the beginning of 2017, 510 forest fires have been registered in the Nizhnevartovsky district on an area of 52685.3 Ha, and 69 fires on the Samotlor field on an area of 2611.6 Ha [14]

It is very difficult to avoid fires in this area, since there is a high swampiness, and therefore a large accumulation of peat deposits.

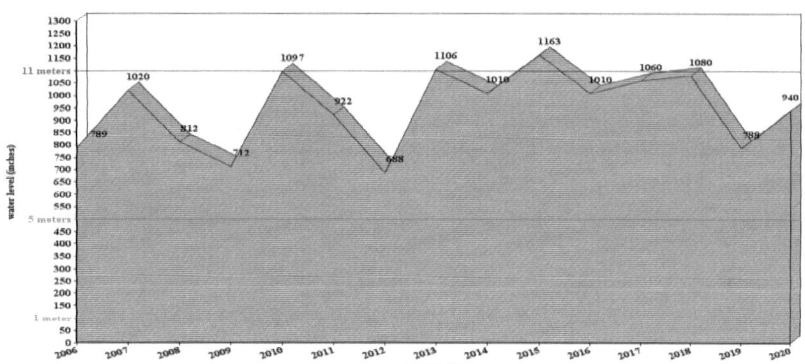

Figure 2. Increasing flood peaks of the river Ob in Nizhnevartovsk

In the case of oil waste ignition at the polygon, the plumes will take on terrifying dimensions. Nizhnevartovsk, Izluchinsk village, Bolshetarkhovo village, as well as many others, will be included in the zone of dangerous environmental impact. Even if there is no direct contact, there will be very strong smog and smoke, making it impossible for people to live and function equipment.

Another characteristic natural feature of the Nizhnevartovsky district is the long spring-summer flood. Its duration is about 100 days. Based on the topographical basis, as well as on hydrological data, it is possible to predict what will happen when the water rises in the river Ob relative to the maximum level of 1979 year.

Based on the graph data in Figure 2, we can say that the peaks of spring-summer flooding are increasing, which is influenced by the global warming trend. The greater the sum of positive air temperatures during the snowmelt period, the more intense the melting of snow cover and ice is. Also, the period of rise becomes shorter and the maximum flood occurs faster [15].

An excess of 12-13 m can lead to almost complete flooding of the Nizhnevartovsk's territory. An excess of 14-15 m will lead to flooding of a large part of the Samotlor field. As a result of the analysis, an increase in flood peaks on the Ob River in the area of Nizhnevartovsk was revealed. It should also be taken into account that in the case of flooding, there will be a chain of interconnected water objects connecting the oil waste polygon, Nizhnevartovsk, and Izluchinsk urban settlement, primarily through the Vakh river. When the water level rises, another complex of processes is expected, related to the aggravation of swamping processes and the restructuring of swamp systems.

For environmental analysis and evaluation of the optimal placement of the described polygon, we offer the following criteria:

1) reducing the level of danger and harmfulness of the negative factors source by improving its construction and working process implemented in it;

2) increasing the distance from the source of danger to the object of protection;

3) installation of means that reduce the level of dangerous and harmful factors between the source of danger or damage effect and the object of protection.

2.2. Zoning of territory at project location of oil waste polygon

Based on the previously developed tables, a comparative analysis of the impact of the possible negative consequences of accidents at the oil waste polygon for Nizhnevartovsk city, Izluchinsk urban settlement, Bolshetarkhovo village. (figure 3) [16].

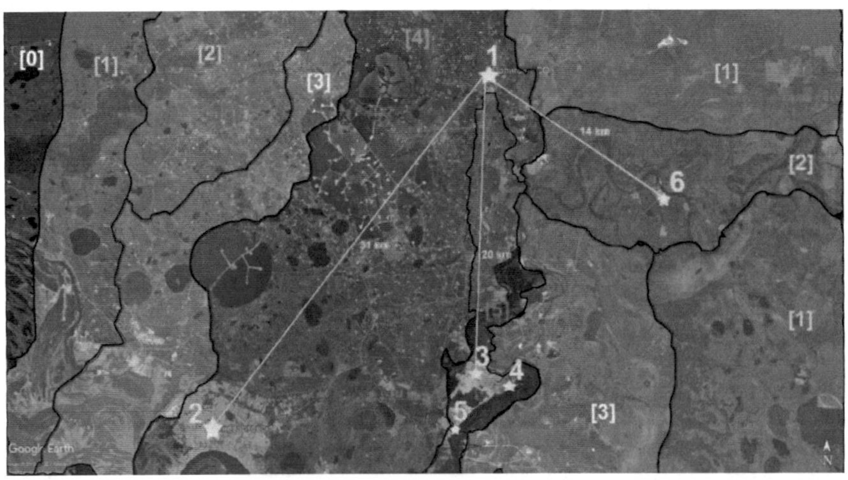

Figure 3. Zones of negative environmental consequences from a polygon for processing oil wastes in the case of project location. Explanation of figure: 1- project location of the polygon; 2 – Nizhnevartovsk city; 3 – Izluchinsk village; 4 – Nizhnevartovsk SDPS; 5 – drinking water intake from the Vakh River; 6 – Bolshetarkhovo village.

According to figure 3, it can be seen that the Bolshetarkhovo village, which is the closest of all the considered objects to the polygon, can only be negatively affected by the latter in the case of fire (through smog). A negative impact on the Nizhnevartovsk city and Izluchinsk village is caused, firstly, by the getting into the radius of a strategic location that provides electricity demand for the entire district - Nizhnevartovsk SDPS. Secondly, in the case of a leak of toxic oil wastes at the polygon, the waste inevitably falls into the Vakh river course, from where water intake is carried out to provide drinking water to Nizhnevartovsk's citizens. We also developed pollution scales, in which the color corresponds to a certain level of pollution, and calculated the area of each zone. (table 1)

Table 1. The level of negative consequences from the oil waste polygon on objects entering the exposure zone in the case of project location

Name	Color	Level of environmental pollution, %	Area of contamination
[0] points		no consequences	32.845 322 sq.km
[1] point		about 20	441.844529 sq.km
[2] points		21-40	301.813913 sq.km
[3] points		41-60	401.381378 sq.km
[4] points		61-80	632.731496 sq.km
[5] points		81-100	129.831483 sq.km

3. Results and Discussion

The proposed criteria were applied to the analysis of the location of toxic oil waste polygon relative to nearby cities and villages (table 2).

Table 2. The intensity of the negative impact from the oil waste polygon in the realization of the following hazards, taking into account the proposed principles that reduce the intensity of the negative impact of the oil waste polygon

Hazards	Nizhnevartovsk city (31.5 km.)			Izluchinsk village (20 km.)			Bolshetarkhovo village (14.5 km.)		
	Factors			Factors			Factors		
	A	B	C	A	B	C	A	B	C
Earthquakes	[1]	[0]	[0]	[0]	[0]	[0]	[0]	[0]	[0]
Fires	[1]	[0]	[0]	[1]	[1]	[0]	[1]	[1]	[0]
Floods	[2]	[1]	[0]	[2]	[1]	[0]	[0]	[0]	[0]

Explanations for table 2:
The distance from the oil waste polygon to the locality is shown in brackets
Criteria that reduce the intensity of the negative impact of the oil waste polygon:
A - reducing the level of danger and harmfulness of the negative factors source by improving its construction and working process implemented in it (reducing production volumes);
B - increasing the distance from the source of danger to the object of protection;
C - placement of means that reduce the level of dangerous and harmful factors between the source of danger or damage effect and the object of protection (placing a polygon behind a natural barrier);
Table 2 shows that subject to the proposed safety criteria (A, B, C), there will be a significant reduction in the negative impact of the oil waste polygon, up to zero values.
Applying these criteria (B,C), we propose the most optimal location for the oil waste polygon (figure 4.)

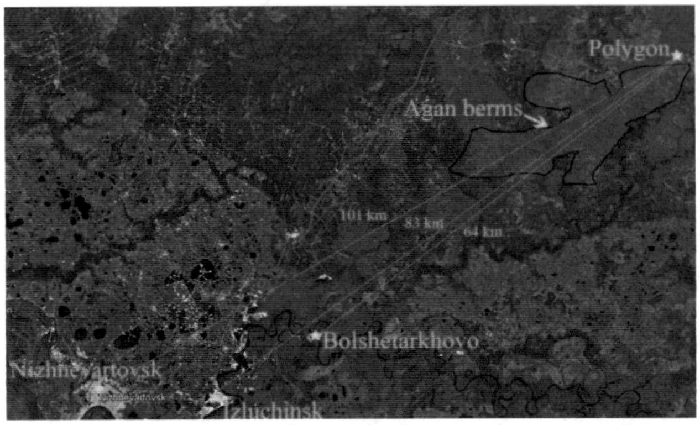

Figure 4. Our proposed location for the oil waste polygon.

Placing the polygon behind the Aganskiy Uvals, we simultaneously apply two proposed criteria: B – increasing the distance from the source of danger to the object of protection, C – placing the source of danger so that there is a natural barrier between the source and the object of protection (elevation - Aganskiy Uvals). We are on the way of creating a unified, universal methodology for evaluating the location of such hazardous industrial facilities. The next stage of work is the implementation of software products with geographic information systems (GIS). The application of GIS programs can speed up and improve the the process of evaluating the environmental hazard of facilities. At the moment, there are no similar software products that will comprehensively consider all factors (relief, hydrogeological terrain structure, wind rose, modeling of various emergencies taking into account local natural hazards).

4. Conclusions

The results obtained in the study show that we have achieved our goal – we have developed criteria for choosing the optimal location of the oil waste polygon, which made it possible to reduce the negative impact of the polygon on Nizhnevartovsk, Izluchinsk and the village Bolshetarkhovo. In addition, we conducted zoning of dangerous environmental consequences from the polygon, calculated the areas of these zones, developed pollution scales, and proposed the most optimal location of the oil waste polygon.

References

[1] Solodovnikov A U and Soromotin A V 2015 Drilling waste utilization experience in the Khanty-Mansi Autonomous Okrug - Yugra *Environmental protection in oil and gas complex* **12** 22-48

[2] Astapenko E O 2015 On certain aspects related to application of systemic environmental monitoring in zones influenced by petrochemical enterprises located in Khanty-Mansi Autonomous Okrug – Yugra *Innovation and Investment* **7** 8-12

[3] Litvinova T A, Tsokur O S, Zubenko U U and Kosulina T P 2012 Problem solution of utilization of oil wastes with their involvement in resource management *Modern problems of science and education* **7** 67

[4] Pashkevich M A 2006 Estimation of the environmental treat of production facilities during mining and processing of mineral resources *Notes of Mining Institute* **168** 29-31

[5] Moroz O V 2018 Problems of legal consolidation of objects of strategic environmental assessment, environmental impact assessment and state ecological expertise *Vesnik of Yanka Kupala State University of Grodno* **5** 29-37

[6] Nikitina A V 2007 Regional environmental aspects of environmental impact assessment in the projecting of oil and gas facilities on the territory of the Primorye territory *Vologdin reading* **64** 31-33

IOP Conf. Series: Materials Science and Engineering **952** (2020) 012028 doi:10.1088/1757-899X/952/1/012028

[7] Nozik M L 2010 Analysis of the basic indices for estimation the condition of the radiation safety and the radiation-and-ecological safety at the units of the fuel-and-power-generating complex *Prospect and protection of mineral resources* **8** 45-50

[8] Amosov P V and Novozhilova N V 2010 Assessment of the environmental risk zone (using the example of an underground radiation-dangerous facility) *Mining informational and analytical bulletin* **54** 366-374

[9] Decree of the Government of the Russian Federation № 555 of 07.07.2011 *On the Federal target program "Reducing risks and mitigating the consequences of natural and man-made emergencies in the Russian Federation until 2015*

[10] Khudaiberdiev A T 2019 About optimization of placement of environmentally dangerous objects in the perspective of global warming *Regional energy and energy saving* **2** 68

[11] Ryansky F N, Korkin S E and Aitov I S 2005 *Analysis of natural and anthropogenic factors of emergency situations* (Nizhnevartovsk: Nizhnevartovsk Humanitarian University Press)

[12] Ryansky F N, Korkin S E, Grebeniuk G N and Aitov I S 2005 Environmental management for sustainable development in the context of climate change in the North of Western Siberia *Polzunovsky vestnik. Ecological and sustainable development issues* **4** 61-64

[13] Efimov V I and Rybak L V 2012 *Production and environment* (Moscow: Federal state budgetary educational institution of higher professional education " Moscow state mining University")

[14] Belousova E P, Latysheva I V, Latyshev S V, Loshchenko K A and Shcheblykin A S 2016 Natural factors of forest fires in Irkutsk Oblast *Interdisciplinary journal of basic and applied sciences "Biosfera"* **4** 390-400

[15] Khudaiberdiev A T 2019 Analysis of the placement the oil waste processing polygon at the Samotlor field *VII regional youth conference named after V. I. Shpilman "Problems of rational nature management and the history of geological search in Western Siberia"* (Khanty-Mansiysk: LLC "Yugorsky format"), pp 182 – 184

[16] Khudaiberdiev A T 2019 Methods for assessing the placement of environmentally dangerous objects on the example of the Samotlor field's toxic waste polygon in the perspective of global warming *IX International scientific and practical conference of students, postgraduates and scientists "Experience, actual problems and perspectives of development of oil and gas complex"* (Tyumen: Industrial University of Tyumen) pp 182 – 184

IOP Conf. Series: Materials Science and Engineering **952** (2020) 012029 doi:10.1088/1757-899X/952/1/012029

Limiting state of plate with hole

Yu I Kazarinov

Tyumen Industrial University, Nizhnevartovsk branch, 38, Volodarskogo St., Tyumen, 625000, Russia

E-mail: kazarinov.yury@yandex.ru

Abstract. The problem of determining the limit state of a plate with hole is considered. Solutions of two other problems are proposed to solve this problem, namely, determination of stress intensity factor in elements of structure with cracks and calculation of stress-strain state of a plate of finite width with hole. The first objective is based on the fracture mechanics positions for the crack plate using the software complex KOMPAS-3D (3D design system). And the second objective is based on the theoretical method of elasticity theory, applied to a plate of finite width with hole. Obtained results of both objectives were compared with known solutions.

1. Introduction

When designing structural elements, their strength calculation is performed. Structural elements must reliably perform certain functions during the defined service life**.** The main criterion for the operability of machine parts is their strength [1]. The structural elements may contain technological stress concentrators due to operational necessity. For such parts, the right choice of strength criteria will be an important point in the strength calculation. When determining the limit state of structural elements, the presence/absence of a crack is important for selecting the strength criterion. Then, in the presence of a crack, the methods of fracture mechanics are applied, and in the presence of a non-sharp-edged defect, classical and new criteria for the strength of a solid material are used.

The development and implementation of the principle of safe damage is possible only with the use of fracture mechanics methods. According to the conventional provisions, the stress state of bodies with cracks is fully characterized by the stress intensity factor (SIF) Almost all currently known criteria for brittle and quasi-brittle fracture, which describe the growth of fatigue cracks, are based on the preliminary determination of the SIF.

Various analytical and numerical methods are proposed for calculating stress intensity factors [2-3]. However, during practical calculations, especially for bodies of complex shapes, structural elements with damage, there is a need for a more convenient method for determining SIF, in order to obtain more accurate results.

The numerical method for determining SIF in structural elements with cracks using the finite element method makes it possible to significantly simplify the calculation method and expand the research.

2. Materials and methods

Strength and fracture criteria for continuum medium mechanics. The stress state of real parts even with the simplest schemes of force application is always complex. The results of the strength calculation largely depend on the accepted strength criterion, on the basis of which it is possible, based on the data on material behavior under the simplest loads, to predict when a dangerous state will occur under the action of any complex load system.

The following four classical strength criteria are most widely known in engineering practice: the highest normal stress, linear strain, and shear stress, as well as the energy criterion [4].

Analysis of material limit state criteria from crack mechanics positions. The presence of sharp-pointed stress concentrators in the real body fundamentally complicates its strength calculation. In such cases, the classical approaches of continuum medium mechanics lead to incorrect results. In the classical theories of strength, for calculating the limiting state of a material, the special stress-strain state of the material near the apex of a sharp-pointed defect – a crack in the process of body deformation. This is due to the fact that the radius of the apex's curvature of such a concentrator is comparable with the parameters of the material's structure itself.

The basic idea of fracture mechanics of materials is based on the following. It is assumed that the transition of the deformable body element from a solid state to a fracture state is accompanied by an intermediate state. The most important feature of areas of deformable solid body, in which a pre-fracture state (pre-fracture areas) occurs, is that the material is always deformed beyond the elastic limit and that the most intense plastic flow occurs in them.

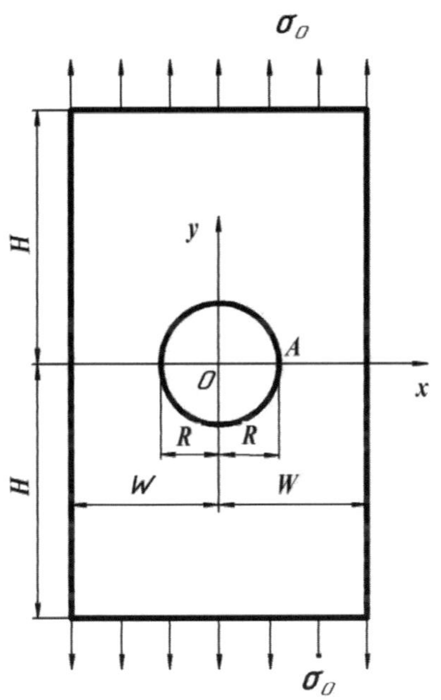

Figure 1. Plate with central circular hole

Therefore, assessing the strength of a solid, it is necessary to take into account its local physical and mechanical properties. Taking into account the states of material pre-fracture within the framework of continuum medium mechanics requires the introduction of new (in comparison with classical) computational models and concepts. Since the main characteristics that control the material behavior at the apex of the crack are stress, strain, and energy, then all the criteria of fracture mechanics similarly to the classical theories of strength, are divided into energy, force, and strain.

3. Characteristics and description of the proposed criterion for static fracture of components.

Consider a loaded plate with hole. It is known from experimental data that at the moment of fracture of a plate with a hole, the limiting stress state is not reached over the entire weakened section. Upon reaching a certain load near the hole in the dangerous section, a zone of the material limiting state appears. In this zone, a discontinuity of the material will be observed, which is equivalent to the formation of a crack. With a further increase in the load, the limit state zone (crack) will increase until the crack reaches the critical length. Then there will be almost instantaneous plate fracture.

The following conditions are accepted as criteria for static fracture of brittle materials:

$$\sigma_{eq} \geq \sigma_{\text{B}} \cup \sigma_1 \geq S_k, \tag{1}$$

where σ_{eq} is determined according to the Pisarenko-Lebedev strength theory; σ_{B} – is the ultimate strength; σ_1 – is the primary stress; $\cup$ – is the logical summation sign; S_k, – is the true resistance to separation.

Irwin condition is used as a fracture condition

$$K_1 = K_c, \tag{2}$$

where K_1 and K_c – are the stress intensity factors (SIF) of normal separation and its critical value in the case of the plane stress condition.

Thus, when using formulae (1) and (2) together, it is possible to determine the limit state of parts with stress concentrators, hat differ from cracks, and their fracture. To apply conditions (1) and (2) to concrete details, it is necessary to know the stress state in the weakened section and the formula for determining SIF near a concentrator with a crack.

Hence, we write down the criterion for static fracture of a part with a non-crack-shaped concentrator:

$$\begin{cases} \sigma_{eq} \geq \sigma_{\text{B}} \cup \sigma_1 \geq S_k \\ K_1 = K_{c.} \end{cases} \tag{3}$$

4. Determination of the stress intensity factor in structural elements with cracks

Let us consider an infinite elastic body under the conditions of the plane problem of elasticity theory. The body has a narrow internal crack with a length of 2a (figure 1).

Let the crack propagate under the pressure effect $p(x)$, which can vary along the crack.

We assume that there is symmetry about the x axis, so the problem is reduced to determining displacements and stresses in the elastic half-plane $y \geq 0$.

Boundary conditions in this case have the form, if $y = 0$:

$$\tau_{xy} = 0, \ -\infty < x < \infty;$$
$$\sigma_y = -p(x), \qquad |x| \leq a;$$
$$v = 0, \qquad |x| \geq a,$$

where v – is the projection of the displacement vector onto the y axis.

Solving the mixed problem of elasticity theory, we find, if $p(x) = p = const$:

$$\sigma_y\,(x,0) = -p\,, if\ |x| < a,$$
$$\sigma_y\,(x,0) = p\left[\frac{|x|}{(x^2 - a^2)^{\frac{1}{2}}} - 1\right], if\ |x| > a. \tag{4}$$

The stress curve σ_y, defined by the formula (4), is shown in figure 1. The stress σ_y approaches infinity at the crack apex, that is, it has a singularity at the tip.

To establish the nature of the normal stress singularity at the crack tip, we find the asymptotic image for σ_y, $if\ x \to \text{a} + 0$. Letting in formula (4) x tend to $a + 0$ and discarding the constant summand, we find

$$\sigma_y\,(x,0) = \frac{pa^{1/2}}{2^{1/2}(x-a)^{1/2}}, \qquad x \to \text{a} + 0. \tag{5}$$

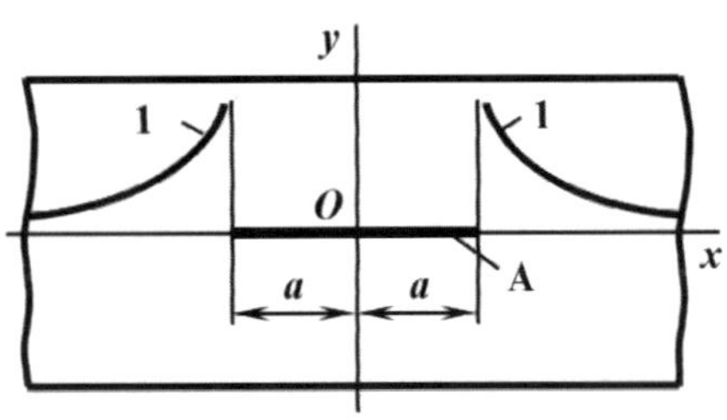

Figure 2. Elastic body with a crack **A**: 1 – curve σ_y.

If we introduce the notation

$$K_1 = p(\pi a)^{1/2}, \tag{6}$$

then formula (5) is:

$$\sigma_y\,(x,0) = \frac{K_1}{(2\pi)^{\frac{1}{2}}(x-a)^{\frac{1}{2}}}, \qquad x \to \text{a} + 0, \tag{7}$$

where K_1 – SIF of normal separation.

Using formula (6), we determine SIF for a plane with a single crack, when uniform pressure p is applied to the crack boundaries, or when the plane is subjected to uniform stretching along the normal to the crack line. In this case, SIF depends on the pressure and the crack size. Formula (7) is maintained when the crack is in an elastic body of finite sizes, while the pressure p (x) may not be constant. However, in such cases, the expression for K_1 will differ from formula (6). For a finite body it also depends on its size.

The introduction of the concept of "SIF" has proven to be very useful in linear fracture mechanics. The stress intensity factor can be determined from both known stresses and from displacements. From formula (7) we find

$$K_1 = lim\left[2\pi^{1/2}(x-a)^{1/2}\sigma_y\,(x,0)\right]. \tag{8}$$

Formula (8) allows you to determine SIF of normal separation by known stresses. The K_1 factor can also be found from the known displacements v:

$$K_1 = \frac{(2\pi)^{1/2}\mu}{1+\chi}\lim_{x\to a-0}\left[\frac{v(x,+0)-v(x,-0)}{(a-x)^{1/2}}\right]; \tag{9}$$

$$K_1 = \frac{2\mu}{1+\chi}\lim\left((2\pi)^{1/2}(a-x)^{1/2}\frac{\partial}{\partial x}[v(x,+0)-v(x,-0)]\right), \tag{10}$$

where μ – is the shear modulus; $\chi = 3 - 4v$ – for plane strain;

$\chi = \frac{3-v}{1+v}$ – for the generalized plane stress state;

v – Poisson's ratio.

For the determination of SIF of transverse K_2 and longitudinal shift K_3 there are formulas similar to the formulas (8)…(10) [5].

5. Numerical method for determining the stress intensity factor

Within the framework of linear fracture mechanics, which considers a model of an ideally elastic body and depicts a crack as a zero-thickness section, the surfaces of which are stress-free, the problem under consideration is reduced to the boundary value problem of elasticity theory.

The K_1 factor is the main basic characteristic of the stress-strain state of the material in the vicinity of a crack, which is extremely important in fracture mechanics.

Although real materials do not show perfectly elastic behavior during fracture, nevertheless, the models of linear fracture mechanics continue to be correct for them, provided that the zone of plastic processes is limited (implementing a quasi-brittle fracture scheme). This allows us to assume that the size of the pre-fracture zone and the state of the material in it as a whole is controlled by SIF [5].

For the problems under consideration, it is almost impossible to obtain an accurate analytical solution, therefore, so the numerical finite element method (FEM) in the form of the displacement method, implemented in the KOMPAS-3D software package, was used to calculate the K_1 factor.

For a generalized plane stress state, formula (6) is converted to a form that is convenient for use in the case of FEM applying:

$$K_1 = \frac{(2\pi)^{1/2}(1+v)\mu}{4}\cdot\frac{v\left(x,+\frac{\delta}{2}\right)-v\left(x,-\frac{\delta}{2}\right)}{(a-x)^{1/2}}, \tag{11}$$

where $v\left(x,+\frac{\delta}{2}\right)$ and $v\left(x,-\frac{\delta}{2}\right)$ – displacement at points near the crack apex in the y-axis direction (figure 2); δ – crack thickness.

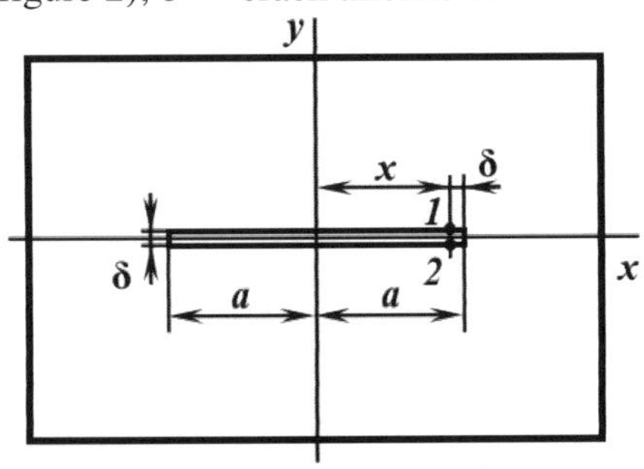

Figure 3. Study of body displacements at the crack apex

In some software complexes (KOMPAS-3D - 3D design system), SIF is calculated from several points at the crack apex.

The computational task was also to study the optimal number of points and their placement at the crack apex of the strained body during the determination of the K_1 factor and its comparison with the known solutions.

Two types of bodies with cracks are considered to solve the problem and to test the proposed method of determining SIF K_1:

– rectangular plate with a central crack (figure 4, a);

– strip with two symmetrical edge cracks under axial tension with normal stress σ = 100 MPa (figure 4, b).

Body thickness t = 3 mm, width W = 150 mm. For all variants, the crack thickness δ = 0.1 mm is acceptable.

Body material – mild steel with characteristics:

$$\nu = 0.28;$$

$$E = 2{,}1 \cdot 10^5 \text{ MPa};$$

$$\mu = 7{,}9 \cdot 10^4 \text{ MPa}.$$

Analysis of stress-strain state of plate and strip is performed using FEM and software complex.

The plate and strip were divided into finite elements with automatic grid clustering around the crack apex and had such sizes that allowed us to study the strain behavior of the material bodies at the specified points at the crack tip.

The K_1 factors for normal separation cracks were determined by the formula (11) and compared with the solution [6] for the plate, obtained by the joint method of decomposition of complex stress potentials and boundary collocation

$$K_1 = \sigma(\pi a)^{1/2} F_1\,(\alpha, \beta), \tag{12}$$

and for the strip obtained by the volumetric force method: $K_1 = \sigma(\pi a)^{1/2} F_1\,(\alpha)$;

$$F(\alpha) = 1{,}122 — 0{,}154\alpha + 0{,}807\alpha^2 - 1{,}894\alpha^3 + 2{,}494\alpha^4;$$

$$\alpha = 2a/W;\ \beta = 2H/W,$$

where $F_1\,(\alpha, \beta)$, $F(\alpha)$ – functions that depend on the geometry of bodies and are determined according to the work [6].

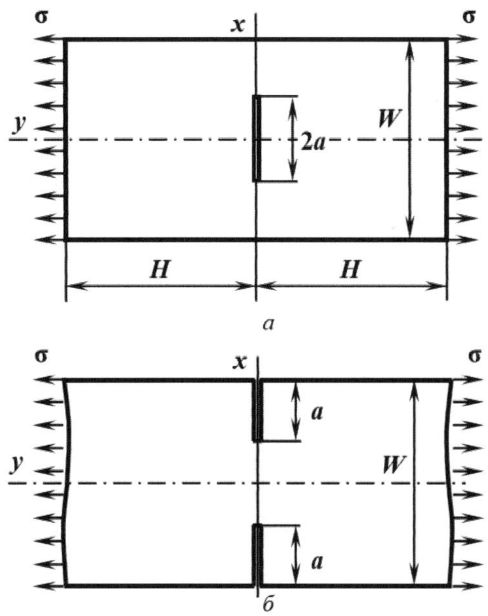

Figure 4. Calculation schemes of the studied bodies under axial tension:
a – rectangular plate with a central crack;
b – strip with two symmetrical edge cracks.

The results of studying the influence of the number of points and their placement at the crack apex of the plate on the SIF value showed, that the optimal sufficient option is to determine the displacements along the *y* axis at two points located at a distance δ from the crack tip (figure 3).

The obtained results of comparison of the K_1 value indicate a greater accuracy in determining SIF values at two points at the crack apex. The discrepancy does not exceed 4 % [6].

6. Comparison of calculated and experimental data

Using experimental data [7], we show the application of the fracture criterion (3) for the plate of final width ($2W$ = 140 mm, H = 200 mm, t = 1,83 mm, where t – plate thickness), weakened by a central circular hole of radius R = 14 mm, and subjected to uniform stretching (figure 1). The D16AT plate material has the following mechanical characteristics: true resistance to separation S_k = 700 MPa, ultimate strength σ_B = 450 MPa, fracture toughness K_c = 37 MN/m$^{3/2}$. To determine the stress state in a weakened section, we apply the formula 2.24 obtained for a plate of finite sizes with a hole in [7] for figure 1:

$$\sigma_y(x,0) = \sigma_0\left(1 + \frac{1}{2}\cdot\frac{R^2}{x^2} + \frac{3}{2}\cdot\frac{1+\varepsilon}{1-\varepsilon^2}\cdot\frac{R^4}{x^4}\right),$$

where $\varepsilon = R/w$, $R \leq x \leq W$.

To determine SIF near a concentrator with a crack - formula (12).

The application of criterion (3) and the work data [7] showed that the largest discrepancy between experimental and theoretical data obtained using the criterion $\sigma_I \geq S_k$ does not exceed 7%, and obtained using the criterion $\sigma_{eq} \geq \sigma_B$ – no more than 9%.

7. Conclusions

A numerical method for determining the SIF for structural elements with cracks is proposed. Several options for the placement of points were considered when determining the displacement field on the crack surface. Comparative analysis and numerical calculations show that the optimal case is when displacements are determined at two points (figure 3, points 1 and 2), located from the crack apex at a distance of its thickness δ. In this case, the error usually does not exceed 5%. The proposed method of calculating SIF (8) can be widely used during the study of the limiting state of elements with cracks.

As can be seen from the example considered above, the error of the proposed static fracture criterion is relatively small and is quite acceptable for engineering calculations.

References

[1] Kogaev V P, Makhutov N A, Gusenkov A P 1985 Calculations of machine parts and structures for strength and durability: Handbook, Moscow. Mechanical engineering
[2] Parton V Z 2008 Elastic-plastic fracture mechanics: Fundamentals of fracture mechanics: manual. Moscow. LKI publishing
[3] Siratori M 1986 Computational failure mechanics. Moscow: Mir
[4] Pisarenko G S 1976 Deformation and strength of materials at complex stress state. Kiev. Nauk. Dumka
[5] Panasyuk V V 1988 Fracture mechanics and strength of materials: in 4 vol. Kiev: Nauk. Dumka

[6] Murakami Y 1990 Stress Intensity Factors Handbook: in 2 vol. Moscow. Mir
[7] Kazarinov Yu I 2017 Strength of structural elements with cutouts and damages. Tyumen, IUT

IOP Conf. Series: Materials Science and Engineering **952** (2020) 012030 doi:10.1088/1757-899X/952/1/012030

Application of PCIU for protection of emergency pipelines against corrosion

N N Saveleva, A R Salyakhova, K E Koretskiy, P E Koretskiy

Tyumen Industrial University, 38, Volodarsky St., Tyumen, 625000, Russia

E-mail: nnsavelieva@yandex.ru, angelinasalihova@mail.ru everlast86rus@bk.ru loullli@bk.ru

Abstract. The field pipelines at the Samotlor field are the main means of oil and gas transportation. Corrosion causes premature wear of pipes and piping equipment and can cause breakdowns. To prevent emergencies, it is necessary to carry out repair and technical work timely, using chemical dosing units. To protect the field pipelines at the Samotlor field, stationary reagent dosing units are used. All of them are installed at well pads, closer to automated group metering units. They have a number of significant disadvantages: difficulty in installing; difficulty of the repair work; impossibility to remotely control the process of pumping a chemical reagent in real time; long idle time; outdated security system. A modern solution to the listed above problems for the oil and gas industry is the use of portable chemical injection units (PCIU). A PCIU is a state-of-the-art trailer-based system, and it can also be installed permanently on a well pad. The advantage of this installation is its high mobility, ease of installing, minimal installation and maintenance costs, the ability to automate.

1. Introduction

Currently, pipelines are the main type of oil and gas transportation in Russia. Corrosion leads to premature wear-out of pipes and pipeline equipment, can cause accidents, thereby causing spills of oil products, harming nature, and increasing transportation costs [1]. To prevent emergency situations, it is necessary to carry out repair and maintenance works in a timely manner using chemical dosing skids (CDS).

The purpose of this study is to compare portable CDS with stationary CDS currently operating at Samotlor field. We will also determine the economic benefits of using portable dosing skids.

The Samotlor oil and gas field was discovered in 1965 and brought into development in 1969. It is located in the Nizhnevartovsky district of the Khanty-Mansi Autonomous Okrug of the Tyumen region, 750 km North-East of Tyumen, near Nizhnevartovsk.

The area of the Samotlor license site, which is being developed by Samotlorneftegaz, is 2463.2 sq. km. The field has 8902 producing and 3901 injection wells equipped with the latest high-tech equipment. The length of oil pipelines is 2490 km, water pipelines – 2422 km, and other pipelines – 445 km. An extensive network of hard-surface roads with a total length of 1,923 km has been laid throughout the field. Figure 1 shows an overview map of the Samotlor field with a pipelines' layout scheme.

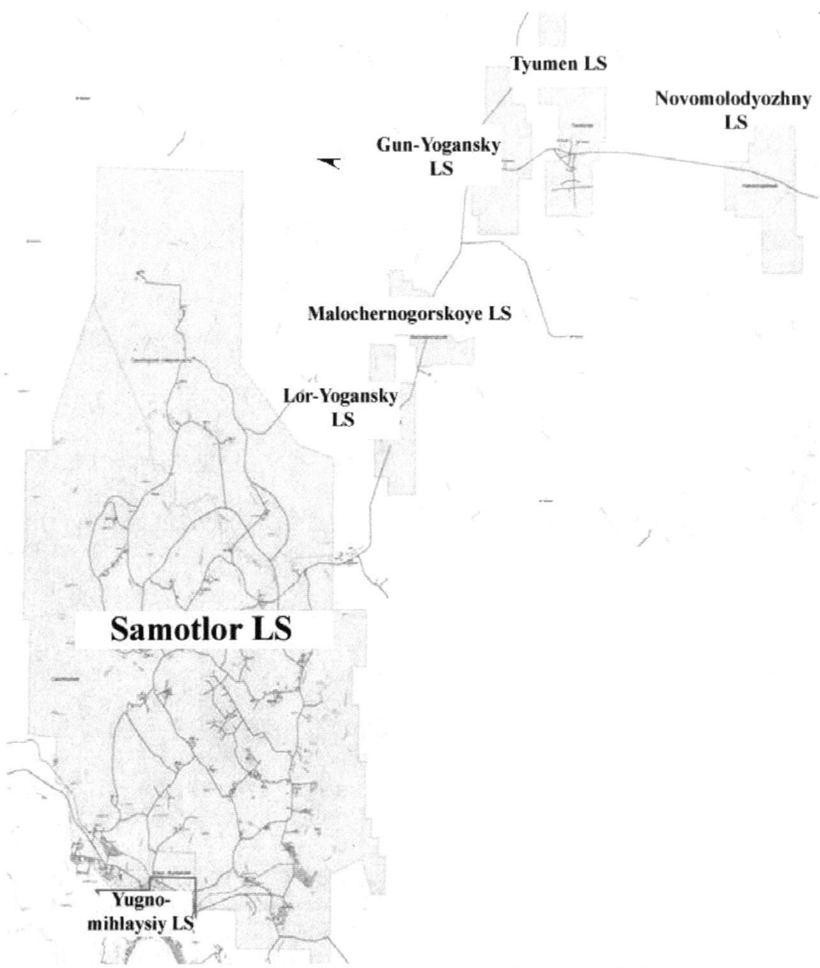

Figure 1. Overview map of the Samotlor field.

2. Problem statement

The current issue is corrosion control, which affects most pipelines and pipe network equipment. Due to its effects, Corrosion complicates the process of transporting fluid along the pipeline, after which there are emergencies, including breaks that completely stop work for an extended period of time.

We propose to use stationary chemical dosing skids to protect the field pipelines of the Samotlor field. All of them are installed on multiple-well platforms, closer to automated group measure units (AGMU). Further, the injection line should be connected from the chemical dosing skids (CDS) to the point of the main collector in order to protect the linear part of the oil collection.

IOP Conf. Series: Materials Science and Engineering **952** (2020) 012030 doi:10.1088/1757-899X/952/1/012030

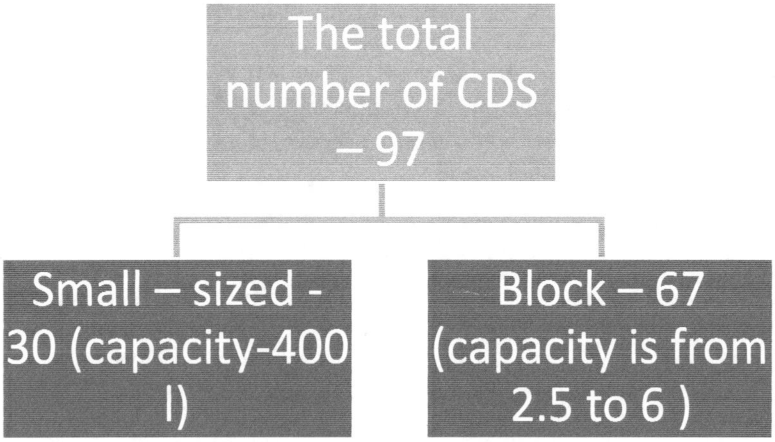

Figure 2. Number of CDS at the Samotlor field.

The total number of CDS in the emergency areas of the Samotlor field is 97 and is shown in figure 2. 30 units of them are fine dimensional blocks and their capacity is 400 l, the remaining 67 units are block objects installed according to the project for the arrangement of the multiple-well platform and their capacity is from 2.5 to 6 m^3.

The diagram in figure 3 shows that more than 25% of the total number of all CDS at the Samotlor field is occupied by fine dimensional blocks, which were installed after the implementation of the project for the arrangement of the multiple-well platform, in order to protect emergency pipeline sections at the oilfield.

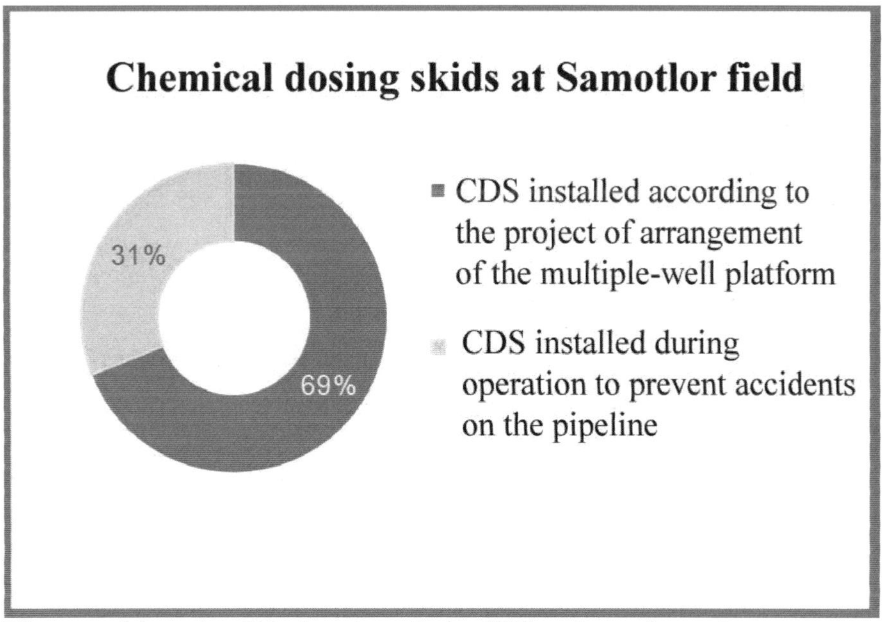

Figure 3. Diagram of the ratio of block and small-size CDS to the total quantity.

In addition to the fact that the existing CDS are stationary, they also have a number of significant disadvantages:

1. Installation complexity (many services are involved in order to complete all installation and dismantling cycles);
2. The complexity of repair work, including coordination of hot works, takes a lot of time;
3. For example, to remove a container from a chemical reagent unit, it is necessary to use the following components:
    3.1. Agree on the order for hot works
    3.2. Apply for a power outage
    3.3. Pump out the remaining chemical reagent from the container with the involvement of special equipment from the customer.
    3.4. Dismantling the installation with the involvement of the welding course from the customer
    3.5. Loading of the chemical reagents unit with the involvement of special equipment from the customer.
    3.6. Removal of the installation to the workshop territory
    3.7. And so, to repair one container, you will have to use the following special equipment: crane, trawl, fire brigade, vacuum booster, steam truck, welding course.
4. Inability to remotely control the process of chemical reagent injection in the real time;
5. After scheduled work, stationary CDS are preserved, which leads to non-productive delay;
6. Security system that does not meet modern requirements.

Also, a large number of CDS at the Samotlor field have been operating for more than 10 years, so they are under increased safety control. Due to the short service life, these units will soon need to be replaced.

Based on this, there are doubts about the rationality of using stationary CDS both from a logistical and economic point of view.

Therefore, oil and gas enterprises are faced with the issue of developing innovative dosing units that can minimize all the costs, related to existing chemical dosing units.

3. Problem solution

A modern solution to these problems for the oil and gas industry is the use of portable chemical injection unit (PCIU). PCIU is the latest trailer-based system and can also be installed stationary on the multiple well platform.

The unit is equipped with electric metering pump units in the amount of 3 pieces. This unit can operate from the main electrical networks or independently from the Geko power station with a voltage of 380 V, the power consumption of which is no more than 4 kW.

The unit also has antistatic plastic containers in the amount of 3 pieces. The volume of one container is 3000l, the total volume is 9000l. Containers are intended for storing any chemical reagents, including methanol. If necessary, the PCIU unit can simultaneously pump three different chemical reagents. The IBRC installation can supply the reagent simultaneously to three wells, or supply the reagent to the well, pipeline, and AGMU.

The following are 3D models of PCIU (Figure 4,5,6).

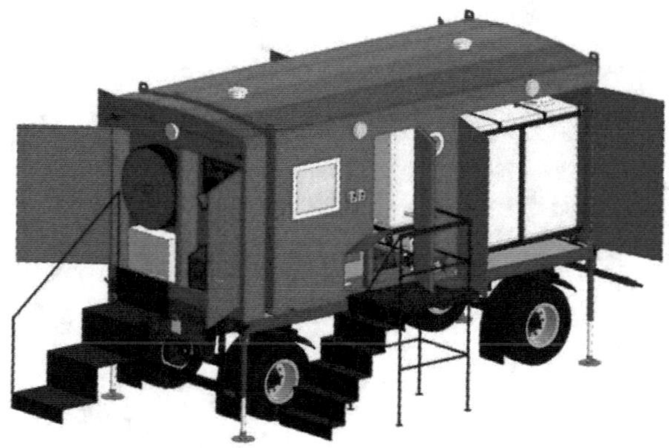

Figure 4. Unit in assembly. General view of PCIU

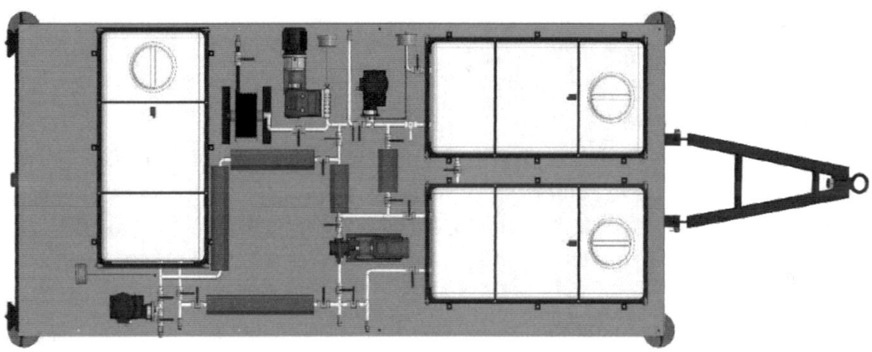

Figure 5. Unit in assembly. Top view.

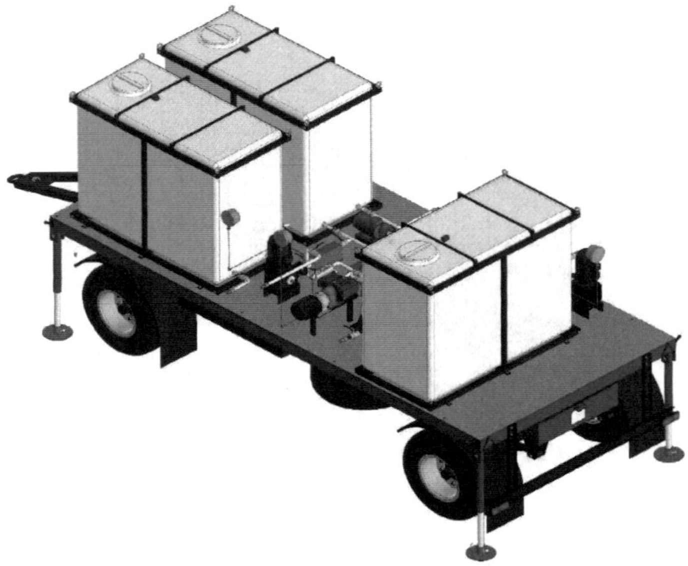

Figure 6. Unit in assembly.

The portable block includes:

- electric metering pump unit
- electric metering glandless pump unit
- electric gear oil pump unit
- 3-plunger electric pump unit
- additional capacity for 3000 liters
- control cabinet;
- pipe manifold and isolation valve;
- pressure, temperature, liquid flow rate and liquid level monitoring devices;
- remote control of metering pumps to increase or decrease chemical supply;
- control position;
- fire-fighting board;
- ventilation;
- interior lighting;
- exterior lighting;
- energy-saving block compartment heaters;
- terminal boxes, switches;
- electrical and control cable;
- unauthorized access alarm with output to the dispatcher
- real-time video surveillance gives the customer the ability to monitor the timely and safe performance of service organizations.
- and also, there is no need to coordinate hot works

4. Discussion of results

Advantages of this installation:

- Easy to install. (Maximum 2 specialists involved);
- Repair works are carried out immediately on site (including removal of containers in case of their failure);
- Monitoring and controlling PCIU remotely in the real time;

IOP Conf. Series: Materials Science and Engineering **952** (2020) 012030 doi:10.1088/1757-899X/952/1/012030

- Due to the portability of this unit, the need for conservation is eliminated;
- High security against unauthorized access;
- Full-service life more than 10 years;
- Minimal installation and maintenance costs.

It should also be noted that all work on the production process automation can be carried out using the considered equipment [3].

Economic efficiency:

The economic efficiency of these installations is great. Due to PCIU, it is possible to significantly reduce the cost and time for its maintenance and repair without involving special equipment from the customer [4].

For the year, at the Samotlor field, the cost of servicing all CDS is 17 million rubles. Approximately for a month the cost is 1 million 400 thousand rubles. As mentioned earlier, most of the installations at the field have been in service for more than 5 years, which in turn requires even more repair costs or complete replacement of old CDS with new ones.

With a similar cost of stationary and portable CDS, it is more profitable to install portable installations. As costs can be reduced due to the portability of PCIU, longer service life and ease of installation, repair, and maintenance. For example, 2 specialists are enough when dismantling this unit (replacing the container with a chemical reagent).

5. Conclusion

Performance of the oil and gas pipeline is one of the most important aspects of the oil and gas industry [5]. The correct tool is needed to solve this problem. The portable chemical injection unit from LLC "YugraChemService" will be a worthy replacement for the stationary chemical dosing skids available at the Samotlor field. PCIU will reduce the economic costs and time spent on repairs and maintenance, thereby increasing the productivity of the production process.

References

[1] Saveleva N N 2019 Improvement of technological equipment of the system of gathering and welding preparation. *Modern High Technologies* **2** 138-142

[2] Karakulov I I 2012 Patent 2456435 Russian Federation, IPC E21B37/06. Mobile chemical injection unit (MCIU) for submitting chemical reagents for the processing of oil and gas wells and pipelines: № 2010138170

[3] Saveleva N N 2018 Creation of an automation system for engineering calculation of preparation for the production at high-technology enterprises of mechanical engineering. *IOP Conference Series: Earth and Environmental Science* 012-029

[4] Saliakhova A R 2019 Analysis of anti-corrosion protection of oil and gas pipelines, in: *Collection: Experience, current issues and prospects for the development of the oil and gas complex* (Nizhnevartovsk) pp 70-73

[5] Khaidersbakh R 2014 *Corrosion protection and metallurgy of equipment for oil and gas production* (Vologda: Infra-Inzheneriya)

Improvement of pin flexible coupling construction

N N Saveleva and Y V Savelev

Tyumen Industrial University, Bld 9, 2/P, Lenina St., Nizhnevartovsk, 628616, Russia

E-mail: nnsavelieva@yandex.ru

Abstract. The purpose of the study is to increase the service life of coupling joints of centrifugal pumping units by designing a modernized coupling joint. During the operation of centrifugal units, there is a problem associated with a short service life of the couplings due to the imbalance of the half couplings, improper balancing of the rotor, etc. The authors conducted an objective and comprehensive analysis of the flexible coupling operation. Based on the results of the study of the existing construction, the coupling joint was modernized using the example of the pin flexible coupling for a centrifugal pump 200D90. The flexible coupling construction, which has no analogues in Russia, has been created. Due to the proposed upgraded construction, the coupling provides increased service life and more reliable operability in service.

1. Introduction

During the operation of centrifugal pumping units for transporting well products and process water, a problem arises due to the short service life of the couplings, which are used to transfer torque from the electric motor shaft to the shaft of the electric centrifugal pump.

In general, the coupling is the most worn-out part of a centrifugal pumping unit. Currently, the following couplings are used to connect centrifugal pumps with electric motors: PFC CP 60/198 State Standard 21424-75, PFC 200D/90, PFC CP 1000/180 State Standard 21424-93 [1,5]. The presented coupling joints have a number of disadvantages: it takes a long time to install couplings; installing and removing couplings, it is possible to deform the joint parts, violate the integrity of the coupling joint structure; get more vibrations during operation and, as a result, reduce the mean-time-between-failures; increased risk during hot works in the pumping block at process installations; outdated construction of used coupling joints.

Also, the main reason for reducing the service life and premature failure of the couplings is the pumping unit vibration [2]. List parameters affecting the level of pumping unit vibration:

- misalignment of the drive and pump;
- rotor imbalance;
- incorrect landing of the rotor on the shaft;
- incorrect connection of pipelines;
- impact of couplings;
- violation of bearing elements geometry;
- defects of bearing assembly;
- cavitation of the pumped medium;
- loosening of parts attachments;
- thermal expansion of pump construction elements;

- damage to foundations and supports;
- insufficient or poor quality lubricants, etc.

The purpose of the study is to increase the service life of coupling joints of centrifugal pumping units by designing a modernized coupling joint. Research objectives: reduction of non-production time during work on replacing the electric motor of pumping units; reducing the risk level of installation and dismantling of coupling joints; elimination of possible coupling joint destruction; reducing costs related to the use of welding engineering; simplification of coupling joints replacement procedure; possibility of restoration due to the modularity of the proposed product construction; simplification of repair work for routine repair and maintenance, as well as bearing assembly replacement operations; reducing the cost of repair and maintenance work.

2. Materials and methods

The vibration level was measured in accordance with the international standard ISO 10816-3:1998 "Evaluation of machine vibration by measurements on non-rotating parts" [8]. Also, the operation of centrifugal pumps was monitored based on the condition of bearings, couplings, shafts in accordance with the standards for flexible couplings [3].

Factors affecting pumping unit vibration level. As a result of field test of the centrifugal pumping unit, the following factors were identified, that reduce the operational reliability of the equipment.

✓ Incorrect alignment with the pump;

✓ Half couplings imbalance (worn-out pins, misalignment of pins holes or misalignment of half coupling);

✓ Clearances of bearing assemblies (bearing defects);

✓ Balancing the pump and electric motor rotors [4];

✓ Frame base abutment and attachments.

3. Results and discussion

In case of deviation of one or several factors of pump operation, the bearings fail, both in the pump and in the electric motor.

Measures for extension of coupling service life. The authors proposed measures to improve the efficiency and safe operation of pumping units. The vibration state of pumping units was monitored by determining the zones and boundaries of the vibration state, measuring the vibration velocity, vibration acceleration, and constructing their spectral graphs. The operation of pumping units in non-nominal operating modes was also studied. The installation of elements of a vibration isolating compensating system was carried out and bearing capacity of pumping blocks foundations was increased. The study initially considered the existing construction of the used coupling joint shown in Figure 1.

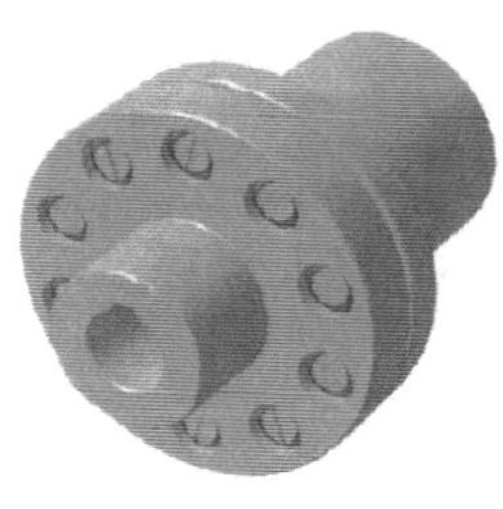

Figure 1. Standard factory-made couplings

The following disadvantages were identified:

1. Non-production time spending on the installation and dismantling of coupling joints [6,7];
2. The use of pullers during disassembly leads to the formation of scorings along the inner diameter of the half coupling, on the pump shaft and the electric motor rotor;
3. Coupling joints are supplied unbalanced which increases the vibration coefficient, and as a result, reduces the time to failure;
4. The service life is reduced, and the imbalance of the coupling joints increases due to coupling installation and dismantling operations;
5. Increased risk during hot works in the pumping unit at process installations;
6. Short service life from 3 to 5 years.

Modernized version of the flexible coupling. As part of the ongoing construction work, a modernized coupling joint was created, shown in figure 2, 3. Construction work was carried out using the professional application program Kompas 3D V18 [9]. Dowel pins have been added to the existing construction, which are installed in special milled holes. They allow performing routine repair operations, quick installation, dismantling and replacement of coupling joints.

Modernized version of coupling:

Figure 2. Modernized coupling assembly

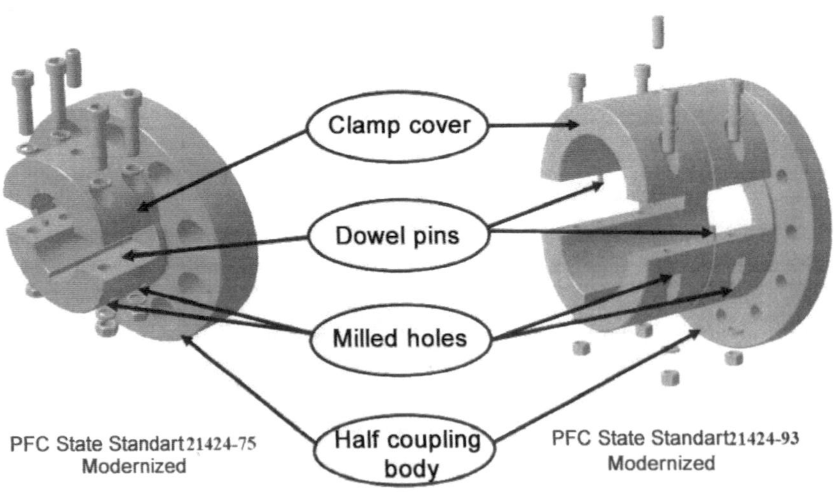

Figure 3. Modernized pin flexible coupling

Modernized coupling construction. Figure 3 shows a disassembled modernized half coupling of an electric motor PFC State Standard 21424-93. The considered construction of the coupling joint makes it possible to exclude incorrect alignment of the pump and the electric motor, and as a result, reliable operation of bearing assemblies will be ensured, and bending of shafts is excluded. Thus, we simplify the methods of installation and dismantling of this type of PFC (Pin Flexible Coupling State Standard-21424-93 for a centrifugal pump 200D90).

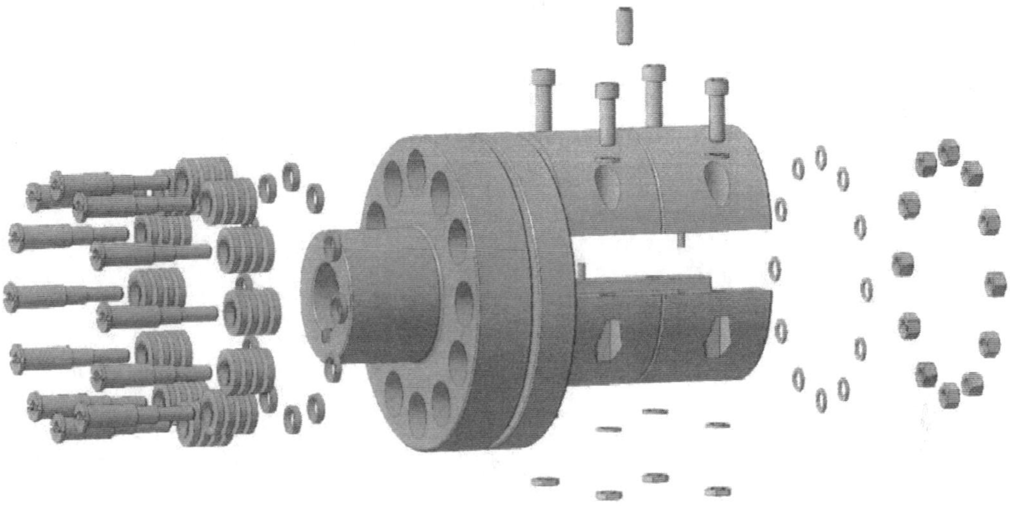

Figure 4. Modernized pin flexible coupling

Figure 4 shows the modernized pin flexible coupling. The advantages of the proposed model are: simplicity of construction, it is also possible to offset the shafts relative to the horizontal axis from 1 to 5 mm, the radial offset of the shafts is from 0.1 to 0.4 mm. This is achieved by using compensators made of rubber that make it possible to compensate for the above offsets. The most significant advantage is the possibility of installing the coupling without pressing operation, which greatly simplifies the installation and dismantling of the coupling joint. Construction disadvantage is balancing on false shafts during coupling installation.

4. Conclusions

Tests have shown that the use of a modernized pin flexible coupling increases its service life, operability operating safety of centrifugal pumping units. The results of the prototypes confirm the increase of vibration speed level to the level of "good", an increase in the bearing capacity of pile footings, foundations of pumping blocks.

The following advantages of the modernized construction should be noted:

1. Reducing the time spent on the installation and dismantling of coupling joints due to the use of a detachable construction with locating pins;
2. Reducing the dismantling time of the modernized coupling due to the use of the modernized construction;
3. Reduction of non-production time and cost for routine repairs and overhaul of pumping units due to reduced time for disassembly and assembly of the modernized pin flexible coupling;
4. Ease of installation and dismantling of the half coupling during maintenance and repair operations of centrifugal units due to the removal of the half coupling's upper part;
5. Reducing the vibration level of pumping units due to the balancing of the coupling joint, and as a result, increasing the interrepair time;
6. Reducing the risk due to the refusal of hot works in the pumping block at process installations;

The construction of a pin flexible coupling, proposed in the work, allows improving the technological maintenance of pumping units, simplifying the routine repair overhaul. As a result, the reliability of pumping units increases, which is very relevant for oil complex facilities pumping large volumes of liquid.

References

[1] Anuryev V I 2001 *Handbook of the constructor-machinist: in 3 vol.* (Moscow: Mechanical engineering) 2

[2] Vladimirov A I 2016 *Testing of oil and gas equipment and their metrological support. Manual.* (Moscow: Prospect)

[3] State Standart 21424-93. *Flexible couplings with rubber-bushed studs. Parameters and dimensions.*

[4] Lasukov R Y 2015 Analysis of the causes of premature failures in the operation of the electrical centrifugal pump in the YuS formations of the East Surgut fields and methods of dealing with them *Earth science* **11**

[5] Polyakov V S, Barbash I D and Riakhovskiy O A 1974 *Coupling reference* (Moscow: Mechanical engineering)

[6] Savelyeva N N Sokolova I U and Belyaev O V 2018 *Oil and gas field equipment guidance manual* (Tyumen: IUT)

[7] Safiullin R R, Matveev U G and Burtsev E A 2011 *Analysis of operation of electric centrifugal pumps units and technical methods of improving their reliability* (Ufa: USPTU publishing)

[8] Graham Kelly S 2012 *Mechanical vibrations.* Six editions (The university of Akron. Seala)

[9] Savelyeva N N 2018 Creation of an automation system for engineering calculation of preparation for the production at high-technology enterprises of mechanical engineering. In the collection *IOP Conference Series: Earth and Environmental Science* 12-29

IOP Conf. Series: Materials Science and Engineering **952** (2020) 012032 doi:10.1088/1757-899X/952/1/012032

Methods for illegal tapping detection using non-destructive testing tools

A R Salyakhova, K E Koretskiy, P E Koretskiy

Tyumen Industrial University, Lenin Street, 2P, 9, Nizhnevartovsk, 628616, Russia

E-mail: angelinasalihova@mail.ru; everlast86rus@bk.ru; loullli@bk.ru

Abstract. The article discusses the statistics of accidents occurred as a result of illegal tapping or sabotage at the sections of the main oil pipeline. The authors analyzed existing methods for determining illegal tapping, revealed the need to develop such an advanced method for detecting tapping, which would combine several predominantly promising properties for the oil and gas industry. One of the main conditions for the development prospects is the environmental and economic feasibility of the development introduction. Significant advantages of the impulse reflectometry method are established, which consists in oil pipeline sounding, presented in the form of a two-wire electric line and detecting the wave resistance deviation at heterogeneous sections of oil pipeline. Various defects can present such deviations in comparison with the already existing control methods, which mainly depend on human factor during patrolling along the main oil pipelines.

1. Introduction

Nowadays in oil and gas complex, there is an acute problem of the detection and prevention of illegal tapping and sabotage at sections of the main oil pipeline. This is primarily explained by the perfection of tapping equipment methods, as well as human factor when patrolling pipeline sections [3].

According to the statistics of accidents shown in Table 1 [2] we can conclude that the causes of most accidents and as a result environmental pollution are tapping and sabotage. Since individual parts of main oil pipelines sometimes reach several thousand kilometers in length, the number of illegal tapping can reach ten, or even hundreds. As a result, there are the huge losses from the budget of the operating pipeline organizations, associated with both the loss of raw materials and environmental consequences in the case of poor-quality tapping.

The important difference between illegal tapping and emergency leak is the absence of traces of raw material leakage in case of illegal tapping exactly at the tapping point next to the oil pipeline. However, in the case of emergency leak of raw materials, an oil spill or traces of oil leakage are present near the oil pipeline.

Table 1. Statistics of accidents on the main oil pipeline

Cause of accident	Number of accidents (in a year)						Share, %
	2013	2014	2015	2016	2017	In total	
Defect in construction and installation activities	0	2	2	0	0	4	2,68
Organizational reasons	0	0	3	0	2	5	3,35
Mechanical impact during groundworks	0	0	1	0	0	1	0,67
Corrosion	0	0	2	1	0	3	2,01
Illegal tapping	15	8	12	10	2	47	31,49
Production defect	2	2	0	3	0	7	4,69
Other issues	0	0	0	0	0	0	0
Total	17	12	20	14	4	67	100

According to the annual data of the Federal Service for Ecological, Technological and Atomic Inspection, it is possible to estimate the rate of accidents along the route of oil trunk pipelines. The average statistical rate of accidents in recent years is 0.27 accidents per 1000 kilometers of the pipeline, 58% of which occur due to illegal tapping and sabotage [1].

2. Problem statement

In order to solve this problem, it is necessary to provide appropriate preventive measures.

Methods: The existing today methods for detecting illegal tapping are either insufficiently effective or require huge capital investments for to be implemented in production.

In order to solve the problem of detecting illegal tapping, it is proposed to use impulsive reflectometry.

The principle of operation is similar to the determination of the coordinates of a cable break section, thus, a main pipeline with anticorrosive insulation is an equivalent two-wire electrical line with different values of wave resistance in such sections as a weld seam or illegal trapping.

The method of impulse reflectometry is based on the property of an infinitely long uniform line, from which it follows that the ratio between currents and voltages introduced into the electromagnetic wave line is equal at any point of the uniform line.

This relation has the following form (1)

$$W = \frac{I}{U} \tag{1}$$

And it is called the line resistance.

To implement this method of pulse reflectometry, it is necessary to send a probe pulse to the line in order to measure the interval tx - time during which the pulse will pass twice to the source of damage or interference (the place where the heterogeneity of the wave resistance is detected). In order to

IOP Conf. Series: Materials Science and Engineering **952** (2020) 012032 doi:10.1088/1757-899X/952/1/012032

understand at what the distance the heterogeneity of the wave resistance from the impulse source is, the following formula is used (2)

$$Lx = tx * \frac{V}{2} \tag{2}$$

where V - impulse velocity

The ratio of the amplitudes of the reflected (U_0) to the sounding (U_3) pulse is called the pulse reflection coefficient (K_{pr})

$$K_{pr} = \frac{U_0}{U3} = \frac{(W_1 - W)}{(W_1 + W)}$$

Where W - line resistance at line homogeneity zone,

W_1 - line resistance at damage zone or heterogeneity.

A distorted signal appears in the places of a line where the wave resistance deviates from its mean value: in places of raw material leakage, illegal tapping, oil leakage, etc. If the output resistance of reflectometer differs from the wave resistance on the line, then reflections appear in the area where the reflectometer is connected to the oil pipeline, which are reflections from the resistances of the reflectometer input of the reflection signals that occurred in the connection of reflectometer. The input and output resistances of reflectometer are equal.

In order to make the pulse values sufficiently reliable, it is necessary to match the line resistance with the output resistance of reflectometer. The reason for the change in polarity and amplitude of the reflectometer is the ratio of the characteristic resistance to the output resistance of the reflectometer (Figure 1).

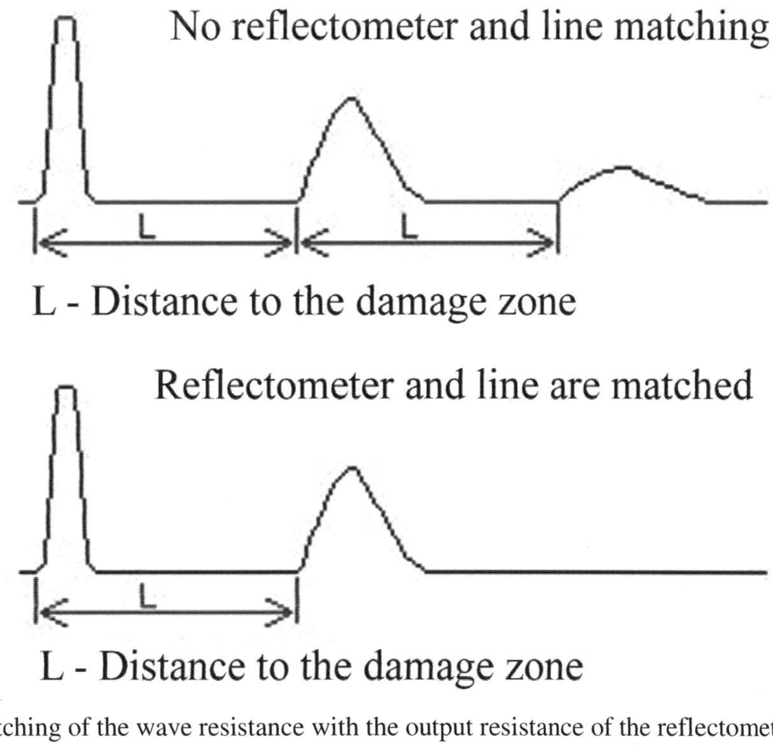

Figure 1. Matching of the wave resistance with the output resistance of the reflectometer

The pulse signal tends to subside when it passes the line. It means that the amplitude decreases.

Line damping is determined by the materials used in the production of the pipeline and the insulation protection. As a result, the pulse signal is frequency-dependent.

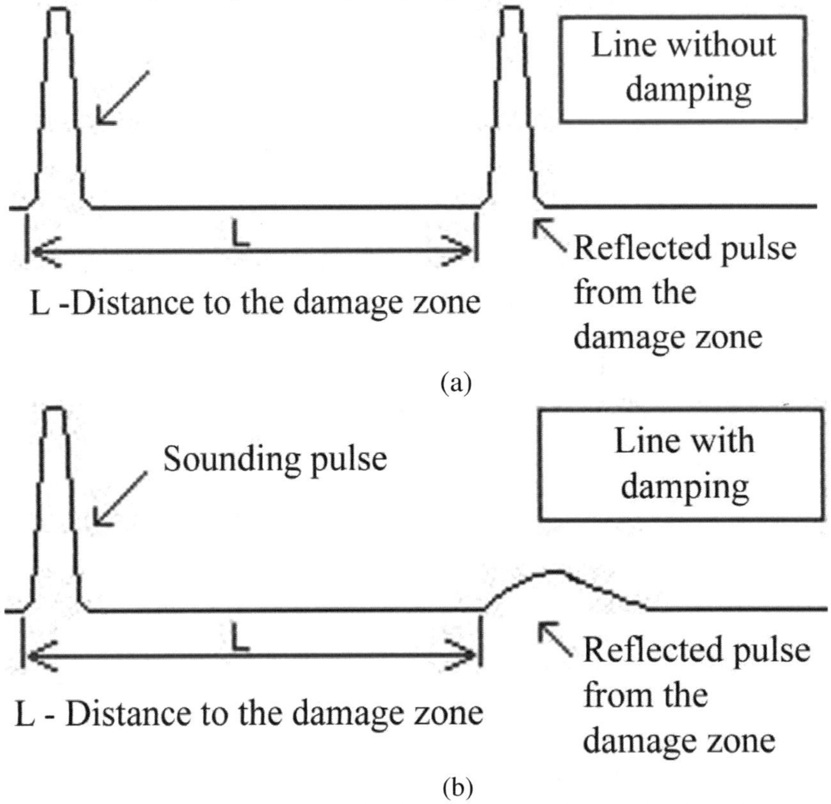

Figure 2. Reflectograms with (a) and without (b) impulse damping

One of the problems in the determination of the exact distance to the tapping point is the change in the sounding pulses, since not only the pulse resistance changes, but also its form, which characterizes the duration of the pulse front and its cut. This phenomenon is called "pulse spreading". The greater the distance to the defect, the more diffuse the signal and the smaller the pulse amplitude. The examples of traces with and without damping (ideal line) are shown (Figure 2.)

3. Conclusion

During the sounding of such an electrical circuit with impulse signals, electromagnetic energy is reflected from heterogeneous internal sections of oil pipelines. These pulses are fed back to the reflectometer, which is used to compare this value with a standard. If a mismatch is found, the coordinate is determined and the corresponding technical operations are performed in order to eliminate the tapping.

This method is convenient due to the simplicity of equipment installation. It can be installed at any end of the pipeline.

References

[1] Mitropolskiy Yu I 2016 *Multi-architectural computational supersystems. Development prospects.* Moscow: Technosphere

[2] Nekrasova A P 2000 On the statistics of accidents and unauthorized tie-ins on main oil product pipelines and measures to reduce their number. *Transport and storage of oil products* **8-9** 9-11

[3] Shestakov R A 2015 On the question of methods of detecting leaks and unauthorized tie-ins on main oil pipelines. *Proceedings of the Gubkin Russian State University of Oil and Gas* **1** 85-94

[4] Kharenko D N 2004 *A system for detecting unauthorized tie-ins into the main oil pipeline* (Kazan District Oil Pipeline Administration)
[5] Nekrasova A P 2007 Methods of detecting leaks and unauthorized tie-ins on main oil pipelines *Transportation and storage of oil products* **8-9** 15-18

IOP Conf. Series: Materials Science and Engineering **952** (2020) 012033 doi:10.1088/1757-899X/952/1/012033

Calculation of the torque on the crank shaft of the low-speed balancing drive sucker rod pumps

Galeev, A.S.[1], Arslanov, R.I.[2], Suleymanov, R.N.[1], Filimonov, O.V.[1]

[1]Ufa State Petroleum Technological University, Branch of the University in the City of Oktyabrsky, 54a, Devonskaya St., Oktyabrsky, Republic of Bashkortostan, 452607, Russian Federation
[2]SBEI of HE "Almetyevsk State Oil Institute", 2, Lenin St., Almetyevsk, Republic of Tatarstan, 423450, Russia

E-mail: ssgaleev@mail.ru

Annotation. The operation of well sucker rod pump installations is the most massive way of mechanized oil production. A significant part of the wells equipped with low-speed plants of this type have a low flow rate and high water cut. Such conditions create certain restrictions on updating and modernization of the production fund of equipment in operation. The life cycle of a rocking machine is usually significant and often exceeds 20 years. During ISPR operation, both ground and downhole equipment will inevitably wear out. A significant proportion of failures of the ground part of low-speed ISPR is associated with the reduction of the rocking machine, which is the most responsible and most expensive drive element. The resource of the gearbox directly depends on its load, which is characterized by the moment on its crank (output) shaft. Reducing the maximum torque and moment fluctuations in one full swing of the rocking machine will significantly increase the reliability of the gearbox and the entire installation. This paper describes the mathematical apparatus for calculating the moment on the crank shaft of the gearbox, based on the exact kinematics of the drive and the principle of possible movements. The proposed methodology will allow to solve the problems of designing a balancing drive of a low-speed ISPR, to determine the requirements for the engine and transmission.

1. Introduction

A significant part of the world oil well fund is operated mechanically using installations with sucker-rod pump (ISPR). Moreover, in recent years, the proportion of wells producing highly viscous products has been steadily growing. Taking into account equipping a large part of such wells with low-speed ISPR, improving their efficiency becomes one of the priority tasks [1-4]. When producing high-viscosity oil, complications arise that adversely affect the operation of downhole equipment and entail increased wear and premature failure of ISPR equipment [5]. ISPR can be divided into three parts: a downhole pump, sucker rods and ground drive. The most conservative part of the entire installation, which has been in operation for the longest time, is the conversion mechanism of the ground drive, in most cases presented in the form of a beam-pumping unit. The actual service life, based on international experience, often exceeds 20 years [6,7].

2. Methods and materials

This work was carried out in the framework of the Federal Target Program "Research and Development in Priority Directions for the Development of the Russian Science and Technology Complex for 2014 - 2020" under the Agreement on the provision of subsidy No. 14.610.21.0019 of 10.23.17 on the topic "Creation of a set of technological solutions for increasing oil recovery containing highly viscous oil ", a unique identifier for the work RFMEFI61017X0019.

3. Results and Discussion

The balancing drive of a borehole sucker rod pump installation is a mechanism that converts the rotational movement of an electric motor into reciprocating motion of a string of pump rods [8].

The mechanism of the rocking machine is a four-link and has one degree of freedom. The kinematic diagram of the rocking machine is shown in Figure 1. The angle of rotation of the crank O_1O_2 of length l_1 is taken as the generalized coordinate. The axis of the crank O_1 (x_1, y_1) is stationary. The movement of the hinge O_2 (x_2, y_2) occurs in a circle:

$$x_2 = x_1 + l_1 \cos\phi \tag{1}$$

$$y_2 = y_1 + l_1 \sin\phi \tag{2}$$

Rocker (balancer) rotates around a fixed center O_4 (x_4, y_4).

The hinge O_3 (x_3, y_3) connecting the balancer to the connecting rod also moves along an arc of a circle of radius l_3 and its motion is described by the equations:

$$x_3 = x_4 + l_3 \cos\psi \tag{3}$$

$$y_3 = y_4 + l_3 \sin\psi \tag{4}$$

whereψ is the angle between the balancer and the X axis. The distance between the hinges is equal to the length of the O_2O_3 connecting rod (l_2):

$$(x_3 - x_2)^2 + (y_3 - y_2)^2 = l_2^2 \tag{5}$$

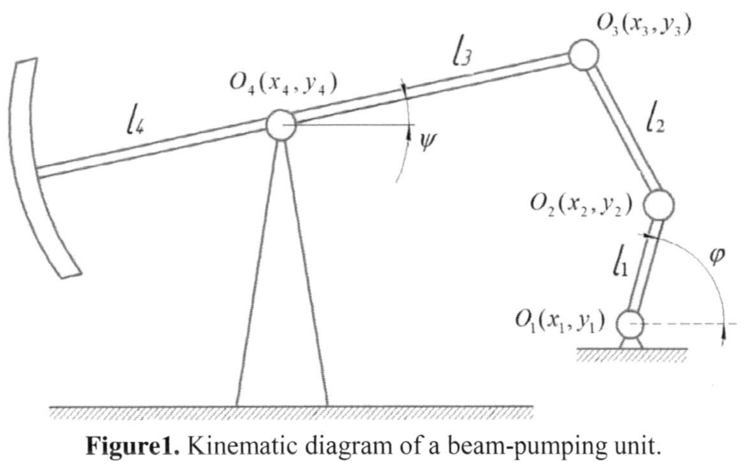

Figure1. Kinematic diagram of a beam-pumping unit.

and auxiliary angle θ by the formula:

$$\theta = \arcsin \frac{a}{\sqrt{a^2 + b^2}} \tag{6}$$

we get:

$$\psi = (-1)^n \arcsin\left(\frac{l_2^2 - l_3^2 - a^2 - b^2}{2l_3\sqrt{a^2+b^2}}\right) - \theta + \pi n,\ n \in \mathrm{Z} \quad (7)$$

The movement of the ISPR drive is determined by the following external forces: gravity of the rod string, gravity of the liquid column, drag force of the column against the borehole wall (boundary friction), viscous friction force of the rod string against the fluid, friction forces in the gearbox and a couple of forces from the electric motor [11, 12].

In slow-moving rocking machines, the dynamic loads are negligible and, accordingly, the moment on the crank shaft of the gearbox is determined only by the load on the traverse and the balancing counterweight.

To calculate the moment on the crank shaft, the principle of possible movements is used, according to which the sum of the work of active forces and at any possible movement of the system is zero. The design scheme of the ISPR drive is shown in Figure 2. The active forces of the ISPR drive are:

1) Torque M_{cr} applied to the crank (transmitted from the engine through the gearbox);
2) The gravity of the crank G_{cr};
3) Gravity forces of counterweights G_1 and G_2;
4) The gravity of the rocker G_3;
5) Column weight G_{quan};
6) The weight of the liquid column (only when moving up) G_{liq};
7) Column resistance force F_{re} (directed all the time against the movement of the column).

Consider an arbitrary position of the system (the crank is rotated through an arbitrary angle φ). We give the crank an infinitesimal displacement $\delta\varphi$. Then the balancer will rotate a certain angle $\delta\psi$ and the column will move a distance δS.

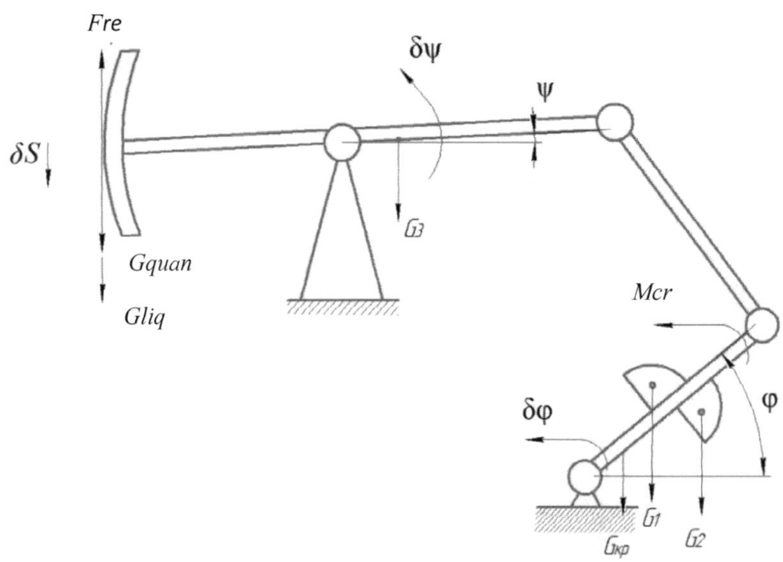

Figure 2. The design scheme of the ISPR drive.

The work of active forces on this movement will be:

$$\begin{aligned}\delta \mathbf{A} = \mathbf{M}_{cr}\delta\varphi - G_{cr}\cdot\delta\varphi\cdot r_3 cos\varphi - G_1(r_1 cos\varphi - h_1 sin\varphi) - G_2(r_2 cos\varphi + \\ + h_2 sin\varphi) - G_3\cdot r_4 cos\psi\delta\psi + (G_{quan} + G_{liq} - F_{re})\delta S,\end{aligned} \tag{8}$$

where r_3 is the distance from the axis of the crank to its center of gravity;
r_1,h_1 and r_2, h_2 – radial and tangential displacements of the counterweights relative to the crank;
r_4 is the distance from the axis of the balancer to its center of gravity (taking into account the traverse and connecting rods);

In accordance with the principle of possible movements, this work should be zero. Expressing the increment of the angle δψ and the small displacement of the column δS through the increment of the angle δφ [13-18], we obtain an equation from which we can find the moment on the crank shaft, depending on the design of the ISPR, and the position of the crank:

$$\begin{aligned}\mathrm{M}_{cr} = G_{cr} r_3 cos\varphi + G_1(r_1 cos\varphi - h_1 sin\varphi) + \\ + G_2(r_2 cos\varphi + h_2 sin\varphi) + G_3\cdot r_4 cos\psi\frac{\partial\psi}{\partial\varphi} - \left(G_{quan} + G_{liq} - F_{re}\right)\frac{\partial\psi}{\partial\varphi} l_4\end{aligned} \tag{9}$$

$$\frac{\partial\psi}{\partial\varphi} = \psi' = \frac{l_1 l_3 \sin(\varphi-\psi) + l_1(x_4 - x_1)\sin\varphi - l_1(y_4 - y_1)\cos\varphi}{l_3[(x_4 - x_1 - l_1\cos\varphi)\sin\psi - (y_4 - y_1 - l_1\sin\varphi)\cos\psi]} \tag{10}$$

4. Conclusion

The calculation is implemented in the form of a software product, in which, after entering the main parameters of the ISPR drive, we calculate the work performed by the acting forces, the degree of balance, and the change in the effective torque on the crank. The input window for the main parameters of the ISPR drive is shown in Figure 3.

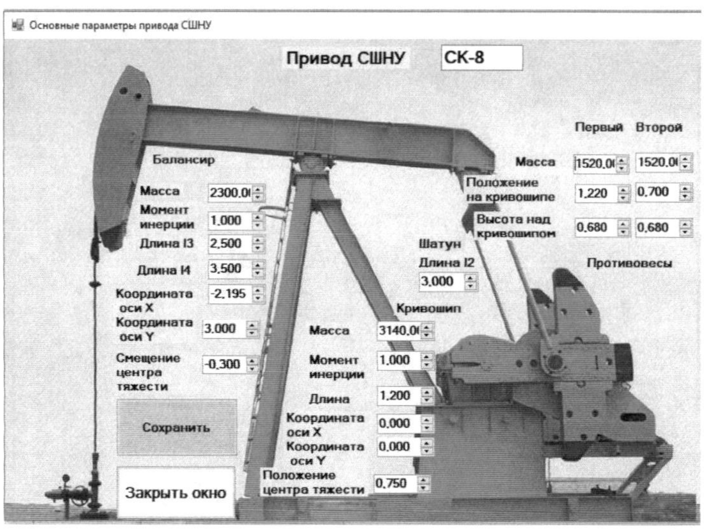

Figure 3. The input graphical user interface for the basic parameters of the ISPR drive.

Translation (from left to right, up to bottom): ISPR drive, balancer, mass, inertia moment, length 13, length 14, X-axis coordinate, Y-axis coordinate, centre of gravity shift, save, close the window, connecting rod, crank, mass, inertia moment, length, -axis coordinate, Y-axis coordinate, center of gravity position, first, second, mass, on crank position, height above crank, counterweights.

The program window with an example of the calculation is shown in Figure 4.

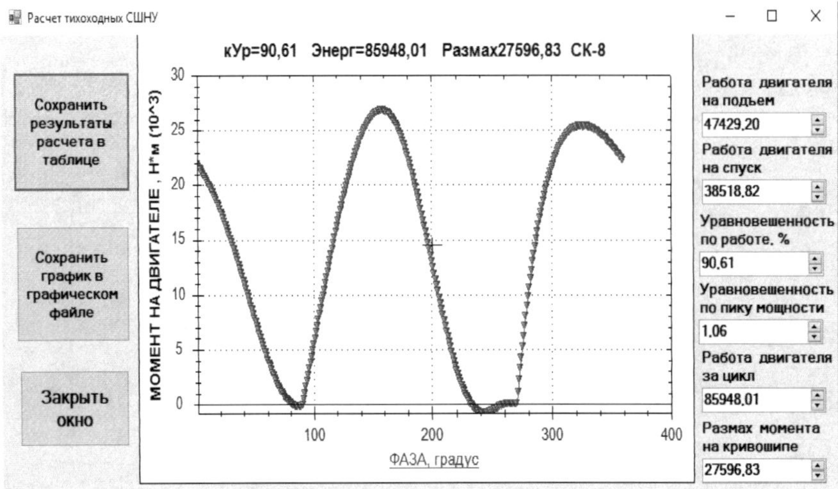

Figure 4. The output graphical user interface of the developed software.

Translation (from left to right, up to bottom): Save the results of the calculation in the table, save the graph in the graph file, close the window, engine torque, engine run up, engine run down, operating balance, power peak balance, engine run per cycle, torque swing on crank.

The resulting equation for the dependence of the moment on the crank shaft of the gearbox on the mass of the drive elements, geometric parameters and the load on the cable suspension will allow us to solve direct and inverse problems of designing the balancing drive of the low-speed ISPR, to determine the requirements for the gearbox of the rocking machine, belt drive and engine.

References

[1] Ogarkov Y M, Tsylev P N, Korotayev A D, Burmakin A M 2005 Improving the efficiency of oil production from low-rate wells. *Bulletin of PNRPU. Geology. Oil and gas and mining* **6** pp.172-178. (in Russ.).

[2] Beliaev E F, Tashkinov A A, Tsyliov P N 2012 Improvements of the beam pumping unit electrical drive of the dripper. *Bulletin of PNRPU. Geology. Oil and gas and mining* **4**. pp. 91-103. (in Russ.).

[3] Bikbulatova G I, Galeev A S, Boltneva Y A, Larin P A, Suleymanov R N, Filimonov O V 2019 Optimization of pumping fixed volume of liquid on two directions *Bulletin of the Tomsk Polytechnic University* **330**(1) pp. 134-144.

[4] Khabibullin M Y 2019 Development of the design of the sucker-rod pump for sandy wells [IOP Conference Series: Materials Science and Engineering **560**(1) 012065. DOI: 10.1088/1757-899X/560/1/012065.

[5] Urazakov K R, Nurgaliev R Z, Latypov B M, Bikbulatova G I and Boltneva Y A 2018 Selection of the design of the weighted bottom of the pump rod column for high-viscosity oil production *Oil industry* **10** pp. 111-113.

[6] Bikbulatova G I, Isaev A A, Boltneva Y A 2018 Investigation of the influence of complicating factors on the efficiency of screw rod pumps. *Proceedings of Tomsk Polytechnic University. Geo-resource engineering* **329**(4) pp.21-29.

[7] Urazakov K R, Nurgaliev R Z, Belov A E, Bikbulatova G I, Davletshin F F 2019 Investigation of

complications in the operation of downhole rod pumps during simultaneous and separate operation *Oil industry* **7** pp. 114-117.

[8] Gabzalilova A K, Soloviev N N, Garifullina Z A Effect of changes in the crank radius on the kinematics of the pumping unit when regulating the dynamic level of the fluid in the *well IOP Conference Series: Earth and Environmental Science* **378**(1) 012112. DOI: 10.1088/1755-1315/378/1/012112.

[9] Galeev A S, Filimonov O V, Suleymanov R N 2019 Development of a laser beam to the line in measuring systems *IOP Conf. Series: Materials Science and Engineering* **560** 012125.

[10] Rahman T, Ramanathan R, Seliktar R, Harwin W 1995 A simple technique to passively gravity balance articulated mechanisms. *Mech. J. Des.* 117 (4) pp. 655-658.

[11] Johnson W, Laithwaite E, Slater R 1964 An experimental impact – extrusion machine driver by a linear induction motor *Proc. Instn. Mech. Engrs.* **179** pp. 15–35.

[12] Williams F, Laithwaite E, Piggott L 1957 Bruchless variable – speed Induction. Motors *Proceedings of IEE* **14**(A) pp.102–118.

[13] Jean M, Gosselin C M 1996 Static balancing of planar parallel manipulators. *Proc. IEEE Int. Conf. on Robotics and Automation 1996* pp. 3732-3737. DOI: 10.1109/ROBOT.1996.509149.

[14] Van Dam P L, Herder J L 2011 Static balancing of translational parallel mechanisms. *Proceedings of the ASME 2011 International Design Engineering Technical Conferences & Computers and Information in Engineering Conference IDETC/CIE* DETC 2011-47525.

[15] Baradat C, Arakelian V, Briot S, Guégan S 2008 Design and prototyping of a new balancing mechanism for spatial parallel manipulators. *Mech. J. Des.* **130**(7) pp. 072305-1-072305-13.

[16] Russo A, Sinatra R, Xi F 2005 Static balancing of parallel robots. *Mech. Mach. Theory* **40**(2) pp. 191-202.

[17] Uicker J J, Denavit J J, Hartenberg R S 1964 An iterative method for the displacement analysis of spatial mechanisms *J. Appl. Mech. Trans. ASME Ser. E* .**86**(2) pp.309-314.

[18] Walsh G J, Streit D A, Gilmore B J 1991 Spatial spring equilibrator theory. *Mech. Mach. Theory* **26**(2) pp.155-170.

IOP Conf. Series: Materials Science and Engineering **952** (2020) 012034 doi:10.1088/1757-899X/952/1/012034

Development of a hydraulic classifier used to obtain various sand fractions

A E Verisokin[1], A P Yanukyan[2] and P M Sorokin[2]

[1] North Caucasian Federal University, Institute of Oil and Gas, 1, Pushkin Str., Stavropol, 355009, Russia
[2] Tyumen industrial University, 49/3, Respubliki Str., Tyumen, 625000, Russia

E-mail: verisokin.aleksandr@mail.ru

Abstract. The principle of crack filler selection depending on the crack depth and closing stress is presented. The study of existing classifiers and proposed a more effective device for sand sieving. The main elements of the classifier are presented and the principle of its operation is described.

1. Introduction

Due to the increase in the proportion of oil reserves in low-permeability reservoirs with adverse physical and geological conditions its extraction requires the use of new technologies, materials and techniques [1]. One effective way of increasing the production of oil from low permeable collectors is hydraulic fracturing [2]. To carry out efficient fracturing under low-permeability reservoirs, permeability less than 0.01 μm is necessary to create the greatest possible extent of cracks to expand the field of drainage wells by increasing the mass of injected proppant material.

2. Materials and methods

As a granular proppant, sand or ceramic proppant is often used. The proppant has high strength, sphericity and roundness, conductivity and permeability. In turn, sand is much cheaper. The analysis showed that a well calibrated sand may be used during hydraulic fracturing at a depth of 3000 m, where the compression stress does not exceed 40 MPa. If sand is ground, it can withstand closure pressures of up to 55 MPa. Such a natural crack filler is called "super sand". Based on numerous studies, the principle of choosing a filler of a crack was developed, which is presented in Figure 1.

The main advantage of sand is its low cost. Russian oil companies have used ceramic proppant even at shallow depths where white sand can be used. This leads to a rise in the cost of the process of hydraulic fracturing. In our country, there are grades of sand of various roundedness and strength. To study the conductivity of sand of different fractional composition allows the installation presented in [3]. To conduct research on a test bench, it is necessary to break the sand into fractions of 0.2 - 0.4 mm, 0.4 - 0.8 mm, 0.7 - 1.2 mm, 0.8 - 1.6 mm. To develop an effective classifier, existing devices were investigated and their shortcomings revealed.

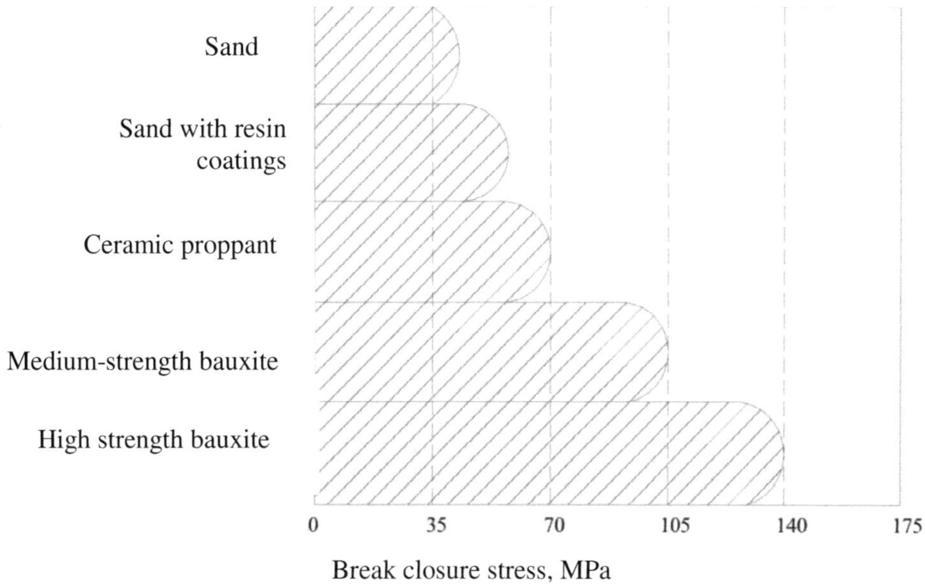

Figure 1. The principle of selecting the type of proppants

3. Research results

The characteristics of a stationary hydraulic classifier have been studied, which has the following disadvantages: high metal consumption, complexity in manufacturing and installation [4].

A centrifugal hydraulic classifier was studied, consisting of a cylindrical body turning into a conical one with inlet and drain pipes, which does not allow obtaining sand of the exact fraction [5]:

-it is possible to obtain only a large fraction with the discharge of the remaining part into the drain, which reduces the efficiency of the process;

- it is necessary to measure the flow rate to the feeding rate of the condensed liquid through the bottom and the drain pipes to maintain the optimum water level inside the housing and ensure the continuity of the process.

Known hydraulic classifier cone, consisting of a housing tube for supplying the initial mass, a nozzle for discharging fine product and a pipe for discharging condensed product [5]. The disadvantages of the design include the following:

- classification of granular material occurs only for one fraction, which is impractical because in some cases smaller fractions are also used, for example, in the oil and gas industry for installing gravel filters or for hydraulic fracturing.

The analysis of the hydraulic classifier of fine-grained materials, reveals the following disadvantage [5]:

-the presence of two independent streams supplied to different chambers causes uncertainty in the operation of the device.

A more advanced device for the hydraulic classification of fine-grained materials was developed, which is illustrated in the figure. 2 [5].

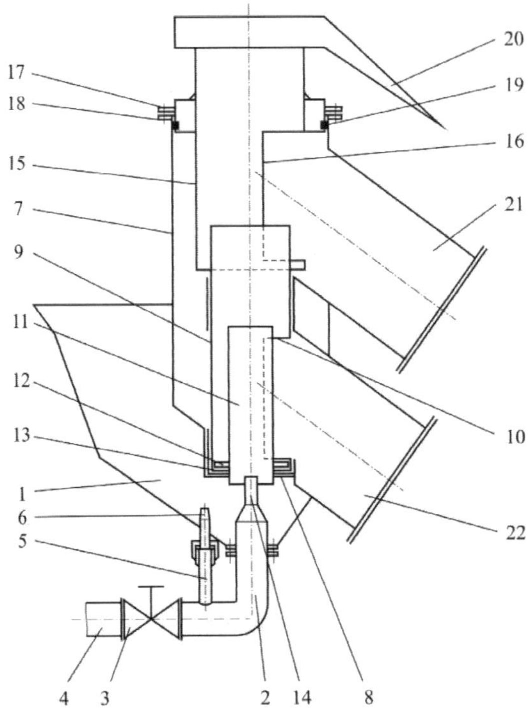

Figure 2. Device for hydraulic classification of fine-grained material

The device consists of a loading hopper 1, on the bottom of which a water supply pipe 2 is installed, connected through a valve 3 with a supply pipe 4. A drain 5 is additionally connected to the inlet pipe 2 with a nozzle 6 extending into the cavity of the loading hopper 1, which is connected to the housing 7. At the lower end constriction is provided, at the bottom 8. In the constriction, there is a classification chamber 9 with a window 10, inside which a receiving and separating chamber 11 with a flange 12 is mounted, extending at the bottom 13 in the classification chamber 9 and exiting into the cavity of the feed hopper 1. The inlet pipe 2 of the water supply is equipped with a jet nozzle 14, which extends into the axial channel of the receiving and separation chamber 11. Axially symmetrically to the casing 7, an enrichment chamber 15 is installed, covering with a gap the upper end of the classification chamber 9 and provided with a cut-out 16. A flange 17 is mounted on the outside of the enrichment chamber 15 for connection with the counter flange 18 of the casing 7. The annular gap between the housing 7 and the enrichment chamber 15 is closed by the seal 19. The enrichment chamber 15 is equipped with a discharge pipe 20. The housing 7 is equipped with a soil collector 21 for the middle fraction, the cavity of which through the cut-out 16 in the body of the enrichment chamber 15 is connected with its axial channel. Below is a soil collector 22 for a large fraction, the axial channel of which through the window 10 in the classification chamber 9 is constantly connected with its axial channel. The design of the device provides for the possibility of the sequential dismantling of enrichment 15, reception and separation 11 and classification 9 chambers. The inner diameter of the axial channel of the receiving and separating chamber 11 is consistent with the flow rate of the supplied water and the known diametrical sizes of the grains of sand. The inner diameter of the axial channel of the classification chamber 9 is designed with a prerequisite for lifting average particle sand fraction which is discharged through the cut-out 16 into the soil collector 21 The diameter of the axial channel of the enrichment chamber 15 is selected from the condition of maintaining the speed of the upward flow, sufficient to transport small fractions to the outlet pipe 20.

Apparatus for fractional sieving and hydraulic classification of fine-grained materials is as follows. Open the valve 3 to supply water to the cavity of the device. Sand is fed into the loading hopper 1 with a different fraction and a stream of water flowing from the bubbler nozzle 6 into the cavity of the loading hopper 1, a pulp is created which enters the axial channel of the receiving and separation chamber 11 through the annular gap between its walls and the body of the jet nozzle 14. The pulp through this annular gap enters the axial channel of the receiving and separation chamber 11 and is transported upward at a certain upward flow rate. The upstream velocity should prevent large fractions from settling down. The pulp stream leaves the axial channel of the receiving and separation chamber 11 into the axial channel of the classification chamber 9, having a larger diameter. Particles of coarse sand come out of the pulp stream and through the window 10 enter the soil collector 22. Then the pulp stream, freed from the coarse fraction, moves into the axial channel of the enrichment chamber 15 with a decrease in the flow rate. Particles of sand of medium fraction through the cut-out 16 in the body of the enrichment chamber 15 fall into the soil collector 21, where they accumulate. The depleted pulp stream, having a fine fraction, enters the outlet pipe 20 through the axial channel of the enrichment chamber 15. The upward flow rate must be sufficient to move all the small particles to be discharged to the fine fraction dump. The sand of different fractional composition is supplied to the loading hopper 1 periodically in separate portions as it is fed for sorting into the receiving and separation chamber 11. When filling the soil collectors 21 and 22 with medium and large fractions of sand, the sorting process is stopped by closing the valve 3, the water is drained from the device through crane in the housing 7 (not shown in the figure). The soil collector 21 is opened sequentially and the middle fraction is removed, then the soil collector 22 is opened and a large fraction of sand is taken. Close the lids on the soil collectors 21 and 22 and repeat the sorting process by analogy with the previously described [9].

The overall dimensions of soil collectors 21 and 22 make it possible to obtain medium and large fractions in a volume of about 0.5 - 0.7 tons of each of them in one cycle.

4. Conclusion

Using the proposed classifier will provide advantages over existing analogues:

-the ability to provide a technological process with sieving sand into several fractions;

-the ability to obtain fractions of various sizes depending on the fractional composition of the initial product by increasing or decreasing the flow of water;

-the possibility of replacing worn-out cells with new ones;

-the possibility of switching to sieving sand of various fractional composition by changing the diameters of the chambers and changing the flow rate of the supplied water.

References

[1] Zakirov A, Fadeev A, Gabidullin Sh, Cherepanov D, Petrov A, Zotov A, Arzhevitin D and Garifullin R 2019 New fluid for hydraulic fracturing in complicated geological conditions *Oil industry* **1** 58 - 60

[2] Shipilov A, Babkina N, Menshikov I 2018 Study of the properties of a process fluid for hydraulic fracturing based on a viscoelastic surfactant *Oil industry* **3** 30 - 33

[3] Verisokin A 2018 The proppant test procedure for hydraulic fracturing *Science and Technology in the gas industry* **2** 62-69

[4] Kupavikh K S, Nutskova M V 2016 Ecological features of oil well repair at low - permeability reservoir. *International Journal of Applied Engineering Research*

[5] Verisokin A, Petrenko V, Varyagov S and Fedorova N 2019 Development of a hydraulic classifier used to obtain various sand fractions *Science and Technology in the gas industry* **4** 66-70

[6] Verisokin A, Vasil'yev V and Gun'kina T 2019 Packer design research used in hydraulic fracturing *IOP Conf. Series: Earth and Environmental Science* **378** 7

IOP Conf. Series: Materials Science and Engineering **952** (2020) 012035 doi:10.1088/1757-899X/952/1/012035

Experience in developing oil and gas deposits with horizontal wells located near the gas processing plant

E V Vaganov[1], V F Tomskaya[2], I I Krasnov[3] and M J Z Alsheikhly[4]

[1] NOVATEK-Pur JSC, Urengoy settlement, 5 microdistrict, 22, Novy Urengoy, 629860, Russia
[2] Industrial University of Tyumen", 36, Volodarskogo Str., Tyumen, 625000, Russia
[3] FSBEI HE North-Eastern Federal University M.K. Ammosov, Polytechnic Institute (branch), 14 Oyunskogo Str., Mirny, 678175, Republic of Sakha (Yakutia)
[4] University of Kirkuk, Al-Sayada St., Kirkuk, 36001, Iraq.

E-mail: vanessatomskaya@gmail.com

Abstract. In this work, four wells were analyzed at the Srednebotuobinskoye oil, gas and condensate field located in the Republic of Sakha (Yakutia). Production of oil reserves is complicated by the problem of gas breakthrough from the gas cap into the oil area of the reservoir, which leads to gasification of production wells up to their full shutdown. The field is characterized by the presence of a thin oil rim. That is why there is an urgent need to develop innovative technologies capable of ensuring higher oil recovery rates. The result of the conducted analysis is the object of implementation of the research results of this work.

1. Introduction

The object of research in the work is oil and gas deposits of Botuobinskoye horizon, located in Srednebotuobinskoye field. The peculiarity of oil and gas condensate field development is the direct dependence of hydrodynamic connection between oil and gas condensate objects. In accordance with the adopted project decision of the current field development project, oil reserves development of the Bt facility implies operation with horizontal wells with the length of more than 1250 meters. The geometry of the grid reflects the condition of in-line placement of wells: distance between rows of 500 meters, between wells in a row of 1000 meters. The rows of wells are shifted relative to each other, the gentle section of which is confined to the oil-saturated intervals of the deposit.

Well operation indicators, modes and conditions of their operation are essentially determined by the tasks of field studies, which are formulated as follows: justification of the optimal profile of horizontal well site location, taking into account the specific character of oil and gas saturation of the horizon. In the course of field development, the resource and energy potential of the formation is the uniform commissioning of objects in development.

2. Results and Discussion

The Srednebotuobinskoye oil and gas condensate field is located in Western Yakutia. In close proximity to the Srednebotuobinskoye field, the Taas-Yuryakhskoye, Bes-Yuryakhskoye, Irelyakhskoye, Mirninskoye, Severo-Nilbinskoye, Khoto-Murbaiskoye oil and gas condensate fields, which are at various stages of development (exploration, construction or development), have been explored and are listed on the state balance sheet. According to the adopted scheme of oil and gas geologic zoning of the territory of the Siberian platform, the Srednebotuobinskoye field is timed to the Nepsko-Botuobinskoye oil and gas bearing region of the Leno-Tunguskaya oil and gas bearing province. A number of oil and gas fields have been discovered within the oil and gas bearing area, including the Chayandinskoye field, unique in terms of gas reserves with condensate. The Verkhnechonskoye, Talakanskoye, Chayandinskoye, Yaraktinskoye, Severo-Talakanskoye and Verkhnevilyuchanskoye fields are the largest in terms of oil reserves [1].

The Srednebotuobinskoye field is large in terms of both oil and gas reserves. The field is characterized by an intensive manifestation of block tectonics, which led to the formation of four large tectonic blocks: Central, Western, Eastern and Northern, complicated, in turn, by smaller order faults. All identified faults are considered hydrodynamic barriers. The proven oil and gas bearing capacity of the field is primarily associated with sediments of the Botuobin horizon, which contains the majority of oil and gas reserves. Gas deposits with condensate have been discovered in sediments of the Ulahan and Talakh horizons of the Kursk Vendian and Osin horizons of the Cambrian Bilir Formation. The prospects of oil and gas bearing are also connected with deposits of the Upper-Byuk Subformation, the Uspun Formation's Transfiguration Horizon, and the Jurassic-I-III Yuriakh Formation's horizons. The proven productivity of the Central Tectonic Block of the Srednebotuobinskoye field is associated with sandstones of the Botuobinskoye horizon and with the Lower Cambrian carbonate sediments of the Bilir Formation - the Osinskoye horizon, the O-I and O-II formations. Oil and gas bearing capacity within the Northern Block was found in four horizons - Botuobin, Ulakhana I, Ulakhana II and Talakh. Within the Eastern Block III only the Botuobin production horizon was discovered, the Ulakhani horizon has a local distribution within the field and there is no oil content within the Eastern blocks [1]. The Talakh horizon has not been sufficiently explored within the northern block and its prospects have not been assessed within Eastern block III. The Botuobinskiy horizon is the main one at Srednebotuobinskiy field. Maximum oil flow rate was received in square No. Sbt-53 - 192 m^3/day, maximum gas flow rate in square No. Sbt-86 - 1616 thousand m^3/day, maximum water flow rate in square No. Sbt-7 - 144 m^3/day.

When deciding on the field development strategy, one should take into account the fact that the practice of oil rim development is based on the organization of barrier flooding, regulation of oil and gas withdrawal rates and water injection in order to minimize losses associated with the introduction of oil into the gas part of deposits. The modern practice of oil rim development, which is covered with gas cap and underfloor water, presupposes wide application of horizontal wells. Important is the economic result, which predetermines the order of hydrocarbon reserves development, the order and rate of formation drilling, their status: main object or returnable object.

Thus, from the point of view of geological and technological conditions, within the limits of the Central Tectonic Block of Srednebotuobinskoye field the objects of development are Bt and O-I+II layers.

Extracting reserves from gas-oil and gas-condensate deposits is generally considered more difficult than developing oil fields. Specific features of these deposits (complex geological structure, presence of oil, gas, gas condensate and water in the reservoir) to a certain extent complicate the complex of problems associated with their development. It is especially typical for large gas and oilfields of Western Siberia (Samotlor, Fedorovskoye, Lyantorskoye, Varioganskoye, etc.), including the area of Surgut arch, where 48 % of current oil reserves of the fields under development fall on the share of oil and gas deposits. Analyses show that due to the high production of more geologically simple oil deposits, the prospects for oil recovery in this area will be primarily related to the development of complex gas and oil deposits. The main oil and gas reserves of the Surgut region's gas and oil deposits are concentrated at two sites - the Lyantorskoye and Fedorovskoye fields, which are characterized by a complex geological structure, small oil layer thicknesses and a large share of pore volume of oil-saturated layer. Extraction of these reserves with the highest oil recovery and lowest costs is one of the main problems in the development of fields of this type.

The tasks of rational development of gas-condensate (GKZ), gas-condensate oil (GKNZ) and oil and gas (NGZ) deposits with bottom water are given, the generalization of world experience in the development of these deposits is given, the design theory is developed taking into account the geological and hydrogasodynamic features of NGZ and GKNZ and their classification, current problems of design and rational development of these objects are considered. The main difficulties in the development of oil and gas deposits are associated with technological difficulties of oil recovery, depending on the mode of their development. In this case, mainly show the modes of dissolved gas and the elastic-water pressure; the first has a dominant value and determines the final coefficient of oil recovery, in most cases insignificant. Due to the increasing world demand for oil and the depletion of large oil fields, there is increased interest in the development of oil and gas deposits and oil rims. As it is known, a rational method of oil rim recovery is considered to be the advanced oil rim development with saving of gas cap energy. However, as the world practice shows, sometimes the method of simultaneous extraction of oil and gas reserves from oil and gas deposits with preservation of gas and oil contact immovability is useful [2].

The oil and gas deposit of the Botuobinsky horizon is characterized by a complex geological structure, abnormally low formation pressure and temperature. Low reservoir pressure (50 kg/cm^2 less than hydrostatic pressure) [3] indicates that the reservoir has limited reservoir energy reserves and its long-term operation in the depletion mode is not expedient. Based on the results of field works performed during the development of Srednebotuobinskoye field we can conclude that geological conditions of hydrocarbons occurrence have an adverse effect on the efficiency of oil recovery process. Oil production under such conditions is inevitably accompanied by the extraction of significant volumes of associated formation water and top gas breakthrough [4].

Efficiency of gas-insulation measures in development of oil and gas deposits of Botuobinskoye horizon, as well as for other fields with hard-to-recover oil reserves, is an important problem for the near future. The regulation and control of oil reserves development processes and limitation of inefficient utilization of associated gas for flaring are of great importance. The assessment of the prospects of applying gas-insulation methods is connected with the creation of unconventional technologies, the physical nature of which is not only high technological efficiency, but also resource saving with a significant expansion of geological criteria of their applicability.

Analysis of the nature of gasification of oil producing wells and study of the causes shows that the main ones are:

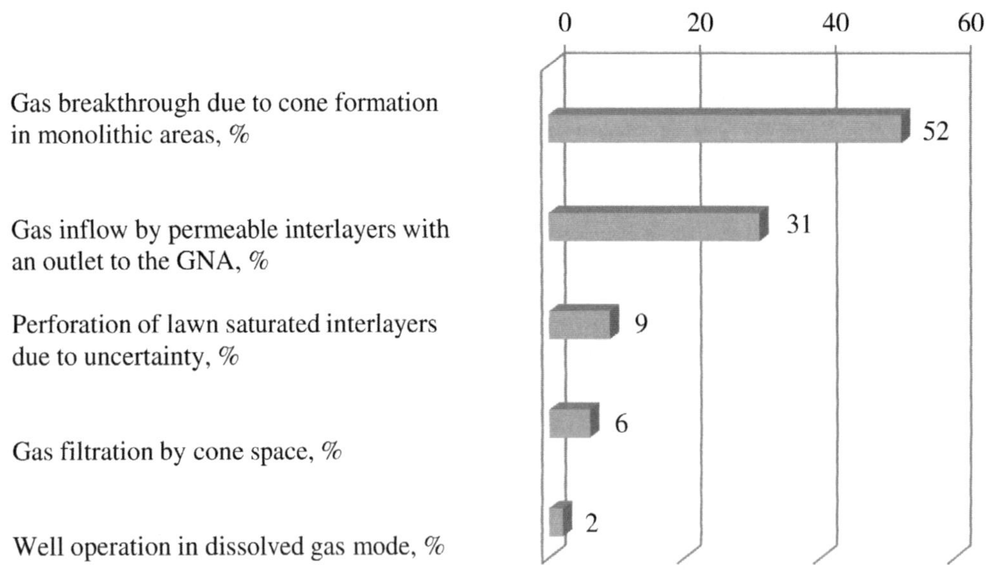

Figure 1. Reasons for gasification of wells under conditions of gas and oil deposits development

The primitiveness of the process of gas and oil reservoirs development is determined by a 2-phase system, which is in equilibrium at initial formation conditions. In the process of development thermobaric parameters of the deposit are changed and the equilibrium is broken. General decrease of reservoir pressure (or decrease in some parts of the reservoir) leads to movement of CPC. Due to viscosity instability, the movement of the gas condensate downstream contributes to the breakthrough of gas tongue into the oil wells and the formation of cones [5].

Development of gas and oil deposits is complicated:

- by the difficulty of regulating the movement of G.N.A;
- equal saturation pressure (Pnas) and initial formation pressure (Rnach.pl.);
- the unconditional proximity of the bottom of the GNK well when draining the oil rim;
- rapid gas breakthrough to the bottom of production wells and instability of oil displacement by gas, resulting in loss of reservoir energy;
- Possible SSC mobility in the vicinity of the well in the process of development and full hydrodynamic connection of the oil bed with the gas cap.

The growth of the gas factor is directly related to the location of wells relative to the gas content, the closer a well is to the gas content contour, the shorter is the gas-free period of well operation and the faster is the tightening of the gas cone. Only well #124 operates with relatively high (70-80 t/day) oil flow rate and stable gas factor (400 m3/t). For the rest of the wells the time of gas breakthrough from the wellhead (over 300 m^3/t) is 1-2 days. Well '99 is an exception - where the breakthrough time was 18 days. The most probable reason is the influence of clay lintel, which is installed in close proximity to the HSS. Also it should be noted that the gas factor measured at wells (including during hydrodynamic studies), can not be the basis for the justification of parameters of the initial formation oil, because these parameters should be taken on the basis of experimental measurements (PVT) on the depth samples in the laboratory.

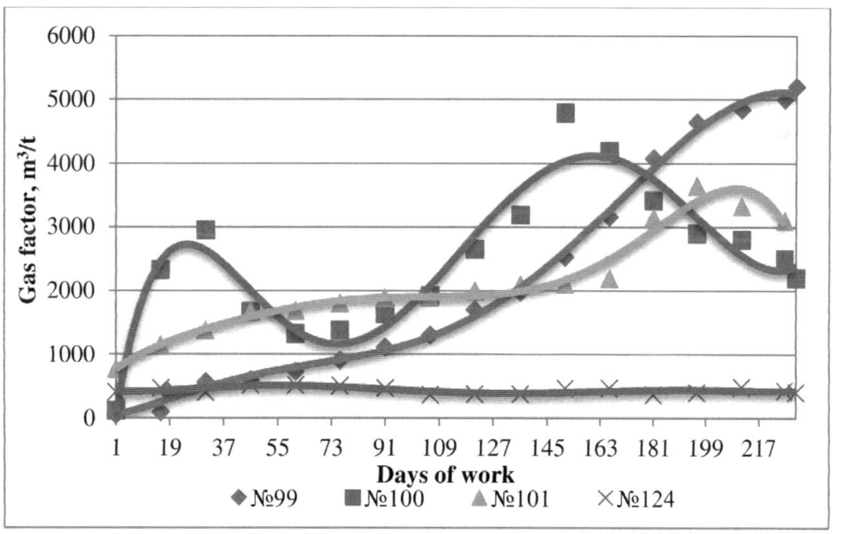

Figure 2. Dynamics of Gas Factor of BT formation by wells located near the gas processing plant.

According to the analysis of field data, the current well flow rate for oil varies in the range from 3.5 to 204.6 tons per day, for liquids from 4 to 236 m^3/day. The average oil flow rate is 90 tpd, fluid flow rate is 109.5 m^3/day. Such a significant dispersion is related to the well construction and the wiring profile. Horizontal wells certainly justify themselves both from the point of view of productivity, which is four times higher than that of directional wells, and oil flow rate (HS - 110 tons per day, NPS - 30-40 tons per day).

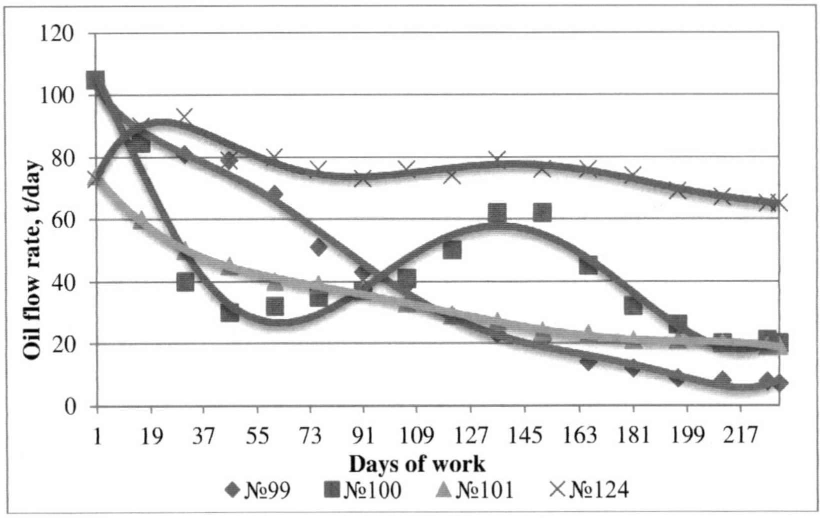

Figure 3. Dynamics of oil flow rate of Bt formation through wells located near the gas processing plant.

Analysis of BT reservoir development showed that failure to achieve the design indicators for oil production and oil flow rates in 2017 becomes more significant: at actual values of 228.9 thousand tons, 101.4 tons per day and design values of 597 thousand tons and 42.0 tons per day respectively. The reason for the discrepancy between the design and actual indicators is the difference in the current production fund of about 30 % (project - 41, fact – 28), due to the disconnection of wells due to high gas factor and the absence of new wells from drilling. Under the project, 39 new wells were planned to be commissioned, 9 wells were drilled, and 27 wells were commissioned (due to inaction and development of previous years). The water content of production wells under the project is 5.2 %, actually from 5 % to 25 %. The wells are operated with high gas factor (≈ 2000 m^3/t), which is caused by imperfect technology of well construction and installation [6].

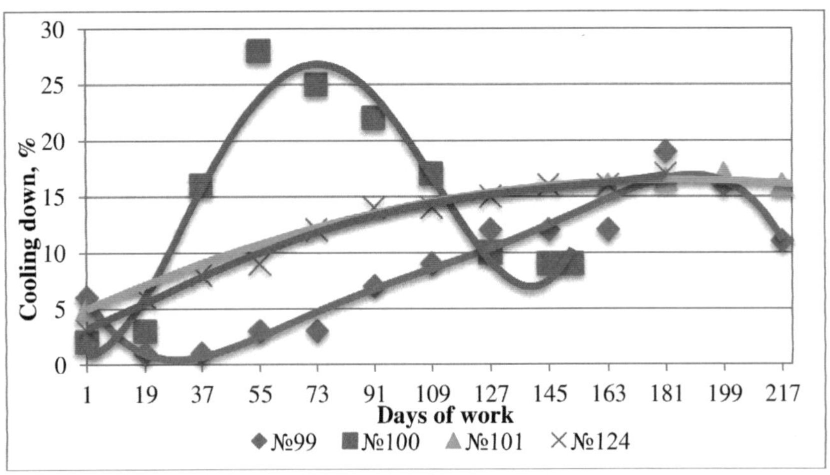

Figure 4. Dynamics of water encroachment of BT formation by wells located near the gas processing plant.

Despite the operation of wells on limited modes (underbalanced 1.0-3.0 MPa, downhole pressures up to 9.0-12.0 MPa at design Rzab – 4.0 MPa), unsatisfactory quality of cementing, fastening of couplings and packers led to the advance gas breakthrough of gas cap. The current gas factor is considerable 328 m^3/t and exceeds the calculated value by 4 times. Since the beginning of development, 1106 thousand tons of oil and liquid have been produced at the field. The analysis of well operation indicators and field research results reflects a number of development process regularities, which were not confirmed before.

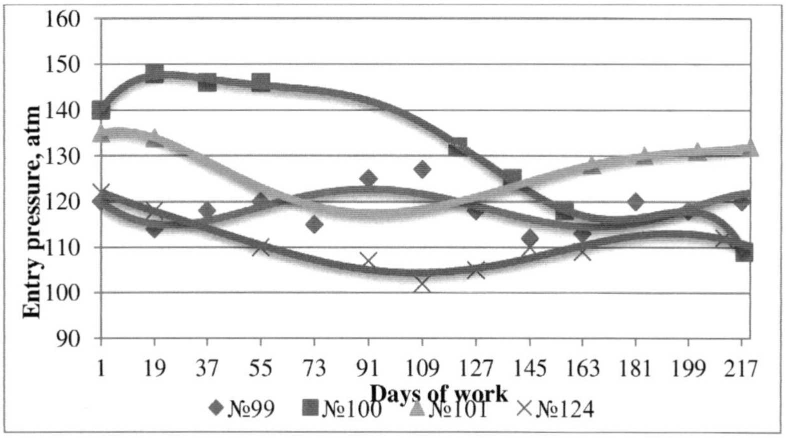

Figure 5. Dynamics of bottomhole formation pressure BT through the wells located near the SOC.

3. Conclusion

On the basis of the performed analysis of the experience in developing oil and gas deposits with a large gas cap and bedding bottom water, the reasons affecting the effective production of oil reserves were identified. The main reasons for high gas factor are gas breakthrough in horizontal wells behind the column space due to poor quality cementing and tightening of the upper gas cone in the areas of monolithic structure of the deposit. For the conditions of development of low-temperature deposits of the Srednebotuobinskoye field the main development indicators were analyzed during commissioning of new wells. It was found out that the optimal location of the horizontal well section should be closer to the water-oil contact. In addition, in the under-gas zones, it is necessary to reduce underbalance and optimize the parameters and profile of the hollow well. In consequence of operation of wells with increased underbalance up to 3.0-5.0 MPa leads to rapid growth of HF gas factor up to 1000 m^3/m^3 and higher. Based on the analysis of field data to ensure longer operation without gas breakthrough, the optimal underbalance is not more than 1.0 MPa. It is recommended to use the results of the study when developing oil reserves from under-gas zones of oil and gas fields.

References

[1] Tomskaya V F, Inyakin V V, Tomskiy K O and Krasnov I I 2018 Features of oil reserves development at the oilfields of the Republic of Sakha (Yakutia) *Collection: State, trends and problems of Western Siberia oil and gas potential development* pp. 79-86

[2] Inyakina E I, Tomskaya V F, Shavaleyeva A A and Varlamov V V 2018 Estimation of development efficiency of the oil and gas deposits in Srednebotuobinskoye field *Scientific forum. Siberia* **1** 26-27

[3] Krasnov I I, Tomskaya V F, Inyakina E I, Tomskiy K O, Ivanova M S and Katanova R K 2019 Restriction of the gas flows under the Botuobinskiy horizon development *Higher education news. Oil and gas* **6** 159-166

[4] Grachev S I, Strekalov A V and Samoilov A S 20106 *Increase of the oil fields development efficiency by the horizontal wells* (Tyumen)

[5] Krasnova T L 1997 Cone formation control at the oil and gas deposits development with the underflow water G*eology, geophysics and oil and gas fields development* **4** 38

[6] Tomskaya V F, Gracheva S K, Krasnov I I and Vaganov E V 2019 Prediction of the oil and gas deposits development with application of the gas flow limitation technology in the wells *Oil and gas: scientific innovations* **2** 3-19

IOP Conf. Series: Materials Science and Engineering **952** (2020) 012036 doi:10.1088/1757-899X/952/1/012036

An analysis of the effectiveness of hydraulic fracturing at YS1 of the Northern field

A V Karpikov[1], R I Aliev[2], N V Babyr[3]

[1] Irkutsk National Research Technical University, 83, Lermontov str., Irkutsk, 664074, Russia
[2] Chechen State University, 17a, Dudaev Boulevard, Groznyy, 364015, Russia
[3] Saint-Petersburg Mining University, 21 liniya 2, Vasilievsky-Island, 199106, Russia

E-mail: mamadaliev_it@mail.ru

Abstract. There are many methods of enhanced oil recovery and stimulation of oil inflow, and in order to find their most effective application, it is necessary to study the object of works, candidate wells, as well as their impacts on these wells. One of the methods of production intensification, which makes it possible to improve profitability of oil extraction, is hydraulic fracturing.

1. Introduction

For the first time, the hydraulic fracturing method was applied in the Northern field in 1993. In 2009-2013, the BS11 and YS_1 facilities were commissioned. The share of the YS_1 in the field's production has significantly increased and amounts to 28.1%, while the share of the current recoverable reserves is 27.1%. This was achieved due to advanced technologies used for drilling horizontal wells using the multi-zone hydraulic fracturing method. As of 01.01.2014, 2,467 wells were put into operation after their hydraulic fracturing (Figure 1, Figure 2):

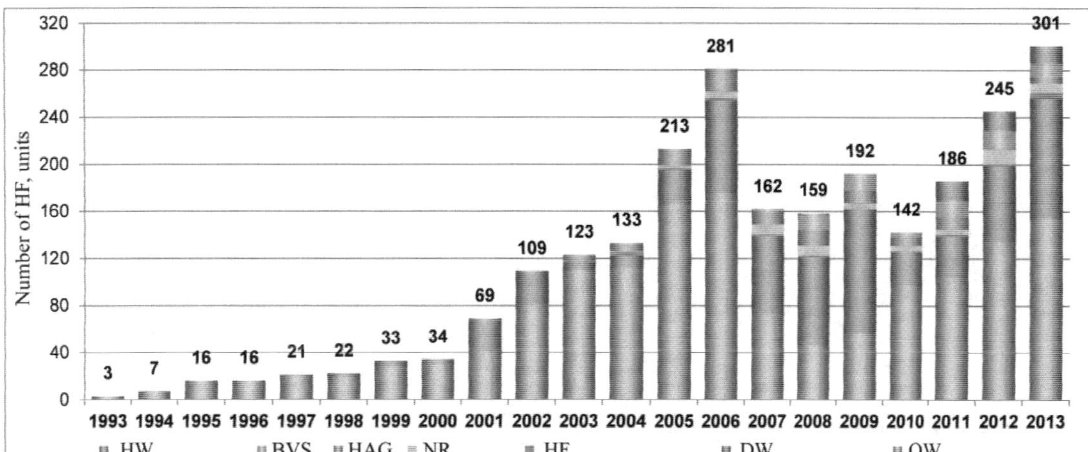

Figure 1. Distribution of hydraulic fracturing volumes by years and categories of wells. Northern field

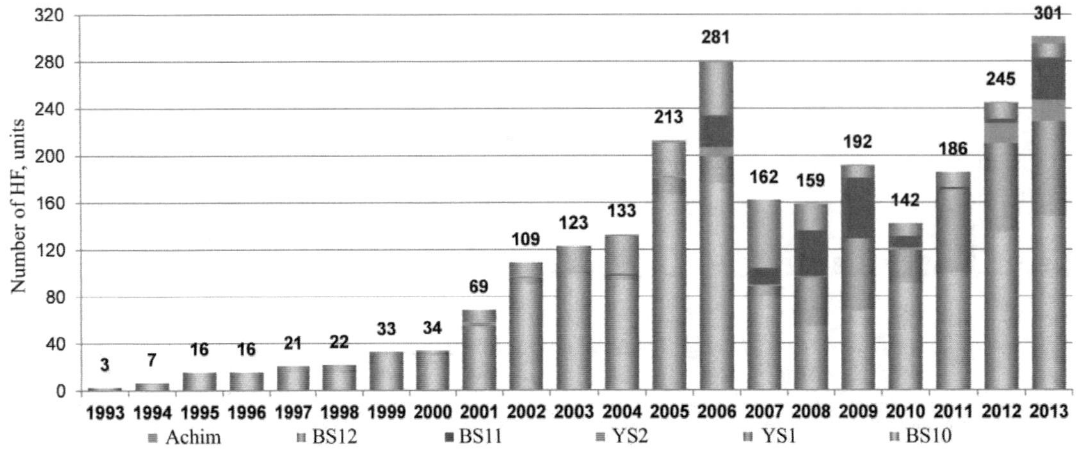

Figure 2. Distribution of hydraulic fracturing volumes by years and objects. Northern field

YS$_1$. The hydraulic fracturing began in 1993. As of 01.01.2014, 429 treatments were performed (including 96 - at the operating fund, 227 - at directional wells from drilling, 24 - during sidetracking of the second wellbore, 47 - at horizontal wells, 7 - when transferring to another horizon, 1 - with the addition of a new object to previously operated formations, 18 - at injection wells);

An increase in the number of treatments has been observed since 2005; in 2013, the number reached 81. The dynamics of hydraulic fracturing operations is shown in Figure 3. The well stock coverage was 69% (Figure 4).

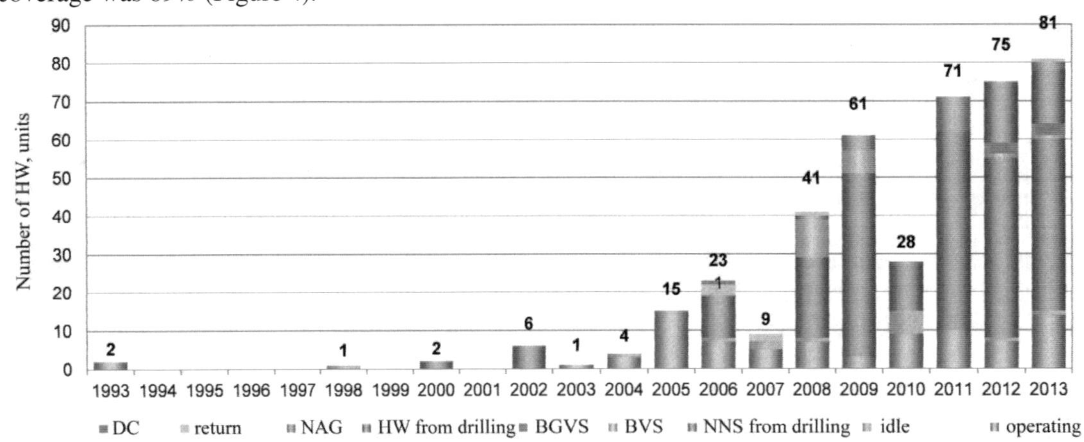

Figure 3. Dynamics of hydraulic fracturing at YUS1 by years

The average flow rate of fluid after hydraulic fracturing was 38.7 t / day, oil - 26.9 t / day, at the operating wells - 26 and 10.9 t / day, at drilled wells - 42.9 and 32.5 t / day, respectively. Additional oil production amounted to 5,671.3 thousand tons or 14.1 thousand tons per well (Figure 5), including:

- at operating wells - 697.9 thousand tons or 7.3 thousand tons / well, respectively;
- at WEM wells - 20.4 thousand tons per well;
- at new recompletion wells - 27.2 thousand tons or 3.9 thousand tons per well;
-at drilled wells - 4925.9 thousand tons or 16.5 thousand tons / well, respectively.

To predict the flow rate, empirical dependences on the effective reservoir thickness were used; the water cut was set according to the results of a geological analysis. For some wells, the indicators were adjusted based on the actual experience of using the hydraulic fracturing method at offset wells. When calculating additional oil production, we used the actual rate of a decrease in the effect for hydraulic fractured wells in 2010-2013.

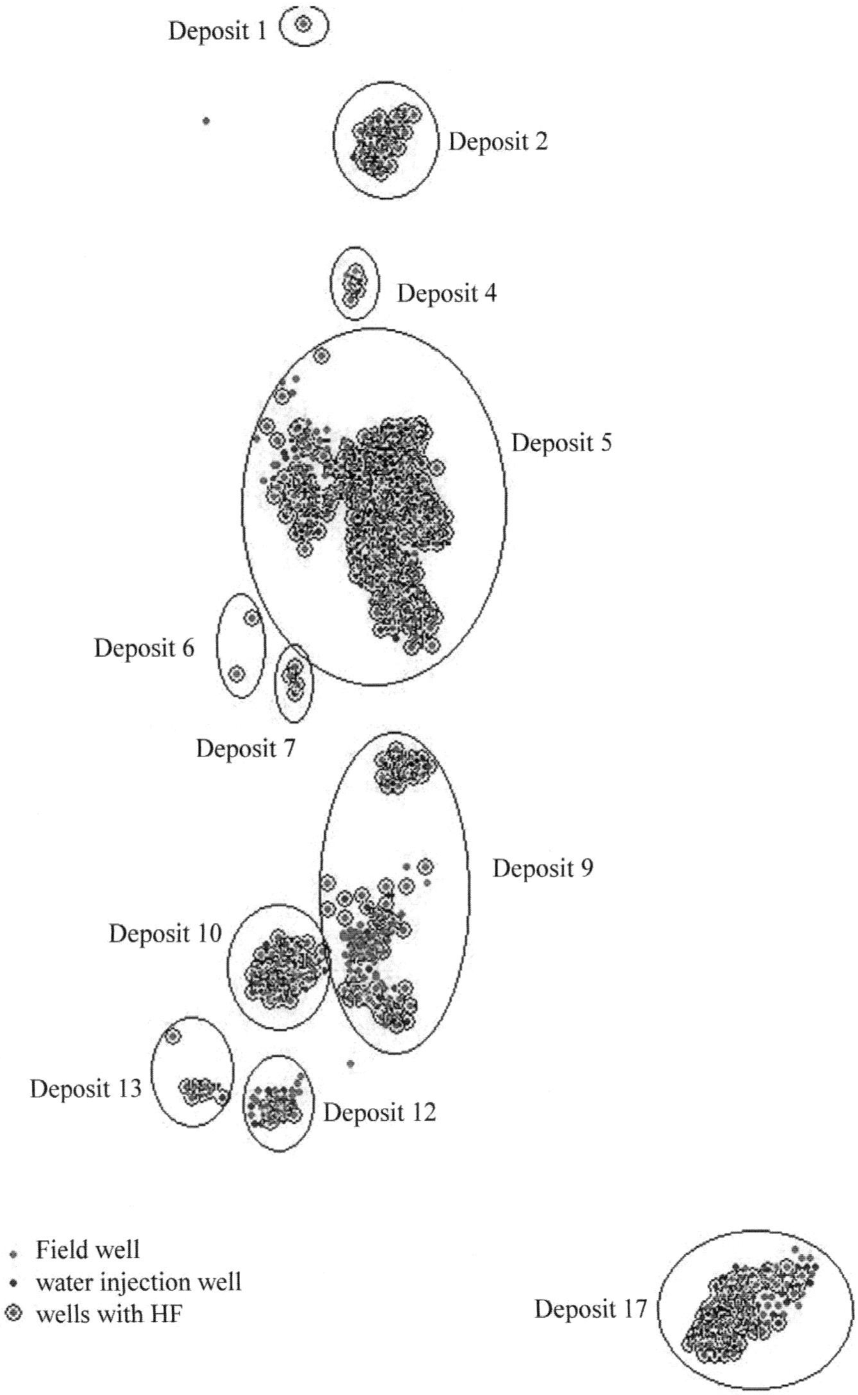

Figure 4. Well stock coverage with hydraulic fracturing operations YUS_1

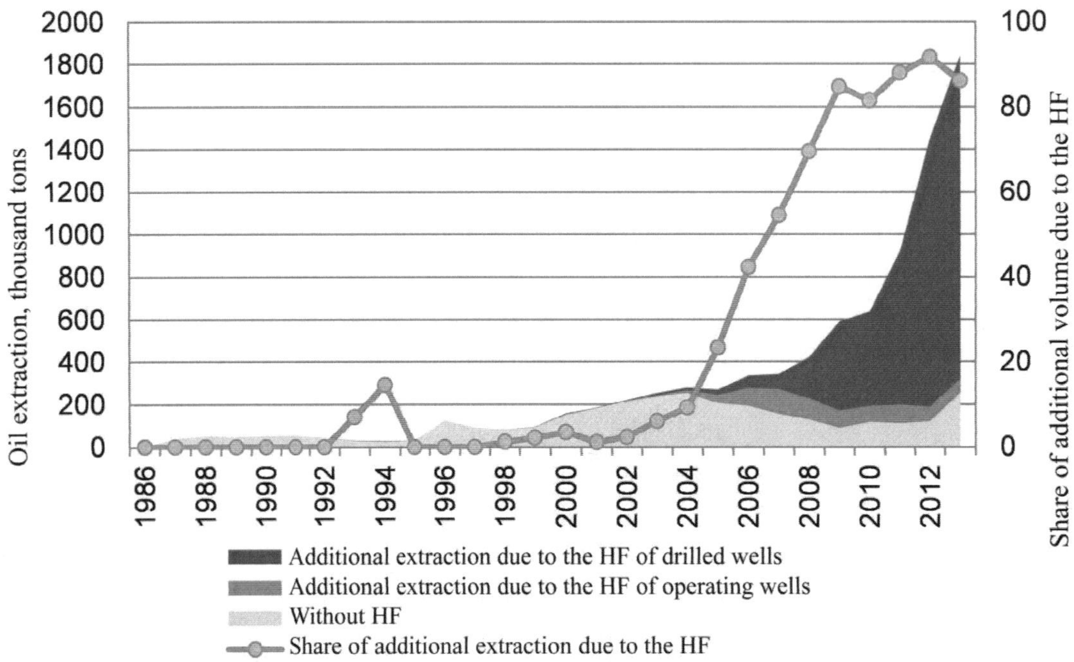

Figure 5. Dynamics of changes in additional oil production by years. YS1

2. MES.

It is proposed to conduct pilot works at horizontal well No. 2010G using SurgiFrac technology - multi-zone hydraulic fracturing using a coiled tubing unit (fracturing fluid with proppant is supplied through the Hydra-Jet system: a high-pressure jet initiates a fracture, while fluid is fed through the annulus, which helps to expand the fracture) ...

This technology has a high accident rate. In fact, a more reliable multi-zone hydraulic fracturing (MHF) technology was applied at t YUS1 in horizontal wells, which implies the wellbore arrangement with a special multistage assembly at the construction stage. The liner arrangement allows us to divide the horizontal part of the well into several sections in an open hole and affect each interval.

The main element of the system is circulation valves or couplings for interval fracturing, which were opened using a ball release system. The balls are dropped from the second interval; the first hydraulic fracturing is carried out without dropping the ball through the already pre-perforated pipe.

The HF technology by dropping balls is as follows: a device is installed at the wellhead, the device starts the first ball (of the smallest diameter), installed on the cylinder and activates the coupling, establishing a hydraulic connection with the annulus, hydraulic fracturing is carried out; the second ball is released (larger diameter), which simultaneously overlaps the previous interval (where the fracturing was carried out) and actuates its sleeve. In this sequence, all couplings are triggered towards the wellhead, and multi-point hydraulic fracturing is carried out along the entire wellbore. The annulus between two adjacent couplings is sealed with packers that swell in the presence of hydrocarbon or water.

For 2011-2013, at YS_1, multi-zone 47 horizontal wells were fractured. The number of fracturing ports was 4-5. The average oil production rate was 78 t / day, which is three times higher compared to fracturing from drilling directional wells.

Thus, the multi-zone hydraulic fracturing technology is recommended for commercial implementation at the YS_1 of the Northern field.

In addition, in 2012-2013, 4 wells (sidetracks with horizontal ends) No. 2108L, 2206L, 8333L, 6397L underwent multi-zone hydraulic fracturing using the Schlumberger AbrasiFRAC technology (hydrosand-jet perforation and stimulation of several intervals in one operation using coiled tubing, which allows the multistage fracturing in small-diameter barrels and reduces time input). Hydraulic fracturing was performed in 3-4 intervals.

The length of one perforation interval was 5 m, the distance between the perforations was 65 - 100 m. When performing hydraulic fracturing using this technology, STOPs often occur (due to the complexity of hydrosand-jet perforation in the horizontal wellbore). The average oil production rate of wells No. 2108L, 2206L was 7.6 t / day, water cut was 73 and 90%, respectively; for wells No. 6397L, the average oil production rate was 8333L - 42 t / day and water cut was less than 10%.

Thus, the AbrasiFRAC multi-zone hydraulic fracturing technology is recommended in case of confident predictions about the nature of oil saturation and not advisable when drilling into zones with unclear oil saturation (high water-cut and for surrounding wells, the presence of technogenic fracturing due to previously performed hydraulic fracturing); in case of a low productivity factor along the sidetrack from drilling, it is more expedient to perform a less expensive "blind" hydraulic fracturing with a packer installed in the vertical part of the wellbore.

In addition, at YS_1, nitrogen-foam hydraulic fracturing technologies and fracturing fluid with MPF were tested.

Nitrogen fracturing was performed on two wells, both operations were successful. After nitrogen fracturing with a foam rate of 30% (proppant weight 48 tons), Well No. 7621N yielded 22 tons / day for liquid and 19.3 tons / day for oil; an increase was 13.6 tons / day; the duration of the effect was one and a half years. At the same time, in 2009, the well was fractured during commissioning: 29 tons of proppant were injected, the input flow rate of fluid was 24 tons / day, oil - 12.5 tons / day; the less stable dynamics of work was observed (a decrease was 4 times per six months). After nitrogen fracturing with a foam rate of 50% (proppant weight 40 tons) Well No. 3627 yielded 51 tons / day for liquid and 31.9 tons / day for oil; an increase was 22, 6 tons / day, the duration of the effect was more than 3 years; at currently, the oil production rate is 24 tons / day.

3. Conclusion

Thus, the nitrogen hydraulic fracturing technology with a foam rate of at least 60% is recommended for the industrial use at YS_1 of the Northern field.

Analyzing the results of non-standard hydraulic fracturing technologies at the Northern field, the following recommendations were proposed for their industrial implementation. It is not recommended to limit the crack height by using low-viscosity surfactant-based fluids due to the lack of experience and inadequacy of the object. It is not recommend to reduce the relative phase permeability to water using a fracturing fluid with the addition of MPP (CW-Frac, Harriberton; PPM-1, solvent U-066 from Nalco Petroalliance; WCA-1, TricanVellServich; WLP-3700, KATConfet) or proppants with hydrophobic coating (ForeMPP, Schlumbrezhe) due to negative results of the laboratory tests. Reagents have a beneficial effect on fracture conductivity and matrix permeability. The use of reagents has a positive effect on conductivity of the fracture and permeability of the rock matrix if the fracturing fluid is foamed (nitrogen foam fracturing). It is not recommended if the fracturing fluid has a reduced polymer loading and self-dissolving organic fibers due to negative results of the laboratory tests. It is also not recommended to use preliminary isolation of water-cut intervals by pumping plugging compounds due to the high water cut before and after fracturing (more than 95%).

Despite the fact that measures taken at YS_1 in order to increase oil recovery and stimulate oil inflow have a positive effect, the most effective method is hydraulic fracturing. High rates of fluid and oil production rates were obtained for wells with a pronounced reservoir thickness (more than 7 m), high permeability and a low partition coefficient (less than 6 units). An important criterion for the successful hydraulic fracturing is the energy state of the area and a good hydrodynamic connection with injection wells.

At directional wells, an increase in fluid and oil production rates were observed with an effective reservoir thickness of more than 5 m, a proppant weight of over 50 tons and a proppant specific gravity of over 7 t / m.

The efficiency of multistage fracturing is influenced by the length of the horizontal section of the wellbore and the number of frac-ports. The recommended number of stages is five or more frac-ports.

IOP Conf. Series: Materials Science and Engineering **952** (2020) 012036 doi:10.1088/1757-899X/952/1/012036

References

[1] *Minutes No. 1860-dsp of 27.02.2009 of the meeting of the State Commission for Mineral Reserves.* GKZ Rosnedra.

[2] GOST R53710-2009. *National standard of the Russian Federation: Oil and gas-oil fields.* Development design rules. - Introduction. 2011-01-07. Reference and legal system "Garant" NPP "Garant-Service". - 06 July 2015

[3] Rules for geophysical research and work in oil and gas wells: approved. M-vom fuel and energy Ros. Federation and the Ministry of Natural Resources Ros. Federation 28.12.99

[4] Usachev P M 1986 *Hydraulic fracturing.* Moscow. Nedra

[5] Mishchenko I T 1989 *Calculations in oil production.* Moscow. Nedra

[6] Surguchev M L 1995 *Secondary and tertiary methods of enhanced oil recovery* Moscow. Nedra

[7] Ivanov M I 2008 Hydraulic fracturing. *Oil Industry* **01** 12-15

[8] *About the subsoil: Feder. the law Ros.* Federation of February 21, 1992 № 2395-1. Reference and legal system "Garant": NPP "Garant-Service"

[9] Broslavsky L I 2019 ENVIRONMENTAL SAFETY: GLOBAL WARMING AND CLIMATE CHANGE (LEGAL PROBLEMS). *Business, management and law* **4** 15-20

[10] RepinaV.A., GalkinV.I., GalkinS.V. Complex Petrophysical Correction in the Adaptation of Geological Hydrodynamic Models (on the example of Visean pool of Gondyrev oil field).Journal of Mining Institute.2018. Vol.231,p.268-274. DOI: 10.25515/PMI.2018.3.268

IOP Conf. Series: Materials Science and Engineering **952** (2020) 012037 doi:10.1088/1757-899X/952/1/012037

Results of the experimental study of the fatigue characteristics of the repaired welded joint of the marine stationary oil and gas producing platform

C Teodoriu[1], I V Starokon[2] and Ch S Guseynov[2]

[1] The University of Oklahoma, 660, Parrington Oval, Norman, OK, 73019-0390, USA
[2] National University of Oil and Gas «Gubkin University», 65, Leninsky Prospekt, Moscow, 119991, Russia

Abstract. Offshore stationary platforms are actively exploited in the world. The lifespan of some of these platforms exceeds 20 years. Such a significant period of operation leads to the fact that the platforms are exposed to various loads for a long time, causing alternating stresses and, as a result, fatigue damage, including fatigue cracks in welded joints and the base metal of the platform. When fatigue cracks are identified, defect joints are repaired commonly using welding techniques. However, the duration of the subsequent operation of such repaired welded joints is not currently known. The authors conducted an extensive experimental study to understand the life span of welding repair technologies. This approach was for the first time used to repair the welded joint of a platform located in the White Tiger. White Tiger is the Vietnamese offshore oil field, located 200 km east of Ho Chi Minh City on the shelf of the South China Sea. The result of this study was the fatigue diagram and its equation, which were first obtained by the authors.

Email: starokon@mail.ru

1. Introduction

Fixed offshore platform, commissioned more than 20 years ago, are still operated in the world. In connection with this significant period of operation during which the platform was exposed to various loads that cause alternating stresses and as a result the fatigue damage, the problem of estimation of residual resource of welded joints of offshore platforms after repair becomes extremely urgent. In accordance with Russian Federal law No. 116 "On industrial safety of hazardous production facilities" [1] after 20 years of operation the platform should be subjected to comprehensive technical diagnostics, based on which part of the welded joints with unacceptable defects should be repaired. To confirm the reliability of such repaired joints, residual resource calculation must be made in the framework of the preparation of the conclusions of industrial safety expertise. In preparing these opinions, the resource of welded compounds of the platform is determined from various conditions (for corrosion rates, compliance strength, etc.), including assessment of the resource for each welded joint based on the conditions of the fatigue life under the action of cyclic loads [2, 3]. The residual resource of the long maintained welded joints of platforms differs from a recently commissioned. However, determination of the impact of variable stresses on such elements is possible only by defining in any way the limit of endurance of these elements [4, 5]. In this article, we will talk about the experimental method of determining the fatigue limit of welded joints of offshore platforms on the example of the technology of repair, designed for the repair of the platforms on the field "White tiger"

[6, 7]. Many authors have investigated the problems of metal fracture in the oil and gas industry [8, 9], including for tubular joints in sea and land conditions [10, 11, 12].

2. Results and discussion

In the case under consideration during the survey, on one of the platforms installed at the White Tiger field, cracks were revealed (Figure 1) at the joint section of the column with a diameter of 2000 mm and a horizontal pipe-shaped element with a diameter of 1000 mm. The wall thickness of the column is 38mm.

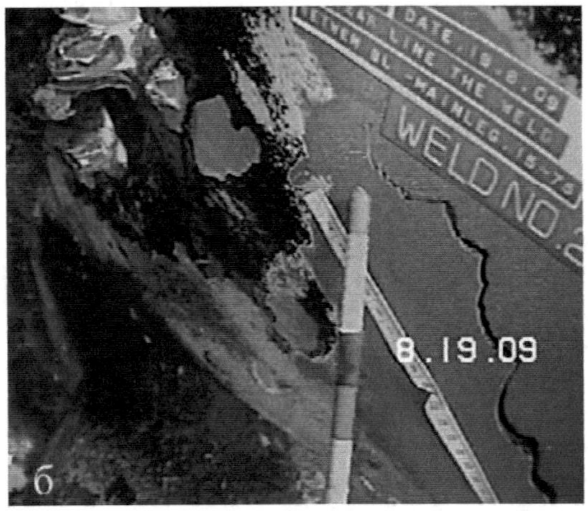

Figure 1. Crackdiscoveredduringthesurveyoftheoffshoreplatform

The crack has the form described in [7] and formed in the transition zone from the weld to the base metal and subsequently developed during operation along the fusion line. After the crack circled the horizontal element, its top continued its development into the base metal of the column wall.

To stop the development of the crack, it was decided to weld the crack itself and install a reinforcing pad on the wall of the column. It should be noted that when using arc welding to repair supports with a wall thickness of 38 mm, there is a high probability of the formation of cold cracks in the heat-affected zone, which is caused by the high cooling rate of the surrounding base material. When welding structures with such a significant wall thickness, it is recommended to preheat to reduce the cooling rate, but when welding under water in limited conditions of the caisson, it is extremely difficult to perform heating. To exclude heating during welding of low alloy steels, the following technological method was used. Before welding, a layer of austenitic metal was deposited on the welded edges, then the welded edges were welded together. This technique allows you to create a buffer layer of limy austenitic material, which reduces the likelihood of cracking during welding. The welding itself was carried out by the so-called dry method. For this, a caisson was used, i.e. a special chamber filled either with air (for depths up to 20 meters) or with a mixture of helium and oxygen at depths greater than 20 meters, in which, regardless of depth, a partial oxygen pressure of 29.4 kPa is maintained. In this case, the welder is also placed in this sealed chamber. The experience of using this technology shows that welded joints made in this way do not differ in quality from those made under ordinary atmospheric conditions. At the same time, the caisson should be hermetically adjacent to the structural section being repaired under water and be spacious enough for welders to work in it.

Remedial works of the resulting crack were carried out in the following sequence. The crack was either welded or, if a through hole was formed, which had to be sealed, a plug was introduced in the form of a curved plate of small thickness, the second was welded to the pipe wall with an angular seam in a wet way. Since the plug is not part of the supporting structure, the use of a wet welding method is acceptable. For welding, austenitic electrodes of the SoufTouch brand were used. It should be borne in mind that the repair of the supporting structure of oil platforms is allowed only by dry method. After welding the crack, a reinforcing pad was installed outside of the column. Then, the lining was assembled at the tack in the caisson and the sections of the column surface 30–40 mm wide were cleaned, adjacent to the end of the lining along its entire contour to a metallic luster. Then, a buffer layer along the contour of the patch is deposited on the surface of the support body by austenitic electrodes. The junction of the buffer layer to the end of the lining is cleaned to remove slag inclusions and prevent the formation of pockets. After that, the main fillet welds are welded. To ensure a smooth transition from the weld to the base metal, the external contour of the weld must be treated with an abrasive tool. The presence of austenitic weld support on the surface of a pearlite body can contribute to an increase in the rate of electrochemical corrosion. To prevent corrosion at the repair site, it is necessary to carry out a protective painting of the welding zone and install protectors.

To assess the resource of the welded joint, repaired as described above, an experiment was designed and conducted. Experimental installations were built in such a way that a welded joint is mimicked in the laboratory scale and cyclic loaded until a crack is observed. The laboratory scale setup consists of a 7” inch vertical pipe to which a 4” horizontal pipe is welded. The weld is first inspected to show that is free of defects and then cyclic loaded using a hydraulic cylinder controlled by a computer until the first crack is observed. Once the crack is measured, the crack is repaired, and the cyclic loading is restarted until a new crack is detected.

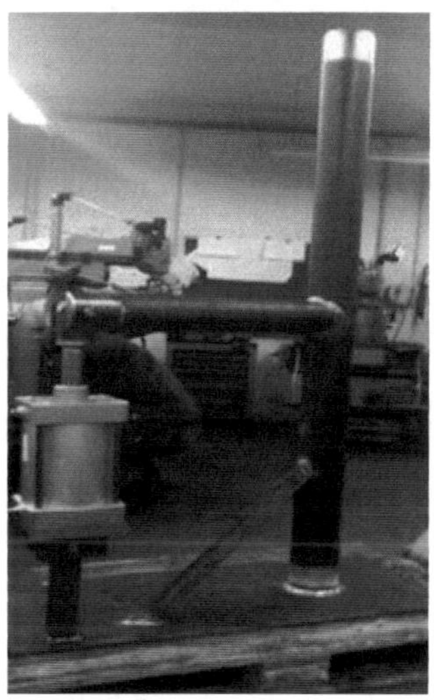

Figure 2. The photo of the experimental setup

An algorithm implemented in the LabVIEW software package was used as a control module. The experiment was carried out as follows. From the high-pressure pipeline with compressed air, a retraction was conducted to the module for monitoring the values of the applied force. In this module, the pressure which controlles the force applied to the welded joint was decreased to a predetermined value set in the control module of the experiment. Then, pressure entered the switching unit of the experimental load vector. And then it entered the pneumatic cylinder in which it was transformed by virtue of a certain value, which, under the control of the switching unit, developed sequentially along the vertical axis in different directions. This load created a destructive experimental moment in the studied welded joint. The general installation diagram is shown in Figure 3. The author carried out numerical and analytical modelling of the stress state of welded joints of the offshore platform located at the Black Sea fields. As a result of modelling, it was found that under various conditions of wave action, the maximum amplitude of alternating stresses is in the range from 50 to 90 MPa.

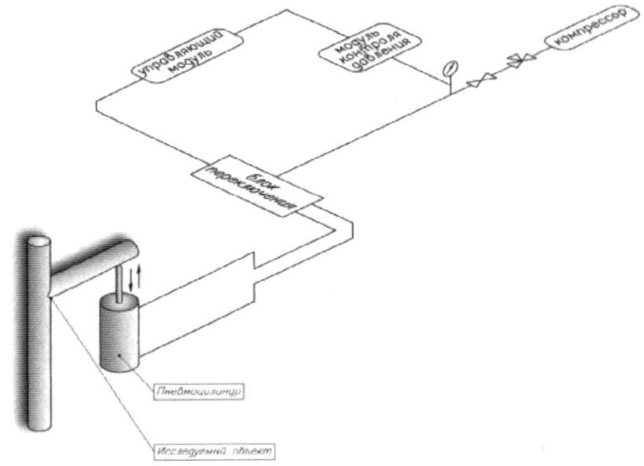

Figure 3. Schematic diagram of the experimental setup

The results of the experiment are shown in table 1.

Table 1. The number of cycles N obtained from the experimental study of the resource of the restored welded joint

Stress amplitude σ_a, MPa	**Number of cycles, N**						**The resulting N indicating the magnitude of the variance**
	Sample №1	**Sample №2**	**Sample №3**	**Sample №4**	**Sample №5**	**Sample №6**	**N±Δx**
90	81538	75557	74514	79348	77516	73243	76953±4964
51	1041861	1167943	1324561	1287892	1165742	1216567	1200761±160142

The calculation of the endurance limit was carried out by the method of D.I. Goltseva [10] according to the formula:

$$\sigma_R = \sqrt[2]{(\sigma_1^2 N_1 - \sigma_2^2 N_2)/(N_1 - N_2)}, \quad (1)$$

As a result, the endurance limit of 47 MPa was calculated. However, the endurance limit obtained as a result of experiments must be correlated with the actual operating conditions of the welded joints of the offshore platform in the conditions of the shelf. In [10,11], the following main parameters were identified that should be taken into account: the ratio of sample sizes to real objects, roughness, asymmetry of stress cycles, residual stresses, and other factors. Using this technique, we compare the obtained experimental result with a real welded joint.

Table 2. Results of processing the experimental endurance limit of a welded joint, taking into account real conditions of operation on the shelf

Name of the parameter[1]	**Result**
The limiting amplitude of the first main stress, σ_{1a}	90,3
Ошибка! Закладка не определена. Residual stress asymmetry coefficient, R^{res}	0,65
Tensilestresscoefficient, η_1	0,1
Compressive stresses coefficient, η_2**Ошибка! Закладка не определена.**	0,4
Normalstresscoefficient, η_3	0,12
Endurance limit, σ_{-1}	47
Endurance limit based on actual operating conditions, σ^{act}_{-1}	15

[1]-other values are given in the work [7].

After the necessary calculations, we obtain the endurance limit of 15 MPa, and the slope of the fatigue curve m = 3.6. Calculation in accordance with the Baskvin equation made it possible to determine the N_G fracture point at the level of $3,1*10^7$ cycles. The approximating equation of the fatigue curve and the value of N_G are calculated in the Wolpram software package. As a result, the following equation was obtained:

$$\sigma_a = 224{,}395 - 12{,}1201 \cdot \ln(N), \tag{2}$$

where: σ_a- amplitude of alternating stresses acting in a welded joint; N-number of cycles.

The fatigue diagram of the restored welded joint by the repair method developed at the White Tiger deposit was built as shown in figure 4.

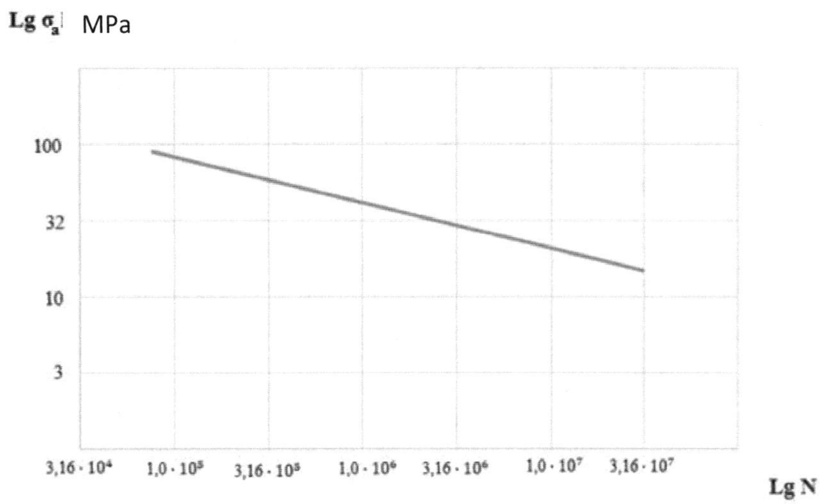

Figure 4. Fatigue diagram of reconditioned welded joint repaired according to the repair technology developed during the repair of the offshore platform at the White Tiger field

Analyzing it, we can say that the value of the endurance limit according to the results of an experimental-analytical study turned out to be higher than for a new compound, but the dependence of the amplitudes of the alternating stresses on the number of cycles shifted to the right, which indicates that the destruction of the compound repaired by this method will occur later than described in the technology of welding cracks with the installation of crack traps.

3. Conclusions

An experimental setup was built to simulate the fatigue of welded joints commonly used for offshore structures.

The experimental investigations have shown that the number of cycles to failure are in the range of 77000 cycles for 90 MPa induced stress and 1.2 million cycles at a stress level of 50 MPa.

Using the experimental data a new equation was derived to assess the fatigue resistance for the White Tiger field selected platform.

The value of the endurance limit according to the results of an experimental-analytical study turned out to be higher than for a new compound, but the dependence of the amplitudes of the alternating stresses on the number of cycles shifted to the right.

Acknowledgement
The authors would like to thank the Technical University of Clausthal for the laboratory support offered for the experimental investigations.

References

[1] Bär F and Overmeyer T 2013 Approaches for determination and reduction of non-productive times of drilling rigs for deep wells *Logistics Journal* **7** 2192-9084

[2] Topchiev A 2020 Design and monitoring of oil and gas industry facilities based on the use of ultra-light aviation and digital technologies *ICMSIT 2020 Journal of Physics: Conference Series* 1515 042030 IOP Publishing doi:10.1088/1742-6596/1515/4/042030

[3] Bezkorovayniy V, Bayazitov V and Bobov D 2018 Management of the Design and Construction of Offshore Oil and Gas Facilities with Bim Base *IOP Conference Series: Materials Science and Engineering* **463 (4)** 042056

[4] Reitenbach V, Ganzer L, Albrecht D and Hagemann B 2015 Influence of added hydrogen on underground gas storage: a review of key issues *IOP Conference Series: Environmental Earth Sciences* **73(10)** 6927–37

[5] Hagemann B, Rasoulzadeh M, Panfilov M, Ganzer L and Reitenbach V 2016 Hydrogenization of underground storage of natural gas. Impact of hydrogen on the hydrodynamic and biochemical behavior *IOP Conference Series: Environmental Earth Sciences* **20** 595–606

[6] Efimenko L A, Elagina O Yu and Kapustin O E 2017 Equipment for determining the carbon equivalent of low-carbon pipe steels of a new generation *Welding production* **10** 3-7.

[7] Lukyanov V F, Rogozin D V and Gritsykhin V A 2011 Repair of metal structures of offshore drilling and oil production platforms *Vestnik DGTU* **9(60)** 1630-1636

[8] Yuan Z, Schubert J, Esteban U, Chantose P and Teodoriu C 2013 Casing failure mechanism and characterization under HPHT conditions in south texas Society of Petroleum Engineers - *International Petroleum Technology Conference 2013, IPTC 2013: Challenging Technology and Economic Limits to Meet the Global Energy Demand* vol 3 pp 2207-17

[9] Teodoriu C 2012 Selection criteria for tubular connection used for shale and tight gas applications *Society of Petroleum Engineers - SPE/EAGE European Unconventional Resources Conference and Exhibition* pp 865-70

[10] Starokon I 2019 Methods for solving the problems of extending the resource of offshore stationary platforms *Journal of Physics* 1399 055087

[11] Starokon I and Ermakov A 2019 Assessment of jacket-type platform stress state in corrosion environment *Materials Science and Engineering* **700** 012018

[12] Giotis T, Pavlou D and Belibassakis K 2019 Flexural dynamic behavior of submerged cylindrical structures under wave loads *IOP Conference Series: Materials Science and Engineering* **700** 012022

IOP Conf. Series: Materials Science and Engineering **952** (2020) 012038 doi:10.1088/1757-899X/952/1/012038

Development of the non-stationary waterflooding technology

E M Almukhametova, E M Veliev, A M Puchkov

Ufa State Petroleum Technological University, Branch of the University in the City of Oktyabrsky, 54a, Devonskaya St., Oktyabrsky, Republic of Bashkortostan, 452607, Russia

E-mail: elikaza@mail.ru

Abstract. The article analyzes the non-stationary waterflooding technology. The basic physical principles of the technology are described. The reasons for its low popularity among subsoil users are revealed. The article aims to substantiate the relevance of non-stationary technologies, conduct experimental-industrial works to adapt and create new non-stationary technologies, and determine conditions for their effective application. In the 1990s, Western technologies were widely used in the oil industry. Non-stationary waterflooding was not included in the list of these technologies. Despite its relative simplicity, the practical implementation of the non-stationary waterflooding technology requires preparation works. Non-stationary waterflooding is one of the most knowledge-intensive methods of enhanced oil recovery. Non-stationary waterflooding tends to "age", i.e. its effectiveness decreases over time, which requires constant changes in the technology and its combination with other methods.

1. Introduction

When developing highly productive heterogeneous reservoirs, due to the advanced waterflooding of highly permeable zones (layers, interlayers), oil reserves are uneven. At the same time, depending on the ratio of low-permeability and high-permeability layers, the ratio of viscosities of oil and water and other factors, the water cut can reach limiting values with low production rates from the initial recoverable oil reserves. The situation when the water cut reaches and even exceeds limiting values, and significant volumes of mobile oil reserves remain in the reservoirs is common in the development of oil fields. This is especially evident in high-viscosity oil deposits.

This paper discusses the development of non-stationary waterflooding technology.

Non-stationary waterflooding is a common sweep enhancement technology in the development of reservoirs with heterogeneous permeability. The implementation of non-stationary waterflooding does not require additional costs, and the effect of the technology is significant. Therefore, this technology is widely used in the development of oil fields. To date, extensive experience in the use of non-stationary waterflooding under various geological and technological conditions has been accumulated. Basically, non-stationary waterflooding has been applied for oil reservoirs with low and medium viscosity. For example, the Romashkinskoe field has been a testing ground for various types of non-stationary impacts. There are also reports on the use of the NT on oil deposits with an increased viscosity.

2. Materials and methods

The problem of developing oil reserves from a heterogeneous formation has been well studied in underground hydrodynamics. A sector model of a heterogeneous reservoir was built. The information was processed using modern methods. An analysis of the non-stationary waterflooding technology showed that it has been implemented in practice and the results of implementation have confirmed its effectiveness.

3. Results

The model is described in [1]. The situation when viscosities of the displaced agents were comparable by $\mu o / \mu w = 1.25$ is analyzed.

Figure 1 shows the waterflooding of the model formation. As expected, there is a rapid breakthrough of water through the highly permeable interlayer. With full flooding of a highly permeable interlayer, the maximum water cut is achieved. At the same time, residual oil reserves are concentrated in low-permeability intervals of the formation.

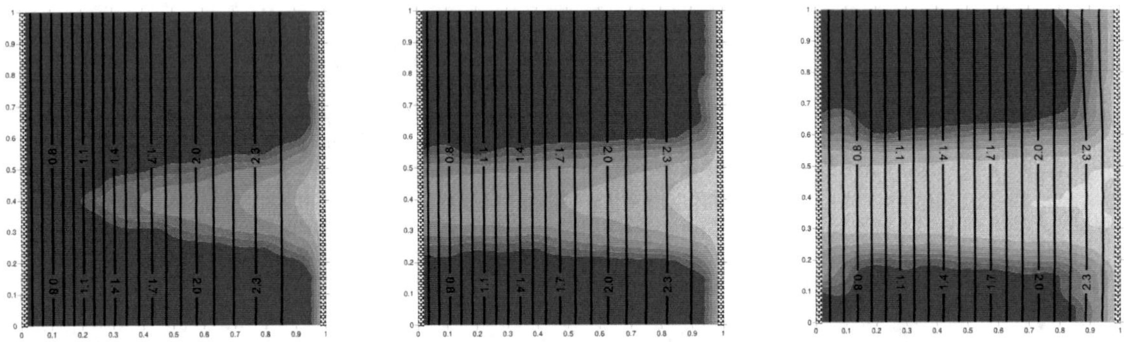

Figure 1. Changes in water saturation and pressure during flooding of a heterogeneous formation (according to [1])

As a result of the advanced flooding of the highly permeable layer, a technogenic oil-water contact is formed. It has a significant strike area. By creating conditions for the exchange of fluids between the flooded and oil areas, it is possible to use residual reserves of the partially flooded formation. In this case, the high-permeability layer becomes a transport channel for the oil flowing from the low-permeability layers.

The ability to create pressure drops between the water-flooded and oil intervals is inherent in the non-stationary (waterflooding) technology. The literature also refers to it as the "elastic-capillary cyclic development method" [3,12].

The physical foundations of unsteady flooding were developed in the 1960s. In 1965, a number of scientists from the All-Russian Research Institute (Surguchev M.L., Zheltov Yu.P., Boxerman A.A., Gubanov A.I., Ogandzhanyants V.G., Kocheshkov A.A.) formulated the basic principle of the NS technology: "The method involves increasing the elastic reserve of the reservoir system by increasing and decreasing the water injection pressure. This is a prerequisite for the unsteady pressure drops within the formation and the corresponding unsteady fluid flows between layers (areas) of different permeability. In the half-cycle of increasing the injection pressure, water from the layers with higher permeability penetrates into the low-permeability layers, and during the half-cycle of pressure reduction, oil from the low-permeability layers moves to the highly permeable part of the reservoir. "

As a result of waterflooding of heterogeneous layers, undrained residual oil reserves are located in low-permeability layers, zones, areas. Initially, a fairly homogeneous field of the oil-saturated reservoir is cut by flooding zones; "under such conditions, the formation is a random alternation of flooded and oil-saturated areas" [5]. The transition to the non-stationary water injection will allow the remaining oil reserves to be involved in the drainage process.

Studies of the physical processes occurring during periodic water injection are summarized by V.E. Gavura. According to the researcher, "the effectiveness of the elastic-capillary cyclic method of waterflooding of heterogeneous formations is determined by two inextricably linked processes: the introduction of water into low-permeability zones of the formation due to pressure drops arising from an uneven distribution of pressures caused by the heterogeneity of the formation, and capillary retention in low-permeability zones of the formation of penetrated water" [2].

The NST is often combined with the flow reversal (FR) technologies. In this case, the combination of technologies makes it possible to obtain an effect determined by the non-stationary processes in the formation and changes in the direction of the predominant flows of formation fluids. The integrated technology increases the coverage of the reservoir by acting along the section and the lateral [4,9].

In practice, an unsteady pressure field is created by regular shutdowns (or changes in injectivity) of injection wells and / or high-water-cut production wells. Methods for creating regular fluctuations of the pressure field can be combined. The main advantage of non-stationary waterflooding is its low cost; no capital costs are required to implement the technology, and there is no increase in operating costs. The non-stationary waterflooding allows us to obtain a significant effect. According to the expert assessment, the effect value varies from 8% to 14% of the cumulative oil production. For high-viscosity oil deposits, an increase in oil production is 12 ... 20% [11], for low-viscosity oil deposits, it is 6 ... 14% [6].

The effect value is determined by many parameters: the permeable heterogeneity of the reservoir, the initial water cut, the ratio of viscosities of oil and water, the impact amplitude (from the side of injection wells), the duration of shutdowns and operation of injection and production wells, the type of a waterflooding system, etc.

According to Vladimirov, "the use of non-stationary waterflooding technologies has provided a significant effect. However, non-stationary waterflooding technologies have a decreasing efficiency during their long-term use. It is necessary to adapt the NS technology to the changing geological and technological conditions. Therefore, the applied technologies undergo continuous changes.

According to [8], non-stationary stimulation technologies multiply their efficiency when rebuilding the reservoir pressure maintenance system (for example, when transferring water injection foci). However, this way of increasing the efficiency of the NS technology is costly and requires a significant amount of time to be implemented. However, [7] has shown that other ways of increasing the efficiency of non-stationary technologies can be used. They involve the use of other modes of regular operation of injection wells.

The commonly used method for reducing the water cut are technologies limiting the water inflow and flow diverting technologies [10]. These types of technologies limit the movement of water through highly permeable intervals of the reservoir due to the injection of special plugging compounds. A sustainable positive effect can be achieved by pumping large volumes of plugging compounds. Otherwise, the impact area is limited to the near-wellbore zone, which is often insufficient to obtain a positive economic effect.

Unlike costly technologies for limiting water production and flow diversion technologies, non-stationary waterflooding (NSWF) is a free technology. The use of non-stationary waterflooding technologies at the fields of the former Soviet Union is techonologically and economically efficient. At the same time, it should be noted that the theory and practice of non-stationary waterflooding were developed for conditions when the water cut is significantly lower than the limit values. However, currently, the conditions for the development of these fields have changed, which requires further studies of non-stationary processes and adaptation of the NSWF technology to the current conditions.

4. Conclusion

The data obtained demonstrates the attractiveness of non-stationary waterflooding technologies in the development of oil deposits, especially those that are at the final stage of development. However, the NSWF technology is not widely used. After its rapid development in the 1970-1980s, the oil recovery technology was beyond the scope of the most popular enhanced oil recovery (EOR) technologies. This is due to a number of circumstances. Firstly, in the 1990s, specialists used Western technologies. Despite the successful use of the non-stationary waterflooding, it was an "exotic" method of enhanced oil recovery for Western specialists. There was a gap between the rich experience of the Soviet school of petroleum science and the current understanding of oil field development. Secondly, despite its relative simplicity, the practical implementation of the technology requires preparation works. Non-stationary waterflooding is one of the most knowledge-intensive methods of enhanced oil recovery. Therefore, the NSWF programs are being developed by specialized scientific and design organizations. It is possible to distinguish between three directions in preparing this technology for implementation: selection of a site, selection of a NS technology and its parameters, coordination of the technology with capabilities of reservoir pressure maintenance and collection systems. Thirdly, the NS technology tends to "age", i.e. its effectiveness decreases over time. The oil recovery technology is quickly aging at high-viscosity oil deposits. This requires constant changes in the technology, its combination with other methods.

References

[1] Almukhametova E M 2018 Recommendations on thermal non-stationary waterflooding when modeling oil field development *IOP Conference Series: Earth and Environmental Science* **194 (6)** 062002

[2] Beliveau D 2008 Waterflooding Viscous Oil Reservoirs *SPE* 113132

[3] Colonomos P A Feasibility study of Cyclic Steam Injection in a Deep Heavy Oil Reservoir in Western Venezuela *SPE* 15091

[4] Das S K 1997 Vapex: An Efficient Process for the Recovery of Heavy Oil and Bitumen *SPE* 50941

[5] Kuleshova L S, Mukhametshin V V and Safiullina A R 2019 Applying information technologies in identifying the features of deposit identification under conditions of different oil-and gas provinces *Journal of Physics: Conference Series (ITBI 2019 – International Conference "Information Technologies in Business and Industry")* **1333(7)** 1-5 DOI: 10.1088/1742-6596/1333/7/072012

[6] Batalov S A, Andreev V E, Lobankov V M and Mukhametshin V Sh 2019 Numerical simulation of oil formation with regulated disturbances. Oil recovery quality simulation *Journal of Physics: Conference Series (ITBI 2019 – International Conference "Information Technologies in Business and Industry")* **1333(3)** 1-6 DOI: 10.1088/1742-6596/1333/3/032006

[7] Akhmetov R T, Mukhametshin V V and Kuleshova L S 2019 Simulation of the absolute permeability based on the capillary pressure curves using the dumbbell model *Journal of Physics: Conference Series (ITBI 2019 – International Conference "Information Technologies in Business and Industry")* **1333(3)** 1-8 DOI: 10.1088/1742-6596/1333/3/032001

[8] Batalov S A, Andreev V E, Lobankov V M and Mukhametshin V Sh 2019 Numerical simulation of the oil reservoir with regulated disturbances. Oil recovery stability simulation *Journal of Physics: Conference Series (ITBI 2019 – International Conference "Information Technologies in Business and Industry")* **1333(3)** 1-6 DOI:10.1088/1742-6596/1333/3/032007

[9] Mustoni J L, Norman C A and Denyer P 2010 Deep Conformance Control by a Novel Thermally Activated Particle System to Improve Sweep Efficiency in Mature Waterfloods of the San Jorge Basin *SPE* 129732

[10] Kuleshova L S, Mukhametshin V V 2019 Elimination of uncertainties in predicting well interaction using indirect geological field information *IOP Conference Series: Earth and Environmental Science (IPDME 2019 – International Workshop on Innovations and Prospects of Development of Mining Machinery and Electrical Engineering)* **378 (1) (012115)** 1-8 DOI: 10.1088/1755-1315/378/1/012115

[11] Rocha de Farias M L, et al. 2012 A Comparative Study of Emulsion Flooding and other IOR Methods for Heavy Oil *SPE* 152290

[12] Vladimirov I V, Almuhametova E M, Veliyev E M 2016 Conditions of the effective use of hot water flooding for development of heavy oil deposits with non-uniform permeability *Oil Industry* **9** 62-65

IOP Conf. Series: Materials Science and Engineering **952** (2020) 012039 doi:10.1088/1757-899X/952/1/012039

Approbation of the system technology for improving well drilling under difficult mining conditions of Bashkortostan fields

A P Chizhov[1,2], V E Andreev[1,2], V V Mukhametshin[1] and L S Kuleshova[3]

[1] Ufa State Petroleum Technological University, 1, Kosmonavtov st., Ufa, Republic of Bashkortostan, 450062, Russian Federation
[2] SASI "Institute of strategic research of the Republic of Bashkortostan", 129/3, October avenue, Ufa, Republic of Bashkortostan, 450075, Russian Federation
[3] Ufa State Petroleum Technological University, Branch of the University in the City of Oktyabrsky, 54a, Devonskaya St., Oktyabrsky, Republic of Bashkortostan, 452607, Russian Federation

E-mail: vv@of.ugntu.ru

Abstract. A method for systematic improvement of the drilling technology in difficult mining conditions has been suggested. The study deals with unstable rocks of the sedimentary cover of the eastern edge of the Russian platform and conditions for well construction. Wells are directional with horizontal ends. Field tests have shown that technical and technological solutions can solve scientific and technical problems. It is recommended to test these solutions in the conditions of other Russian regions and abroad.

1. Introduction

Currently, there is an increase in the complexity of mining conditions for well construction. The increasing complexity of conditions is associated with both natural and man-made factors. The increasing complexity of conditions decreases quality indicators for well construction, and deteriorates the efficiency of drilling technologies [1–15]. This is due to the abnormality of changing thermodynamic and geological and technical drilling conditions and oil and gas well completion. In addition, there are no methods for monitoring and controlling well drilling technological processes.

The experience of using traditional construction technologies in existing conditions shows that a decrease in the main indicators of drilling efficiency is due to unsteady hydraulic conditions of drilling, irreversible violations of initial properties of technological solutions in the wellbore, lack of control and regulation of interaction of technological solutions with fluids of the rock mass.

2. Materials and methods

Complex field problems, such as stabilization of the hydraulic state and well behavior, preservation of natural reservoir properties of productive formations and their isolation from fluid-saturated formations of the productive strata, improvement of tightness of the annular space of the support, can be solved based on a systematic approach.

The systematic approach involves the implementation of the following scientific principles: information support, organization and management of drilling technological processes.

Information support provides data about difficulties arising when deepening the well bottom. For example, in order to understand the characteristics of the penetrated loss interval, tightness and strength of the open hole are assessed by pressure testing from the wellhead.

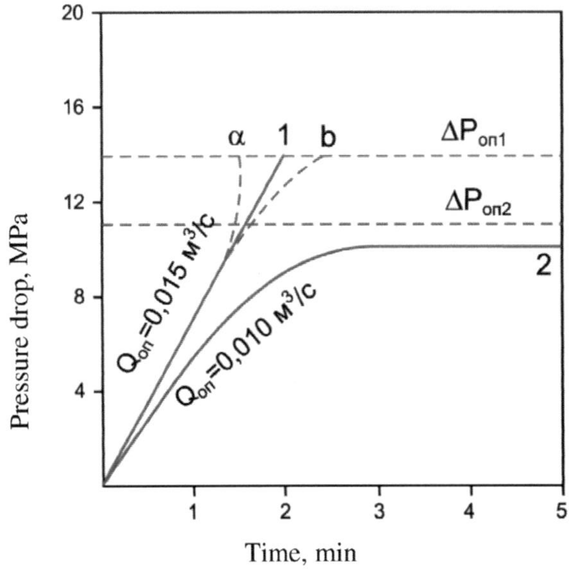

Figure 1. Hydromechanical strength of the wellbore: 1 - pressure testing in the absence or unsteady fluid filtration; 2 - pressure testing with established fluid filtration; a, b - possible deviations of the crimping pressure; Cp1, Cp2 - calculated permissible pressure of crimping

Determined and controlled:

- tightness index — $К = \frac{Q}{\Delta P}$, (m³/s/MPa);
- barrel strength index — $\nabla P = \frac{\Delta P}{H}$, MPa/m;
- fluid filtration start pressure gradient — $\nabla P_P = \frac{\Delta P_1}{H_1}$;

The organization and management of drilling technological processes is impossible without the methods based on hydromechanical strengthening of the wellbore by water jet drilling fluids. The technology is implemented simultaneously with the well drilling process and ensures the formation of a near-wellbore waterproofing screen. The near-wellbore waterproofing screen that is formed under these thermodynamic conditions is characterized by thickness of the adhesive layer on the borehole wall - 2–3 mm; depth of the colmatation zone in the matrix of permeable rocks - 10–30 mm; hydromechanical strength of the well under repression - 0.018-0.025 MPa / m (rock pressure gradient); shaft tightness in case of depression - 0.010–0.015 MPa / m [1].

In order to increase the efficiency of the screen and improve the hydromechanical strength of the walls of wells drilled in unstable and unstable rocks, gel-cement drilling solution that was tested can be used. The development of a new drilling solution was carried out in collaboration with scientists and specialists of NPP "BURINTEH" LLC at the company's laboratory facilities. Gel-cement drilling solution "Gel-Drill" [2, 4] has good thixotropic, waterproofing, inhibiting parameters and carbonic aggression.

3. Results and Discussion

In 2016-2017, with the participation of BURINTEH and TEHGEOSERVICE, field tests (FT) of the wellbore hardening technology were carried out using new drilling fluid.

The pilot project was aimed at assessing the compliance of technological parameters and operational characteristics of drilling fluids with their purpose when drilling deep wells in difficult geological and thermodynamic conditions, as well as improving quality, efficiency and environmental

IOP Conf. Series: Materials Science and Engineering **952** (2020) 012039 doi:10.1088/1757-899X/952/1/012039

safety of drilling operations and extending the operating time of oil and gas wells at maximum limited volumes of produced formation waters.

Pilot tests were carried out on 17 inclined horizontal wells with a depth of 1430–2050 m. The testing interval was 220–1780 m (pilot, transport and production casing boreholes).

Typical for the drilling depth is the presence of fluid-saturated wells in the section of wells with reservoir pressure differentiated by the section and area of deposits, as well as unstable rocks prone to hydraulic fracturing. They cause violations and complications of the drilling technology and associated negative consequences. With the traditional drilling technology, this happens: when the Kashir horizon is opened in the range of 959–1010 m (integral values); the Verey horizon is opened in the range of 1040–1098 m; the Bashkirian stage - in the range of 1098-1155m; the Serpukhovian stage – in the range of 1155–1350m.

The technological complexity of geological and technical conditions for well drilling is characterized by the following integral indicators: intensity of most (≤80%) losses - 30-50 m^3/h, water showings - 0.5 to 1-12 m^3/h, gas showings - background or short-term - 0.015–0.6 vol. /%, talus and landslides - from weak to intense (landslide) ones.

Drilling of pilot wells was adapted to serial equipment and design technological solutions, providing a synchronous and coordinated mode of hydromechanical strengthening of the wellbore by quantum impact.

The tool consisted of a three-cone chisel 215.9 RDS US 616 I430 with six nozzles with a diameter of 11.1 mm, an above-bit sub - a colmatator with a side nozzle diameter of 12.7 mm, a drill collar. The drilling method is rotary with a rotation frequency of 30–35 rpm, the contact of the jet with the borehole wall is 0.013–0.004 s, the wellhead pressure is 9–13 MPa. Flushing fluid: polymer clay drilling mud (BR) "Gel-DRILL" (patent of LLC NPP "BURINTECH" [4]) with the following parameters: ρ = 1190-1230 kg / m3, SNS 1 min / 10 min, HC + 50-70 s, Ø = 5 ml / 30 min, c ≥ 16% (bentonin + cement).

Analytical assessment of pilot testing results was carried out according to the reports of service support of LLC NPP "BURINTEH" and reports on the services provided for telemetric and technological support of LLC "TECHGEOSERVICE". Well logging methods were also used (cavernous and thermometry, SGDT, cementometry).

Thus, the pilot test based on the hydromechanical strengthening method identified that

1. "Technology" corresponds to its purpose and ensures the achievement of the goals. Wells are drilled without complications and disruption of technological processes in the intervals for technical (pilot, transport) and production strings.

2. Similar results were obtained when using the "technology" instead of "Gel-Drill" polymer clay drilling mud.

3. Of six wells drilled without the use of an over-bit sub-collator, a positive result was obtained for two wells (No. 1200G and No. 13479G). The result is accidental and insufficiently effective since the method does not stabilize the initial properties of the cement slurry-stone, does not improve tightness of the cemented casing annulus and does not protect productive formations from pollution.

The advantages of the "technology" are: a non-linear increase in quality, technical, economic, and environmental indicators of drilling operations, and deterioration of production safety.

Table 1. Technical and economic indicators when drilling wells No. 453g and 454g of the Tatyshlin field

Technical and economic indicator	No 453g (experimental)	No 4504g (serial)
Time spent on additional operations (planned), hour:	0.20	39.91 (1.7 days)
cementing of the Verey horizon	-	32.33 (1.3 days)
transfer of a well from one type of solution to another one	0.20	7.58 (0.3 days)
Cost of 1 m^3 of solution, s.u.	0.88 (GEL-DRILL)	1.00 (clay-emulsion)

Reducing the total time spent on drilling wells. 453g (experimental) - 79.4 hours (3.3 days). Total: 3.4 + 3.3 +7.7 days

Formation of a sealed waterproofing near-wellbore screen isolates the wellbore from the rock mass; possibility of their active interaction during the drilling process.

Generalization and analytical evaluation of the data and results of pilot tests of "Gel-DRILL" and "Polcarb-BIO" during hydraulic hardening of the wellbore allow us to make a well-founded conclusion on improving the quality and efficiency of drilling operations in difficult geological and technogenic conditions.

4. Conclusion

The current level of quality, technical, economic and economic safety achieved by using traditional well drilling technologies excludes the likelihood of their further improvement and development without switching to systemic technological developments that ensure nonlinear growth of the baseline well construction indicators. In general, the results of the pilot study allowed for the following conclusions:

1. The use of "Gel-Drill" in combination with hydromechanical strengthening of the wellbore improved quality and efficiency of drilling of nine experimental wells with horizontal completion at "BASHNEFT" fields.

As a result, the technical and economic indicators of drilling operations increased (for example, well No. 453g Tatyshly):

- ROP - by 13.6%;
- technical drilling speed - by 8.6%;
- commercial drilling speed - by 6.5%

The technological indicators changed in the following way:

- the coefficient of cavernousness of the Verey horizon decreased;
- the time for additional operations (isolation operations, borehole reaming, drilling fluid replacement) decreased by 97.5%.
- the prime cost of 1 m^3 of drilling mud decreased by 12%.

To continue the technological stagnation, improve technological processes and efficiency of development of oil and gas deposits, one of the promising scientific and technical directions is a synchronous and coordinated regime of the quantum impact on the formation of a near-wellbore waterproofing screen during well drilling.

References

[1] Polyakov V N, Zeigman Yu V, Kotenev Yu A, Mukhametshin V V, Sultanov Sh Kh, and Chizhov A P 2018 System solution for technological problems of well construction completion *Nanotechnologies in Construction* **10(1)** 72–87 DOI: 10.15828/2075-8545-2018-10-1-72-87

[2] Polyakov V N, Chizhov A P, Kotenev Yu A and Mukhametshin V Sh 2019 Results of System Drilling Techniques and Completion of Oil and Gas Wells *IOP Conference Series: Earth and Environmental Science (IPDME 2019 – International Workshop on Innovations and Prospects of Development of Mining Machinery and Electrical Engineering)* **378(1) 012119** 1–7 DOI: 10.1088/1755-1315/378/1/012119

[3] Malyarenko A M, Bogdan V A, Kotenev Yu A, Mukhametshin V Sh, Umetbaev V G 2019 Wettability and formation conditions of reservoirs *IOP Conference Series: Earth and Environmental Science (IPDME 2019 – International Workshop on Innovations and Prospects of Development of Mining Machinery and Electrical Engineering)* **378(1) 012040** 1–6 DOI: 10.1088/1755-1315/378/1/012040

[4] Khokhlov V I, Galimov Sh S, Devyatkova S G, Kotenev Yu A, Sultanov Sh Kh and Mukhametshin V Sh 2019 Justification of impact and planning of technology efficiency on the basis of limy-emulsion formulation in low-permeability highly-rugged reservoirs of Tyumen deposits *IOP Conference Series: Earth and Environmental Science (IPDME 2019 –*

International Workshop on Innovations and Prospects of Development of Mining Machinery and Electrical Engineering) **378(1) 012114** 1–6 DOI: 10.1088/1755-1315/378/1/012114

[5] Bone K, Bradley B, and Kale I 2019 Next-Generation Hybrid Drill Bit Produces Exceptional Drilling Dynamics and Time-Saving Improvements *Offshore Technology Conference (6-9 May Houston Texas)* 1–18 DOI: 10.4043/29440-MS

[6] Dolezal T, Felderhoff F, and Holliday A 2011 Expansion of Field Testing and Application of New Hybrid Drill Bit *SPE Annual Technical Conference and Exhibition, Denver, Colorado, 30 October-02 November* 1–12 DOI: 10.2118/146737-MS

[7] Dupriest F E, and Koederitz W L 2005 Maximizing Drill Rates with Real-Time Surveillance of Mechanical Specific Energy *SPE/IADC Drilling Conference (Amsterdam, The Netherlands, 23-25 February)* 1–10 DOI: 10.2118/92194-MS

[8] Dykstra M W, Armenta M A, Fitzerald A M, Adesokan O, Schornick T L, Chowdhury A R, and Allain M D 2018 Converting Power to Performance: Gulf of Mexico Examples of an Optimization Workflow for Bit Selection, Drilling System Design and Operation *Offshore Technology Conference (Houston, Texas, 30 April–3 May)* 1–29 DOI: 10.4043/29065-MS

[9] Romero S, Monroy R R, Johnson C, Cardenas F, and Abraham G A T 2006 Preventing Lost Circulation by Use of Lightweight Slurries With Reticular Systems: Depleted Reservoirs in Southern Mexico *SPE Drilling & Completions Journal* **21(3)** 185-192 DOI: 10.2118/92187-PA

[10] Roth J, Reeves C, Johnson C R, De Bruijn G, Bellabarba M, Le Roy-Delage S, and Bulte-Loyer H 2008 Innovative Hydraulic Isolation Material Preserves Well Integrity *IADC/SPE Drilling Conference (Orlando, Florida, USA, March 4-6)* 1–14 DOI:10.2118/112715-MS

[11] Rodgers J D 2006 *Assessments of Technologies for Environmentally Friendly Drilling Project: Land-Based Operations Houston Advanced Research Center* (Houston, Texas, Draft 3, Version 2, March 23)

[12] Brown T J, Idoine N E, Raycraft E R, Shaw R A, Hobbs S F, Everett P, Deady E A, and Bide T World Mineral Production 2012-2016 *British Geological Society*

[13] Valeev A S, Kotenev Yu A, and Mukhametshin V Sh 2018 Evaluation of Water-Alternating-Gas Efficiency when Using Wide Range of Gas Composition *IOP Conference Series: Earth and Environmental Science (IPDME 2018 – International Conference on Innovations and Prospects of Development of Mining Machinery and Electrical Engineering)* **194(8)** 1-8 DOI: 10.1088/1755-1315/194/8/082042

[14] Mukhametshin V Sh, Kotenev Yu A and Sultanov Sh Kh 2018 Assessment of the Need to Stimulate the Development of Hard-to-Recover Reserves in Carbonate Reservoirs *IOP Conference Series: Earth and Environmental Science (IPDME 2018 – International Conference on Innovations and Prospects of Development of Mining Machinery and Electrical Engineering)* **194(8)** 1-4 DOI: 10.1088/1755-1315/194/8/082027

[15] Kotenev Yu A, Mukhametshin V Sh, and Sultanov Sh Kh 2018 Energy-efficient technology for recovery of oil reserves with gas injection *IOP Conference Series: Earth and Environmental Science (IPDME 2018 – International Conference on Innovations and Prospects of Development of Mining Machinery and Electrical Engineering)* **194(8)** 1-6 DOI: 10.1088/1755-1315/194/8/082019

IOP Conf. Series: Materials Science and Engineering **952** (2020) 012040 doi:10.1088/1757-899X/952/1/012040

Hydraulically perfect modes of injection of grouting mixtures when isolating absorbing formations

A P Chizhov[1,2], V E Andreev[1,2], A V Chibisov[1,2], V V Mukhametshin[1] and L S Kuleshova[3]

[1] Ufa State Petroleum Technological University, 1, Kosmonavtov st., Ufa, Republic of Bashkortostan, 450062, Russian Federation
[2] SASI "Institute of strategic research of the Republic of Bashkortostan", 129/3, October avenue, Ufa, Republic of Bashkortostan, 450075, Russian Federation
[3] Ufa State Petroleum Technological University, Branch of the University in the City of Oktyabrsky, 54a, Devonskaya St., Oktyabrsky, Republic of Bashkortostan, 452607, Russian Federation

E-mail: vv@of.ugntu.ru

Abstract. The article deals with a mechanism of isolation of absorbing intervals during the construction of oil and gas wells. Research is based on systemic approaches to complex scientific and technical problems. The results of laboratory and technological experiments aimed at improving the hydromechanical processes of cementation of absorbing rocks are presented. Various technological schemes for injecting hardening solutions into intervals of absorbing formations, the mechanism of interaction between solutions and permeable rocks, hydraulically perfect modes of injection of backfill mixtures and quality and efficiency of insulation works are analyzed.

1. Introduction

The main qualitative and technical and economic indicators of drilling and completing deep wells depend on the effectiveness of methods used to deal with complications (losses, gas and oil showings, hydraulic fracturing of rocks, rock falls, interstratal flows, etc.) [1–13]. However, the greatest amount of time (22–23%) and material and financial resources (4–8%) are spent on the fight against losses of drilling and grouting fluids [14–16].

To solve this complex technical problem, various technological methods are used: regulation of the parameters of drilling and grouting fluids, the supply of mud pumps, the concentration of clogging fillers, the injection of highly structured clay solutions of pastes and paste plugs, viscoelastic systems (VES), injection of cement, gel-cement and polymer-cement mixtures, etc. All these methods have the same significant drawbacks - low efficiency (efficiency rate - 20-50%) and quality of insulation works [2, 6, 14]. This situation is due to underdeveloped technologies and the lack of systemic approaches and solutions, methods for monitoring and regulating the hydromechanical processes of interaction between cement slurries and absorbing rocks during the implementation of various mechanisms of their isolation [15].

In practice, this leads to an arbitrary and unjustified choice of technological parameters of the mode of injection of plugging mixtures into the channels of absorbing formations (without taking into account the mechanisms for reducing the intensity of absorbing rocks), types, plugging-technical properties and volumes of waterproofing solutions and hardening mixtures, fillers and chemicals, which are not adequate to the geological and physical conditions of insulation works [16]. As a result, the quality and technical and economic indicators of insulation works become worse, and he area of their effective application is reduced.

2. Materials and methods

Waterproofing of absorbing formations is an unsteady process of hydromechanical interaction of injected viscous-plastic fluids (HPL) and permeable rocks at the radius of penetration into filtration channels under the action of external factors: fluid flow rate and injection pressure. Despite the widespread use of this technology, the hydromechanical process of isolation of highly permeable absorbing rocks is understudied [1-14].

Therefore, the purpose of this article is to provide examples of implementation of system approaches and solutions in the drilling process.

3. Results and Discussion

Injection of cement slurries into the bottomhole zone of absorbing formations can occur under different injection modes - lateral transition and bottom ones [14]. Each mode has own mechanism of impact on permeable rocks, parameters and an area of effective application.

The mode of lateral injection of grouting mixtures into the lost circulation zone is implemented by creating repression on the formation top by regulating the initial structural and mechanical properties of the grouting mixtures and modes of their injection from the wellbore into the isolated interval (pump flow and injection pressure) (Figure 1).

Under hydrodynamic conditions, the movement of plugging mixtures in the bottomhole zone occurs simultaneously throughout the entire thickness of absorbing rocks. In terms of effectiveness of the influence of this injection scheme on the absorption intensity reduction, it is hydraulically perfect. It allows you to implement a number of important technological effects:

- "piston" displacement and replacement at the radius of penetration of formation fluid by cement slurries simultaneously throughout the entire thickness of the absorption interval, excluding negative changes in the initial properties of cement slurries and occurrence of intra-formation flows during and after the isolation operation;
- intensification of the dehydration process of hardening grout mixtures and their structural-mechanical and grout-technical properties (Figures 2, 3, Table 1). The mechanism of dehydration of hardening mixtures accelerates the process of structure formation in cement slurries, in which within 2–3 minutes the initial plastic viscosity of cement slurry increases 4–6 times (Figure 2), and the ultimate shear stress - 6.0–7.5 times (Figure 3). According to the results of laboratory studies, the time of the beginning of setting of the viscoplastic fluid and the water-cement ratio of 0.35–0.40 are reduced 1.5–1.9 times, and at the end of setting - 1.4–1.5 times.
- stable control and operational regulation of the technological process of isolation during the injection of solidifying solutions into the absorption channels.

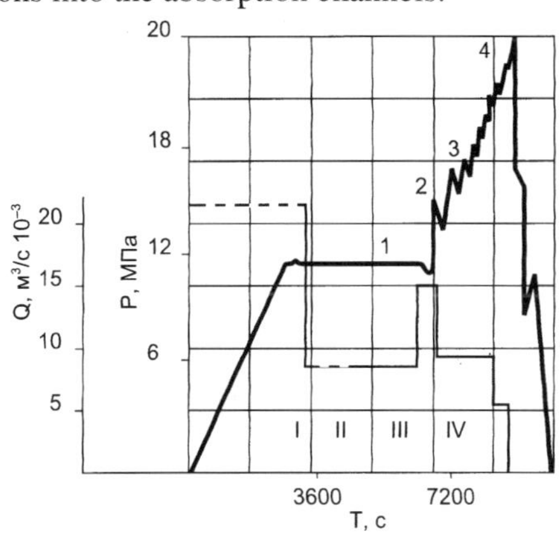

Figure 1. Isolation of absorption intensity up to 50 m^3/h in the mode of lateral injection of grouting mixtures: 1-2 - clay paste; 2-4 - cement mortar

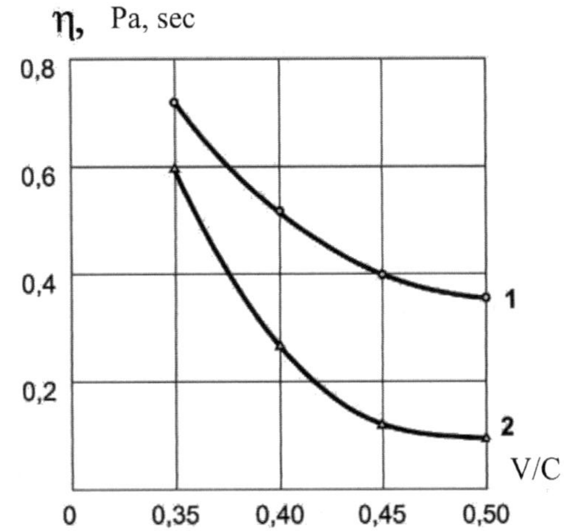

Figure 2. Dependence of plastic viscosity on the water-cement ratio: 1 - 20 minutes after mixing; 2 - 4 minutes after mixing:

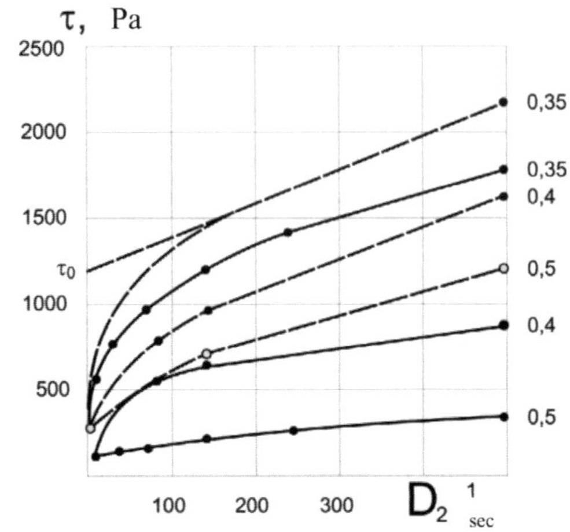

Figure 3. Dependence of the shear stress on the velocity gradient at different water-cement ratios: ——4 minutes after mixing; - - - - - 20 minutes after mixing

Table 1. Dependence of the grouting properties of the cement mortar-stone on the water-cement ratio

Water-cement ratio, units.	Spreadability, cm	Density, kg / m^3	Plastic viscosity, Pa s	Shear stress, Pa	Setting time, hour: min		Ultimate strength, MPa	
					start	end	When bending	When compressing
0.50	22	1830	0.10	300	6:19	2:06	2.22	4.5
0.45	20	1880	0.12	-	5:05	2:05	2.63	7.5
0.40	15	1950	0.26	800	4:12	1:45	3.50	11.0
0.35	12	2030	0.60	1750	3:20	1:33	4.60	18.3

The practical implementation of the above-mentioned intrasystem effects in the side injection scheme allows the isolation of losses with an intensity of 10 to 70 m3 / h and reduction of material and financial costs for combating losses by 1.5–1.85 times. The comparative technological efficiency of traditional and improved methods of absorption isolation is presented in Table 2.

Table 2. Performance of advanced reservoir isolation techniques

Isolation technology	Volume		Success rate	Time spent on one absorption, hour	Consumption of grouting materials and reagents per absorption		
	number of wells	number of isolations			Cement, t	Clay mortar, m3	Fillers, t
Serial	74	156	0.474	28.53	24.4	11.6	6.5
Recommended	87	104	0.836	15.42	16.4	10.3	1.0
Performance indicators			+1.76	-1.85	-1.49	-1.13	-5.9

The transient injection mode (Figure 4) is characterized by two sequential processes. At the beginning of the inflow of cement slurry into the bottomhole formation zone, this flow mode is formed during repression in the top of a formation close to or smaller than the formation (position 2 in Figure 4). At the second stage, as part of the filtration channels (cracks) is isolated and repression occurs in the top of absorbing strata, this flow mode switches to the mode of lateral injection of grouting mixture (positions 3-4 in Figure 4), where the indicator of hydraulic perfection of the injection mode δ is determined as:

$$\delta = \Delta P_r/\Delta P_b,$$

where ΔP_r – pressure drop on the roof of the block during injection of the cement slurry, MPa; ΔP_b – pressure drop at the bottom of the absorbing formation during the injection of the cement slurry, MPa.

A characteristic feature of the technological process is the periodic supply of absorbing rocks from the top in the interval of isolation of formation fluids (water, gas). This leads to the dilution of grouting solutions and violation of their original functional properties. The indicator of the hydraulic perfection of the modes of injection is 0.4-0.6. The area of effective application of the transient HPF injection mode is associated with catastrophic losses of more than 80–100 m^3/h.

The mode of bottom injection of HPF into an absorbing formation occurs when applied equipment (pumps, packers, etc.) and technological solutions (parameters of HPF injection, types and properties of backfill mixtures, plugging fillers, methods of regulating the setting and hardening of backfill mixtures etc.) do not create repressions on the top of the absorbing rocks (Figure 5). In such hydraulic conditions, the movement of grouting solutions is accompanied by a constant inflow of formation fluids from the roof part (intra-formation crossflow). The index of hydraulic perfection of the isolation mode does not exceed 0.4.

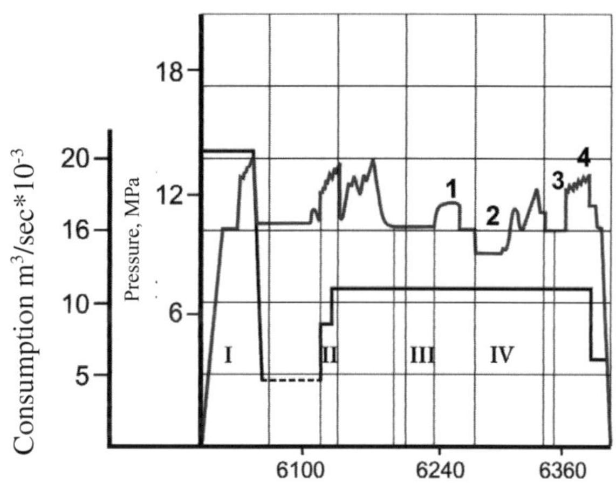

Figure 4. Isolation of absorption with an intensity of 50–80 m3 / h in the transient injection mode of grouting mixtures: 1-3 - non-hardening and hardening FWP; 3-4 - hardening FWP

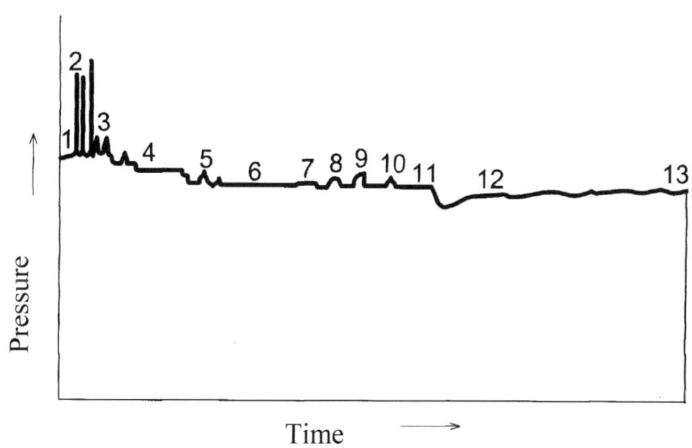

Figure 5. Hydraulic pressure in time.

4. Conclusion

The massive use of the improved technology of isolation of absorbing formations under hydraulically perfect modes of injection of backfill mixtures in Bashkortostan, Tatarstan, Eastern Siberia, and Orenburg region with the use of domestic non-scarce materials (clay powder, backfill cement, polymers and chemicals) contributed to the nonlinear improvement of quality and efficiency of insulation works (1.5-2.0 times higher than when using traditional methods) and reduction of financial costs and time to combat absorption with an intensity from 10 m^3/h to 150 m^3/h by 30-50%.

References

[1] Clayton C and Power P 2002 Managing Geotechnical Risk in Deepwater *Offshore Site Investigation and Geotechnics "Diversity and Sustainability" Proceedings of an International Conference (26-28 November, London, UK)* 1–19

[2] Roberts T S, Schen A E, and Wise J L 2005 Optimization of PDC Drill Bit Performance Utilizing High-Speed, Real-Time Downhole Data Acquired under a Cooperative Research and Development Agreement *SPE/IADC Drilling Conference (Amsterdam, The Netherlands, 23–25 February)* 1–14 DOI: 10.2118/91782-MS

[3] Cayeux E and Daireaux B 2009 Early Detection of Drilling Conditions Deterioration Using Real-Time Calibration of Computer Models: Field Example from North Sea Drilling Operations *SPE/IADC Drilling Conference and Exhibition (17-19 March, Amsterdam, The Netherlands)* 1–13 DOI: 10.2118/119435-MS

[4] Dupriest F E, and Koederitz W L 2005 Maximizing Drill Rates with Real-Time Surveillance of Mechanical Specific Energy *SPE/IADC Drilling Conference (23-25 February, Amsterdam, Netherlands)* 1–10 DOI: 10.2118/92194-MS

[5] Mukhametshin V V and Andreev V E 2017 Search and argumentation of decisions aimed at increasing the efficiency of bottom-hole zone stimulation in oil accumulations with challenged reserves *SPE Russian Petroleum Technology Conference (16-18 October 2017, Moscow)* 1–23 DOI: 10.2118/187785-MS

[6] Pink A P, Harrell L L, Perez G D, Carrico J, and Mackinnon P 2011 Integrated Drilling Optimization and Downhole Drilling Dynamics Enables PDC Bits to Drill in the Colorado Region of the San Juan Unconventional Gas Basin *SPE/IADC Drilling Conference and Exhibition (1-3 March, Amsterdam, The Netherlands)* 1–12 DOI: 10.2118/139848-MS

[7] Logan W D 2015 Engineered Shale Completions Based on Common Drilling Data *SPE Annual Technical Conference and Exhibition (Houston, Texas, USA, 28–30 September)* 1–15 DOI: 10.2118/174839-MS

[8] Yakupov R F, Gimazov A A, Mukhametshin V Sh, and Makaev R I 2018 Analytical method for estimating efficiency of oil recovery technology in case of bottom water-drive reservoir, verified on the hydrodynamic model *Oil Industry* **6** 66-69 DOI: 10.24887/0028-2448-2018-6-66-69

[9] Ejofodomi E, Baihly J, Malpani R, Altman R, Huchton T, Welch D and Zieche J 2011 Integrating All Available Data to Improve Production in the Marcellus Shale *North American Unconventional Gas Conference and Exhibition (14-16 June, The Woodlands, Texas, USA)* 1–37 DOI: 10.2118/144321-MS

[10] Khokhlov V I, Galimov Sh S, Devyatkova S G, Kotenev Yu A, Sultanov Sh Kh and Mukhametshin V Sh 2019 Justification of impact and planning of technology efficiency on the basis of limy-emulsion formulation in low-permeability highly-rugged reservoirs of Tyumen deposits *IOP Conference Series: Earth and Environmental Science (IPDME 2019 – International Workshop on Innovations and Prospects of Development of Mining Machinery and Electrical Engineering)* **378(1) 012114** 1–6 DOI: 10.1088/1755-1315/378/1/012114

[11] Bevilacqua M, Ciarapica F and Marchetti 2013 Bcquisition, Processing and Evaluation of Down Hole Data for Monitoring Efficiency of Drilling Processes *Journal of Petroleum Science Research (JPSR)* **2(2)** 49–56

[12] Malyarenko A M, Bogdan V A, Kotenev Yu A, Mukhametshin V Sh, Umetbaev V G 2019 Wettability and formation conditions of reservoirs *IOP Conference Series: Earth and*

Environmental Science (IPDME 2019 – International Workshop on Innovations and Prospects of Development of Mining Machinery and Electrical Engineering) **378(1) 012040** 1–6 DOI: 10.1088/1755-1315/378/1/012040

[13] Dwars S 2015 Recent Advances in Soft Torque Rotary Systems *SPE/IADC Drilling Conference and Exhibition (17-19 March, London, England, UK)* 1-12 DOI: 10.2118/173037-MS

[14] Polyakov V N, Mavlyutov M R, Alekseev L A, and Kolodkin V A 1998 *Technology and techniques for fighting absorption in the construction of wells* Ufa: Kitap 192 p

[15] Polyakov V N, Chizhov A P, Kotenev Yu A and Mukhametshin V Sh 2019 Results of System Drilling Techniques and Completion of Oil and Gas Wells *IOP Conference Series: Earth and Environmental Science (IPDME 2019 – International Workshop on Innovations and Prospects of Development of Mining Machinery and Electrical Engineering)* **378(1) 012119** 1–7 DOI: 10.1088/1755-1315/378/1/012119

[16] Polyakov V N, Sayapin E D, and Valyamov R G 1977 Technological influence of the flow rate and pressure of injection of a tamponage mixture on the effectiveness of operations for isolation of absorbing layers Drilling: scientific and technical collection **9** 20-22

Experience in using the Russian LWD complex for well drilling

A Y Davydov[1] and I A Yakhina[2]

[1] Ufa State Petroleum Technological University, Branch of the University in the City of Oktyabrsky, 54a, Devonskaya St., Oktyabrsky, Republic of Bashkortostan, 452607, Russian Federation
[2] LLC SPC «VNIIGIS-ZTK», 16A, Sadovoekoltso St., Oktyabrsky, Republic of Bashkortostan, 452607, Russian Federation

E-mail: alex-dy@yandex.ru

Abstract. The experience of using the LWD production complex of Research and Production Company VNIIGIS-ZTK LLC, which allows you to track the lithology and thickness of the layers in the direction of drilling, to predict the point of entry into the reservoir, to conduct the well on the most productive part of the reservoir, is considered.

1. Introduction

Despite the sufficient knowledge of the fields of the Volga-Ural oil and gas province as a whole, the reservoirs of many of them can be classified as complex, characterized by a high proportion of heterogeneity of the structure and reservoir properties. Currently, for the effective development of such fields at a late stage of operation, it is necessary to use modern means and methods of designing, researching and operating to achieve the maximum result on the way to increase the flow rate of wells. Collective work in the field of using modern field development tools combines a wide range of specialists with key skills in the field of knowledge of geology, geophysics, petrophysics, mechanics, etc. [1].

2. LWD complex of Russian production

Since 2012, Research and Production Company VNIIGIS-ZTK LLC has been actively developing in the field of creating logging equipment for the drilling process - LWD (logging while drilling). LWD technologies allow real-time adjustment of the position of the wellbore in the formation with optimal filtration and capacitive properties. In Russia, this market began to develop relatively recently. The complexes implemented on the basis of telemetry with a hydraulic communication channel LWD, represented by a wide range of geophysical modules, are successfully used by foreign companies (Schlumberger, Halliburton, Baker Huges, Weatherford, etc.). A significant amount of LWD work at the moment relates primarily to the drilling of horizontal sections of wells, which are a means of access to the development of hard-to-recover reserves and are widely used in offshore field development. Horizontal technologies contribute to an increase in oil recovery, and, consequently, well production due to an increase in drainage area and play a decisive role in well construction. However, the services provided with the LWD complex of foreign companies have one significant drawback - the very high cost of the work. This leads to a significant increase in the cost of well construction. And such additional costs, even in spite of all their efficiency, are not economically justified for deposits that are at a late stage of development [2, 3].

To solve the problem of determining the formation resistance and its reservoir properties during drilling, the Research and Production Company VNIIGIS-ZTK specialists developed a complex of geophysical methods of electromagnetic and radioactive logging: GRL gamma logging methods, IL induction logging, NNL neutron neutron logging, neutron logging gamma ray logging NGL. The NNL, NGL radioactive logging complex is used only as part of the extracted ZTK-42KK downhole telemetry system. Figure 1 shows the layout option of the bottom of the drill string (BHA) with LWD modules and provides indicative non-measurement data (distance from the bit to the measurement sensors of each module).

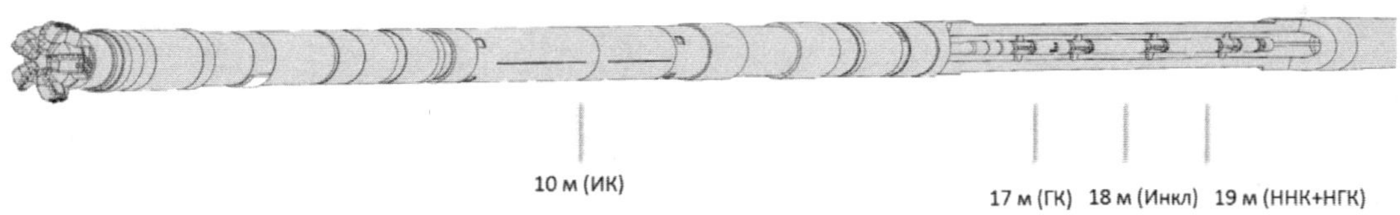

Figure 1. Drill string bottom layout with LWD modules

The neutron logging module (NLM) manufactured by VNIIGIS-ZTK NPF LLC has two thermal neutron neutron-neutron logging (NNL) probes and one gamma-ray neutron (NGL) probe (Figure 2, Table 1). It has end-to-end design, that is, it is installed inside a non-magnetic pipe and has overall dimensions corresponding to the dimensions of other nodal modules of the tele-system. It is used to determine the porosity (hydrogen content) of rocks during drilling. The module also has an internal flash memory for reading data after drilling [4, 5].

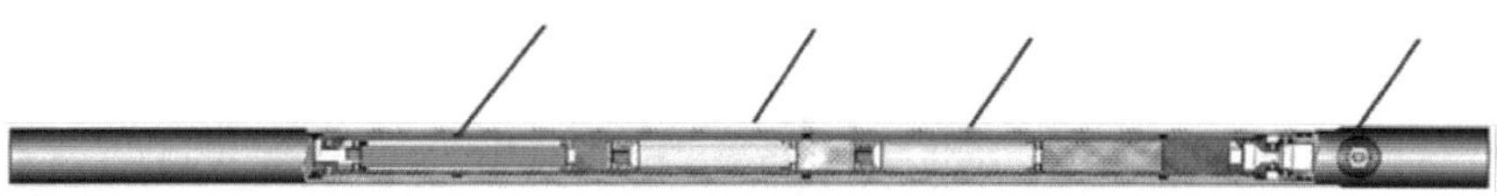

Figure 2. Location of sensors in the neutron logging module: NGL, NNL2, NNL1, III

Table 1.NLM Specifications

Options	Values
The range of measurement of porosity (volumetric hydrogen content),%	1-40
Pressure, MPa	0-60
Maximum working temperature, ° C	100
Overall dimensions, mm:	
diameter	42
length	1120

Ampoules containing Pu-Be powder are used as a source of ionizing radiation. The ampoule source is installed by specially trained personnel at the rig just before the telesystem is launched to the face.

The induction logging module is designed to differentiate rocks by resistance (conductivity) during drilling, and is used on any type of drilling fluid, including saline and invert emulsion. The induction logging module provides measurement and calculation of electrical resistivity by phase shift and amplitude attenuation using two current generation frequencies - 2 MHz and 400 kHz. 4 resistance curves are transmitted to the surface in real time (table 2). The module consists of the most measuring multi-coil device located inside a non-magnetic pipe (casing), which has radio-transparent windows. To work with the induction logging module, a plug-in autonomous power supply and a transceiver for docking with a telemetry system are used [6].

Table 2. Induction Log Module Specifications

Options	Values
Nomenclature of transmitted curves, Ohm · m:	
- P06h	0.1-2800
- P06L	0.1-2800
- P10H	0.1-900
- A06H	0.1-120
Maximum working temperature, ° C	120
Maximum hydrostatic pressure, MPa	80
Continuous work hours	150
Overall dimensions, mm:	
- diameter of the measuring module	48
- diameters of casings	108; 120; 178
- the total length of the module assembly with power and a transmitter-receiver unit	5500

The LWD operating mode provides for the simultaneous recording of GRL, IL, NNL, NGL logs and the transfer of all data, including from the over-bit module (or engine with telemetry), to the surface in real time via a wireless communication channel at a speed of up to 5 bits/sec. Logging during drilling makes it possible to quickly correct the wellbore trajectory depending on changing geological conditions, and allows to abandon additional intermediate and georeferenced logs. A comprehensive interpretation of the GRL, IL, NNL methods allows online to determine the reservoir properties of the reservoir and to calculate the petrophysical parameters already at the stage of initial opening.

3. Experience with the use of the LWD complex

Industrial implementation of the LWD complex produced by Research and Production Company VNIIGIS-ZTK LLC began in 2015 at the facilities of PJSC Tatneft. In total, for the period 2015-2019, with the full LWD complex, 7 horizontal wells were drilled in PJSC TATNEFT. As part of the BHA, in addition to improving the quality of the wellbore wiring in the reservoir, measurement systems on the bit were used - an over-bit module or an engine with integrated sensors. The data transmitted from the LWD modules helps to conduct the drilling of wells inside the reservoir in a more efficient way. Neutron-neutron logging data allows you to select dense rocks in the section (Figure 3) [7, 8].

In 2018, the joint work of Permneftegeofizika PJSC (Perm) and Research and Production Company VNIIGIS-ZTK (Oktyabrsky) LLC on providing service support for drilling at the facilities of LUKOIL Perm LLC (Perm). The main purpose of drilling was to restore the old well stock by drilling sidetracks, which at the end have an extended horizontal section. The objects of research were terrigenous reservoirs of the Bobrikov and Tula horizons and carbonate rocks of the Tournaisian stage. At the development stage, in terms of drilling support for LUKOIL-Perm LLC, a comprehensive approach was applied, including telemetric, technological, geo-navigation support services provided by Research and Production Company VNIIGIS-ZTK. The telemetry complex with a combined communication channel (ZTK-42KK) included the full LWD complex: gamma-ray logging (GRL), neutron gamma-ray logging (NGL), neutron-neutron logging (NNL), induction logging (IL). Based on the data on resistance and porosity obtained at the stage of primary opening, the main filtration-capacitive properties of rocks were calculated [9, 10].

The specialists of Permneftegeofizika PJSC based on the data of radioactive and electromagnetic logs obtained during the drilling process made the final interpretation of the well material. Correction of the wellbore during drilling was carried out on the basis of a dynamic geological model, which is built according to the correlation of the set of logging while drilling (LWD) methods of the reference and drilling wells. Geo-navigation support for drilling was carried out by geo-navigators of Research and Production Company VNIIGIS-ZTK LLC (Figure 4).

IOP Conf. Series: Materials Science and Engineering **952** (2020) 012041 doi:10.1088/1757-899X/952/1/012041

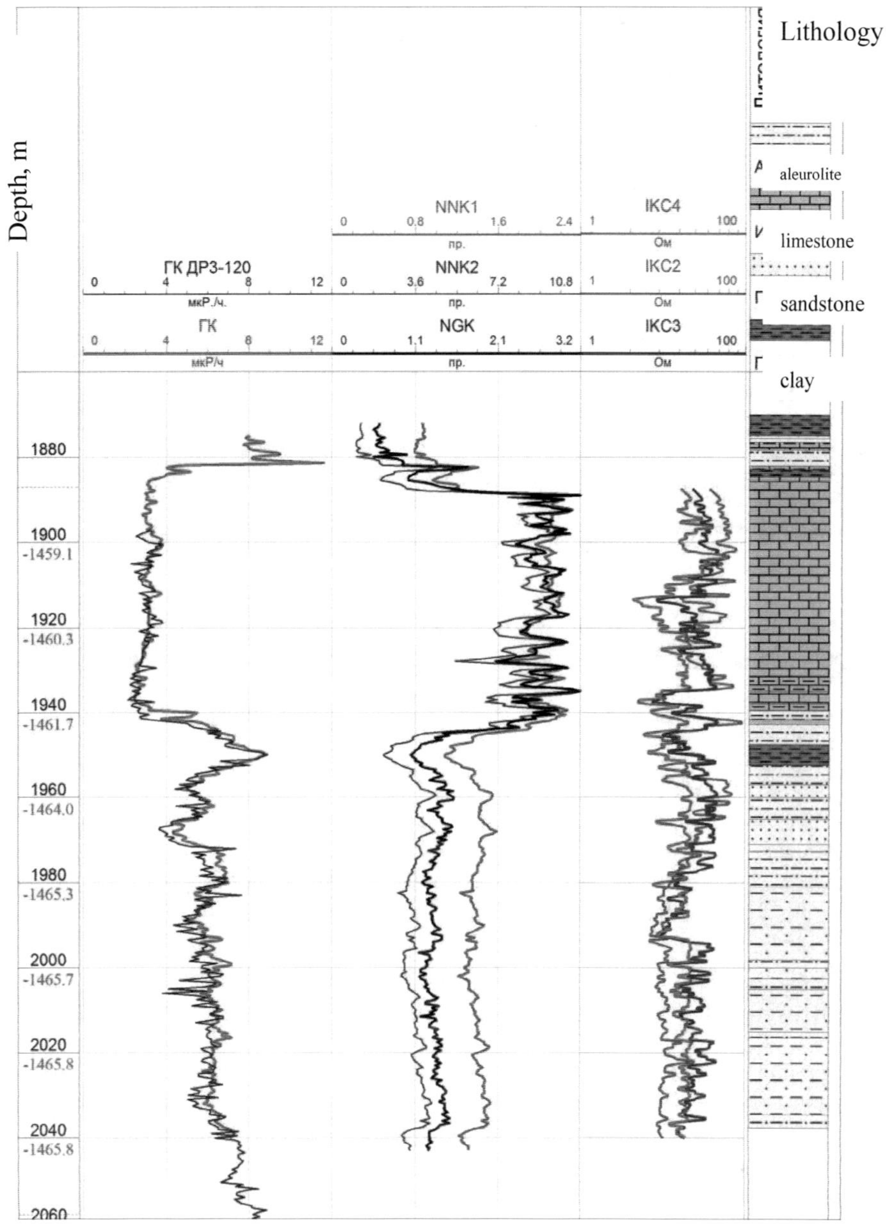

Figure 3. Determination of the interval of dense rocks according to the NNL method of the LWD complex

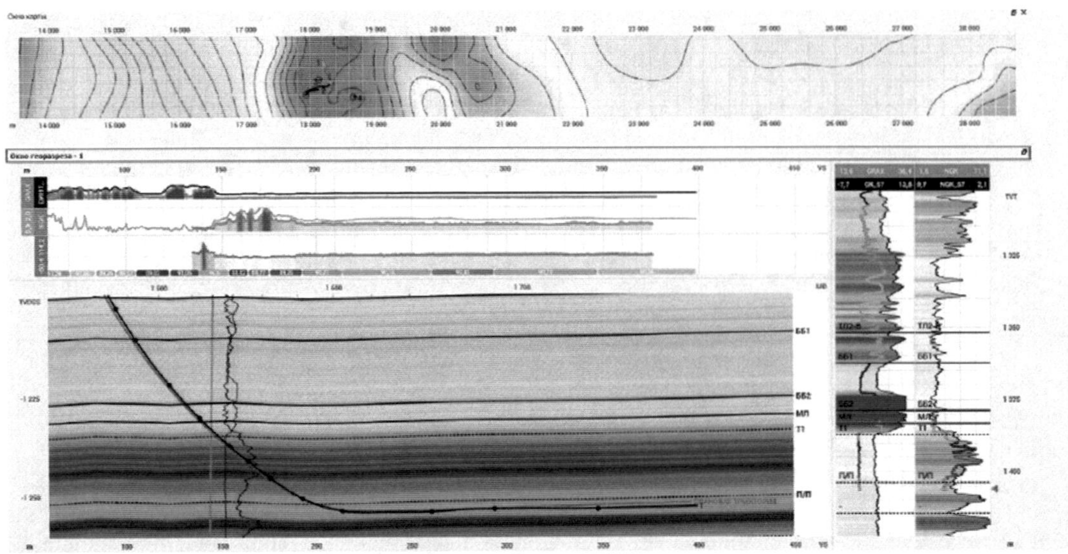

Figure 4. Geosteering support for drilling according to LWD

Between July 2018 and December 2019, 10 horizontal lateral shafts were drilled with the LWD complex of Research and Production Company VNIIGIS-ZTK LLC in Perm at the facilities of Lukoil-Perm LLC.

In 2019, the LWD complex was successfully widely used at one of the fields owned by the small oil company in Tatarstan, Okhtin-Oil JSC, where well support services using the LWD complex were provided at 7 horizontal wells [11, 12].

4. Conclusion

With a full range of LWD (including radioactive and electromagnetic logs) for the period 2015-2019. more than 30 wells were drilled in various mining and geological conditions. The use of LWD systems significantly reduced the time for well construction, as it allowed to abandon the final logging. The availability of online data from the logging complex and geo-navigation support services allowed us to quickly respond to changes in the geological situation at the bottom, reduce risks and uncertainties during the drilling process (control of the OWC zone, trace lithology and power in the direction of drilling), and predict the position of the wellbore relative to structural boundaries, to maneuver in the interbedded stratum of thin rocks. The integrated approach used made it possible to conduct wells in the most efficient way, having the ability to control the position of the productive part of the reservoir zone. The subsequent development of the drilled wells confirmed the presence in the interval of penetration of the horizontal section of oil-saturated reservoirs identified earlier during the drilling process.

References

[1] Andreev V E, Chizhov A P and Chibisov A V and Mukhametshin V S 2019. Forecasting the use of enhanced oil recovery methods in oilfields of Bashkortostan. *IOP Conference Series: Earth and Environmental Science (International Symposium «Earth sciences: history, contemporary issues and prospects»)* **350(1)** 1–6 DOI: 10.1088/1755-1315/350/1/012025.

[2] Davydov A Y, Yakhina I A 2019 Bit measuring systems *IOP Conference Series: Earth and Environmental Science (International Conference on Innovations and Prospects of Development of Mining Machinery and Electrical Engineering. 2019, IPDME 2019)* **378(1)** 012078 DOI: 10.1088/1755-1315/378/1/012078.

[3] Davydov A Y 2018 Utilization of expert systems for screw pump sets with surface drive management A*dvances in Engineering Research (AER) (International conference "Actual issues of mechanical engineering" (AIME 2018))* **157** 129-133

[4] Davydov A Y 2019 Monitoring of the operation of surface-driven screw pumping units using

expert control systems *Journal of Physics: Conference Series,* **1333(4)** 042005 DOI: 10.1088/1742-6596/1333/4/042005

[5] Zaripova L M and Gabdrakhimov M S 2019 Layout for drilling seismic wells by vibration method *IOP Conference Series: Earth and Environmental Science, 2019. International Symposium on Earth Sciences: History, Contemporary Issues and Prospects, ESHCIP* **350**(1) 012048. DOI: 10.1088/1755-1315/350/1/012048

[6] Zaripova L M and Gabdrakhimov M S 2019 Restoration of intake capacity of injection well by vibrations *IOP Conference Series: Earth and Environmental Science (International Conference on Innovations and Prospects of Development of Mining Machinery and Electrical Engineering 2019, IPDME* **378(1)** 012120. DOI: 10.1088/1755-1315/378/1/012120

[7] Zainagalina L Z, Suleimanov R I, Gabdrakhimov M S and Khabibullin M Y 2018 Determining oscillating system dynamic parameters of a near-bit junk pulper A*dvances in Engineering Research (AER) (International conference "Actual issues of mechanical engineering" (AIME 2018)* **157** 642-645

[8] Giltner, M., Earle, L., Willis, J., Tellez, D., Neel, R.Performance impact of downhole data from wired drill pipe and downhole sensors // 2019. SPE/IADC Drilling Conference, Proceedings 2019.

[9] Cohen J H, Maurer W C, Evans C R, Westcott P A 1995 High-power downhole motor *American Society of Mechanical Engineers, Petroleum Division (Publication) PD* **65** 265-271

[10] Hohl A, Herbig C, Arevalo P, (...), Macpherson J, Hughes B 2018 Measurement of dynamics phenomena in downhole tools - requirements, theory and interpretation *Society of Petroleum Engineers - IADC/SPE Drilling Conference and Exhibition, DC,* 2018, March.

[11] Al-Shuker N, Kirby C, Hughes B and Brinsdon M 2018 The application of real time downhole drilling dynamic signatures as a possible early indicator of lithology changes *Society of Petroleum Engineers - SPE/DGS Saudi Arabia Section Technical Symposium and Exhibition,* 2011, pp. 100-101.

[12] Wilson J K, Whitacre T and Heisig G 2019 High-frequency at-bit measurements provide new insights into torsional dynamics when drilling with steerable mud motors in unconventional horizontal wells *SPE/IADC Drilling Conference, Proceedings,* 2019, March.

IOP Conf. Series: Materials Science and Engineering **952** (2020) 012042 doi:10.1088/1757-899X/952/1/012042

Intensification of the gas inflow when bringing wells into production

G S Dubinskiy[1,2], V E Andreev[1,2], L S Kuleshova[3], V V Mukhametshin[1]

[1] Ufa State Petroleum Technological University, 1, Kosmonavtov st., Ufa, Republic of Bashkortostan, 450062, Russian Federation
[2] SASI "Institute of strategic research of the Republic of Bashkortostan", 129/3, October avenue, Ufa, Republic of Bashkortostan, 450075, Russian Federation
[3] Ufa State Petroleum Technological University, Branch of the University in the City of Oktyabrsky, 54a, Devonskaya St., Oktyabrsky, Republic of Bashkortostan, 452607, Russian Federation

E-mail: vv@of.ugntu.ru

Abstract. In wells brought into production, the bottomhole formation zone is contaminated. Contamination must be eliminated by acid treatment. The degree of contamination depends on reservoir characteristics and technological parameters of the primary and secondary opening. The intervals of formation permeability were identified; additional measures aimed at cleaning the reservoir from contamination and restoring well productivity should be implemented. Multi-solution acid treatment of the bottomhole formation zone improves productive characteristics of wells with deteriorated properties of the bottomhole reservoir.

1. Introduction

In the bottomhole formation zone (BFZ) of wells, contamination (clogging) of the reservoir occurs due to the penetration of filtrate and solid particles from drilling and other technological solutions [1, 2]. To reduce and / or remove clogging, a number of known methods are used [3–5]. The degree of contamination depends on the reservoir characteristics and technological parameters of the primary and secondary penetration of productive reservoirs [1, 6, 7]. When opening a productive formation by drilling and creating a sump, the time of contact of the drilling fluid with the borehole walls in the oil and gas-saturated reservoir increases for technological reasons and with an increasing sump depth. A long duration of the contact increases the contamination of the near-wellbore zone of the formation with drilling mud filtrate and mechanical particles [8–10]. When drilling with a high-viscosity mud, the repression on the formation increases. In such cases, the correct opening technology, quality of drilling fluids, active chemical agents have a tangible effect on the quality of the formation opening. During the reopening of the productive formation, perforation fluid penetrates into the reservoir, which also degrades the reservoir properties. With good reservoir properties of the perforated productive formation, the above factors contaminate the formation, and a zone of penetration of filtrate and drilling mud particles with a large radius is formed [11-14].

2. Materials and methods

Information from several fields of Western Siberia and measures taken to increase the inflow of fluids from the reservoir were used to assess the quality of the opening of productive formations, the need for hydrocarbon stimulation.

In order to determine the quality of penetration of productive layers, the indicator "relative permeability" (RP) is often used [1, 6]:

$$ОК= k_{bhf}/k_r , \quad (1)$$

where k_{bhf} – permeability of the bottomhole formation zone with reservoir properties that differ from the natural ones in the remote zone, μm^2; k_r – permeability of a remote formation zone with natural reservoir properties, μm^2.

To assess the quality of reservoir penetration and results of stimulation measures, the results of geological and physical studies (GPS) and hydrodynamic studies of wells (HSW) were used.

3. Results and Discussion

To determine the quality of penetration, calculations were performed for 48 wells of several fields in Western Siberia. In these wells, productive sandstone beds with similar characteristics were drilled (the results of studies and calculations in Table 1 and Figure 1).

The formations of the wells of group 7 have the lowest "relative permeability", the average permeability of the productive formations is $k_{av.}$ = × 10-3 μm^2, and the maximum one reaches 645 × 10-3 μm^2. The average permeability of the reservoir in group 7 is higher by 29.2 times than in group 1 (wells that penetrated the reservoirs with the lowest permeability), and RP_{av} in group 7 is 13.8 times less than in group 1. The opening of the productive formation (primary and secondary) in wells with such parameters is not performed well enough. Therefore, in such wells, it is necessary to plan activities to restore productivity and stimulate the flow of hydrocarbons.

Table 1. Permeability values of the bottomhole formation zone in different groups of wells by the value of relative permeability of the OC

Group of wells	Average OC_{av} in the group, %	Limits of changes in OC in the group, %	Average permeability k_{av} in the group, ×10^{-3} μm^2	Limits of changes in permeability k in the group, ×10^{-3} μm^2
1	162.7	44.4–760	10.0	4–14
2	208.8	3.8–889	20.3	15–33
3	18.4	11.5–50	53.0	40–78
4	38.5	6.5–147.5	117.4	104–134
5	31.5	2.1–68.8	152.9	140–169
6	13.5	3.3–24	277.0	173–195
7	11.8	2.8–22	291.8	200–645

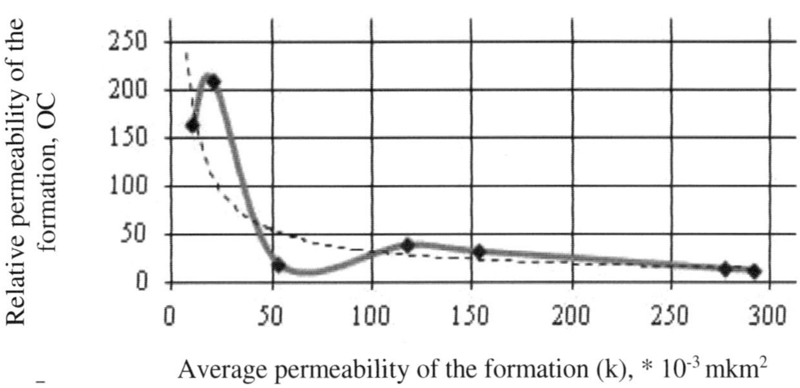

Figure 1. Dependence of the OC parameter on permeability of the bottomhole formation zone (k). The dashed line is a trend line.

The geological and production information indicates a tangible effect of the permeability value on the quality of opening of the reservoir. The following conclusions can be drawn:

- formations with relatively low permeability during opening by drilling and perforation are less prone to mudding, in comparison with highly permeable formations, which allows for self-cleaning of the bottomhole formation zone during standard development, without additional measures;
- during the secondary opening of the productive formation with permeability in the range from 40 × 10-3 μm² to 100 × 10-3 μm² and more than 120–150 × 10-3 μm², additional measures are required to clean up the bottomhole formation zone from contaminants, stimulate the flow of hydrocarbons and restore the well productivity;
- it is very important to predict the value of permeability and porosity of the productive formations in the wells being drilled in order to correctly select the technology for the primary and secondary opening; this will ensure a higher quality of the productive formation opening.

A number of wells after drilling and re-opening of productive formations do not reach the operating mode and / or do not reach the design level of productivity. For example, the wells No. XXX of the Yamburg gas condensate field (YGCF) have a different and heterogeneous section of the productive formations (Table 2). Some of the wells could not be put into operation using a standard inflow challenge; the planned productivity was not achieved. The best reservoir properties are typical of the strata drilled in wells 4, 5, 9. The strata with a high productivity were opened in wells 1, 3, 6, 7, and in well 8, the most dissected section (17 reservoirs with an average effective thickness of 1.4 m) with low permeability was opened.

Table 2. Reservoir characteristics of productive formations, which were penetrated in the wells # XXX YGKM

No of the well	Summary characteristics of formations according to the GPS data, penetrated by perforation			
	Recovered interlayers and layers by perforation, units	Effective formation thickness, m	Permeability, $\times 10^{-3}$ μm^2	Porosity, %
1	15	28.4	27.6	15.9
3	9	16.4	44.2	16.5
3*	3	5.2	123.9	17.1
Total for 3	12	21.6	63.4	16.6
4	17	24.2	106.0	15.4
5	11	16.6	125.3	16.4
6	11	20.0	60.3	16.2
7	8	15.1	60.2	15.8
8	17	23.8	5.7	15.1
9	9	24.4	63.3	16.5

No of the well	The worst penetrated formation by the GPS data			The best penetrated formation by the GPS data		
	Effective thickness, m	Permeability, $\times 10^{-3}$ μm^2	Porosity, %	Effective thickness, m	Permeability, $\times 10^{-3}$ μm^2	Porosity, %
1	2.6	0.4	13.0	6.0 12.6	82.5 23.6	17.2 16.7
3	2.2	0.6	13.6	1.2	188.5	16.9
3*	1.2	60.2	16.7	2.0	188.5	16.9
Total for 3	2.2	0.6	13.6	3.2	188.5	16.9
4	2.0	0.2	13.5	9.0 1.6	250.0 152.7	17.5 16.6
5	1.6	0.4	13.0	7.4	250.0	17.8
6	4.4	0.6	14.0	3.4	250.0	17.6
7	1.0	0.4	13.0	3.3	250.0	17.8
8	1.6	0.4	13.2	1.2 4.0	17.2 13.2	16.0 15.5
9	2.4	3.9	15.2	7.8	153.5	18.3

Note: * - after acid treatment of the exposed layers, BU_9^1.

After the introduction of highly productive layers BU91 in well No. 3, its general characteristics have significantly improved.

In wells No. 3, 5, 6 of the YGCF, prior to commissioning, stimulation of the inflow of hydrocarbons was carried out by the method of multi-solution acid treatment using acid compositions based on hydrochloric and hydrofluoric acids with the addition of surfactants, drying agents and solution stabilizers with reaction products (to prevent precipitation). The results of multi-solution acid treatments and a comparison of the operating parameters of wells after development (before acid treatments) and gas inflow stimulation into wells No. XXX are presented in Tables 3 and 4.

Table 3. Operating modes of wells No. XXX YGKM

No of the well	After well completion (before acidizing)							
	$\frac{Q}{\Delta P^2}$	Absolute free flow rate, thousand m ³	Mode 1			Mode 2		
			Washer diameter, mm	Temperature ^{0}C	Gas flow rate, thousand m ³	Washer diameter, mm	Temperature ^{0}C	Gas flow rate, thousand m ³
1	5.69	559	11.0	24.6	309	17.5	+26.0	465
3	3.27	345	11.0	18.9	241	17.4	+21.6	302
3*	BU_9^1 was added after acidizing							
4	3.77	338	11.0	24.4	241	17.0	+25.0	294
5	The well has not been explored, because it worked in a hydration mode							
6	3.16	301	11.0	21.4	227	17.0	+23.7	283
7	4.53	420	11.0	23.7	278	17.0	+24.0	364
8	3.80	372	11.0	20.5	261	17.0	+20.5	334
9	4.68	447	10.9	25.2	283	17.0	+25.3	404

No of the well	After acid treatment							
	$\frac{Q}{\Delta P^2}$	Absolute free flow rate, thousand m ³	Mode 1			Mode 2		
			Washer diameter, mm	Temperature ^{0}C	Gas flow rate, thousand m ³	Washer diameter, mm	Temperature ^{0}C	Gas flow rate, thousand m ³
1	acid treatment was not carried out							
3	3.73	339	11.0	22.6	253	17.0	+22.9	309
3*	3.89	346	11.0	23.5	241	17.0	+23.0	318
4	acid treatment was not carried out							
5	5.82	511	11.0	27.7	302	17.0	+27.5	419
6	3.37	333	11.0	22.5	232	15.0	+23.5	292
7	acid treatment was not carried out							
8	acid treatment was not carried out							
9	acid treatment was not carried out							

Table 4. Results of hydrodynamic studies of wells No. XXX YGKM

No of the well		1	3	3*	4	5
After development (before acid treatment)	Filtration coefficient A	189.23	405.66	BU_9^1 was added	The well worked in a hydration mode, therefore it has not been investigated	251.87
	Filtration coefficient B	-0.0363	-0.3375			0.0518
After acid treatment	Filtration coefficient A	Acid treatment was not carried out	265,00	206.19	169.76	Acid treatment was not carried out
	Filtration coefficient B		0,0115	0.1854	0.0066	

Table 4. Results of hydrodynamic studies of wells No. XXX YGKM (continue)

No of the well		6	7	8	9
After development (before acid treatment)	Filtration coefficient A	308.17	219.92	259.09	185.90
	Filtration coefficient B	0.0334	0.0028	0.0145	0.0910
After acid treatment	Filtration coefficient A	290.69	Acid treatment was not carried out	Acid treatment was not carried out	Acid treatment was not carried out
	Filtration coefficient B	0.0205			

The results show that the expected inflows were not achieved, primarily in wells No. 5 and No. 3, 6, 4. In well. 4; the gas inflow made it possible to operate the well, but the characteristics of the layers opened are better than in wells No. 1, 8, 9, and the productive performance (Q / ΔP2 parameter in Table 3) is the worst among these four wells. This is due to the clogging of the bottomhole formation zone by deep penetration of drilling mud filtrate and perforation fluid due to the good reservoir properties of the formations. In well No. 5, which has characteristics close to well No. 4, clogging has led to the impossibility of bringing the well to the operating mode without additional works.

Successful acidizing of the bottomhole formation zone in well No. 5 (Table 3) suggests that a similar impact in well No. 4 with similar reservoir properties (Table 2) could give similar results. The bottomhole formation zone in wells No. 1 and 8 was polluted to a shallower depth, which is confirmed by the coefficients "A" and "B" (Table 4); therefore, self-cleaning by the gas flowing to the bottom occurred. Wells No. 7 and 9, which have good reservoir properties of the bottomhole formation zone (Table 2), were brought into production independently (as a result of self-cleaning and good permeability and low clogging of the bottomhole zone) (coefficients "A "And" B "in Table 3).

After perforation and development, wells No. 3 and No 6 had the lowest productivity (Q / ΔP2 parameter in Table 3), despite the fact that the reservoir characteristics of their BHP are comparable to the characteristics of the formations in well No. 7. Higher bollard formation is confirmed by the values of coefficients "A" and "B" (see Table 4). The acid treatment carried out on the bottomhole formation zone of wells No. 3 and 6 allowed them to reach the level of productive characteristics of the operation mode of well No. 4 (see Tables 3 and 4); their productivity increased by 14.1% and 6.6%, respectively. The introduction of highly productive formations BU91 in well No. 3 increased productivity by 4.3%, but it was less than expected. The additionally opened formations have better productivity (see table 2). This can be explained by the deep clogging of the BU91 formations. If perforation followed by acid stimulation was made throughout the interval, the result could have been better.

After the gas production intensification, all three treated wells (3, 5, 6) were brought into operation.

4. Conclusion

1. Acid stimulation made it possible to improve the productive characteristics of wells No. 3 and 6, while well No. 5 achieved the highest productivity value for cluster No. XXX.

2. For well No. 3, it was possible to achieve a better result if BU91 was penetrated simultaneously with other formations and acid stimulation was carried out on all formations (in this case, BU91 clogging can be reduced).

3. In all wells No XXX, the bottomhole formation zone is clogged. The contamination must be eliminated by acid stimulation or using another method.

4. The degree of clogging of the bottomhole formation zone is associated with the technological parameters of the primary (drilling) and secondary (perforation) opening of the productive formations, as well as reservoir characteristics of the formations. The results of well stimulation works are consistent with the conclusions made in the first part of the article, which gives grounds to speak of the need to predict the permeability and porosity of productive formations in drilled wells in order to select a more correct technology for formation opening.

5. When opening in wells formations with permeability from 40 · 10-3 μm^2 to 100 · 10-3 μm^2 and more than 140–150 · 10-3 μm^2, additional measures should be planned to clean up the bottomhole formation zone and restore well productivity. All wells with a permeability of more than 40 × 10-3 μm^2 should be investigated, the quality of penetration should be determined, and if self-cleaning of the formation does not occur, measures should be taken to restore productivity.

References

[1] Ganiev R R, Sultanov V G, Dubinsky G S, and Shamaev G A 1998 Analysis of Some Factors Affecting on the Quality of the Secondary Opening of Productive Layers *Geology, Geophysics and Development of Oil and Gas Fields* **11** 34-36

[2] Liang T, Gu F, Yao E, Zhang L, Yang K, Liu G, and Zhou F 2017 Formation Damage due to Drilling and Fracturing Fluids and Its Solution for Tight Naturally Fractured Sandstone Reservoirs *Geofluids* **2017 (9350967)** DOI: 10.1155/2017/9350967

[3] Basarygin Yu M, Budnikov V F, Bulatov A I, and Proselkov Yu M 2001 *Technological bases of development and silencing of oil and gas wells* M.: LLC Nedra-Biznestsentr 543 p

[4] Rodney L B, and William S B 2004 Production improvement from increased permeability using Engineered Biohemical *Secjndary Recovery Methodology in Marginal Wells of the East Texas Fields* p 132

[5] Horev V S, Dmitriev A Y, Boyko I A, Rakhimov T R, and Kumar P S 2016 The integrated method to select drilling muds for well construction in difficult geological conditions *IOP Conf. Series: Earth and Environmental Science* **33 (012038)** DOI:10.1088/1755-1315/33/1/012038

[6] Lyons C 2010 *Working guide to drilling equipment and operations* Houston: Gulf Publishing 617 p

[7] Agarwal R G, Al-Hussainy R, and Ramey H J 1970 An investigation of wellbore storage and skin effect in unsteady liquid flow *Analytical Treatment, Society of Petroleum Engineers Journal* **10(3)** 279–290 DOI: 10.2118/2466-PA

[8] Ganiev R R, Sultanov V G, Dubinsky G S, and Shamaev G A 1999 About the Need to Limit the Depth of the Sump to Increase the Quality of Reservoir Opening *Geology, Geophysics and Development of Oil and Gas Fields* **4** 28-30

[9] Zeigman Yu V, Mukhametshin V Sh, Khafizov A R, Kharina S B, Abutalipova E M, and Avrenyuk A N 2017 Peculiarities of selecting well-killng fluids composition for difficult conditions *Oil Industry* **1** 66–69

[10] Andreev V E, Chizhov A P, Chibisov A V and Mukhametshin V Sh 2019 Forecasting the use of

enhanced oil recovery methods in oilfields of Bashkortostan *IOP Conference Series: Earth and Environmental Science (International Symposium «Earth sciences: history, contemporary issues and prospects»)* **350(1)** 1–6 DOI: 10.1088/1755-1315/350/1/012025

[11] Kotenev Yu A, Mukhametshin V Sh, and Sultanov Sh Kh 2018 Energy-efficient technology for recovery of oil reserves with gas injection *IOP Conference Series: Earth and Environmental Science (IPDME 2018 – International Conference on Innovations and Prospects of Development of Mining Machinery and Electrical Engineering)* **194(8)** 1-6 DOI: 10.1088/1755-1315/194/8/082019

[12] Chudinova D Yu, Kotenev Yu A, Sultanov Sh Kh, and Mukhametshin V Sh 2018 The neural network for grouping wells of different facies zones of productive layer *IOP Conference Series: Earth and Environmental Science (IPDME 2018 – International Conference on Innovations and Prospects of Development of Mining Machinery and Electrical Engineering)* **194(8)** 1-6 DOI: 10.1088/1755-1315/194/8/082008

[13] Yakupov R F, Gimazov A A, Mukhametshin V Sh, and Makaev R I 2018 Analytical method for estimating efficiency of oil recovery technology in case of bottom water-drive reservoir, verified on the hydrodynamic model *Oil Industry* **6** 66-69 DOI: 10.24887/0028-2448-2018-6-66-69

[14] Mukhametshin V V, and Kuleshova L S 2020 On uncertainty level reduction in managing waterflooding of the deposits with hard to extract reserves *Bulletin of the Tomsk Polytechnic University. Geo Assets Engineering* **331(5)** 140–146 DOI 10.18799/24131830/2020/5/2644

IOP Conf. Series: Materials Science and Engineering **952** (2020) 012043 doi:10.1088/1757-899X/952/1/012043

Wave thermal flooding when designing high-viscosity oil reservoirs

G S Dubinskiy[1,2], V E Andreev[1,2], V V Mukhametshin[1], L S Kuleshova[3], I Sh Mingulov[3]

[1] Ufa State Petroleum Technological University, 1, Kosmonavtov st., Ufa, Republic of Bashkortostan, 450062, Russian Federation
[2] SASI "Institute of strategic research of the Republic of Bashkortostan", 129/3, October avenue, Ufa, Republic of Bashkortostan, 450075, Russian Federation
[3] Ufa State Petroleum Technological University, Branch of the University in the City of Oktyabrsky, 54a, Devonskaya St., Oktyabrsky, Republic of Bashkortostan, 452607, Russian Federation

E-mail: vv@of.ugntu.ru

Abstract. The article describes an environmentally friendly and energy-saving method for increasing the volume of oil recovery - thermal flooding of a high-viscosity oil reservoir integrated with wave actions in injection wells. Heating of a remote formation zone to the effective temperature and wave impacts improve the conditions for effective oil displacement. The integrated use of associated gas and waste water from the oil treatment plant increases the energy efficiency and improves the environmental friendliness of small high-viscosity oil fields. For geological and physical conditions of high-viscosity oil deposits, it is recommended to employ the method of wave thermal flooding.

1. Introduction

The water cut of produced fluid by oil producing wells of "old" fields and a large share of conventional oil reserves production make to pay e attention to hard-to-recover reserves (HTRR), associated with high-viscosity oils (HVO), natural bitumen (NB), low-permeability reservoirs, residual reserves, etc. [1-6]. Thus, the role of small deposits, deposits associated with HTRRs, is increasing; to maintain and stabilize the level of oil production, it is necessary to apply new environmentally friendly and energy-saving technologies [1, 2].

Enhanced oil recovery (EOR) methods have been used for more than half a century in a large number of oil-producing countries and regions [1]. In recent decades, modern technologies of enhanced oil recovery and production intensification have been developed [2, 7, 8]. EORs have been used at fields with different geological and physical conditions, and this made it possible to evaluate their effectiveness and predict effective use for various hydrocarbon fields. The effect of the same enhanced oil recovery technologies at one of the fields decreased. With a decrease in the efficiency of an EOR and an increase in the water cut of fluid and reference of residual oil to the HTRR category, the volume of application decreased. At this time, fields with a complicated geological structure containing explosives began to be exploited. Their reserves belong to the HTRR [1, 9]. This caused the need to search for energy-saving and innovative methods for the development of fields with HTRR under different geological conditions [2, 7–12]. A technical and economic assessment of the use of various oil recovery processes is required [13].

The aim of the study is to determine the effectiveness of maintaining the reservoir temperature in combination with wave action on the productive reservoir for small deposits of high-viscosity oil and to develop recommendation for using the method.

2. Materials and methods

This goal was achieved by analyzing the temperature regime of the productive formation, studying viscosity properties of oil, measuring the flow rates of liquid and oil in wells of the experimental section of the Soldatovskoye field.

3. Results and Discussion

A significant number of studies on the applicability of various EOR for the development of HTRR have been carried out [1, 2, 9, 10]; criteria for the applicability of EOR have been found [1, 2, 11]. Using modern techniques, several types of EOR that are applicable to high-viscosity oil deposits confined to the South Tatar arch and Melekess depression have been determined [10, 12]. In world practice, a significant number of technologies are used for the development of HTRR [1, 13]. Among other methods of well stimulation and EOR, a wave action was recommended. It can be used for the Bobrikovskaya high-viscosity oil reservoir at the Soldatskoye field (oil viscosity is more than 200 mPa s).

At the Soldatskoye field, thermal flooding was implemented based on the scheme that can reduce the technogenic load on the environment [11]. Hot waste water (temperature up to 60–65°C) obtained from the oil treatment unit was used for injection into the formation. Energy for technological heating was consumed using associated petroleum gas. This scheme made it possible to avoid emissions into the atmosphere and, accordingly, environmental payments (fines) for such emissions. The laboratory experiments determined the effective temperature for the field. It is +29.7^{0}C. Thus, in order to achieve the best oil displacement effect, hot water injection temperature must be +29.7^{o}C.

Field studies have shown that in the reservoir conditions, there was a decrease in oil viscosity by about 30 mPa·s. This means that the development indicators for the Bobrikovskaya deposit of the Soldatskoye field have been obtained due to the thermal effect on this deposit. The forecast of changes in viscosity made on the basis of laboratory studies shows (the correlation coefficient is close to "1") that oil heated to more than 40–50 °C has a lower energy efficiency than oil heated to 30–40 °C. In fact, the temperature was 29.5–30 °C.

This mode provided an increase in oil production and achievement of the oil recovery factor design (ORF).

Injection well No. 4 (figure) was selected for the wave action (WA), into which hot water is injected to maintain reservoir pressure and temperature. It was expected that the wave action would increase the injectivity of the well due to the inclusion of idle intervals of the productive formation in the filtration process and cleaning the bottomhole zone from clogging material.

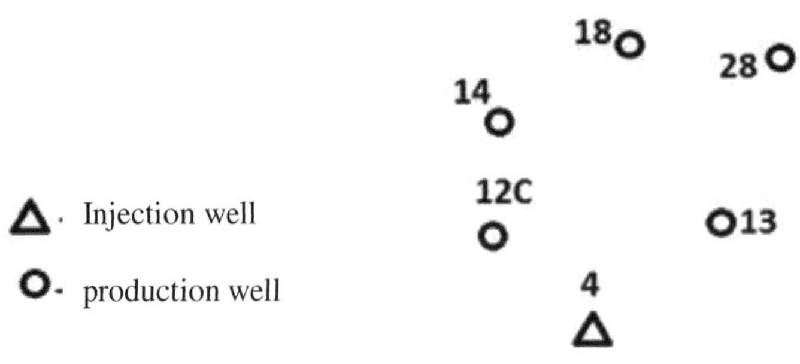

Figure 1. Scheme of the wave thermal flooding section

The bottom hole was treated with a hydraulic generator using the cavitation method. The generator has a design that allows you to create radiation in high and low frequency ranges. BHZ treatment is

IOP Conf. Series: Materials Science and Engineering **952** (2020) 012043 doi:10.1088/1757-899X/952/1/012043

carried out in each perforation interval, i.e. the generator worked at two points. The operation of the generator was confirmed by recording with a deep self-contained pressure gauge. The processing was carried out for more than 8 hours. Then injection well # 4 was launched with a wave hydraulic generator installed on the tubing shoe above the upper perforation interval.

After the wave impact on the bottomhole formation zone and the start of the wave thermal flooding (WTF) process, the injectivity of the well exceeded the initial one by 25%. In the reacting wells, 2018 tons of oil were additionally produced after the beginning of the wave thermal flooding (table)

Table 1. Results of implementation of the wave thermal flooding method at the site of the Soldatskoye deposit

No of wells	Additional oil production per year after wave treatment, tons	Average current gain in oil production rate, t / day
12C	453	1.3
13	459	1.3
14	821	4.1
18	218	0.7
28	67	0.2
Total	**2018**	**1.5**

The average monthly technological efficiency of WTF (in comparison with simple thermal flooding) per one production well was 33.6 tons. The smallest increase in oil production was observed in wells # 18 and # 28, which are the farthest from the stimulating injection well # 4.

Experimental works on the thermal wave water flooding technology confirm theoretical and experimental studies, which identified and described the processes and effects arising from wave and resonance impacts on porous and fractured oil gas reservoirs [12, 14]. The movement of fluids is accelerated. There appear new ones. There are differences between porous blocks-fractals and cracks, due to which the previously immobile fluid is involved. It is obvious that the use of wave technologies has significant potential and can allow one to involve in the cost-effective development of HTRR.

4. Conclusion

The complex system of RPM and PPT allows solving environmental and energy saving issues. The reservoir pressure and temperature maintenance system is able to maintain the initial reservoir temperature and prevent complications due to the cool reservoir containing high-viscosity oil. The use of associated gas as an energy carrier increases the energy efficiency and improves the environmental friendliness of high-viscosity oil fields.

Thermal wave waterflooding, which is a modification of thermal flooding, provides an increase in production and final oil recovery.

For the geological and physical conditions of the HVO deposits, the method of wave thermal flooding can be used. It will increase the energy efficiency of the HTRR recovery and oil recovery.

References

[1] Muslimov R Kh 2005 *Modern methods of oil recovery increasing: design, optimization and performance evaluation* (Kazan': FEN)

[2] Druetta P, Tesi P, De Persis C, Picchioni F 2016 Methods in Oil Recovery Processes and Reservoir Simulation *Advances in Chemical Engineering and Science* **6** 399-435 DOI: 10.4236/aces.2016.64039

[3] Andreev V E, Chizhov A P, Chibisov A V and Mukhametshin V Sh 2019 Forecasting the use of enhanced oil recovery methods in oilfields of Bashkortostan *IOP Conference Series: Earth and Environmental Science (International Symposium «Earth sciences: history, contemporary issues and prospects»)* **350(1)** 1–6 DOI: 10.1088/1755-1315/350/1/012025

[4] Valeev A S, Kotenev Yu A, and Mukhametshin V Sh 2018 Evaluation of Water-Alternating-Gas Efficiency when Using Wide Range of Gas Composition *IOP Conference Series: Earth and

Environmental Science (IPDME 2018 – International Conference on Innovations and Prospects of Development of Mining Machinery and Electrical Engineering) **194(8)** 1-8 DOI: 10.1088/1755-1315/194/8/082042

[5] Mukhametshin V Sh, Kotenev Yu A and Sultanov Sh Kh 2018 Assessment of the Need to Stimulate the Development of Hard-to-Recover Reserves in Carbonate Reservoirs *IOP Conference Series: Earth and Environmental Science (IPDME 2018 – International Conference on Innovations and Prospects of Development of Mining Machinery and Electrical Engineering)* **194(8)** 1-4 DOI: 10.1088/1755-1315/194/8/082027

[6] Yakupov R F, Gimazov A A, Mukhametshin V Sh, and Makaev R I 2018 Analytical method for estimating efficiency of oil recovery technology in case of bottom water-drive reservoir, verified on the hydrodynamic model *Oil Industry* **6** 66-69 DOI: 10.24887/0028-2448-2018-6-66-69

[7] Druetta P, Tesi P, De Persis C, and Picchioni F 2016 Methods in Oil Recovery Processes and Reservoir Simulation *Advances in Chemical Engineering and Science* 399-435 DOI: 10.4236/aces.2016.64039

[8] Afeez O G, Junin R, Manan M A, Agi A, Yusuff A S 2019 An overview of chemical enhanced oil recovery recent advanes and prospects *International Nano Letters* **9** 171-202 DOI: 10.1007/s40089-019-0272-8

[9] Lipaev A A 2013 *Development of heavy oil and natural bitumen deposits* (Moscow-Izhevsk: Institute of computer research)

[10] Tyncherov K T, Mukhametshin V Sh, Paderin M G, Selivanova M V, Shokurov I V, Almukhametova E M 2018 Thermoacoustic inductor for heavy oil extraction *IOP Conference Series: Materials Science and Engineering (MEACS 2017 – International Conference on Mechanical Engineering, Automation and Control Systems)* **327 (4) (042111)** 1–8 DOI:10.1088/1757-899X/327/4/042111

[11] Khisamov R S, Khuzin R R, Andreev V E, Dubinsky G S, and Miyassarov A Sh 2015 Prospects of increase in efficiency of development of reservoirs containing high-viscosity oil with the use of energy-saving and resource-saving technologies *Oil Industry* **4** 52–55

[12] Ganiev O R, Ganiev R F, and Ukrainskiy L E 2014 *Resonant macro-and micromechanics of an oil reservoir. Intensification of oil production and enhanced oil recovery. Science and practice* (Moscow-Izhevsk: Institute of computer research)

[13] Satter A, Iqbal G M 2016 17-Enhanced oil recovery processes: thermal, chemical, and miscible floods *Reservoir Engineering. The Fundamentals, Simulation, and Management of Conventional and Unconventional Recoveries* 313-337 DOI: 10.1016/B978-0-12-800219-3.00017-6

[14] Dubinsky G S 2015 About the response of fractal structures of fluid-saturated reservoir rocks under wave impact on them *The Collection of Scientific Papers. The Development of Science in the 21st Century: Natural and Technical Sciences* 51–56 DOI: 10.17809/06(2015)-06

IOP Conf. Series: Materials Science and Engineering **952** (2020) 012044 doi:10.1088/1757-899X/952/1/012044

The use of a liquid destructor with flow into the stream during the injection of the fracturing mixture

I G Fattakhov[1,2], L S Kuleshova[2], AVKochetkov[1], I F Galiullina[2], I G Andaryanov[1], A R Safiullina[2] and I R Gizatullin[2]

[1] PJSC TATNEFT named after V.D. Shashin, 75, Lenin St., Almetyevsk, Republic of Tatarstan, 423450, Russian Federation
[2]Ufa State Petroleum Technological University, Branch of the University in the City of Oktyabrsky, 54a, Devonskaya St., Oktyabrsky, Republic of Bashkortostan, 452607, Russian Federation

E-mail: i-fattakhov@rambler.ru

Abstract. For most modern development systems, hydraulic fracturing is the main method of intensification and maintenance of hydrocarbon production. Hydraulic fracturing is achieved with the volume of communicating voids between the granules of fractional proppant. The volume of voids should be maximized, which is achieved by selecting the optimal proppant fraction composition, and in addition, hydraulic fracturing fluid, the choice of which is based on a number of competing factors: high viscosity, which provides the necessary hydraulic width created at the buffer stage of the crack, as well as prediction of subsidence proppant in the fracture and subsequent premature stop of the injection, which creates additional costs. One of the new solutions, which will increase the efficiency and expand potential objects for hydraulic fracturing, is the use of a liquid destructor with a flow into the stream during the injection of the fracturing mixture.

1. Introduction

For most modern development systems, hydraulic fracturing is the main method of intensification and maintenance of hydrocarbon production. Non-traditional collectors are commissioned, the industrial development of which is not economically feasible without the use of relevant intensification methods. Proppant ones are distinguished among the types of hydraulic fracturing, the effectiveness of which is ensured by the proppant delivery to the fracture. Proppant stages can be both basic for traditional treatments of terrigenous strata and part of complex hybrid processes [1, 2].

The efficiency of a hydraulic fracture fixed by a proppant is achieved by creating a volume of communicating voids between the granules of fractional proppant. To achieve the greatest conductivity of the fracture, the volume of voids should be maximum, which is ensured by selecting the optimal fractional composition of proppant, as well as hydraulic fracturing fluid. Against the background of an obvious correlation between the conductivity and the fractional composition of the proppant, the significance of hydraulic fracturing fluid is often underestimated, which can lead to to failure when trying to achieve the planned rates of flow rate growth. The conductivity is achieved by the proppant fixed in the formation and can be significantly reduced by the remains of the undamaged fluid.

IOP Conf. Series: Materials Science and Engineering **952** (2020) 012044 doi:10.1088/1757-899X/952/1/012044

2. Materials and methods

Hydraulic fracture fluid selection is based on a number of competing factors. So, to ensure the technological feasibility of using proppant of a larger fraction, it is necessary to use a more viscous liquid to prevent technological complications during work. Insufficient viscosity can lead to the loss of hydraulic width created at the buffer stage of the hydraulic fracture process, as well as to unpredictable proppant sedimentation in the fracture and subsequent premature injection stop. In this case, the planned proppant volume is not placed in the fracture, the proppant pumped into the formation is not efficiently fixed, and the proppant remaining in the technological tubing must be removed, which can lead to additional costs [3–5]. On the other hand, an unreasonably high viscosity of the fluid used can lead to the formation of a crack with non-optimal geometry,. In particular, it can lead to a decrease in the residual conductivity due to an increase in the mass of the gel reaction products with the destructor, partially clogging the void space of the proppant crack (figure).

A tool for controlling the liquid viscosity is a way to control the concentration of applied chemicals, including a gelling agent. After creating, fixing, and closing the fracture, the fluid viscosity required for proppant transportation becomes a negative factor preventing the flow of fluid through the fracture created by hydraulic fracture. Slow destructors are used to “activate” the operation of the fracture, which allow the proppant to be transported into the fracture, and then release the pore space of the proppant pack for fluid flow. Increasing the load of the gelling agent leads to serious requirements for the quality of the destructor, which ultimately imposes a restriction on the use of hydraulic fracture technology in general.

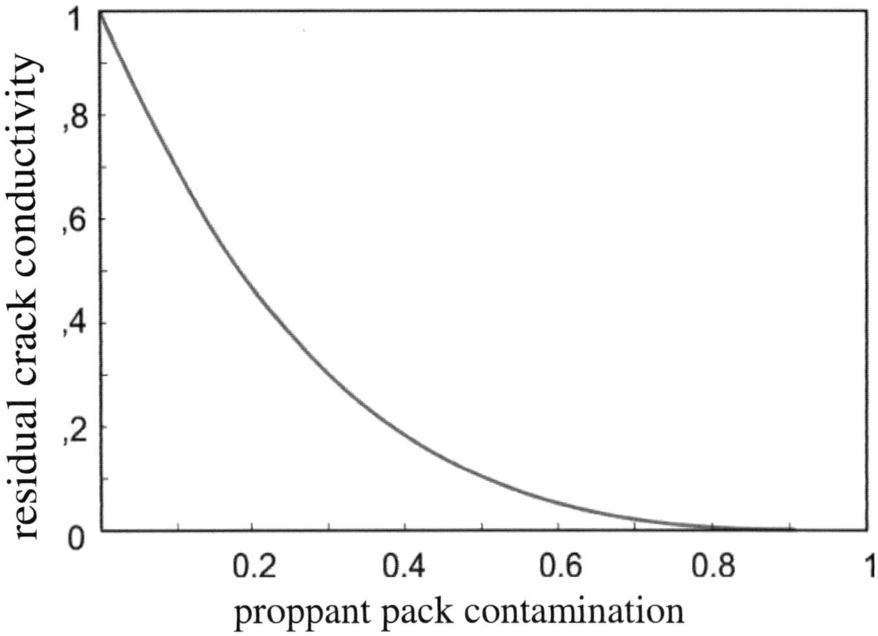

Figure 1. The effect of non-destroyed gel on conductivity

3. Results and Discussion

Thus, the success of the hydraulic fracture process depends on the quality of the gel destruction. The aim of this work is to increase the efficiency of proppant hydraulic fracture by modernizing the breaker system.

Previously used breakers are solid. Adding a breaker to the gel is done by mixing a powdery destructor with a liquid gel. For mixing, a dry chemical reagent feeder was used, as a result of which the concentration of the reagent could not be properly controlled. We monitored the operation of the mechanism, but the amount of reagent passing through the mechanism could not be fixed. In the absence of control, the actual breaker concentration could be either overstated or underestimated. A reduced concentration of the breaker leads to a significant decrease in the effect of hydraulic fracture, while an increased concentration can lead to technological complications during the process [6–8].

One of the new solutions that made it possible to increase the efficiency and expand potential objects for hydraulic fracture was the use of a liquid destructor with feeding into the stream during the injection of the fracturing mixture. To ensure correct control, as well as for uniform distribution of the reagent in the hydraulic fracture mixture, the use of a liquid destructor was introduced. The new destructor is also an oxidizing destructor, as well as a dry analogue. The effectiveness of its application is achieved due to a more correct and confirmed application during proppant hydraulic fracture.

The ready-to-use destructor is supplied as a liquid in a sealed container, which eliminates incorrect proportions when preparing the fluid at the well. Using a separate liquid chemical feed pump, the destructor is fed into the hydraulic fracture mixture stream. In this case, the flow control is carried out by direct measurement using flowmeters at the pump outlet, in front of the main line. The implemented solution allows not only to control the supply of the destructor in real time, but also to carry out claims or research work using digital data records.

Thus, we introduced a new tool for monitoring and controlling both classical and hybrid hydraulic fracture processes into the study. This technology is most relevant for traditional proppant hydraulic fracture in terrigenous reservoirs, where the conductivity of the created fracture is the main criterion for the success of the whole process.

The introduction of this technology allowed expanding the scope of well selection for hydraulic fracture. It became possible to ensure the destruction of a denser viscous gel with a high load of the gel-forming agent necessary for hydraulic fracture under conditions of increased geomechanical stresses. The sampling results indicate the successful cleansing of the created crack from the remnants of a highly viscous gel.

An important positive factor in the application of this technology is the additional control of hydraulic fracture contractors. Additional material is provided for the research, in the event of a flow decrease after hydraulic fracture. A common cause of this complication can be the absence or improper supply of the destructor [9–12].

4. Conclusion

Based on the results of technology implementation, the following conclusions are formulated:

1) for all proppant hydraulic fracture it is advisable to use chemical reagent systems involving the use of a liquid destructor with a feed into the stream;

2) it is necessary to develop breaker systems. Under conditions of controlled use of a liquid destructor, it becomes possible to select reagents with an increased reaction rate and even more complete destruction of the gel.

The search for the reasons for the failure of the geological and technical measures is one of the tools for improvement and further development, without which it is impossible to maintain production with the involvement of hydrocarbon reserves previously considered unconventional. Another important step has been taken towards the control of high-tech methods of well stimulation.

References

[1] Davide S C, Lounganib F P, Marcos S.M. and Poplawski-Ribeirob 2018 Oil prices and inflation dynamics: Evidence from advanced and developing economies *Journal of International Money and Finance* **82** 71-96

[2] Kozikhin R A, Daminov A M, Fattakhov I G, Kuleshova L S and Gabbasov A K 2018. Identifying the efficiency factors on the basis of evaluation of acidizing of carbonate reservoirs *IOP Conference Series: Earth and Environmental Science* **194(6)** 1-6 DOI: 10.1088/1755-1315/194/6/062013

IOP Conf. Series: Materials Science and Engineering **952** (2020) 012044 doi:10.1088/1757-899X/952/1/012044

[3] Bin L, Shiqing C, Xiangrong N and Yongjie Z 2013 Evaluation of damage to horizontal wells through equivalent horizontal well length *Petroleum Exploration and Development* **40(3)** 378-382

[4] Khisamov R S, Huzin R R, Miyassarov A S, Salikhov D A, Andreev V E and Galikeev I A 2017 The experience of heavy oil deposits development using multilateral wells with simultaneous-separate operation with two elevators *Oil Industry* 496-99. DOI:10.24887/0028-2448-2017-4-96-99

[5] Babichev I N, Fattakhov I G, Kuleshova L S, Zaripov L F and Morozov M A 2019 Potential for the wellbore zone development using the dynamic impact. *IOP Conference Series: Earth and Environmental Science (IPDME 2019 – International Workshop on Innovations and Prospects of Development of Mining Machinery and Electrical Engineering)* **378(1)** 1-5 DOI: 10.1088/1755-1315/378/1/012111

[6] Márton L S, Casper S H, Wei Y, Jens H W and Stefan L G 2018 Near-wellbore modeling of a horizontal well with Computational Fluid Dynamics *Journal of Petroleum Science and Engineering* **160** 119-128

[7] Bahtizin R N, Nurgaliev R Z, Fattakhov I G, Andreev V E and Safiullina A R 2018 On the question of the efficiency analysis of the bottom-hole area stimulation method *International Journal of Mechanical Engineering and Technology (IJMET)* **9(6)** 1035–1044.

[8] Nianyin Li, Jinxin Dai, Pingli Liu, Zhifeng Luo and Liqiang Zhao 2015. Experimental study on influencing factors of acid-fracturing effect for carbonate reservoirs *Petroleum* **1(2)** 146-153

[9] Xu B 2018. Influencing factors governing paraffin wax deposition of heavy oil and research on wellbore paraffin remover *Petroleum Science and Technology* **36(20)** 1635-1641.

[10] Maganov N U, Khisamov R S, Bazarevskaya V G and Podavalov V B 2016. Evaluation of natural fracturing to develop induced fracturing methods best suited for low-permeability reservoirs *Society of Petroleum Engineers (SPE Russian Petroleum Technology Conference and Exhibition, 24-26 October, Moscow, Russia)* 764–774 DOI: 10.2118/181911-MS

[11] Shakurova A F and Shakurova A F 2018. The influence of hydraulic fracturing on the estimated ultimate recovery *IOP Conference Series: Earth and Environmental Science* **194(8)**1-4 DOI: 10.1088/1755-1315/194/8/082039.

[12] Baykov N M 2003 New technologies of acid treatment of the productive strata *Oil Industry* **3** 114-116

IOP Conf. Series: Materials Science and Engineering **952** (2020) 012045 doi:10.1088/1757-899X/952/1/012045

Probabilistic statistical model for predicting the effectiveness of hydraulic fracturing

K F Gabdrakhmanova, G R Izmaylova and L Z Samigullina

Ufa State Petroleum Technological University, Branch of the University in the City of Oktyabrsky, 54a, Devonskaya St., Oktyabrsky, Republic of Bashkortostan, 452607, Russian Federation

E-mail: klara47@mail.ru

Abstract. The article studies the effectiveness of single-stage and multi-stage hydraulic fracturing technologies using the mathematical statistics method. The studies are based on the geological and physical characteristics of the Shirotny Priob fields which makes it possible to substantiate the methods for choosing the hydraulic fracturing technology. Based on the probabilistic and statistical analysis of field efficiency data, quantitative criteria for the selection of fracture parameters by length, height and amount of injected propane have been established. They can be used to predict the use of multi-stage and single-stage hydraulic fracturing methods. The productive horizon AS12-3 has a complex geological and physical structure and low filtration and storage properties, which complicates its development and contributes to the active formation of hard-to-recover reserves. The one-stage hydraulic fracturing technology is efficient only for 3-4 years. The multistage hydraulic fracturing technology showed is more efficient in comparison with the single-stage one.

1. Introduction

Russian and foreign technologies used to increase production efficiency are based on hydraulic fracturing which increases oil flow. The technology is based on the mechanism of rock crack formation and propagation. In foreign and Russian works on hydraulic fracturing, the effectiveness of hydraulic fracturing is studied using the probabilistic-statistical approach. Few works determine the optimal number of hydraulic fracturing stages, with an estimate of distances between the fractures. Therefore, it is necessary to develop new criteria for multiple hydraulic fracturing in integrated reservoir stimulation using the probabilistic-statistical approach

2. Literature review

Many foreign and Russian scientists deal with the efficiency of hydraulic fracturing, especially in the conditions of hard-to-recover oil reserves. Many scientists [1–4] describe ways to improve the multistage hydraulic fracturing technology in anisotropic heterogeneous oil-saturated reservoirs.

[3–6] describe the possibility of assessing the productivity index of deposits based on indirect data when drawing up the first design documents and modeling oil recovery processes using the experience of developing fields that have been in operation for a long time.

The study described in [10] is based on the method of determining the amount of injected propane. It uses new equations that describe the fracture closure process. An analysis of the influence of fluid viscosity, its density, size and formation permeability was conducted. These studies will be useful when choosing a propane composition.

The experiment described in [10] is aimed at improving the efficiency of hydraulic fracturing using the pulse-technology of hydraulic fracturing of sand. The authors introduced the concept of discrete multi-layer grinding inside cracks. They proved that the use of pulse duration per formation can increase the productivity of wells, reduce the volume of hydraulic fracturing of the propane.

Despite some studies on reservoir productivity which can be improved using the single-stage and multi-stage hydraulic fracturing technologies, there are some unsolved issues. On the basis of publications and textbooks [14, 15], research using the apparatus of mathematical statistics, the authors determined the purpose and objectives of the study.

3. Research purpose and objectives

To achieve this purpose, the following tasks will be solved:

1) Statistical analysis of existing hydraulic fracturing technologies in Russia and abroad;

2) Theoretical studies of the effectiveness of the one-stage and multi-stage hydraulic fracturing technology;

3) Determination of effectiveness criteria for the hydraulic fracturing technology.

By processing statistical data of field efficiency, quantitative criteria for the selection of fracture parameters by length, height and amount of propane injected have been established. It was possible to improve the field efficiency of one-stage hydraulic fracturing and increase the annual cumulative production volume by at least 10-15%. However, 4 years after, the field efficiency of one-stage hydraulic fracturing began to decline, which required new ways to improve it based on the multi-stage modification. An analysis of the results of field testing made it possible to obtain quantitative criteria for the optimization of multistage hydraulic fracturing based on a reasonable choice of both the length and volume of fractures and their number, which made it possible to achieve the planned oil recovery factor during the development of the AS12-3 horizon within the terms established and agreed with the State Reserves Committee of the Russian Federation.

According to the geological and field studies, the oil-saturated thickness of the field has a stratal-garden character and is represented by the main productive horizon of the AS-12-3 formation of the Cherkashinskaya suite, whose thickness increases (up to 16-18 m) towards the top of the field and decreases (up to 4-5 m) in its wing area.

In accordance with the design task, the hydraulic fracturing mode was selected to obtain a single fracture in the productive strata with a height of up to 0.8–0.9 Npl, a length of 70–80 m and an opening (B) of up to 3–5 mm.

The field efficiency of hydraulic fracturing was about 50%. An analysis of the results of unsuccessful hydraulic fracturing showed that it is accompanied by low flow rates and an increase in the water cut compared to the initial one, although the selected wells were located in the zone of high residual reserves.

The study of structural features of the field and their comparison with the direction of the underlying basement faults showed that in the wing zones of the sedimentary strata, there are areas of the stressed state of the rock accompanied by the natural fracturing, which has a predominantly axial (along the swell-like structure of the field) direction.

The comparison of the unsuccessful hydraulic fracturing in the zones of residual reserves concentration and the natural fracturing direction showed that it was caused by penetration of the second fracture wing into the zone of depleted reserves with a high level of water cut [2].

At the stage of pilot testing of the single-stage hydraulic fracturing technology, the wells were selected by the following criteria [7]:

- oil-saturated reservoir thickness is not less than 3 m;
- potential production rate is not less than 10 tons / day;
– initial oil saturation $Кn > 0{,}4 + 0{,}15 \cdot (1 - \sqrt{\alpha_{nc}})$;
- the thickness of the overlapping and underlying screens is at least 3 m.

- the ratio of the current reservoir pressure to the initial one is not less than 0.9.
- the condition of the cement stone in the perforation interval ± 20 m is good;
- the well should not have behind-the-casing flows;
- water cut of products is no more than 50%;
- the angle of deviation of the wellbore from the vertical in the reservoir interval is not more than 10°;
- water cut of surrounding producing wells is not more than 70%.

In order to study the dependence of the technological efficiency of hydraulic fracturing on the implementation technology, the following fracture parameters were studied: length, height, width, amount of propane.

It was found that with an increase in the fracture length, the initial oil production rate increased and the water cut decreased (Figures 1, 2).

At the same time, it was found that with an increase in the fracture height relative to the effective reservoir thickness, the oil production rate decreases, and the water cut increases (Figure 3). This behavior is due to a decrease in the effective oil-saturated reservoir thickness.

It was found that with an increase in the mass of injected propane, the oil production rate also increases (Fig. 4). The dependencies made it possible to correct the one-stage hydraulic fracturing technology by increasing its technological and economic efficiency.

Using the criteria for choosing the optimal parameters of the single-stage hydraulic fracturing technology, it was possible to increase the accumulated current production volume during the operation of the BV8-1 formation.

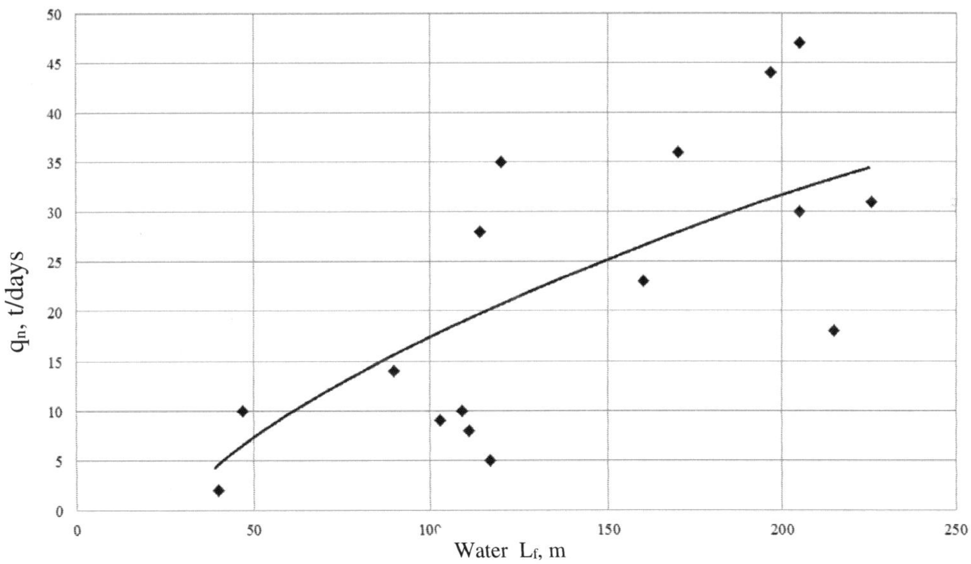

Figure 1. Dependence of the well production rate after single-interval hydraulic fracturing on the fracture length

IOP Conf. Series: Materials Science and Engineering **952** (2020) 012045 doi:10.1088/1757-899X/952/1/012045

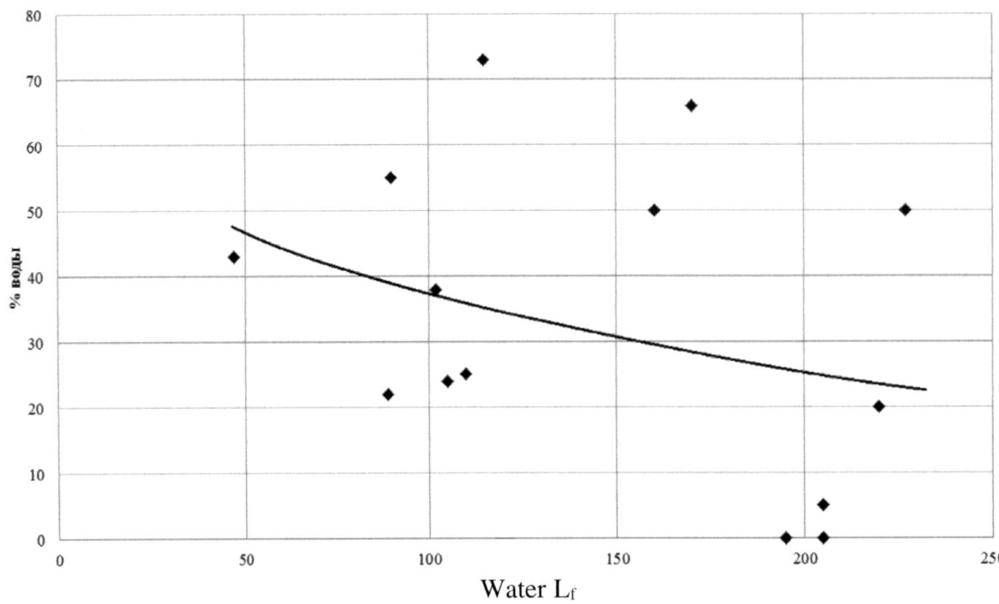

Figure 2. Dependence of the water cut of the well after one-interval (one-stage) hydraulic fracturing on the fracture length

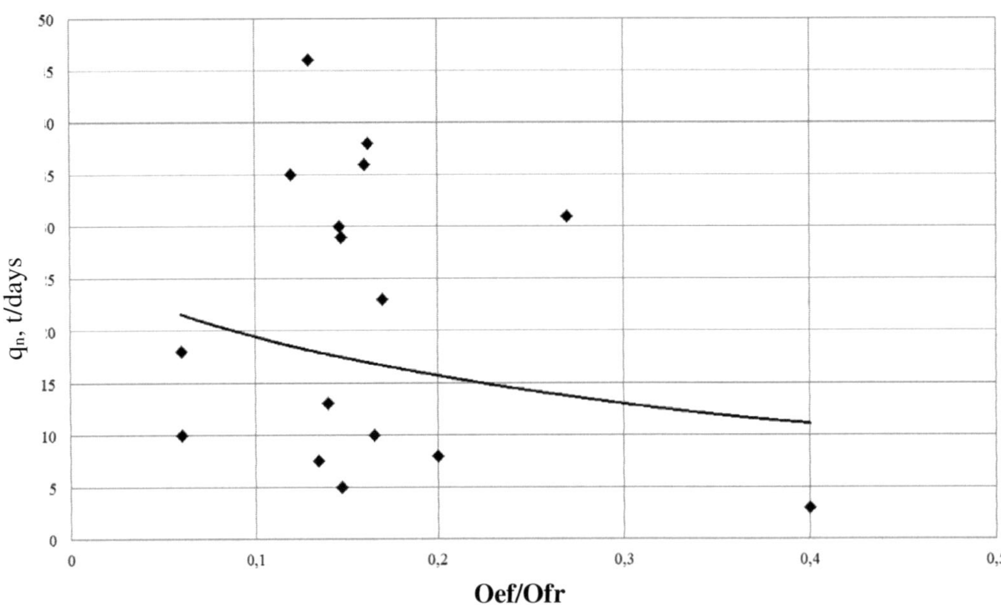

Figure 3. Dependence of the oil production rate after one-stage hydraulic fracturing on the ratio of the effective reservoir thickness to the fracture height

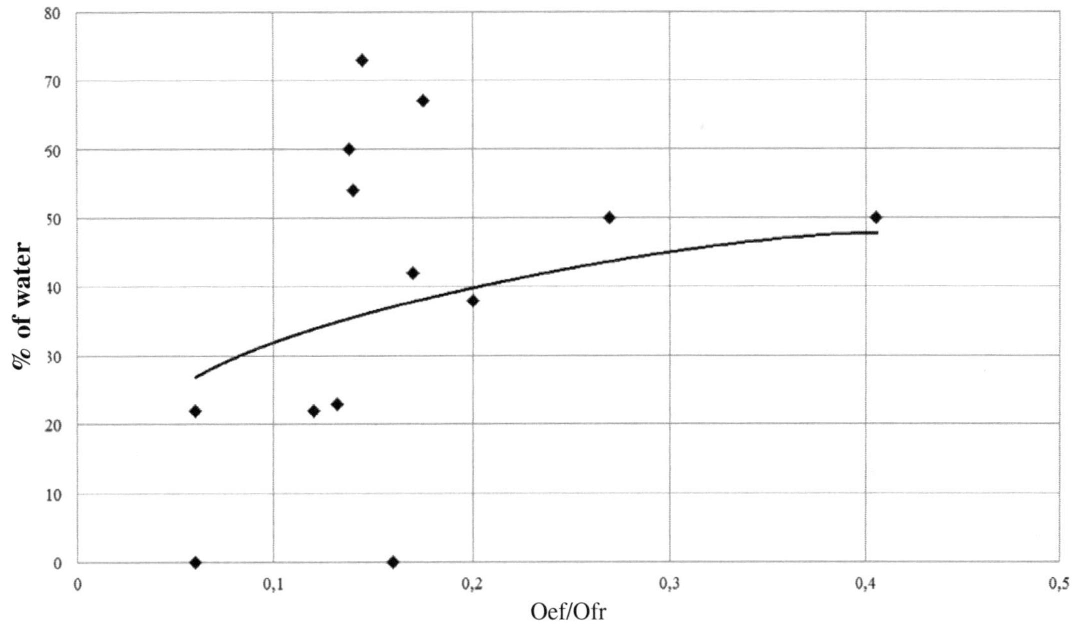

Figure 4. Dependence of the water cut after one-stage hydraulic fracturing on the ratio of the effective reservoir thickness to the fracture height

4. Results

As an alternative to the single-stage hydraulic fracturing technology, the multistage hydraulic fracturing technology has been developed. A horizontal wellbore is drilled in the formation and several fractures located at a certain distance from each other are created [7, 14].

Figure 5 shows the ranges of flow rates: the initial (stop) flow rates before the hydraulic fracturing, the calculated (predicted) and real flow rates, flow rates three and six months after the hydraulic fracturing.

The above dependencies show that the range of calculated and starting flow rates coincides, which indicates a rather perfect method for predicting the field effect from the multistage hydraulic fracturing [7]. However, 3 and 6 months after, the real production rates differ significantly from the calculated data. They can either decrease or increase in comparison with the calculated data. This behavior is due to the systemic water cut calculation error.

We used the statistical data for 2015 and determined the deviations of real flow rates from the predicted ones.

Since when stimulating production using the multistage hydraulic fracturing technology, a gradual decrease in the current flow rate is inevitable due to the depletion of oil reserves, we studied the dependence of differences between the actual decline in the oil production rates and the water cut.

As a result of an analysis of the relationship between the probability of a positive forecast for an increase in oil production and a decrease in its water cut, additional quantitative criteria that determine the optimal properties of the fractures were obtained. They ensure the required technological efficiency of the EOR method.

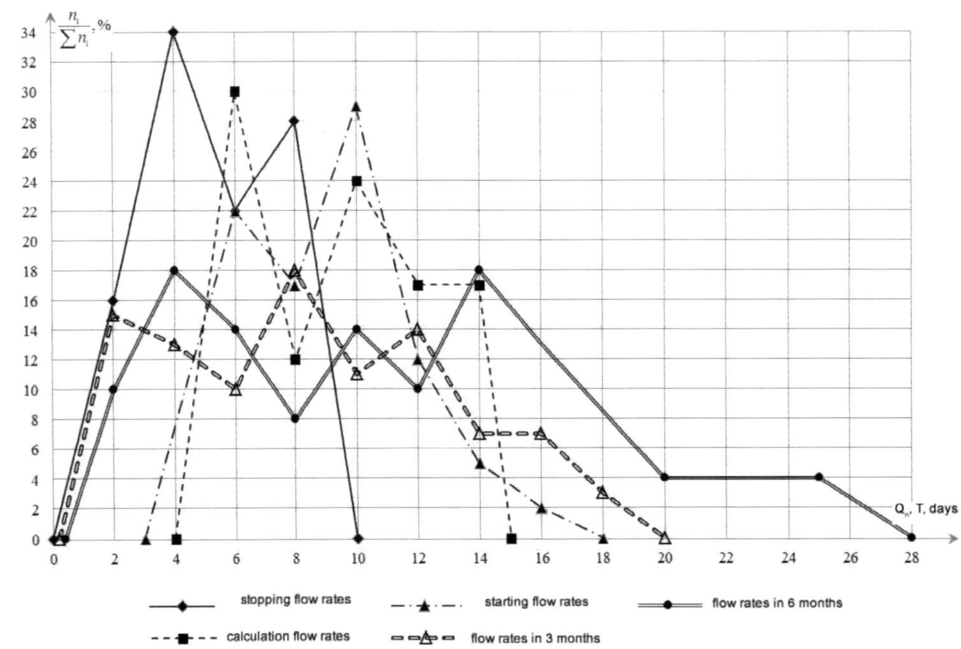

Figure 5. Graphs of well flow rate distribution after multistage hydraulic fracturing of AS12-3 formation

5. Conclusion

1. The BV8 1 horizon development intensification technology based on one-stage hydraulic fracturing is efficient within 3-4 years;
2. The multistage hydraulic fracturing technology applied on the AS12-3 horizon is more efficient. However, the positive predictive indicators did not exceed 50%;
3. An analysis of the statistical indicators for the multistage hydraulic fracturing showed that the geological and physical properties of the AS12-3 horizon were not taken into account;
4. The results of an analysis of statistical indicators of the field results of application of the multistage hydraulic fracturing technology on the AS12-3 horizon made it possible to specify the ranges of the main parameters of hydraulic fractures (length, volume, number of fractures and propane weight).

References

[1] Proskurin V A 2013 Improving the technology of multistage hydraulic fracturing in horizontal wells *Dissertation for the degree of candidate of technical Sciences in spec.* 25.00.17 Ufa IPTER

[2] Muslimov R Kh, Galeev R G, Suleymanov E I and Volkov Yu A 1994 Problems of improving the scientific basis of oil field development *Theses of the scientific and practical conference. "Problems of development of the oil industry of Tatarstan at the late stage of reserves development"* 18-23

[3] Kuleshova L S and Mukhametshin V V 2019 Elimination of uncertainties in predicting well interaction using indirect geological field information *IOP Conference Series: Earth and Environmental Science (IPDME 2019 – International Workshop on Innovations and Prospects of Development of Mining Machinery and Electrical Engineering)* **378 (1) (012115)** 1-8 DOI: 10.1088/1755-1315/378/1/012115

[4] Mukhametshin V V and Kuleshova L S 2019 Prediction of production well flow rates using survey data *IOP Conference Series: Earth and Environmental Science (IPDME 2019 – International*

Workshop on Innovations and Prospects of Development of Mining Machinery and Electrical Engineering) **378(1) 012116** 1-7 DOI: 10.1088/1755-1315/378/1/012116

[5] Yakupov R F, Gimazov A A, Mukhametshin V Sh, and Makaev R I 2018 Analytical method for estimating efficiency of oil recovery technology in case of bottom water-drive reservoir, verified on the hydrodynamic model *Oil Industry* **6** 66-69 DOI: 10.24887/0028-2448-2018-6-66-69

[6] Khokhlov V I, Galimov Sh S, Devyatkova S G, Kotenev Yu A, Sultanov Sh Kh and Mukhametshin V Sh 2019 Justification of impact and planning of technology efficiency on the basis of limy-emulsion formulation in low-permeability highly-rugged reservoirs of Tyumen deposits *IOP Conference Series: Earth and Environmental Science (IPDME 2019 – International Workshop on Innovations and Prospects of Development of Mining Machinery and Electrical Engineering)* **378(1) 012114** 1–6 DOI: 10.1088/1755-1315/378/1/012114

[7] Sogo Shiozawaa, Mark McClurea 2016 Simulation of propane transport with gravitational settling and fracture closure in a three-dimensional hydraulic fracturing simulator *Journal of Petroleum Science and Engineering* DOI:10.1016/J.PETROL.2016.01.002

[8] Stabinskas A P 2014 Evaluation of the efficiency of wells after hydraulic fracturing *Problems of collecting, preparing and transporting oil and petroleum products* **1 (95)** 10-19

[9] Jianchun Guo, Songgen He, Yan Deng, Zhihong Zhao 2015 New stress and initiation model of hydraulic fracturing based on nonlinear constitutive equation *Journal of Natural Gas Science and Engineering* **27 (2)** 666–675

[10] Jinzhou Zhaoa, Xuan Pua, Yongming Lia, Xianjie Heb. A semi-analytical mathematical model for predicting well performance of a multistage hydraulically fractured horizontal well in naturally fractured tight sandstone gas reservoir *Journal of Natural Gas Science and Engineering* **32** 273-291

[11] Bin Qian, Congbin Yin, Juhui Zhu. Research and practice of the impulse sand fracturing technology *Natural Gas Industry* **2 (4)** 334–340.

[12] Branko Damjanac, Peter Cundall 2016 Application of distinct element methods to simulation of hydraulic fracturing in naturally fractured reservoirs **71** 283–294

[13] 2002 *Hydraulic fracturing* Educational and methodological module 1 (Tomsk)

[14] Gabdrakhmanova K, Izmailova G R and Samigullina L Z 2019 Solution of the problem of annular space gas utilization in wells operated by walking-beam pumping unit *IOP Conference Series: Earth and Environmental Science*

[15] Gabdrakhmanova K, Izmailova G R and Samigullina L Z 2019 Task approach in the bachelors' professional competence formation *Advances in Economics, Business and Management Research*

IOP Conf. Series: Materials Science and Engineering **952** (2020) 012046 doi:10.1088/1757-899X/952/1/012046

Stress corrosion cracking in pipelines

A K Gumerov[1] and A R Khasanova[2]

[1] Ufa State Petroleum Technological University, 1, Kosmonavtov st., Ufa, Republic of Bashkortostan, 450062, Russian Federation
[2] Ufa State Petroleum Technological University, Branch of the University in the City of Oktyabrsky, 54a, Devonskaya St., Oktyabrsky, Republic of Bashkortostan, 452607, Russian Federation

E-mail: ajgulkhasanova@mail.ru

Abstract. Stress corrosion is viewed as cracking of a pipe metal in cathodic protection regions of gas transmission pipelines, which are simultaneously affected by various unrelated, at first glance, factors. Stress corrosion is otherwise referred to as stress corrosion cracking or SCC. Stress corrosion cracking of pipelines is the main cause of emergencies in gas transmission pipelines. Currently, this type of corrosion is widespread and very dangerous. The reliability and safety of modern pipelines is of great national importance. Despite being significant, multiple efforts to address this problem have not yet identified the causes and algorithms for handling stress corrosion.

1. Introduction

There are many different considerations and theories to understand the nature of this phenomenon. They are followed by all sorts of implications and recommendations. Some experts recommend individual steel grade for major pipelines, others recommend more highly efficient insulating materials, while others recommend that pipeline construction should be revised and updated. There are even proposals to treat the soil with special compounds entirely throughout the pipeline. However, in fact, none of them will completely address stress corrosion algorithm at large, since they "extract" only one or several involuntary components from the whole gamut of factors and bonds. Meanwhile, the very phenomenon of this type of corrosion still remains unknown. In the fight against stress corrosion, one can hardly rely on success without clarifying such a physical (mechanical, chemical) model [1–4].

Stress corrosion mechanism should explain all facts available and respond to a number of questions, including the following:

- in what way does electrochemical protection (ECP) affect stress corrosion?
- what properties need to be developed to effectively inhibit stress corrosion?
- what is the role of insulation coating in stress corrosion (bitumen, film, factory-made)?
- why does stress corrosion often occur over gas transmission pipelines and does not do at all over oil transmittion pipelines, over pipelines of gas supply systems, over field pipelines?

2. Materials and methods

The research plan is composed of the following:

- addressing the relevance of this problem and streamlining the evolution of stress corrosion over

IOP Conf. Series: Materials Science and Engineering **952** (2020) 012046 doi:10.1088/1757-899X/952/1/012046

major pipelines;

- evaluating all results available and shaping the main prerequisites for stress corrosion cracking of pipelines;
- developing and building up a physical model of pipe stress corrosion, establishing the foundations of this type of corrosion;
- comparing specimens of pipe metals taken from the fracture zone.

3. Results and Discussion

Stress corrosion cracking differs from other types of corrosion. Stress corrosion in pipe walls causes parallel axial cracks. Stress-corrosion cracks in an annular direction occur in those sections of pipelines where there is an elastic bending, and, as a consequence, there are significant bending stresses that are in the longitudinal direction. The outer surface of the metal actually cracks in a stress corroded area and advances deep into the pipe wall, with structural changes of the metal to occur simultaneously. However, the inner surface of the metal still remains plastic.

The elements, including hydrogen, carbon, nitrogen, etc., that can enter from the outer surface, cause the embrittlement of metal, which is possible at high temperatures alone. One of the main elements to induce stress corrosion is hydrogen ($H+$).

The pipe surfaces are negatively charged and therefore a positively charged hydrogen cation will be attracted to the metal. After that, the hydrogen cation is neutralized in the metal surface. Then, the only electron, a hydrogen atom, becomes a part of the electron cloud of metal, and the remaining core is like a proton, i.e. an elementary particle. Further, the proton can easily enter the metal and retain great mobility and activity there. In addition, increased concentrations of the proton in the metal entails an increased internal stress. In turn, the internal stress is combined with external stresses, and such formations cause cracking of the pipeline. Therefore, the stress-corrosion mechanism comprises several stages:

1) incubation period is related to the intrusion of hydrogen atoms deep into the metal;

2) initiation and growth of microcracks, further accumulation of gases, assembling of microcracks, expansion of cracks to the sizes observed through diagnosis methods;

3) crack propagation under internal and external forces;

4) destruction of the major gas pipeline.

Currently, it is believed that many factors influence this type of corrosion: chemical composition and soil moisture, weather conditions, temperature interval, vibrations, residual stresses in the major pipeline, stresses induced by many types of loads, deployment of a pipeline region close to the compressor station, terrain, steel grade, the presence of certain microbes in the soil and many others. Most likely, this is all true. However, all these factors alone will not be able to trigger the stress corrosion mechanism. They only have the ability to enhance or weaken the effects of the main factors: hydrogen and stress [5–7].

Several external factors affect the rate of the first stage, namely the release of atomic hydrogen in the metal surface. The external factors embrace the status of the insulation coating (defect, lack of adhesion), the "ground-pipe" potential difference, the state of metal surface, the chemical composition and saturation of substances in ground water, the pH of the medium (acidity), humidity, vibrations, weather conditions, environmental temperature (soil).

The rate of the second and third stages, namely the propagation of hydrogen into the metal and the processes occurring inside the metal is affected by some internal factors. These include the chemical composition and structure of the metal, the temperature of the metal, the stress state. Moreover, stresses induced by working pressure, as well as external and internal impacts play a crucial role. The latter involves those acting on the crystals and dislocations.

Thus, some factors, like temperature and vibration, fall within both internal and external factors, since they affect all processes that occur both outside and inside the metal.

Unlike other types of corrosion (for example, chloride), corrosion cracking of steels can start and develop not only in the surface, but also inside the metal as such.

A very important conclusion can be derived from this assumption, which may become crucial in ensuring the protection of gas pipelines from stress corrosion. In fact, given that atomic hydrogen is prevented from occurring in the metal surface, all other processes to entail stress corrosion cracking will be stopped as well. These processes will lack a single and principal participant in the process, otherwise known as hydrogen (similar to the fact that if oxygen is excluded, there will be no oxidation) [8, 9].

This assumption is not yet a physical stress corrosion model, but, subject to the outcomes of some further studies in line with the tasks assigned, claims to be.

A comparative analysis shows that gas transmission pipelines (in those sections where stress corrosion is detected) have large diameters and are exploited under high pressures. This means that mechanical stresses in the wall of major gas pipelines are much higher than in all other pipelines. Accordingly, high tensile stresses in the pipeline wall are considered to be the main factor determining the conditions favourable for stress corrosion.

The main stages of the physical model for stress corrosion in major pipelines are:

- Atomic hydrogen is generated in the pipe surface and accumulated in the metal.
- Atomic hydrogen enters the metal in the form of a proton gas (proton).
- Increased tensile stresses result in faster hydrogen propagation into the metal.
- Currently it is proposed to introduce a new concept of "stress corrosion strength". Each metal has stress corrosion strength that characterizes the threshold at which stress corrosion is noticeably accelerated. Below this level, there is almost no stress corrosion.
- Based on this physical model, it was found that the intensity of atomic hydrogen generation in the surface of the pipe metal depends on the quality state of the insulation coating and the operation mode of the electrochemical protection system.

Each individual crack tends to reduce the stress intensity factor about all other cracks. Therefore, a line of parallel cracks is less dangerous than a single crack of the same size. Based on the findings, strength criteria are stated as applied to a stress-corrosion defect [10].

Insulation materials need to be selected properly when it comes to the repairs of existing gas pipelines affected by stress corrosion. In almost all cases, a stress-corrosion defect accompanies the unsatisfactory state of a film insulation coating. In some regions, the adhesion completely disappears, and the film easily moves away from the pipe. In fact, the film coating turns into a kind of shell that contains a pipe within, whereas ground water accumulates between the pipe and the shell. In this case, the pipe does not only have protection against stress corrosion, but partially loses its protection against general and pit corrosion despite the "protection" capacity.

A method is proposed to stop stress corrosion in gas pipelines by restoring the insulation coating through new materials with chemical activity and inhibitory properties.

Being aware of stress corrosion algorithm in major pipelines explains all the above mentioned features and patterns of this phenomenon, and provides the possibility to justify stress corrosion in certain sections of pipelines.

This phenomenon implies that the development of local stress corrosion is possible in oil pipelines in the regions where two conditions appear simultaneously: 1) there is a stress concentration in the form of defects or welded joints; 2) in this region the insulation coating is missing or worn. Stress corrosion is among the deterioration mechanisms (embrittlement and cracking) to affect the metal during long-term operation of pipelines.

4. Conclusion

Thus, hydrogen-induced cracking of a metal of pipelines goes through several stages:

1) generation of active (atomic) hydrogen in a metal surface;
2) atomic hydrogen propagation into a metal;
3) hydrogen-related processes occurring at the level of the microstructure of the metal.

IOP Conf. Series: Materials Science and Engineering **952** (2020) 012046 doi:10.1088/1757-899X/952/1/012046

Actually, these three points are the main elements constituting the physical model of pipeline stress corrosion.

Provided that the mechanism proposed is correct, the protection of the pipeline from stress corrosion cracking should rely on the attempts to remove the source of atomic hydrogen off the surface of the pipeline. Since the insulation coating and ECP cannot be removed, it is worth selecting more effective insulation coatings. Thus, coatings must be highly adaptable so that the insulation of existing gas pipelines can be restored.

References

[1] Barkhatov A F, Gumerov A G, Shammazov A M, Fazletdinov R A and Vafin R V 2019 Implementation and operation of anti-pressure systems of oil pipelines *IOP Conference Series: Earth and Environmental Science (International Conference on Innovations and Prospects of Development of Mining Machinery and Electrical Engineering 2019, IPDME 2019)* **378(1)** DOI: 10.1088/1755-1315/378/1/012077

[2] Batalov S A, Andreev V E, Lobankov V M and Mukhametshin V Sh 2019 Numerical simulation of the oil reservoir with regulated disturbances. Oil recovery stability simulation *Journal of Physics: Conference Series (ITBI 2019 – International Conference "Information Technologies in Business and Industry")* **1333(3)** 1-6 DOI:10.1088/1742-6596/1333/3/032007

[3] Korn G A and Korn T M 1984 *Mathematical Handbook for Scientists and Engineers: Definitions, Theorems, and Formulas for Reference and Review* (Moscow: Nauka)

[4] Aleksandrov A A, Larionov V I, Yuzefovich A V and Gumerov A K Mathematical statement and numerical solving of a problem on three-dimensional pipeline bend under complex geological conditions *Herald of the Bauman Moscow State Technical University, Series Natural Sciences*

[5] Gumerov A K, Khasanova A R, Gumerov A G and Shammazov A M 2019 Applicability of energy flow vector in the mechanics of destruction *Oil and Gas Business* **17(5)** 89-92

[6] Heuzeroth F, Fritzsche J and Werzner E 2015 Viscous force - An important parameter for the modeling of deep bed filtration in liquid media *Powder technology* **283** 190-198

[7] Khabibullin M Ya 2019 Theoretical grounding and controlling optimal parameters for water flooding tests in field pipelines Journal of Physics: Conference Series **1333** 1-5 DOI:10.1088/1742-6596/1333/4/042013

[8] Kuleshova L S, Mukhametshin V V and Safiullina A R 2019 Applying information technologies in identifying the features of deposit identification under conditions of different oil-and gas provinces *Journal of Physics: Conference Series (ITBI 2019 – International Conference "Information Technologies in Business and Industry")* **1333(7)** 1-5 DOI: 10.1088/1742-6596/1333/7/072012

[9] Bykov L I, Shammazov A M, Gumerov A G, Tokarev A P, Safiullina A R 2019 Reduction of vibrations of mechanical and power plant pipelines using vibration insulators with design power characteristics based on guides of a special shape *IOP Conf. Series: Earth and Environmental Science* **378** DOI:10.1088/1755-1315/378/1/012037

[10] Khabibullin M Ya, Suleimanov R I, Sidorkin D I and Arslanov I G 2017 Parameters of damping of vibrations of tubing string in the operation of bottomhole pulse devices *Chem. Petrol. Eng+* **53(5-6)** 378-384

IOP Conf. Series: Materials Science and Engineering **952** (2020) 012047 doi:10.1088/1757-899X/952/1/012047

Set of measures to prevent liner sticking during directional drilling

E L Guseynova[1] and E M Guseynov[2]

[1] Ufa State Petroleum Technological University, Branch of the University in the City of Oktyabrsky, 54a, Devonskaya St., Oktyabrsky, Republic of Bashkortostan, 452607, Russian Federation
[2] LLC Gazprom Export, 2A, Litera A, Ostrovskogo Sq., Saint-Petersburg, 191023, Russian Federation

E-mail: guseinova_elena@hotmail.com

Abstract. The study considers the sticking problems of drilling. The study presents the classification and gives a general characteristic of liners. It is shown that the largest number of troubles occurs due to differential sticking. Differential sticking and methods of its elimination are considered in detail. The set of measures to prevent sticking, including the use of the Rhino XC Tool borehole expander, filters with plugs, dry lubrication, eccentric shoe, is developed and proposed.

1. Introduction

Sticking is the most common type of troubles that occur during directional drilling. Sticking can be divided into two categories: sticking due to mechanical interference, and differential sticking. The majority of sticking that occur during drilling (80%) is differential, the cause of the remaining 20% is mechanical interference or equipment failure. The sticking as a result of mechanical action is divided into two categories: sticking obtained as a result of accumulation of cuttings or due to rock slide in areas with complex wellbore geometry caused by column jamming [1, 2].

In most cases, sticking occurs when the tool is tripped out, and the most common reason for it is the presence of cuttings or rock slide, less often sticking occur during the running of tool into well. Poor borehole cleaning or poor wellbore stability may also cause sticking. Besides, long circulation loss of the washing fluid may cause sticking of a fixed drill string.

Cuttings or rock slide is the most dangerous, since the unfreezing of a column stuck by cuttings or rock slide is more labor-intensive compared to the unfreezing of a column stuck in areas with complex wellbore geometry or stuck by differential sticking. The elimination of such sticking is characterized by the loss of more equipment items and the more frequent need to drill a sidetrack [3].

The indentation of a fixed casing into the filter clay cake formed on the wellbore walls causes differential sticking that occur under the influence of pressure drop in a well and a permeable formation. At the same time the friction force generated between the casing string and the formation rock does not allow moving the drill string out of location. Most often, differential sticking is observed in wells that penetrate depleted productive formations. If the casing remains motionless for a long time, then it is almost impossible to avoid differential sticking [4].

The differential sticking is caused by the presence of permeable formations, repression, thick filter cake, column contact with a wellbore wall, column standstill, time, transverse load, negligence of a drilling crew. The above factors influence the force pressing the string to the wellbore wall with a filter cake and due to differential pressure [5].

Differential sticking occurs only in the permeable formation zone, for example, sand. There are several reasons for sticking (Figure 1):

1. Sticking of the column is possible in case of contact of part of the column with the wellbore wall and its pressing to the filter cake. Hydrostatic pressure of the drilling mud column affects part of the column surface in contact with the filter cake.

2. The pressure gradient of the mud column and in the formation affects the column area in contact with the filter cake of the wellbore wall, and the resulting force holds the string.

Torque caused by differential pressure sticking is calculated as the product of differential pressure, contact area and friction factor [5].

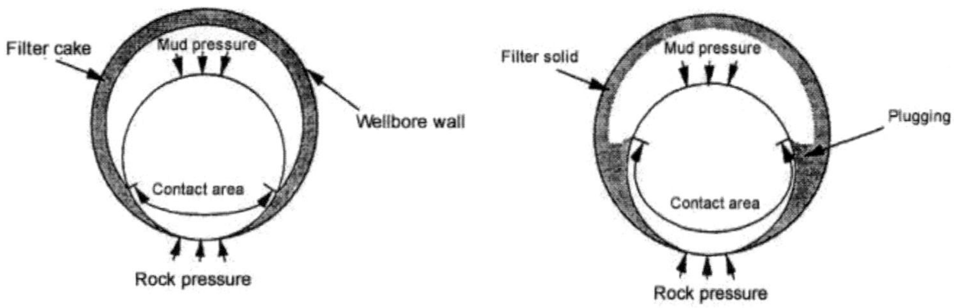

Figure 1. Differential sticking

To successfully eliminate the sticking, it is necessary to determine its cause. Quick and proper actions play a huge role, since over time any situation only becomes more complicated. Statistics shows that within the first four hours after sticking, it is possible to release 50% of all stuck columns, and then this figure is reduced to 10%.

In the event of sticking with cuttings or rock slide, the pressure is released, which is caused by the resulting plug, and then a small pressure is created, since too much pressure can contribute to further movement of the casing into the plug. Low pressure restores the circulation of the washing fluid if the column can be moved out of location.

When trying to lift the casing, it can further advance into the plug. Therefore, it is necessary to move the casing and restore circulation, blur the plug and bring the plug material to the daylight surface. When the circulation is restored, the flow rate of the washing fluid is increased to the maximum value not resulting in absorption. Circulation must be continued until the well is cleaned.

2. Materials and methods

The analysis of experimental data and scientific literature, analytical research methods made it possible to obtain the results presented in the paper.

3. Method procedure and assessment of its efficiency

In the fields of Russia, which have been in operation for more than a year and are in the second and third stages of development, the problem of tailings is quite urgent.

According to the studies, at the deposits of the Krasnoyarsk Territory, the average formation pressure of the Nx-3-4 formation is 160 atm. At the same time, its accepted value is 270 atm. Thus, the coefficient of anomaly of the formation pressure amounts to 0.58. Therefore, the risks of emergency during the construction and reconstruction of wells using side-tracking, the occurrence of differential sticking, sloughing of unstable clays, etc. are obvious.

Considering the need to reduce the idle well stock, using the experience of sidetracking in conditions of low formation pressures of the productive formation Nx 3-4 (up to 60% of hydrostatic pressure), the following set of technical and technological measures are proposed for successful borehole drilling and accident-free running of a tail pipe:

1. Use of the Rhino XC Tool hydraulic expander with the ability to expand the borehole during drilling from 155.6 to 165 mm to the T2 point.
2. Use of a filter with plugs to allow the well to be flushed through the liner bottom, including pumping (if necessary) of calmatation packs of drilling fluid after passing the bottom of the layout through the supercollector interval.
3. Use of STIK LESS dry lubricant. Before running the "liner", a pack with a material that reduces the friction coefficient is installed in the zone of high wellbore intensity to better bring the weight to the liner shoe during running.
4. Use of guide rotating liner shoe.

The Rhino XC Tool developed by Shlumberger is designed to increase the borehole, and its operation is based on the hydraulic activation of additional mills. The expander, passing through the borehole, increases its diameter to 165 mm, which is an advantage of this technology. The ability to activate and deactivate the expander at any stage of well drilling is also a positive factor. The use of this technology reduces the area of liner filter contact with the rock.

The first well in Russia using the Rhino XC expander was drilled in 2017 at the Vankorskoe field. Drilling with an expander was carried out to the target horizon, then the expander was deactivated.

Other companies also have positive experience using the extender. The SOCAR oil company, owned by the Republic of Azerbaijan, used the Rhino XC expander with hydraulic drive in the construction of a well at the Karadagskoe field, the use of the expander allowed drilling and expanding 989 meters of the upper section from 444.5 to 504 mm per run. As a result of using Rhino XC, a well of nominal diameter was drilled with high quality, which made it possible to facilitate the running of the casing as much as possible. The drilling process was accompanied by activation and deactivation of the expander (eight times). SOCAR management plans to expand the experience of using a well-proven expander to drill other wells [6].

Besides, STIK LESS dry lubricant, representing colorless microspheres measuring from 500 to 800 microns, was used to prevent sticking at Russian fields. This technology has several advantages. The addition of microspheres into the drilling fluid significantly improves the lubricity of aqueous, petroleum and synthetic based solutions. Laboratory and practical studies of dry lubrication have proved that its use significantly reduces the shear moment in case of differential sticking. There is also a significant reduction in the likelihood of pipe sticking when pressure drops. In addition, microspheres do not clog the formation and are easily washed out of the well [7]. In May 2018 the liner was successfully lowered using the STIK LESS dry lubricant at the Vankorskoe field.

In 2017, the effect of dry lubricant Alpine drill beard on the drill string mobility was studied at the Vankorskoe field. The 20 kg/m^3 Alpine Drill Beads drilling mud pack was installed in the open hole during the last wiper trip of side-tracking. To assess the efficiency of the Alpine Drill Beads, the weight when lifting the drilling tool before and after installing the pack, as well as the hook weight when running the liner were compared with the calculated parameters. When running the liner, a road map was created allowing clearly seeing a significant decrease in the actual weight of the pack when running the liner compared to the calculated value. As a result of the recalculation, taking into account the actual weight of the pack when running the liner, the friction coefficient made less than 0.3, while the calculated coefficient was 0.4-0.5.

RN-Uvatneftegaz LLC introduced injection of a drilling fluid pack containing "dry lubricant" in the form of calibrated glass microspheres with a diameter of up to 800 μm before running the production casing in wells with vertical deviation of more than 3500 m. The pack was installed at the intervals of the open borehole, namely, in the area of the zenith angle fall in the S-shaped directional wells and in the interval of the zenith angle rise in the J-shaped tangent section of the horizontal well, in which the greatest friction coefficients were observed. Lubrication was used simultaneously with semi-positive centralizer, which made it possible to reduce the friction coefficient when running casing strings in deep wells in a cased hole up to 0.20 and in an open hole up to 0.30. When using standard technologies and tooling the friction coefficients equal 0.25 and 0.40 [8].

Since October 2016, Gazpromneft-Khantos LLC has drilled 10 horizontal wells at the Yuzhno-Priobskoe field, in which drilling mud was used to drill a horizontal hole without the developed active solid phase with the addition of microspheres or "microballs". The use of this technology made it possible to reduce the risk of complications with increased density of the solution caused by sticking [9].

During horizontal drilling, the fixing of the liner, which has a pre-perforated part and is cemented above the filter, may be accompanied by liner failure to reach the bottom without washing through the shoe, and poor-quality removal of clay solution and cuttings may also occur in the filter interval and the borehole annulus of the liner [10-13]. One way to reduce the risk of sticking when running the liner is to use filters with plugs FS114-6000x0.3-10x64-KC132. These filters are used in oil, gas and water wells of any design and operate in the medium of mineralized reservoir water, oil, natural gas, condensate and other well fluids. The filters are designed to prevent destruction of weakly cemented collectors and ingress of sand and other mechanical impurities into the well. The filter is installed at the bottom of the casing string through threading. After cementing and flushing the well, the sealing plugs are cut off the filter of FS114-6000x0.3-10x64-KC132 type with the device for removing the plugs with their tracking at well mouth. The proposed technology allows the following:

- running the tool with the liner to the measured depth with high-quality washing and without complications;
- pumping the colmatation packs.

Another way to reduce the risk of sticking is the use a column rotating shoe – BK-Vr 114. The rotating column shoe BK-Vr is designed for the bottom of the casing string to direct it along the well bore and pass complicated zones (cavities and tight spots) without landing, which is carried out by overcoming ledges and obstacles with the eccentric tip of the shoe. In a horizontal wellbore, the landing and subsequent stoppage of the casing are generally always associated with cuttings accumulated from the lower wellbore area. For better visibility, the movement of the casing can be compared to log skidding. Equipping the log with a rotating hemispherical eccentric head makes it possible to significantly improve its sliding along any uneven surface. A similar principle is implemented in the new design of the rotating shoe. The BK-Vr shoe has structural features: the guide plug of the shoe is made with a bevel of 20-30 ° and an eccentric displacement of the top of the rounded cone relative to the shoe axis. BK-Vr 114 is used in wells with casing columns which nominal diameter is 114 mm. The field of application of the device is vertical and directional horizontal wells. The assembly of the BK-Vr 114 column shoe includes an eccentric aluminum alloy nozzle with one central and four side flushing holes fixed in a coupling and having a sliding bearing.

In May 2018, the technologies proposed in the paper were used at the Vankorskoe field for the successful installation of the borehole and accident-free running of the liner. A lateral borehole with a horizontal end was successfully drilled, during the construction of the well, the experience of drilling boreholes in conditions of low formation pressures of the productive formation Nx 3-4 was used. The use of the Rhino XC expander allowed increasing the diameter of the wellbore, which is reflected in acoustic log data. STICK LESS 20 reagent was used to reduce the friction coefficient when running the liner casing. Besides, the liner arrangement included a guide rotating shoe together with filters with plugs.

4. Conclusion

The well is designed for long-term operation and is an expensive fundamental hydraulic structure. Therefore, the use of advanced modern technologies in well construction will reduce complications and accidents during construction and ensure their better operation. The analysis of the use of the described technologies in various fields suggests that the use of a set of measures to prevent sticking, including the use of the Rhino XC Tool borehole expander, filters with plugs, a rotating shoe guide and a “dry lubricant” makes it possible to recommend these technical and technological measures.

References

[1] Guseynova E L and Guseynov E M 2017 Technology of drilling multi-barrel wells *Modern technologies in the oil and gas industry-2017: proceedings of the international scientific and technical conference* Ufa: usntu Publishing house pp 239-242

[2] Batalov S A, Andreev V E, Lobankov V M and Mukhametshin V Sh 2019 Numerical simulation of oil formation with regulated disturbances. Oil recovery quality simulation *Journal of Physics: Conference Series (ITBI 2019 – International Conference "Information Technologies in Business and Industry")* **1333(3)** 1-6 DOI: 10.1088/1742-6596/1333/3/032006

[3] Guseynova E L and Guseynov E M 2018 Complex of measures to prevent the tailing of the tail *Modern technologies in the oil and gas industry-2018: proceedings of the international scientific and technical conference* (October, March 30, 2018) Ufa: usntu Publishing house, pp. 299-302.

[4] Yangirov F N, Yakhin A R, Mustafin T S and Dikhtyar T D 2018 Substantiation of lubricant selection for drilling technology *Bulletin of the Tomsk Polytechnic University, Geo Assets Engineering* **329 (1)** 51-58

[5] Tretyak A Ya, Rybalchenko Yu M, Lubyanova S I and Turuntaev Yu Y 2016 Ways to combat differential prihvatami *Technologies of production and use of hydrocarbons* **1(6)** available at: http:// http://tp-ning.ru/img/06/09.pdf /

[6] Rhino X C *Expanding the upper section of the well in one chisel* available at: https: //www.slb.ru/library/projects/well_construction_cs/burovoe-oborudovanie-i-servisy/rhino-xc-rasshirenie-verkhney-sektsii-skvazhiny-za-odno-dolblenie/

[7] Makhmutov D, Yakunov A, Lozhkin S, Khristenko A, Cherevko S and Syroegin M 2017 The use of «dry lubricant» for reducing the risk of sticking during liner running for multistage *Drilling and Oil* **2, *available at:*** https://burneft.ru/archive/issues/2017-02/44

[8] Selivanova M V, Tyncherov K T, Ikhsanova F A, Kalmykov I A and Olenev A A 2019 A device for paired zeroing of numbers in a residue system *Journal of Physics: Conference Series (ITBI 2019 – International Conference "Information Technologies in Business and Industry")* **1333(2)** 1-5 DOI: 10.1088/1742-6596/1333/2/022017

[9] *Gazpromneft-Khantos summed up the production results of the first half of 2017,* available at: http://www.ngv.ru/pr/gazpromneft-khantos-podvel-proizvodstvennye-itogi-pervogo-polugodiya-2017-goda/?sphrase_id=9591010

[10] Nikulin V A, Mukhamadiev A M and Galimov R M 2010 Technology of tailings cementation *Specialized journal "Drilling and oil"* **1**, available at: http://burneft.ru/archive/issues/2010-01/15 (accessed: 11.03.2018).

[11] Kuleshova L S, Mukhametshin V V and Safiullina A R 2019 Applying information technologies in identifying the features of deposit identification under conditions of different oil-and gas provinces *Journal of Physics: Conference Series (ITBI 2019 – International Conference "Information Technologies in Business and Industry")* **1333(7)** 1-5 DOI: 10.1088/1742-6596/1333/7/072012

[12] Akhmetov R T, Mukhametshin V V and Kuleshova L S 2019 Simulation of the absolute permeability based on the capillary pressure curves using the dumbbell model *Journal of Physics: Conference Series (ITBI 2019 – International Conference "Information Technologies in Business and Industry")* **1333(3)** 1-8 DOI: 10.1088/1742-6596/1333/3/032001

[13] Batalov S A, Andreev V E, Lobankov V M and Mukhametshin V Sh 2019 Numerical simulation of the oil reservoir with regulated disturbances. Oil recovery stability simulation *Journal of Physics: Conference Series (ITBI 2019 – International Conference "Information Technologies in Business and Industry")* **1333(3)** 1-6 DOI:10.1088/1742-6596/1333/3/032007

IOP Conf. Series: Materials Science and Engineering **952** (2020) 012048 doi:10.1088/1757-899X/952/1/012048

Non-conventional methods for oil production in the conditions of the Abdrakhmanovskaya site of the Romashkino oil field

A Yu Gutorov

Ufa State Petroleum Technological University, Branch of the University in the City of Oktyabrsky, 54a, Devonskaya St., Oktyabrsky, Republic of Bashkortostan, 452607, Russian Federation

E-mail: Gutorov70@mail.ru

Abstract. The paper analyzes the existing non-conventional methods for oil production in the conditions of the Abdrakhmanovskaya site of the Romashkino oil field. It describes the methods for well stimulation by swabbing in tubing strings (TS) and production casing (PC). A number of non-conventional production methods are compared to identify the most innovative. Some technical approaches are outlined once used in specific conditions to effectively deploy the operating well stock. The prospects for their use are evaluated.

1. Introduction
Swabbing can be used as an alternative way to put low-production and poorly-equipped wells into production from an idle well stock without spending much money on infrastructure. The method will help not only involve low-production wells from an inactive well stock (unprofitable wells), but also provide additional oil production from new, temporarily unexplored wells [1].

2. Materials and methods
In 2015, a series of theoretical studies, experimental design and field work carried out to swab oil wells resulted in a set of technical requirements outlined for the above method particularly applied in the conditions of the Abdrakhmanovskaya site of the Romashkino oil field.

Swabbing is used in wells (including those to contain highly viscous products) that lack gas and oil inflow ('brown' fields, abnormal low pressures at oil fields with a negligible gas factor, etc.). No other way to reduce fluids is yet that effective, or impossible at all, as the one proposed herein.

Prior to applying this set of techniques, all equipment is removed from the well followed by wiper trips. As soon as products accumulate in the well, the rig operator backs the machine, consisting of a winch, cable, swap cups, foldable mast and a pulley at the top, up to the edge of the well. Once the equipment is adjusted, the operator, by way of the winch, lowers the cable with a swab cup into the well down to a predetermined depth below the fluid level. Then the swab is pulled up the PC along with a portion of well fluids. The fluids pulled out of the well are drawn to the flow line and then to the collection system or collection tank. Subject to the mission assigned, swabbing cycles, are repeated until the level in the well decreases to a predetermined depth, or until a predetermined volume of fluid is lifted from the well. The schematic for well stimulation by swabbing in the production casing is shown in Figure 1.

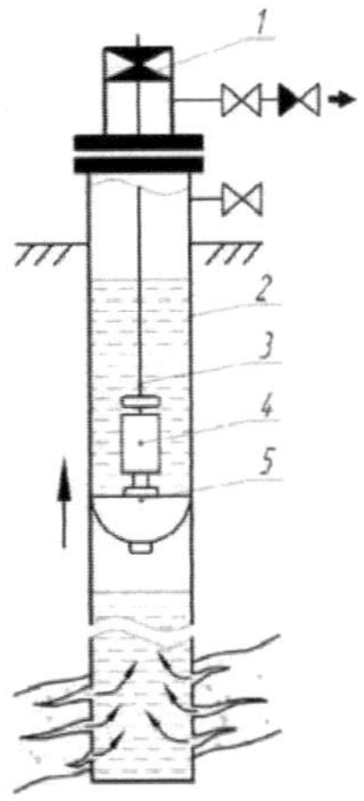

Figure 1. Schematic for well stimulation by swabbing in production casing: 1 – well head; 2 – production casing; 3 – foldable mast; 4 – plummets; 5 – swab cup

PC swabbing benefits:
- high productivity (up to 3 m^3 of liquid per a lift);
- minimum labor intensity for fluid removal;
- opportunity to pull oil from the reservoir above the OWI border;
- improved selection of highly viscous products;
- high-performance fluid withdrawal in small diameter wells;
- low metal consumption (absence of tubing string, sucker rods, pumping equipment).

The drawbacks of PC swabbing include the inability to work in wells with oil and gas inflow, as well as in wells with violated PC integrity or local internal bottlenecks.

Swabbing in tubing strings is applied to wells equipped with tubing strings.

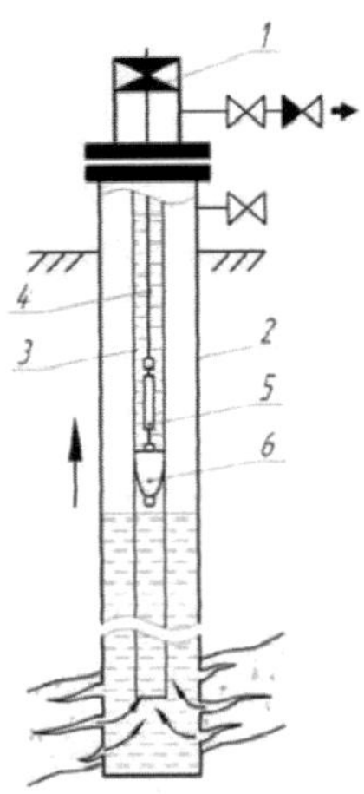

Figure 2. Schematic for well stimulation by swabbing in tubing strings: 1 – well head; 2 – production casing; 3 – tubing string; 4 – foldable mast; 5 – plummets; 6 – swab cup

A wiped tubing string is installed in the well. As soon as the oil accumulates in the well, the rig operator now and then backs the machine accommodating the equipment needed to provide swabbing in a tubing string. Herein, swabbing implies pulling some portions of fluid out of the well with a sequential stepwise decrease in fluids and an appropriately changing depth the swab is lowered with each subsequent run. The swab is lowered into the tubing string down to a predetermined depth below the fluid level (measured in hundreds of meters and limited by the strength characteristics of the swab cups or mast and the characteristics of a winch). After that, it rises up the tubing string along with a portion of the well fluid. The liquid lifted to the wellhead is drawn to the flow line and then to the collection system or collection tank. Subject to the mission assigned, swabbing cycles are repeated until the level in the well decreases to a predetermined depth, or until a predetermined amount of fluid is pulled out of the well. The schematic for well stimulation by swabbing in tubing strings is shown in the Figure 2.

The benefits can include relatively high productivity, well-articulated applications in the wells with oil and gas income, low labor intensity for fluid lifting.

The drawbacks can include the following:

- the lower end of the tubing string is below the oil-water interface (OWI) boundary, hence, oil extraction procedure first involves water arrival in tubing strings, which, along with OWI lowering to the lower end of the tubing string, is subsequently followed by the oil accumulated in the well getting to the tubing string;
- it is impossible to lower the level in the well more than about 100 m to the lower end of the tubing string, which leads to under-extraction of oil from the well in a volume of up to 1.5 m^3;
- reduced selection of highly viscous products (with increasing viscosity it becomes difficult, and sometimes impossible to lower the swab down into the tubing string);
- high metal consumption of well equipment, since the well permanently houses the tubing string.

3. Results

Non-conventional methods are exploited for oil extraction at the Abdrakhmanovskaya site of the Romashkino field. Oil is produced in unequipped, low-production and exploratory wells by way of one-time pumping through swabbing technique in the tubing string and production casing [3, 4]. Actually, swabbing technique applied for pushing products up the production casing often entails significant leaks.

PC swabbing allowed for the average oil production to be 2.5 t/1 per a single run, while the average annual production was 24.4 tons. Total costs amounted to 895 thousand rubles. Net profit made up 51 thousand rubles.

TS swabbing allowed for the average oil production to be 5.3 t/1 per a single run, while the average annual production was 70.8 tons. Total costs amounted to 1704 thousand rubles. Net profit made up 612 thousand rubles.

By way of comparison, as of 2015, TS swabbing was deemed as the most effective non-conventional method of oil production at Abdrakhmanovskaya site, since PC swabbing resulted in total costs exceeding net profit.

TS swabbing provides:

- fluid withdrawal for a single run was 93.4 m^3;
- oil extraction for a single run was 5.3 tons;
- average annual oil production was 70.8 tons;
- total costs amounted to 1704 thousand rubles;
- net profit made up 612 thousand rubles.

The above oil production method reduces operating costs and extends profitable operation terms of low-production, undeveloped and exploratory wells.

With a view to analyzing a non-conventional method of oil production, 5 wells were adequately assessed for PC swabbing throughout 2015 (Table 1).

PC swabbing was deployed in wells No. 10387; 14,129; 6591a; 9293; 3625. The average distance to well sites was 16.6 km.

The maximum fluid withdrawal for a single run from well No. 6591a was 64 m^3, the minimum – 7 m^3 from well No. 3625.

The maximum oil extraction for a single run from well No. 10387, No. 9293 amounted to 3 tons. In wells No. 14129, No. 6591a, oil extraction for a single run was 2 tons, whereas in well No. 3625, oil extraction for a single run was 0 tons.

The maximum number of repeats to produce oil in one year in well No. 6591a was 23 times, the minimum – in well No. 3625 was 1 time.

The annual oil production in wells No. 10387, 14129, 6591a, 9293 and 3625 averaged 24.4 t.

Total costs for wells No. 6591a amounted to 365 thousand rubles, No. 9293 – 258 thousand rubles, No. 10387 – 154 thousand rubles, No. 14129 – 102 thousand rubles, No. 3625 – 16 thousand rubles.

The maximum net profit for well No. 9293 made up 34 thousand rubles, for well No. 6591a – 21 thousand rubles, and for wells No. 10387, No. 14129, and No. 3625, the total costs exceeded net profit.

Table 1. Technical and economic indices of oil production via PC swabbing

Well No.	Distance from well site to oil custody transfer, km	Fluid withdrawal per a single run, m^3	Oil extraction per a single run, t	Number of repeats per year	Annual oil production, t	Total costs, RUB, in thousands	Net profit, RUB, in thousands
10387	5	27	3	4	11	154	-77
14129	18	15	2	7	12	102	-16
6591a	10	64	2	23	56	365	21
9293	15	52	3	14	43	258	34
3625	35	7	0	1	0	16	-13
Total	83	165	10	49	122	895	-51

In 2015, TS swabbing (Table 2) was deployed in wells No. 15439, 20081, 36309, 26508, 147. The average distance to well sites was 9.6 km.

The maximum fluid withdrawal for a single run from well No. 20081 was 184 m^3, the minimum – 39 m^3 from well No. 26508.

The maximum oil extraction for a single run from well No. 15439, No. 36309 amounted to 6 tons. In wells No. 20081, No. 26508, No. 147 oil extraction for a single run was 5 tons.

The maximum number of repeats to produce oil in one year in well No. 20081 was 26 times, the minimum – in well No. 26508 was 6 times.

The annual oil production in wells No. 15439, 20081, 36309, 26508 and 147 averaged 70.8 t.

Total costs for wells No. 20081 amounted to 616 thousand rubles, No. 15439 – 351 thousand rubles, No. 36309 – 375 thousand rubles, No. 147 – 249 thousand rubles and No. 26508 – 163 thousand rubles.

The maximum net profit for well No. 20081 made up 250 thousand rubles, while minimum for well No. 26508 made up 50 thousand rubles.

Table 2. Technical and economic indices of oil production via TS swabbing

Well No.	Distance from well site to oil custody transfer, km	Fluid withdrawal per a single run, m^3	Oil extraction per a single run, t	Number of repeats per year	Annual oil production, t	Total costs, RUB, in thousands	Net profit, RUB, in thousands
15439	7	108	6	13	73	351	125
20081	6	184	5	26	133	616	250
36309	10	78	6	12	67	375	113
26508	15	39	5	6	32	163	50
147	10	58	5	10	49	249	74
Total	48	467	27	67	354	1704	612

4. Conclusion

Technical and economic indices derived from non-conventional oil production at the Abdrakhmanovskaya site provided the following conclusions:

- PC swabbing is not cost effective as total costs exceed net profit;
- TS Swabbing is by far the most effective non-conventional method of oil production in the conditions of the Abdrakhmanovskaya site of the Romashkino oil field, which, with an unequipped, low-yield, inactive well stock, makes this method promising today.

However, swabbing technique applied for pushing products up the production casing often entails significant leaks.

References

[1] Gutorov Yu A 2018 Geological and field preconditions and the possibilities of physical-chemical methods impact on the interwell space to enhance formations oil recovery during water-flooding *Oilfield Engineering* **8** 27-31

[2] Gutorov Yu A 2017 Some information in relation to the efficiency of management of oil-gas fields development based on the well-control intellectual systems *Oilfield Engineering* **8** 10-16

[3] Gutorov Yu A 2013 Improving the development of hard-to-recover oil reserves in the conditions of flooding in the fields of Tatarstan *Gas industry* **696** 61-63

[4] Polyakov V N, Chizhov A P, Kotenev Yu A and Mukhametshin V Sh 2019 Results of System Drilling Techniques and Completion of Oil and Gas Wells *IOP Conference Series: Earth and Environmental Science (IPDME 2019 – International Workshop on Innovations and Prospects of Development of Mining Machinery and Electrical Engineering)* **378(1) 012119** 1–7 DOI: 10.1088/1755-1315/378/1/012119

[5] Shakurova Al F, Shakurova Al F 2018 *IOP Conference Series: Earth and Environmental Science* **8** 194

[6] Akhmetov R T, Kuleshova L S and Mukhametshin V V 2019 Application of the Brooks-Corey model in the conditions of lower cretaceous deposits in terrigenous reservoirs of Western Siberia *IOP Conference Series: Materials Science and Engineering (MEACS 2018 – International Conference on Mechanical Engineering, Automation and Control Systems)* **560(1) 012004** 1-4 DOI: 10.1088/1757-899X/560/1/012004

[7] Kuleshova L S and Mukhametshin V V 2019 Estimation of the wells hydrodynamic drag level based on wells geophysical survey data *Atlantis Highlights in Material Sciences and Technology (ISEES 2019 – International Symposium "Engineering and Earth Sciences: Applied and Fundamental Research" dedicated to the 85th anniversary of H.I. Ibragimov)* **1** 727–729. DOI: 10.2991/isees-19.2019.145

[8] Khokhlov V I, Galimov Sh S, Devyatkova S G, Kotenev Yu A, Sultanov Sh Kh and Mukhametshin V Sh 2019 Justification of impact and planning of technology efficiency on the basis of limy-emulsion formulation in low-permeability highly-rugged reservoirs of Tyumen deposits *IOP Conference Series: Earth and Environmental Science (IPDME 2019 – International Workshop on Innovations and Prospects of Development of Mining Machinery and Electrical Engineering)* **378(1) 012114** 1–6 DOI: 10.1088/1755-1315/378/1/012114

[9] Almukhametova E M 2018 *IOP Conference Series: Earth and Environmental Science* **6** 194

[10] Malyarenko A M, Bogdan V A, Kotenev Yu A, Mukhametshin V Sh, Umetbaev V G 2019 Wettability and formation conditions of reservoirs *IOP Conference Series: Earth and Environmental Science (IPDME 2019 – International Workshop on Innovations and Prospects of Development of Mining Machinery and Electrical Engineering)* **378(1) 012040** 1–6 DOI: 10.1088/1755-1315/378/1/012040

IOP Conf. Series: Materials Science and Engineering **952** (2020) 012049 doi:10.1088/1757-899X/952/1/012049

Development of microcalcite-based drilling fluid

R A Ismakov[1], T D Dikhtyar[2], F N Yangirov[1]

[1] Ufa State Petroleum Technological University, 1, Kosmonavtov St., Ufa, Republic of Bashkortostan, 450062, Russia
[2] Ufa State Petroleum Technological University, Branch of the University in the City of Oktyabrsky, 54a, Devonskaya St., Oktyabrsky, Republic of Bashkortostan, 452607, Russia

E-mail: ismakovrustem@gmail.com

Abstract. One of the most common hindrances to encounter during the construction of oil and gas wells is lost returns. The methods to select prevention and control measures are based on quantitative criteria that reflect the hydrodynamic capacity and geological structure of the strata. One of the ways to prevent this hindrance during the construction of oil and gas wells is pore-plugging – a sophisticated physicomechanical process to reduce the permeability of a reservoir formation, which takes place over time due to mud filtrates, dispersed particles or specially introduced fillers (plugging additives). For this reason, it seems relevant to use domestic microcalcite, as a component of drilling fluids, from the Minyar field. The paper aims to assess the effect of various microcalcite concentrations on the properties of drilling fluids and to assess as to whether this type of material is effective enough for pore-plugging under the specified drilling conditions.

1. Introduction

Well construction is characterized by drilling operations in permeable absorbing formations, which often entails lost returns. Untimely prevention of this type of hindrances can lead to additional technical and technological measures. Therefore, improved means and methods of handling lost circulation is a very urgent challenge to be addressed in drilling wells [1, 2]. One way to prevent lost circulation is to use plugging additives in the drilling fluid.

In addition, through the use of special complexes of different-sized acid-soluble colmatants in drilling fluids, it is possible to develop a thin impermeable pore-plugging screen in the productive matrix to reduce filtrate ingresses into a reservoir formation and preserve the permeability of the near-wellbore area.

A plugging material that meets the described requirements during drilling operations is a powder based on calcium carbonate (microcalcite). Due to its unique physicochemical properties, microcalcite is a highly demanded mineral filler and the most environmentally friendly material, chemically neutral, non-combustible and has low water absorption.

2. Materials and methods

Mud parameters were determined through instrumental methods according to GOST 33213-2014 (ISO 10414-1: 2008) "Monitoring mud parameters in field conditions. Water Based Solutions." The effect of plugging additives and drilling filtrates on the change in rock permeability (core sample) in laboratory settings was studied using the device CPMU-1M.

Microcalcite is a finely divided carbonate product, which is obtained by mechanical grinding of limestone. This product is made of marble that is highly resistant to mechanical destruction, so, unlike other forms of calcium carbonate, it does not disperse. By its physicochemical properties and characteristics, microcalcite varies favorably from other carbonate fillers [3] in that it:

- is high in calcite;
- has high particle strength;
- has low porosity;
- has low moisture absorption;
- has high solubility in mineral acids.

Table 1 presents the technological parameters of the drilling fluid for drilling a borehole section with a possible complication in the interval 5580–5598 m of well No. 1 at the Tsekertinsky license site.

Table 1. Required technological parameters of the drilling fluid

Density (kg/m^3)	Funnel viscosity (s)	Filtration index (cm^3/30min)	pH
1220±0.03	30–36	5–6	10

The above section in the upper part is represented by calcareous dolomites, dolomitic limestones, light gray, dense.

The study implied selecting the composition of drilling fluid and assessing the effect of microcalcite of the Minyar open pit on its properties. The following formulation was selected as a source mud composition:

NaOH (0.1 %) + Na_2CO_3(0.2 %) + PBMA (6 %) + PAC–LV (0.5 %) + PAC–HV (0.2 %) + Lubriol (10 %) + water (balance).

Microcalcite was introduced at a concentration of 50, 100, 150, 200 g/l (5, 10, 15, 20%, respectively):

- experiment No. 1: source solution;
- experiment No. 2: source solution + 5% microcalcite;
- experiment No. 3: source solution + 10% microcalcite;
- experiment No. 4: source solution + 15% microcalcite;
- experiment No. 5: source solution + 20% microcalcite.

The parameters of the above drilling compositions are presented in Table 2.

Table 2. Parameters of microcalcite-based drilling composition

Experiment No.	Density (kg/m^3)	Funnel viscosity (s)	Filtration index (cm^3/30min)	Plastic viscosity (MPa·s)	DSS (dPa)	SSS (Pa)	Friction coefficient	pH
1	1210	35	5.5	12.4	48.0	14/18	0.22	11
2	1220	37	5.0	12.9	50.5	12/17	0.26	11
3	1230	34	4.5	13.4	51.3	15/19	0.31	11
4	1240	36	4.1	14.3	52.1	19/29	0.40	11
5	1250	33	3.8	15.2	52.8	22/31	0.48	11

Due to the higher microcalcite grade in the drilling formulation, as the above table indicates, its density, plastic viscosity, static and dynamic shear stress, and friction coefficient of a mud cake increase. An increased amount of microcalcite introduced leads to a decrease in the filtration rate of the drilling fluid.

The pore-plugging capacity of drilling fluids that contain various microcalcite concentrations was further evaluated through the study of its granularity. The results are presented in Figure 1.

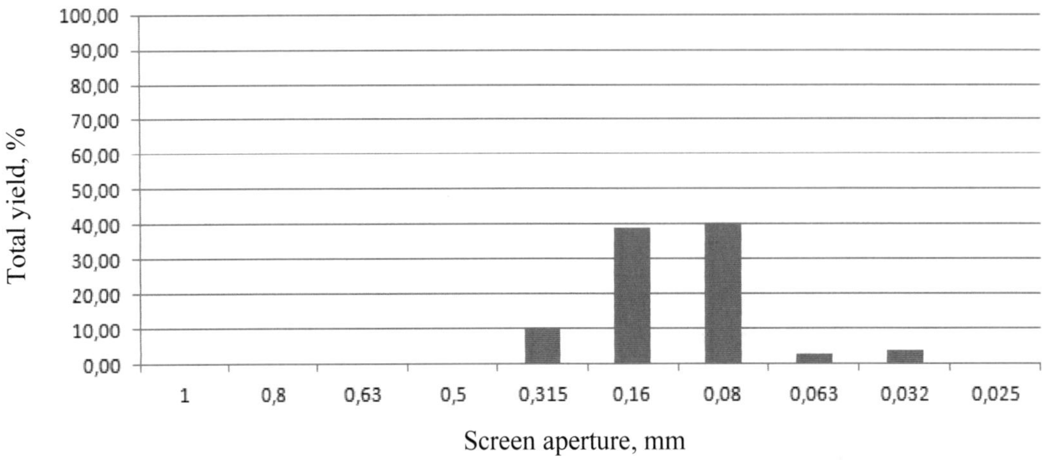

Figure 1. Microcalcite granularity

A fractional pore-plugging substance selected to constitute the drilling fluid will not only reduce lost circulation, but also reduce the rate of filtrate and solid ingresses into the pores and channels of a reservoir. With a view to preserving some reservoir properties, various options are proposed, namely: draw-down or over-balance drilling, high-quality drilling fluids with minimal negative impact on the reservoir properties [4–9], and pore-plugging. Therefore, the next, and most important stage, was the analysis of core samples taken in the interval 5580–5598 m along with an assessment of the effectiveness of the pore-plugging capacity of microcalcite from the Minyar site. A modified installation CPMU-1M was used to study the laws of liquid and gas filtration in a porous medium under the conditions close to the real reservoirs. The dimensions of target core samples were: diameter – 30 mm, height – 50 mm.

Based on the core filtration results (studied drilling compositions), the permeability coefficients of core samples were determined in accordance with the Darcy's linear law before and after being exposed to fluid filtrate. The return permeability coefficient was calculated by the formula:

$$\beta_1 = \frac{k_1}{k_0} \cdot 100\%,$$

where β is the return permeability coefficient; k_1, k_0 are the core permeability for oil before and after drilling fluid being introduced into the core, respectively.

The findings are presented in Table 3 and in Figure 2.

Table 3. Core return permeability coefficient

Sample No.	Core flow rates		Pressure liquid rate	Permeability before c_b (mD)	Permeability after c_a (mD)	Return permeability coefficient, %
	time (s)	volume (m^3)	$Q \cdot 10^{-7}$ (m^3/s)			
1	30	0.0000103	3.42	56	41.0	73.2
2	30	0.0000095	3.17	56	21.4	38.2
3	30	0.0000107	3.57	56	12.45	22.2
4	30	0.0000161	5.36	56	9.58	17.1
5	30	0.0000154	5.13	56	4.87	8.7

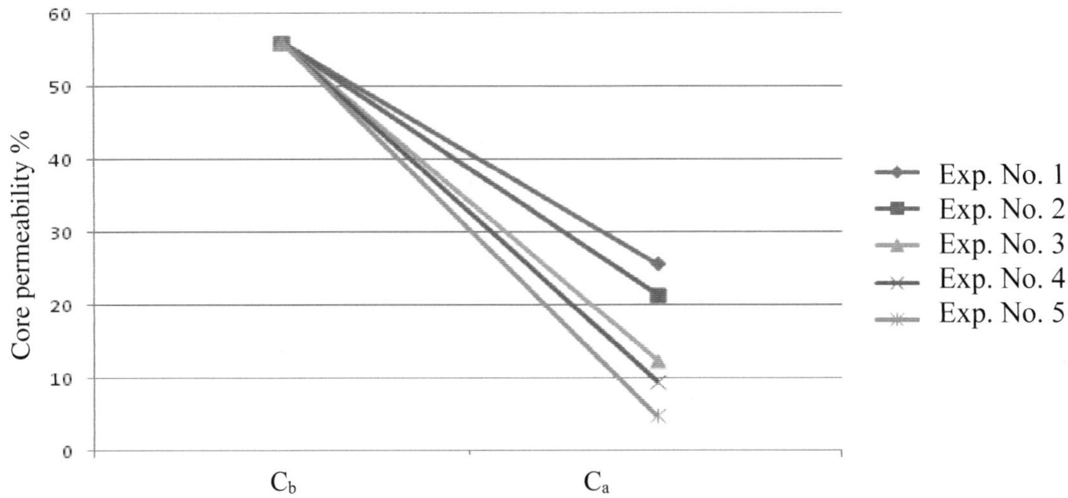

Figure 2. Varying core permeability depending on the type of drilling fluid: respectively, before (c_b) and after (c_a) filtration of the drilling fluid

The permeability of the core samples is reduced when any of the five drilling compositions examined is applied.

3. Results and discussion

The experimental results show the following. Microcalcite is compatible with drilling fluid, but once it increases in the formulation, the density and rheological parameters go up, while the filtration rate of the drilling fluid goes down. An increased friction coefficient of the mud cake produced by the drilling fluid implies a decreased lubricating effect and possible hindrances during drilling operations.

A decrease in the permeability of samples after being exposed to the studied drilling formulations indicates that the microcalcite available in their composition creates a pore-plugging screen in the core. In this respect, the most effective is the fluid with a content of 20% microcalcite (formulation No. 5), since it provides the most significant reduction in core permeability, but, based on geological conditions, the applicability of this formulation is not possible due to the excess of the required density.

The drilling fluid with a content of 5% microcalcite (formulation No. 2) is the most optimal for the initially set geological conditions. Once applied, it reduces the target core permeability and is acceptable for the density required for advancing said density range.

4. Conclusion

One of the ways to prevent lost returns to encounter during the construction of oil and gas wells is the use of pore-plugging additives in drilling fluids. It seems relevant to use a highly resistant to mechanical destruction, non-dispersible, acid-soluble microcalcite of the Minyarsky deposit. Due to the greater microcalcite grade in the drilling fluid, the density, plastic viscosity index, friction coefficient of a mud cake increase, whereas the filtration index decreases. The studies conducted on a modified installation CPMU-1M showed that the most optimal for the geological conditions initially set is the drilling fluid with a content of 5% microcalcite (formulation No. 2), which reduces the permeability of the test core and has the density necessary for advancing said density range.

References

[1] Polyakov V N, Chizhov A P, Kotenev Yu A and Mukhametshin V Sh 2019 Results of System Drilling Techniques and Completion of Oil and Gas Wells *IOP Conference Series: Earth and Environmental Science (IPDME 2019 – International Workshop on Innovations and Prospects of Development of Mining Machinery and Electrical Engineering)* **378(1) 012119** 1–7 DOI:

10.1088/1755-1315/378/1/012119

[2] Teptereva G A, Konesev G V, Ismakov R A, Kantor E A and Dikhtyar T D 2017 Obtaining drill reagents by modification of neutral sulphite alkali phosphonic compounds *Bulletin of the Tomsk Polytechnic University. Geo Assets Engineering* **328(9)** 94-101

[3] Vagizov D I 2017 Fractional analysis of the natural microcalcite of the minyar quarry as a colmatating additive in the drilling washing liquid *Young researcher: challenges and prospects: collection of articles based on the materials of the L International scientific and practical conference* **25 (50)**

[4] Petrov N A and Konesev G V 2017 Research of polymer reagent "gabrose" for use in drilling solutions *Oil and Gas Business* **15(1)** 53-57

[5] Khokhlov V I, Galimov Sh S, Devyatkova S G, Kotenev Yu A, Sultanov Sh Kh and Mukhametshin V Sh 2019 Justification of impact and planning of technology efficiency on the basis of limy-emulsion formulation in low-permeability highly-rugged reservoirs of Tyumen deposits *IOP Conference Series: Earth and Environmental Science (IPDME 2019 – International Workshop on Innovations and Prospects of Development of Mining Machinery and Electrical Engineering)* **378(1) 012114** 1–6 DOI: 10.1088/1755-1315/378/1/012114

[6] Yangirov F N, Yakhin A R, Dikhtyar T D, Loginova M E, Chudnovskaya A V and Shmagel M A 2018 Investigation of surfactants applicable in well drilling *Problems of Gathering, Treatment and Transportation of oil and oil products* **1(111)** 61-68

[7] Yangirov F N, Yakhin A R, Mustafin T S and Dikhtyar T D 2018 Substantiation of lubricant selection for drilling technology *Bulletin of the Tomsk Polytechnic University. Geo Assets Engineering* **329(1)** 51-58

[8] Akhmetov R T, Mukhametshin V V, Andreev A V and Sultanov Sh Kh 2017 Some Testing Results of Productive Strata Wettability Index Forecasting Technique *SOCAR Proceedings* **4** 83–87 DOI: 10.5510/OGP20170400334

[9] Dick M A, Heinz T J, Svoboda C F and Aston M 2000 Optimizing the Selection of Bridging Particles for Reservoir Drilling Fluids *SPE International Symposium on Formation Damage Control, 23-24 February, Lafayette, Louisiana* 1-8

IOP Conf. Series: Materials Science and Engineering **952** (2020) 012050 doi:10.1088/1757-899X/952/1/012050

Investigating the mechanisms of shape distortion of the heat-affected region during electric-arc welding of oilfield pipelines

M Ya Khabibullin[1] **and R I Suleimanov**[1]

[1] Ufa State Petroleum Technological University, Branch of the University in the City of Oktyabrsky, 54a, Devonskaya St., Oktyabrsky, Republic of Bashkortostan, 452607, Russian Federation

E-mail: m-hab@mail.ru

Abstract. The article studies the issue of the influence of the most encountered in practice defects of welded joints of a one-piece joint, formed in the process of applying a weld when connecting oilfield pipelines to a joint by electric arc manual welding, on the strength characteristics of a welded joint by scrutinizing the condition of the pipe material close to the heat-affected zone. As a test sample, a butt welded joint with obvious defects was used. The geometric parameters of the welded joint were measured using standard means and the hardness of the heat-affected zone was determined by the Rockwell method in certain areas with justification of the welding process. The reason for softening is established. The change in the geometry of the heat-affected area is due to the accumulation of filler material during its melting on one side of the weld with a decrease in welding speed. Recommendations are given for the prevention of the destruction of welded structures and the occurrence of emergency situations in oilfield pipelines.

1. Introduction
In the construction of infield pipelines and their subsequent overhaul or maintenance, electric arc manual welding is widely used. During the construction of such structures, due attention is not paid to checking welds for their geometric parameters in the process of connecting pipes in the oil fields, and technical measures are not taken to increase the strength of the welded joint, which often leads to emergency situations caused by a rupture of infield pipelines [1–3]. Consequently, this research will be directly aimed at improving the reliability of welded joints of oil pipelines, repeatedly occurring in the activities of oil companies.

2. Results and discussion
The article considers a variety of the welding process, which is based on the melting of the pipe material when creating high temperature in an electric arc [4–5]. Studies show that the area of the heat-affected area is heated unevenly over the entire width (the maximum temperature occurs in the region of overheating T ≈ 1400–1500 °C). The dependence shown in Figure 1 demonstrates the regions of the heat-affected area that differ in their forming crystal structure with different grain sizes depending on the action [1].

IOP Conf. Series: Materials Science and Engineering **952** (2020) 012050 doi:10.1088/1757-899X/952/1/012050

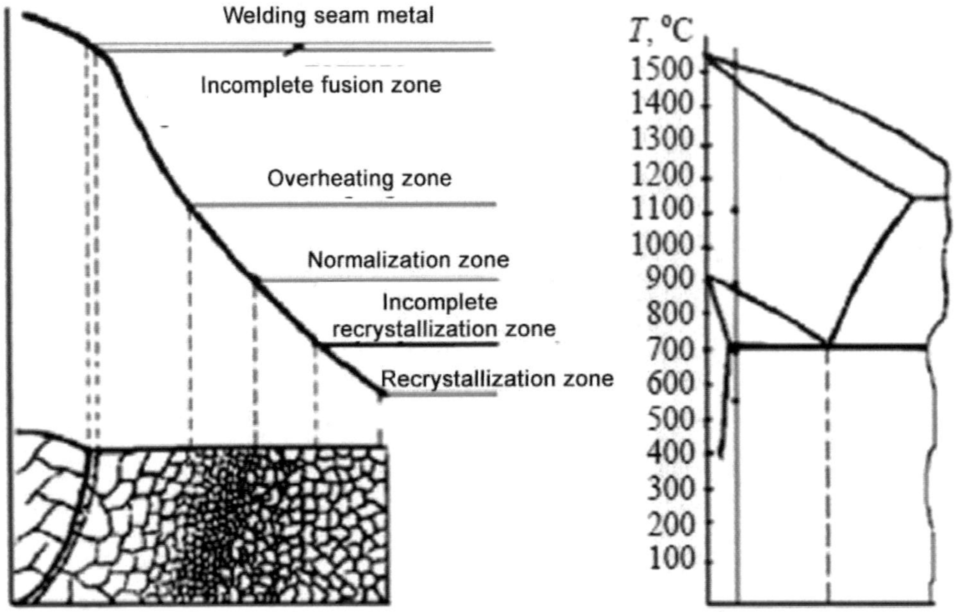

Figure 1. Heat-affected zone

3. Experimental

Figure 2 shows the geometry of the weld under consideration. Figure 3 shows a sample study of a defective welded joint, on which distortions are clearly visible [5].

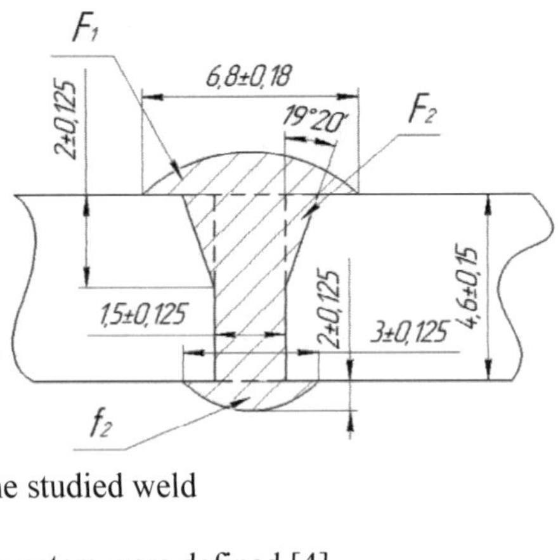

Figure 2. The geometry of the studied weld

The following welding parameters were defined [4]
- welding speed;
- voltage U = 20 V;
- welding current.

To determine the dimensions of the considered thermal-affected zones, Rockwell hardness measurements were carried out in certain areas [6–7]. The obtained data are displayed in the Table.

According to these data, it is clear that the measurements No. 2 and 3 do not coincide, a decrease in the readings on the left side is obvious. The geometric parameters of the heat-affected area are shown in Figure 4, which indicate an increase in the size of the overheating zone. This is due to a change in the thermal conductivity of the pipe material as a result of its recrystallization, which contributed to the transfer of excess heat energy to it, which heated the pipe material to a temperature significantly higher than 830–850 °C. [1, 5], turning it into a molten state with further cooling in the open air.

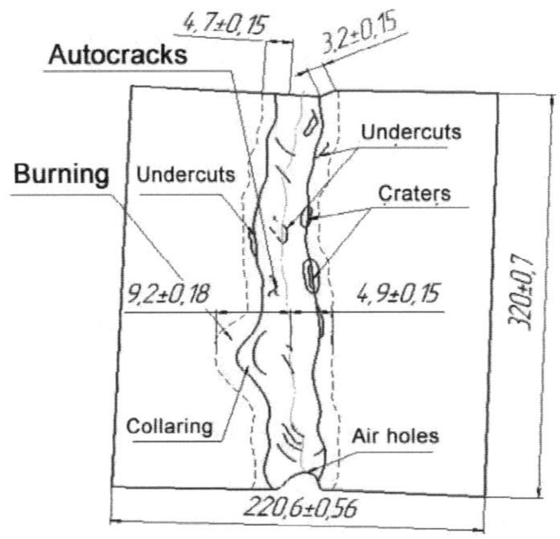

Figure 3. Sample of a defective weld

Table 1 . Hardness values

Measurement	1	2	3	4	5	6
Left side	21 HRC	18 HRC	18 HRC	22 HRC	24 HRC	27 HRC
Right side	21 HRC	18 HRC	22 HRC	23 HRC	24 HRC	26 HRC
25HRC – base metal hardness						

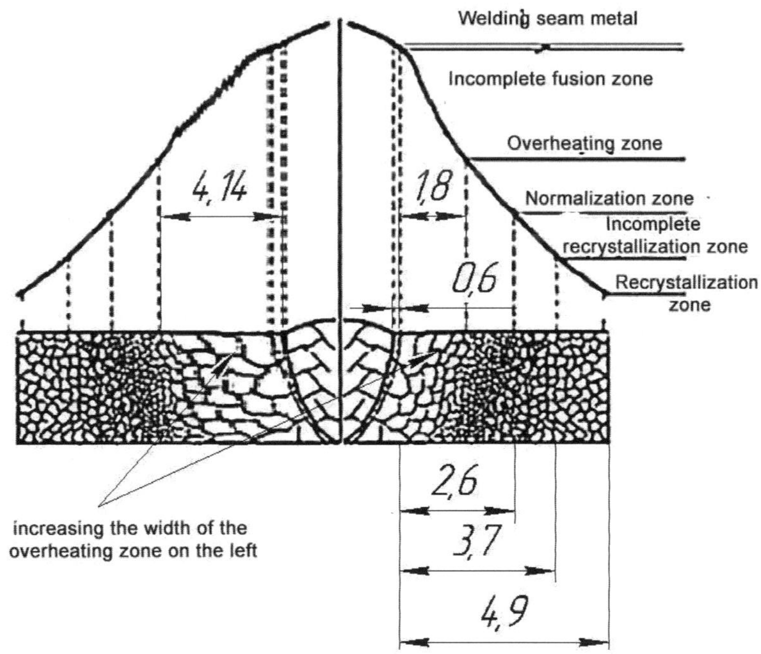

Figure 4. Geometric parameters of the heat-affected zone

The thermal conductivity of the pipe material depends on the grain size of the crystal lattice. Reducing grain sizes reduces thermal conductivity. The initially generated heat propagating from the epicenter of the electric arc has less resistance when passing through a coarse-grained structure, which is caused by high thermal conductivity.

Figure 5 shows the dependence reflecting the ability of various granular structures of the pipe material to transfer thermal energy in the considered areas of thermal impact on different sides of the weld.

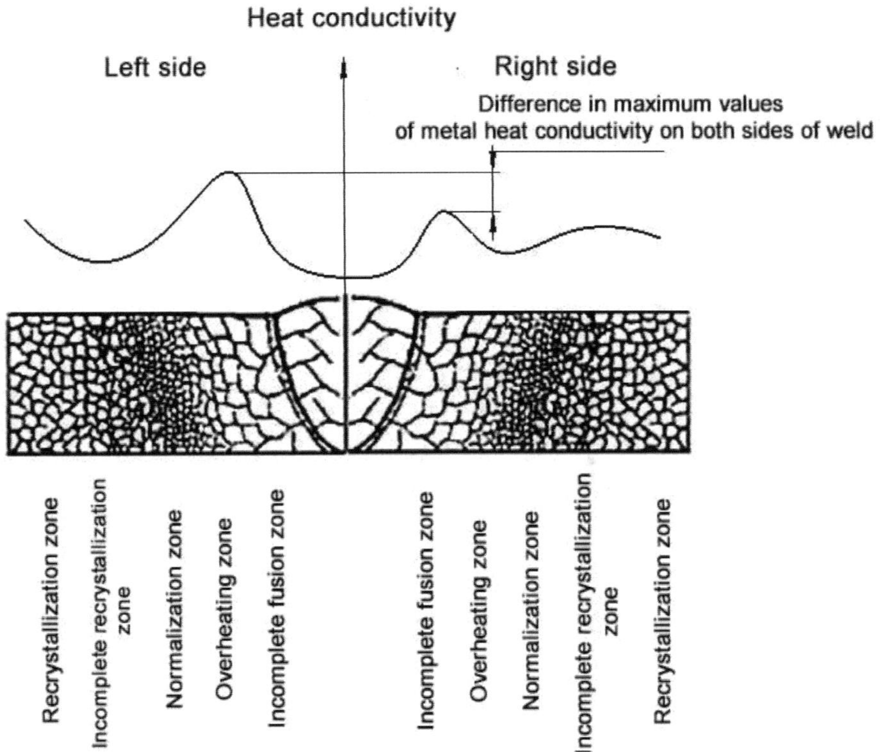

Figure 5. The ability of the pipe material to transfer thermal energy depending on the grain size of the structure

As a result, different regions of the pipe material have different heating, but the completely final structure of the pipe material is formed at t = 723 °C. Peripheral areas demonstrate significant grain growth (Figure 6). A distinctive feature of the overheating region is that the beginning of cooling of the pipe material in the considered region begins with a maximum value, compared with other regions which crystallization occurs at the lowest temperature (red cooling curve). Therefore, the maximum temperature difference is formed in the region of overheating, which creates the condition for the enlargement of the structure of the crystal lattice (Figure 6) [1].

Thus, the reason for the softening of the weld is an increase in the width of the overheating zone. With a decrease in the speed of movement of the electrode, overheating of the metal occurs, followed by a increase in its thermal conductivity. The overheating zone has an enlarged grain structure caused by excessive heat transfer due to the influx of filler material. The results of the study show that even a slight decrease in the speed of electric arc welding leads to a several-fold increase in the width of the most dangerous section of the heat-affected region [8–10]. In such a fragile section of the heat-affected region, the grain sizes are significantly larger than the grain sizes of a metal that is not included in this section, which increases the risk of weld joint failure.

Usually, in the oil and gas fields of Russia, the necessary attention is not paid to correcting and preventing errors that a welder makes during manual electric arc welding. To prevent defects in welded joints, the following control methods are proposed:

- external control of the welded joint and verification of the geometric parameters of the seam;
- physical methods for studying the weld (ultrasonic, magnetographic, X-ray scanning, etc.);
- testing of pressurized infield pipelines for density and strength by gas, hydraulic or pneumatic pressure;
- testing of oil pipelines, the joints of which must be airtight, for density using kerosene, compressed gases, vacuum chambers, ammonia, as well as gas-electric leak detectors.

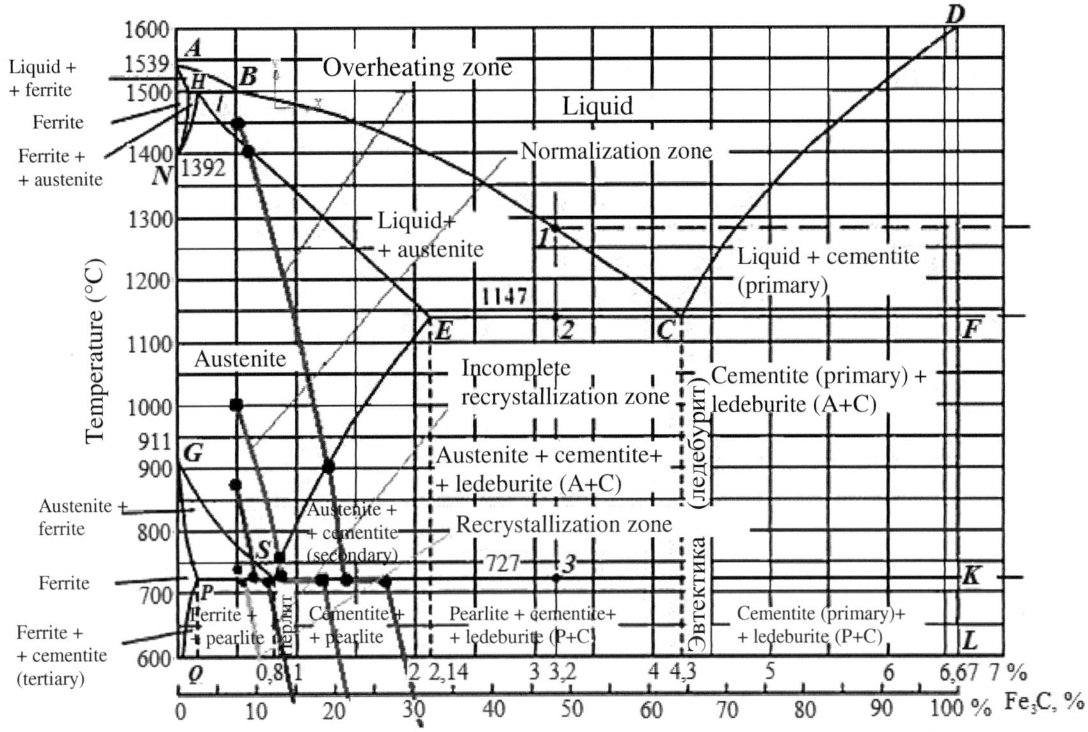

Figure 6. Fe—Fe3C diagram indicating the cooling curves of the areas of the heat-affected area

4. Conclusion

To reduce the residual stresses that form during welding, in practice, immediately after welding, the following methods of processing welds and joints are used: thermal, mechanical, thermomechanical. To prevent the occurrence of corrosion phenomena on welded structures, it is necessary to carry out machining of the welds to remove the remaining slag, and also use various protective coatings, for example, primer paint.

Carrying out a set of measures aimed at preventing the destruction of welded joints of field pipelines is advisable, since when a pipeline ruptures, oil is lost, environmental damage is noticeable and wells are idle during repair work.

References

[1] Meirmanov A and Nekrasova I 2013 Mathematical models of a hydraulic shock *Journal of mathematical analysis and applications* **408(1)** 76-90

[2] Haoran Zh, Yongtu L and Xingyuan Zh 2017 Sensitivity analysis and optimal operation control for large-scale waterflooding pipeline network of oilfield *Journal of petroleum science and engineering* **154** 38-48

[3] Akhmetov R T, Kuleshova L S and Mukhametshin V V 2019 Application of the Brooks-Corey model in the conditions of lower cretaceous deposits in terrigenous reservoirs of Western Siberia *IOP Conference Series: Materials Science and Engineering (MEACS 2018 – International Conference on Mechanical Engineering, Automation and Control Systems)* **560(1) 012004** 1-4 DOI: 10.1088/1757-899X/560/1/012004

[4] Korn G A and Korn T M 1984 *Mathematical Handbook for Scientists and Engineers: Definitions, Theorems, and Formulas for Reference and Review* (Moscow: Nauka)

[5] Heuzeroth F, Fritzsche J and Werzner E 2015 Viscous force - An important parameter for the modeling of deep bed filtration in liquid media *Powder technology* **283** 190-198

[6] Yakupov R F, Mukhametshin V Sh, Khakimzyanov I N and Trofimov V E 2019 Optimization of reserve production from water oil zones of D3ps horizon of Shkapovsky oil field by means of horizontal wells *Georesursy* **21(3)** 55-61. DOI: 10.18599/grs.2019.3.55-61

[7] Polyakov V N, Chizhov A P, Kotenev Yu A and Mukhametshin V Sh 2019 Results of System Drilling Techniques and Completion of Oil and Gas Wells *IOP Conference Series: Earth and Environmental Science (IPDME 2019 – International Workshop on Innovations and Prospects of Development of Mining Machinery and Electrical Engineering)* **378(1) 012119** 1–7 DOI: 10.1088/1755-1315/378/1/012119

[8] Khabibullin M Ya, Suleimanov R I, Sidorkin D I and Arslanov I G 2017 Parameters of damping of vibrations of tubing string in the operation of bottomhole pulse devices *Chemical and Petroleum Engineering* **53(5-6)** 378-384 DOI: 10.1007/s10556-017- 0350-6

[9] Sun W and Mun-Hong H 2017 Forecasting and uncertainty quantification for naturally fractured reservoirs using a new data-space inversion procedure *European Assoc Geoscientists & Engineers Computational geosciences 15th Conference on the Mathematics of Oil Recovery (ECMOR) (Amsterdam, Netherlands)* **21(5-6)** 1443-1458

[10] Khabibullin M Ya and Suleimanov R I 2018 Selection of optimal design of a universal device for nonstationary pulse pumping of liquid in a reservoir pressure maintenance system *Chemical and Petroleum Engineering* **54(3-4)** 225-232 DOI: 10.1007/s10556-018-0467-2

IOP Conf. Series: Materials Science and Engineering **952** (2020) 012051 doi:10.1088/1757-899X/952/1/012051

Improving the design of a downhole filter for fluid production with an increased concentration of suspended particles

M Ya Khabibullin[1], **R I Suleimanov**[1], **Sh G Mingulov**[1], **Yu A Kotenev**[2,3], **Sh Kh Sultanov**[2,3]

[1] Ufa State Petroleum Technological University, Branch of the University in the City of Oktyabrsky, 54a, Devonskaya St., Oktyabrsky, Republic of Bashkortostan, 452607, Russian Federation
[2] Ufa State Petroleum Technological University, 1, Kosmonavtov st., Ufa, Republic of Bashkortostan, 450062, Russian Federation
[3] State autonomous scientific institution "Institute of strategic research of the Republic of Bashkortostan", 129/3, October avenue, Ufa, Republic of Bashkortostan, 450075, Russian Federation

E-mail: m-hab@mail.ru

Abstract. The reliability of the underground equipment of a well sucker rod pump installation will be improved by the use of a new sand filter design, the principle of which is based on the use of centrifugal force of a moving reservoir fluid. To determine the subtlety of cleaning the reservoir fluid from mechanical particles, it should be considered that the fluid velocity is the same as the particle velocity. To do this, we use the universal expression by analogy to our case for the stream function. To develop a specific filter (determination of structural dimensions) taking into account the well conditions, nomograms are developed taking into account all technical and technological conditions. The main advantage of the developed sand filter over other designs is the ability to pass formation fluid even when the cavity for collecting sand and the space between the filter housing and the pipe with radial channels are filled. The presence of the latter also does not affect the performance of the sucker rod pump, in the absence of jamming of the plunger in the cylinder.

1. Introduction

The reliability of the underground equipment of a borehole sucker rod pump installation will be improved by the use of a new sand filter design, the principle of which is based on the use of centrifugal force of a moving reservoir fluid [1].

2. Results and discussion

The structure of the filter design (Figure 1) includes four inlet tubes, which allow creating a swirling liquid flow, as in a hydrocyclone. The reservoir fluid with mechanical impurities flows under pressure from the annulus into the filter. The fluid through the inlet tubes enters the gap between the pipe with radial channels 4 and the central axis 6. Here, the fluid swirls and rises. Further, hitting the walls of the pipe with radial channels, taking into account the friction and gravity forces, the impurities pass through the filtering process through channels 7, which are staggered. The diameters of the filter holes correspond to the size of the impurities. Filtered particles accumulate in the sand collection cavity 2.

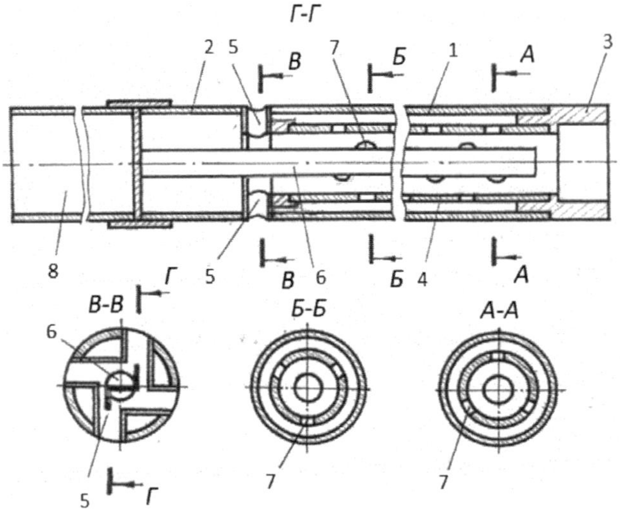

Figure 1. Shortened design of the developed filter: 1 – housing; 2 – sand collection cavity; 3 – sub; 4 – pipe with radial channels; 5 – input tubes; 6 –central axis; 7 – radial channels; 8 – pipe

3. Experimental

To determine the degree of cleaning the reservoir fluid from mechanical particles, it should be considered that the fluid velocity is the same as the particle velocity. To do this, we use the universal expression by analogy to our case for the stream function [2-7]. As a result of transformations of the obtained expressions for the fluid flow rates in the filter [8], we obtain an expression for determining the diameter of the deposited particles:

$$d_ч = \frac{60}{\pi n}\sqrt{\frac{3\mu V_r}{\left(\frac{\rho_ж}{\rho_ч}-1\right)r}}, \tag{1}$$

where μ is the dynamic viscosity of the reservoir fluid, Pa·s; V_r is the radial velocity of the fluid, m/s,

$$V_r = \frac{4Q_0}{\pi(d_1^2-d_2^2)}, \tag{2}$$

where Q_0 is the flow rate of the liquid through the filter, m^3/s; d_1, d_2 are the diameters of the rod and the inner surface of the perforated pipe, m, respectively; n is the frequency of rotation of the liquid in the annular gap, min^{-1}; ρ_l, ρ_p are densities of liquid and particles, kg/m^3; r is the average radius of the annular gap, m.

To develop a specific filter (determination of structural dimensions) taking into account the well conditions, nomograms were developed taking into account all technical conditions.

Sand filters were installed in 2018 at wells Nos. 1246, 1356, 1623 of NGDU Krasnokholmskneft of OAO Bashneft-dobycha. Before the filter was lowered into the wells, samples were taken that established the presence of mechanical impurities in the product. The concentration of suspended particles exceeded the norm by several tens of times, which can lead to premature failure of the pump. In addition, sludge traps were located between the filter and the shank to increase efficiency (not shown in the figure).

Based on the results of sampling, the operation of the sand filter was analyzed. Figures 2 and 3 show the dependences, both for water and for oil, of the concentration of suspended particles. Figure 2 shows that after installing the filters, there was a sharp decrease in the concentration of suspended particles (EHF) in water. Similarly, in Figure 3, the concentration of suspended particles (EHF) in oil also decreased.

IOP Conf. Series: Materials Science and Engineering **952** (2020) 012051 doi:10.1088/1757-899X/952/1/012051

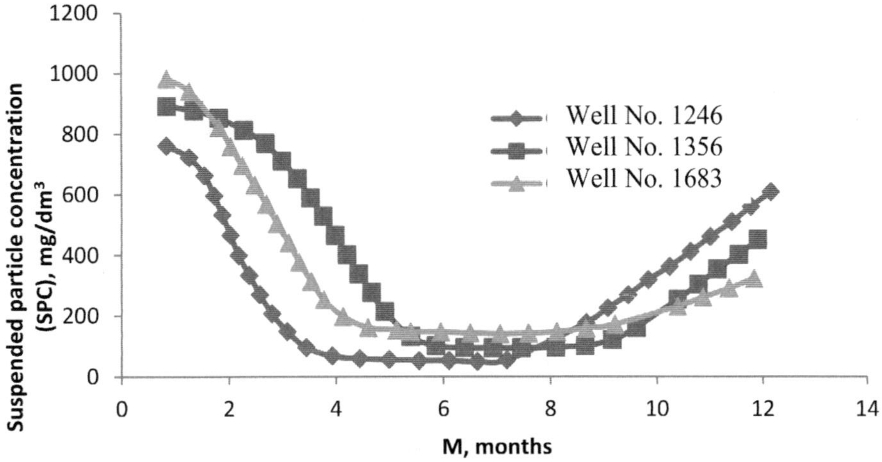

Figure 2. Monthly change in suspended particle concentration in water

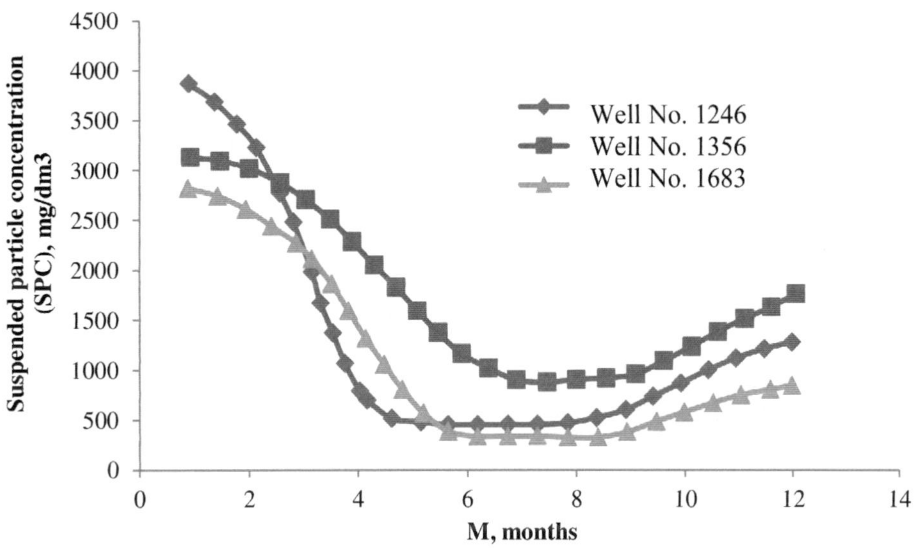

Figure 3. Monthly change in suspended particle concentration in oil

Due to the filtration of the reservoir fluid, no evidence of jamming of the plunger and clogging of the valves was observed. Consequently, the overhaul period increased [9-10]. Figure 3 shows that the SPC in all wells in water and oil grows on average after 8–9 months. This is facilitated by the appearance of larger fractions of other substances that do not pass through the radial channels of the pipe 4. This filter allows the formation fluid to pass through even when the cavity for collecting sand and the space between the filter housing and the pipe with radial channels are filled. The presence of the latter also does not affect the performance of the sucker rod pump.

4. Conclusion

The main advantage of the developed sand filter over other designs is the ability to pass reservoir fluid even when the cavity for collecting sand and the space between the filter housing and the pipe with radial channels are filled. The presence of the latter also does not affect the performance of the sucker rod pump, in the absence of jamming of the plunger in the cylinder.

The use of the new development in conjunction with sludge traps can improve the efficiency of the sucker rod pump in difficult operating conditions.

References

[1] Khabibullin M Ya and Suleimanov R I 2018 Selection of optimal design of a universal device for nonstationary pulse pumping of liquid in a reservoir pressure maintenance system *Chemical and Petroleum Engineering* **54(3-4)** 225-232 DOI: 10.1007/s10556-018-0467-2

[2] Haoran Zh, Yongtu L and Xingyuan Zh 2017 Sensitivity analysis and optimal operation control for large-scale waterflooding pipeline network of oilfield *Journal of petroleum science and engineering* **154** 38-48

[3] Khabibullin M Ya, Suleimanov R I, Sidorkin D I and Arslanov I G 2017 Parameters of damping of vibrations of tubing string in the operation of bottomhole pulse devices *Chemical and Petroleum Engineering* **53(5-6)** 378-384 DOI: 10.1007/s10556-017- 0350-6

[4] Korn G A and Korn T M 1984 *Mathematical Handbook for Scientists and Engineers: Definitions, Theorems, and Formulas for Reference and Review* (Moscow: Nauka)

[5] Mukhametshin V V and Kuleshova L S 2019 Prediction of production well flow rates using survey data *IOP Conference Series: Earth and Environmental Science (IPDME 2019 – International Workshop on Innovations and Prospects of Development of Mining Machinery and Electrical Engineering)* **378(1) 012116** 1-7 DOI: 10.1088/1755-1315/378/1/012116

[6] Welsh E 1987 Borehole Coupling in Porous Media *Ph. d. Thesis, Colorado School of Mines, Golden, Colo* 63 p

[7] Barber A H, George C J, Stiles L H and Thomp-son B B 1983 Infill Drilling to Increase Reserves-Actual Experience in Nine Fields in Texas, Oklahoma and Illinois *J. Pet. Tech* 1530-1538

[8] Sun W and Mun-Hong H 2017 Forecasting and uncertainty quantification for naturally fractured reservoirs using a new data-space inversion procedure *European Assoc Geoscientists & Engineers Computational geosciences 15th Conference on the Mathematics of Oil Recovery (ECMOR) (Amsterdam, Netherlands)* **21(5-6)** 1443-1458

[9] Kuleshova L S and Mukhametshin V V 2019 Estimation of the wells hydrodynamic drag level based on wells geophysical survey data *Atlantis Highlights in Material Sciences and Technology (ISEES 2019 – International Symposium "Engineering and Earth Sciences: Applied and Fundamental Research" dedicated to the 85th anniversary of H.I. Ibragimov)* **1** 727–729. DOI: 10.2991/isees-19.2019.145

[10] Polyakov V N, Chizhov A P, Kotenev Yu A and Mukhametshin V Sh 2019 Results of System Drilling Techniques and Completion of Oil and Gas Wells *IOP Conference Series: Earth and Environmental Science (IPDME 2019 – International Workshop on Innovations and Prospects of Development of Mining Machinery and Electrical Engineering)* **378(1) 012119** 1–7 DOI: 10.1088/1755-1315/378/1/012119

Thermal acid impulse method for enhanced oil recovery

M Ya Khabibullin[1], **N N Soloviev**[2], **A R Safiullina**[1], **A A Karimova**[1]

[1] Ufa State Petroleum Technological University, Branch of the University in the City of Oktyabrsky, 54a, Devonskaya St., Oktyabrsky, Republic of Bashkortostan, 452607, Russian Federation
[2] LLC Gazprom VNIIGAZ, Staraya Basmannaya St., 20, b. 8, Moscow, Russian Federation

E-mail: m-hab@mail.ru

Abstract. The injection rate created by the equipment does not provide uniform heating of the acid solution with thermo-acid samples. The first portion of the solution overheats, and the subsequent ones turn out to be underheated. Therefore, the effectiveness of acid and thermo-acid treatment of highly drained wells with low reservoir pressure is rather low. To increase the efficiency, a method of acid and thermo-acid impulses was proposed, a methodology for its application was developed, and simple designs of the necessary underground and ground equipment for wells were developed (tip reactors and a universal wellhead). If in a well intended for thermoacid treatment by pulsation the filter or some part of it turned out to be blocked by a sand plug, then they were preliminarily cleaned with a bailer without opening the sump. Thermo-acidic (as well as acidic) well treatment by pulsation is carried out without the use of a pumping unit, which can significantly reduce costs. Thus, the most suitable acidic solution for thermoacid pulsation is 15% HCl inhibited by unicol.

1. Introduction

The injection rate created by the units does not provide uniform heating of the acid in thermo-acid samples. The first portion of the solution overheats, and the subsequent ones turn out to be insufficiently heated. Therefore, the efficiency of acidic and thermo-acidic treatments of highly drained wells with low reservoir pressure is rather low [1]. To increase the efficiency, a method of acidic and thermo-acidic pulsation was proposed, a methodology for its application was developed, and simple designs of the necessary underground and ground equipment for wells were developed (tip reactors and a universal wellhead).

2. Results and discussion

The method of thermoacidic pulsation is based on the use of a special tip reactor (Figure 1). A reactor, the main nodes of which are a valve and the reactor itself, filled with magnesium chips, are lowered into the well on tubing until it stops in the bottom with its perforated nozzle tip. At the moment of stop, the valve body rod lifts the ball from the valve seat, and the acid solution, which has been pre-filled into the tubing, penetrates the prefilter zone of the reservoir. The rate of passage of the solution from the pump tubing string through the magnesium-chip reactor is initially much higher than that created by the injection of the acid solution by the aggregate.

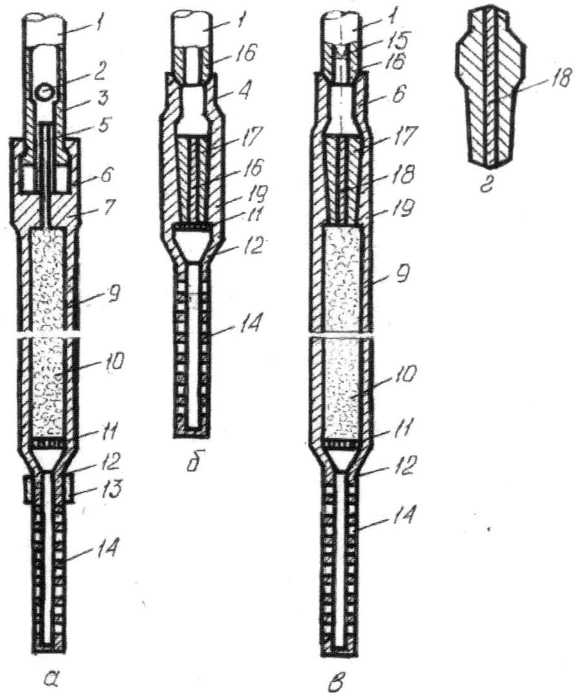

Figure 1. Tip reactors for acid and thermo-acid treatment of wells by pulsation: a – tip reactor for thermo–acid pulsation; b – tip for acid pulsation; c – universal tip reactor for acid and thermo–acid pulsation; g – cone with a magnesium rod (discharged into the well for repeated processing); 1 – pumping compressor pipe; 2 – valve ball; 3 – valve seat; 4 – coupling; 5 – connecting pipe; 6 – persistent branch pipe; 7 – valve body; 8 – coupling; 9 – reactor; 10 – magnesium shavings; 11 – filter (grill); 12 – sub; 13 – coupling; 14 – perforated pipe tip; 15 – saddle for cone with magnesium core; 16 – connecting pipe (adapter sub); 17 – insert fitting; 18 – magnesium rod; 19– fitting body

Then, after the level of the acid solution in the pipes decreases, its flow rate decreases. Thus, uniform heating of the acid solution to the required approximately constant temperature is achieved [2].

It should be noted that in wells prior to thermo-acid treatment by pulsation, conventional acid and thermo-acid treatments were unsuccessfully carried out. The flow rate of the acid solution from the tubing after opening the valve of the reactor tip when conducting thermo-acid pulsation is shown in Figure 2.

The volume of solution poured into the tubing during thermo-acid treatment by pulsation depends on their length and internal diameter. Usually it is taken equal to 1.2–2 m^3.

If in a well planned for thermo-acidic treatment by pulsation, the filter or some part of it turned out to be blocked by a sand plug, then they were preliminarily cleaned with a bailer without opening the sump [3–7].

Thermo-acidic (as well as acidic) well treatment by pulsation was carried out without the use of a pumping unit, which can significantly reduce costs.

Accelerating the movement of the acid solution, heating the entire volume to the desired temperature and creating pressure (due to the weight of the column of solution) increase the depth of penetration and the effectiveness of the effect of acid on the formation.

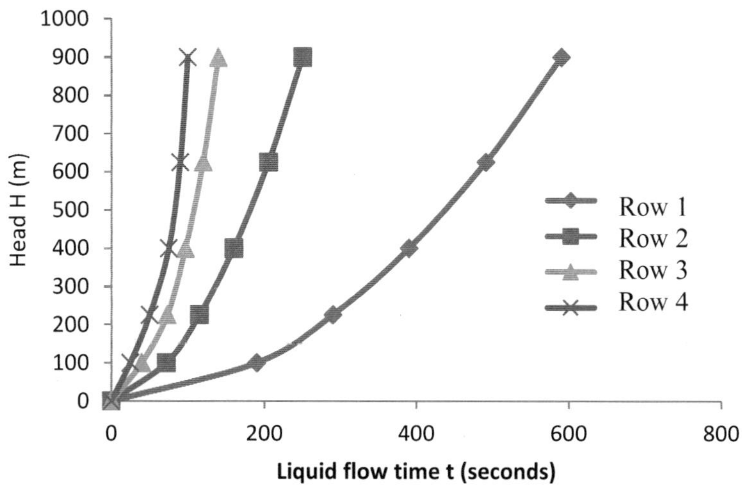

Figure 2. Estimated time I of the fluid flow from 73-mm pipes with a variable head H depending on the diameter of the nozzle d: row 1: d = 10 mm; row 2: d = 15 mm; row 3: d = 20 mm; row 4: d = 25 mm

3. Experimental

Figure 3 shows thermograms taken during pulsed pumping of acid (with a variable pressure) through magnesium chips.

Hydrochloric acid of 15% concentration was prepared with formalin and unicol. It is known that the reaction time for acid salts with magnesium depends on the presence of these inhibitors in the acid solution. The use of unicol to inhibit the acid, which goes to the thermochemical part of the process, inhibits the reaction, especially at elevated pressure, by almost 20 times [8–10].

After each experimental pumping of hydrochloric acid, the residual acidity of the solution was recorded (Table 1).

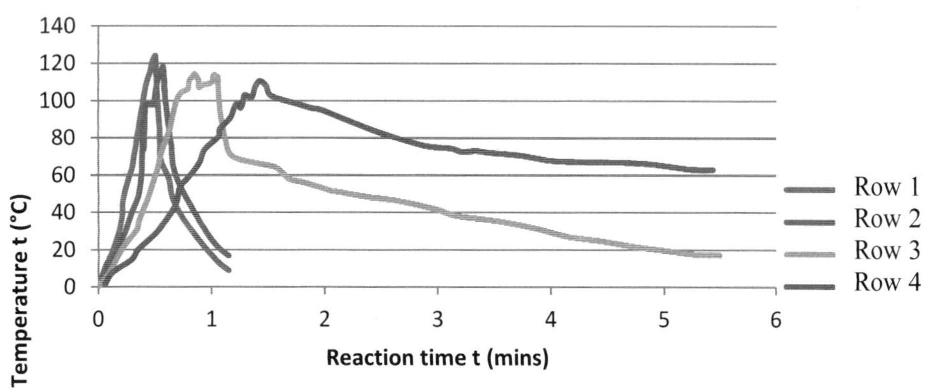

Figure 3. Thermograms taken on the test bench in the process of thermoacid pulsation during the reaction: row 1 – 10% HCI with magnesium shavings; row 2 – 15% HCI with magnesium chips; row 3 – 15% HC1 inhibited by formalin; row 4 – 15% HC1, inhibited by unicol

Table 1. Residual acidity of the solution after thermoacid pulsation

Acidic solution	Residual content of HCl after reaction with magnesium [%]
10%-HCl	9.5
15%-HCl	13.7
15%-HCl inhibited by formalin	12.9
15%-HCl inhibited by unicol	12.4

4. Conclusion

Based on the experiments, the following conclusions can be drawn.

1. The process of thermoacid pulsation to achieve uniform heating of the acid solution is regulated by:

– increase in pumping pressure (pressure is regulated by a column of liquid poured into pipes);

– decrease in pumping speed; the speed of passage of the solution from the column of pumping pipes through the reactor with magnesium chips is initially much higher than the speed created by the injection of the acid solution by the setup; then, as the level of acid in the pipes drops, its outflow rate decreases; as a result, uniform heating of the acid solution to the required temperature is achieved.

2. Inhibition of HCl by formalin and unicol inhibits the reaction. However, only unicol is able to prolong the reaction of HCl with magnesium for the required time, during which the entire process of pumping acid through nozzles proceeds.

Thus, the most suitable acidic solution for thermoacidic pulsation is 15%-HCl inhibited by unicol.

References

[1] Haoran Zh, Yongtu L and Xingyuan Zh 2017 Sensitivity analysis and optimal operation control for large-scale waterflooding pipeline network of oilfield *Journal of petroleum science and engineering* **154** 38-48

[2] Meirmanov A and Nekrasova I 2013 Mathematical models of a hydraulic shock *Journal of mathematical analysis and applications* **408(1)** 76-90

[3] Kuleshova L S and Mukhametshin V V 2019 Estimation of the wells hydrodynamic drag level based on wells geophysical survey data *Atlantis Highlights in Material Sciences and Technology (ISEES 2019 – International Symposium "Engineering and Earth Sciences: Applied and Fundamental Research" dedicated to the 85th anniversary of H.I. Ibragimov)* **1** 727–729. DOI: 10.2991/isees-19.2019.145

[4] Korn G A and Korn T M 1984 Mathematical Handbook for Scientists and Engineers: Definitions, Theorems, and Formulas for Reference and Review (Moscow: Nauka)

[5] Khokhlov V I, Galimov Sh S, Devyatkova S G, Kotenev Yu A, Sultanov Sh Kh and Mukhametshin V Sh 2019 Justification of impact and planning of technology efficiency on the basis of limy-emulsion formulation in low-permeability highly-rugged reservoirs of Tyumen deposits *IOP Conference Series: Earth and Environmental Science (IPDME 2019 – International Workshop on Innovations and Prospects of Development of Mining Machinery and Electrical Engineering)* **378(1) 012114** 1–6 DOI: 10.1088/1755-1315/378/1/012114

[6] Zaichenko A Yu, Glazov S V and Salgansky E A 2017 Filtration combustion of viscous hydrocarbon liquids *Theoretical foundations of chemical engineering* **51(5)** 673-679

[7] Mukhametshin V V and Kuleshova L S 2019 Prediction of production well flow rates using survey data *IOP Conference Series: Earth and Environmental Science (IPDME 2019 – International Workshop on Innovations and Prospects of Development of Mining Machinery and Electrical Engineering)* **378(1) 012116** 1-7 DOI: 10.1088/1755-1315/378/1/012116

[8] Nyemb N V, Jacques H and Elkana P 2017 Numerical Simulation of Heat Distribution with Temperature-Dependent Thermal Conductivity in a Two-Dimensional Liquid Flow *International Journal of nonlinear sciences and numerical simulation* **18(6)** 507-513

[9] Khabibullin M Ya, Suleimanov R I, Sidorkin D I and Arslanov I G 2017 Parameters of damping

of vibrations of tubing string in the operation of bottomhole pulse devices *Chemical and Petroleum Engineering* **53(5-6)** 378-384 DOI: 10.1007/s10556-017- 0350-6
[10] Khabibullin M Ya and Suleimanov R I 2018 Selection of optimal design of a universal device for nonstationary pulse pumping of liquid in a reservoir pressure maintenance system *Chemical and Petroleum Engineering* **54(3-4)** 225-232 DOI: 10.1007/s10556-018-0467-2

Improving the method and equipment of thermo-acidic pulsation treatment

M Ya Khabibullin[1], **G S Dubinskiy**[2,3], **P M Malyshev**[1], **A R Safiullina**[1], **A A Karimova**[1]

[1] Ufa State Petroleum Technological University, Branch of the University in the City of Oktyabrsky, 54a, Devonskaya St., Oktyabrsky, Republic of Bashkortostan, 452607, Russian Federation
[2] Ufa State Petroleum Technological University, 1, Kosmonavtov st., Ufa, Republic of Bashkortostan, 450062, Russian Federation
[3] State autonomous scientific institution "Institute of strategic research of the Republic of Bashkortostan", 129/3, October avenue, Ufa, Republic of Bashkortostan, 450075, Russian Federation

E-mail: m-hab@mail.ru

Abstract. Initially, the method of thermo-acid pulsation was intended for heavily drained wells. But, as practice has shown, it can also be successfully used in the processing of wells of other categories, and depending on the diameter of the production string and the diameter of the reactor (jacket), the acid flow can be changed. A reactor with a diameter of 114 mm is usually used, but for 114 or 127 mm of the production string, a reactor with a diameter of 48 or 88 mm, respectively elongated, can be used to maintain the required capacity in order to accommodate the right amount of magnesium chips. The efficiency of acid treatment operations by pulsing is greater than the efficiency of thermo-acid treatments according to the results of calculations of both the increase in oil production and the duration of the wells at increased production.

1. Introduction
Initially, the method of thermo-acid pulsation was intended for heavily drained wells. But, as practice has shown, it can also be successfully used in the processing of wells of other categories, and depending on the diameter of the production string and the diameter of the reactor (jacket), the acid flow can be changed. A reactor with a diameter of 114 mm is usually used, but for 114 or 127 mm of the production string, a reactor with a diameter of 48 or 88 mm, respectively elongated, can be used to maintain the required capacity in order to accommodate the right amount of magnesium chips.

2. Results and discussion
The tip (Figure 1a), despite the simplicity of its design, nevertheless had a significant drawback: valve opening was associated with the tolerance from the shank to the bottom. In the conditions of sand occurrences, when sand accumulates during the operation of the wells at the bottom, the descent of the liner all the way to the bottom is associated with the danger of sticking pipes. Therefore, a new special tip was proposed and introduced [1-3]. The main part of this tip is a fitting, the channel of which is

closed by a magnesium rod (Figure 1c).
Good results were obtained by the introduction of the acid pulsation method, which involves the use of a special tip in which the nozzle channel is completely closed by a magnesium rod (Figure 1b) [4, 5].

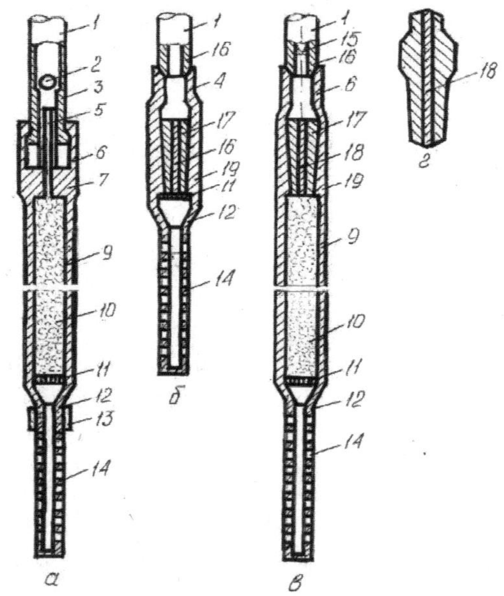

Figure 1. Tip reactors for acidic and thermo-acidic treatment of wells by pulsation: a – tip reactor for thermo–acid pulsation; b – tip for acid impulse; c – universal tip reactor for acidic and thermo–acidic pulsation; g – cone with magnesium rod (discharged into the well during repeated processing); 1 – pump–compressor pipe; 2 – valve ball; 3 – valve seat; 4 – coupling; 5 – connecting pipe; 6 – persistent branch pipe; 7 – valve body; 8 – coupling; 9 – reactor; 10 – magnesium chips; 11 – filter (grill); 12 – pipe sub; 13 – coupling; 14 – perforated pipe tip; 15 – saddle for cone with magnesium core; 16 – connecting pipe (sub); 17 – insert fitting; 18 – magnesium rod; 19– fitting body

The tip is lowered into the well on tubing until the perforated pipe is installed against the production object (in this case, the pipe does not need to abut the bottom). Then, the acid solution is poured into the tubing string, which is retained in it due to the magnesium channel blocking the nozzle channel. As a result of the reaction, the magnesium rod dissolves in the acid solution, after which the latter rushes through the released channel of the nozzle and the perforated nozzle into the filter, creating a so-called acid impulse, i.e. an increase in the injection rate under the influence of hydrostatic pressure is achieved. The pressure created in this case at the bottom of a heavily drained well contributes to a much greater penetration of the active acid solution into the formation compared to conventional acid treatment.

The design of this device has the following advantages:

- the fitting opens without the use of a lifting tractor;
- the acid treatment process can be carried out in wells having a sump (pocket below the filter zone) of considerable length;
- if there is a sand plug or construction of objects on the face there is no danger of sticking pipes.

The length of the magnesium rod is selected according to our nomogram (Figure 2), which was based on laboratory studies of the dissolution of a magnesium rod 10 mm long and 10 mm in diameter (placed in a cartridge-like insert) in a 15% hydrochloric acid solution [6–8] . When constructing the nomograms, we proceeded from the calculation that the complete dissolution of the magnesium rod should occur no earlier than the complete filling of the pipes with an acid solution (at the end of the latter).

Given the length and diameter of the tubing lowered into the well, the volume of the acid solution is determined.

Further, setting the rate of acid solution discharge into the tubing (controlled by the valve opening on the tank truck), the pipe filling time and the length of the magnesium rod are found. So, in this case, at a drain rate of 1 l/s, the pouring time is 53 min, and the rod length is 17 mm.

A significant drawback of the method of acidic pulsation (as well as thermoacid) is the limitation of the volume of the acid solution by the capacity of the deflated tubing.

It should be noted that it is often advisable to use a larger volume of solution. Therefore, a method of multiple acid impulses without extracting tubing from wells was proposed and implemented. For the application of this method, a universal tip reactor is proposed (Figures 1c and 1d).

3. Experimental

Conducting multiple pulsations became possible because the set of the new reactor has a number of cone seats, installed in series in tubing (in Figure 1, only one saddle 15 is shown); moreover, the diameter of the hole in each seat located above is greater than that of the seat located below. After a single acid impulse, a cone is thrown into the tubing (Figure 1, d), which overlaps the seat hole. The hole of the cone is filled with magnesium rod 18. Then one can again fill the acid solution in the tubing, etc. (as with a single acid impulse). This operation can be repeated several times using pre-installed saddles and each time throwing the corresponding cones with magnesium rods (their length in all cases is determined by the nomogram shown in Figure 2) [9].

With repeated pulsation, the number of pulses is determined based on the selected volume of the acid solution and the volume of the solution that can be placed in descended pipes. The latter is determined by the left side of the nomogram, shown in Figure 2. For example, for 500 m of 73-mm pipes, the volume of acid solution that can be used for one pulse will be no more than 1.75 m^3. If during the acid treatment it is advisable to use 3.5 m^3 of the solution, then the pulses will be determined as the quotient of dividing 3.5 by 1.75 and will be equal to two.

In heavily drained wells, it is advisable to first conduct thermo-acid pulsing (especially if paraffin oil was extracted from it). For this, a universal tip reactor can be used (Figure 1c). After thermo-acid pulsation, acid pulsation can be carried out as many times as needed without lifting pipes. To do this, first install the required number of saddles 15 in the tubing (Figure 1c) and cast a cone with a magnesium rod before each acid pulse treatment (Figure 1d).

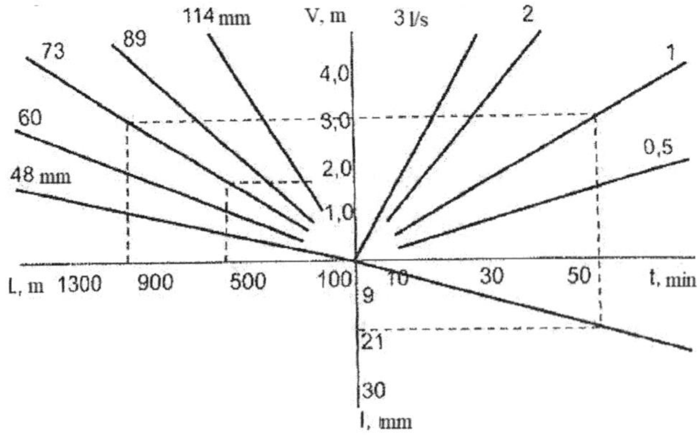

Figure 2. Nomogram for determining the length l of the magnesium rod (for suspension data - the length L and the diameter of the tubing), at which it is completely dissolved only after the filling of the tubing with a given volume of acid solution V (at a certain filling rate of pipes)

When using a conventional filling head, the piston-like movement of the solution in the pipes caused periodic air breakouts to the wellhead. At the same time, the air carried away the acid solution, so it was necessary to periodically stop pouring the solution into the pipes.

A special casting head is proposed (Figure 3). The external 73 mm branch pipe 2 has a branch 3 for draining the acid solution into the tubing 6, and the internal 48 mm pipe 1 serves to evacuate the air displaced into the atmosphere.

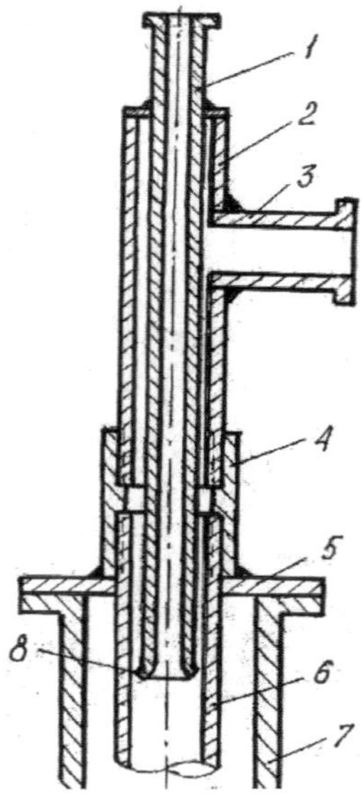

Figure 3. Universal wellhead for acid and thermo-acid pulsation treatment of wells: 1 – pipe with a diameter of 48 mm; 2 – pipe with a diameter of 73 mm; 3 – tap with a diameter of 73 mm; 4 – landing clutch; 5 – faceplate; 6 – tubing; 7 – production casing; 8 – guides for drainage of acid solution

After the pipes are lowered and the said head is installed on them, an acid solution is drained through branch 3. Once in the annular space, the solution first flows along the guides 8, and then along the walls of the tubing 6. The displaced air occupies a central position and is removed through the pipe 1. As a result of the introduction of such a head, there is no loss of time to drain the acid solution into the pipes, as well as loss of the latter due to emissions (together with air); contamination of the wellhead and ensures the safety of workers engaged in pulse acid treatment.

In order to establish the effectiveness of pulsed thermo-acidic treatments in comparison with simple thermo-acidic operations, the probability of successful operations is determined. At the same time, the asymmetric distribution of wells by efficiency was aligned using the Charlier distribution [10].

The results of calculating the probability of successful outcomes for pulsed thermo-acidic treatments are given in the table, where the likelihood of the effectiveness of conventional thermo-acidic treatments calculated above is also indicated for comparison.

Table 1. Operational efficacy probability

Operations	According to calculations of			
	flow rate increase		duration of the well at an increased flow rate	
	Thermo-acid conventional	Thermo-acid pulsed	Thermo-acid conventional	Thermo-acid pulsed
Highly efficient	9	22	7	11
Efficient	11	34	20	23
Ineffective	80	44	73	66

4. Conclusion

It should be noted that a comparison of the two groups of numbers may be valid provided that the difference in their means is not significant. These groups of numbers were processed, as a result of which it was found that the combination represented by them is different. As can be seen from the table, the efficiency of operations on acid treatments by pulsing is greater than the efficiency of thermo-acid treatments according to the results of calculations of both the increase in oil production and the duration of the wells at increased production.

References

[1] Haoran Zh, Yongtu L and Xingyuan Zh 2017 Sensitivity analysis and optimal operation control for large-scale waterflooding pipeline network of oilfield *Journal of petroleum science and engineering* **154** 38-48

[2] Meirmanov A and Nekrasova I 2013 Mathematical models of a hydraulic shock *Journal of mathematical analysis and applications* **408(1)** 76-90

[3] Mukhametshin V V and Kuleshova L S 2019 Prediction of production well flow rates using survey data *IOP Conference Series: Earth and Environmental Science (IPDME 2019 – International Workshop on Innovations and Prospects of Development of Mining Machinery and Electrical Engineering)* **378(1) 012116** 1-7 DOI: 10.1088/1755-1315/378/1/012116

[4] Korn G A and Korn T M 1984 Mathematical Handbook for Scientists and Engineers: Definitions, Theorems, and Formulas for Reference and Review (Moscow: Nauka)

[5] Khokhlov V I, Galimov Sh S, Devyatkova S G, Kotenev Yu A, Sultanov Sh Kh and Mukhametshin V Sh 2019 Justification of impact and planning of technology efficiency on the basis of limy-emulsion formulation in low-permeability highly-rugged reservoirs of Tyumen deposits *IOP Conference Series: Earth and Environmental Science (IPDME 2019 – International Workshop on Innovations and Prospects of Development of Mining Machinery and Electrical Engineering)* **378(1) 012114** 1–6 DOI: 10.1088/1755-1315/378/1/012114

[6] Zaichenko A Yu, Glazov S V and Salgansky E A 2017 Filtration combustion of viscous hydrocarbon liquids *Theoretical foundations of chemical engineering* **51(5)** 673-679

[7] Malyarenko A M, Bogdan V A, Kotenev Yu A, Mukhametshin V Sh, Umetbaev V G 2019 Wettability and formation conditions of reservoirs *IOP Conference Series: Earth and Environmental Science (IPDME 2019 – International Workshop on Innovations and Prospects of Development of Mining Machinery and Electrical Engineering)* **378(1) 012040** 1–6 DOI: 10.1088/1755-1315/378/1/012040

[8] Nyemb N V, Jacques H and Elkana P 2017 Numerical Simulation of Heat Distribution with Temperature-Dependent Thermal Conductivity in a Two-Dimensional Liquid Flow *International Journal of nonlinear sciences and numerical simulation* **18(6)** 507-513

[9] Khabibullin M Ya, Suleimanov R I, Sidorkin D I and Arslanov I G 2017 Parameters of damping of vibrations of tubing string in the operation of bottomhole pulse devices *Chemical and Petroleum Engineering* **53(5-6)** 378-384 DOI: 10.1007/s10556-017- 0350-6

[10] Khabibullin M Ya and Suleimanov R I 2018 Selection of optimal design of a universal device for nonstationary pulse pumping of liquid in a reservoir pressure maintenance system *Chemical and Petroleum Engineering* **54(3-4)** 225-232 DOI: 10.1007/s10556-018-0467-2

IOP Conf. Series: Materials Science and Engineering **952** (2020) 012054 doi:10.1088/1757-899X/952/1/012054

Improving equipment and technology for selective acid treatment of wells

M Ya Khabibullin

Ufa State Petroleum Technological University, Branch of the University in the City of Oktyabrsky, 54a, Devonskaya St., Oktyabrsky, Republic of Bashkortostan, 452607, Russian Federation

E-mail: m-hab@mail.ru

Abstract. This method is based on the following effect. A channel in the selected formation layer or section is formed by sandblasting. The acid solution is pumped into the channel through the nozzle of the perforator. The stream of acidic solution entering the channel prevents this solution from leaving the channel. In order to increase the acid depletion time, i.e. to slow down the reaction rate, it is necessary to add special reagents to the solution. Syntanol DS-10 TU 2483-016-71150986-2012 (a non-ionic surfactant used as an effective surfactant) is a very effective reaction rate reducer. Adding it in an amount of 0.5% (by weight of the volume of the solution) one can reduce the reaction rate by 2.7 times. It is much more difficult to increase the rate of acid solution injection into highly drained formations of low-production wells (up to 5–10 tons of fluid per day) with a low reservoir pressure due to the insufficient capacity of acid aggregates.

1. Introduction

This method is based on the following effect. A channel in the selected formation layer or section is formed by sandblasting. The acid solution is pumped into the channel through the nozzle of the perforator. The stream of acidic solution entering the channel prevents this solution from leaving the channel. As a result, thr abnormal channel pressure is always greater than the reservoir pressure in this zone

$$\Delta p = p_{ch} - p_{sq}, \quad (1)$$

where Δp – difference between channel pressure and reservoir pressure in the zone of channel formation; p_{ch} – channe pressure; p_{sq} – channel formation zone pressure.

Due to the difference Δp, the acid solution in the channel is filtered into its walls and processes the zone around the channel. Only part of the acid solution flows from the channel into the production string, falling into another, more drained zone of the reservoir. Thus, this method allows for strictly selective acid treatment (as well as any other type of treatment) with an accuracy of a few centimeters, determined by the accuracy of the control over the installation of the perforator.

2. Results and discussion

An experimental study of the channel pressure during spot sandblasting was carried out on a bench shown in Figure 1. It consists of a device simulating the channel in the rock, which is enclosed in casing 1, recording pressure gauges 2, pneumatic pressure compensators 3 and pump 4 with a system of manifolds.

The scheme of the device modeling the channel is shown in Figure 2. Thick-walled sub 1 and nut 7 are attached to base 4 by welding. The body of hammer drill 3 with nozzle 2 is screwed into sub 2. 500 mm pipe 8 made of 60-mm pump tube is screwed into nut 7. On the one hand, the pipe is closed by plug 10; on the other hand, it has threaded sleeve 6. Fitting 5 is screwed into the sleeve, simulating the entrance to the channel.

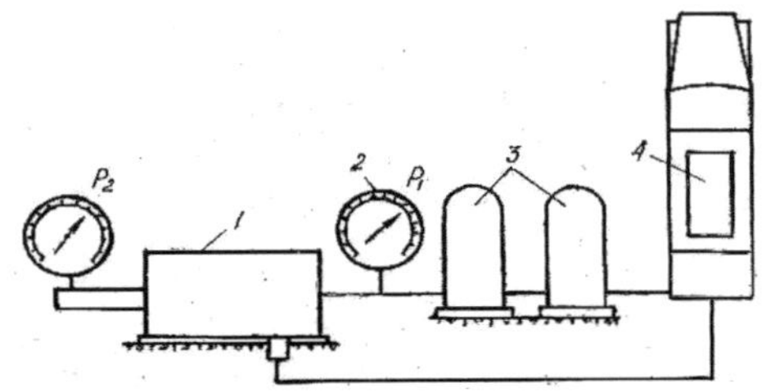

Figure 1. The bench for studying channel pressure changes during hydro-sandblasting of the formation

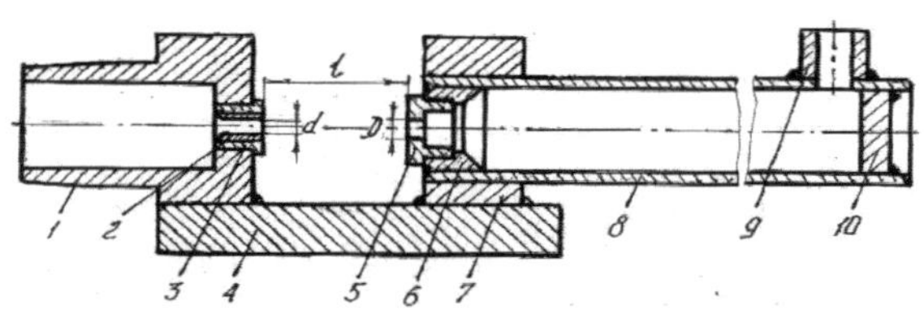

Figure 2. The device simulating processes in the channel

On pipe 8, the pressure gauge is connected to coupling 9. The axis of the nozzle coincides with the axis of the fitting. In order to avoid splashing of the liquid, the device is enclosed in a casing. During the study, nozzles with a diameter of $d_1 = 4.5$ mm and $d_2 = 6$ mm were used.

Since the diameter of the hole in the casing is approximately 2.5–3.0 of the diameters of the nozzle during water-jet sandblasting, nozzles with a diameter of $D_1 = 11.5$ mm, $D_2 = 15$ mm and $D_3 = 22$ mm were used. The distance from the nozzle to the entrance to the channel *l* was varied from 5 to 100 mm by rotating nozzle 8 in nut 7.

3. Experiment

The experimental technique is as follows. In the device, a nozzle and a fitting of diameters d and D were installed at the required distance between them. The pump pumped the liquid at a pressure in front of nozzle p_1 and fixed the pressure inside the nozzle p_2. In the experiments, water and an aqueous solution of sulfite-alcohol stillage (SAS) with a viscosity of 12.65 cPz at a temperature of 20 ° C were used as working fluid.

Table 1 shows that in the studied area of change in the distances from the nozzle to the entrance to channel l from 5 to 100 mm, three features of the change in p_2 are observed depending on the ratio of the diameter of the inlet to the channel to the diameter of the nozzle D/d. At a ratio D / d> 2.55, with an increase in d, p_2 increases. At D/d <2.55, p_2 decreases with increasing d. At D/d ≈2.55 with increasing d, p_2 first increases and then decreases.

The dependence $p_2 = f(p_1)$ is always such that when p_1 increases, p_2 increases. However, the nature of this change depends on the D/d ratio, as can be seen from the following data (Table 2).

It should be noted that during the bench experiments, it was difficult to record p_2 recording. Despite two high-pressure pneumatic compensators and tubular compensators, strong fluctuations in pressure

p_2 were observed. They did not allow us to accurately record its average value (the absence of backpressure at the jet exit from the nozzle was significant).

Table 1. The nature of changes in channel pressure depending on the distance between the nozzle and the inlet $p_2 = f(l)$ for various D / d ratios

No of the experiment	Working liquid	Fittin g diameter D mm	Nozzl e diameter d mm	Ratio *D/d*	Nature of dependence $p_2 = f(p_1)$
1.	water	11.5	4.5	2.55	p_2 first increases and then decreases
2.	water	15.0	4.5	3.33	p_2 first increases and then decreases
3.	water	22.0	4.5	4.88	p_2 decreases
4.	water	15.0	6.0	2.50	p_2 first decreases and then increases
5.	water	22.0	6.0	3.67	p_2 first decreases and then increases
6.	SAS solution	11.5	4.5	2.55	p_2 decreases
7.	SAS solution	15.0	4.5	3.33	p_2 increases
8.	SAS solution	22.0	4.5	4.88	p_2 increases
9.	SAS solution	11.5	6.0	1.91	p_2 decreases
10.	SAS solution	15.0	6.0	2.50	p_2 decreases
11.	SAS solution	22.0	6.0	3.67	p_2 decreases

Table 2. The nature of dependence of $p_2 = f(p)$ on the ratio D/ d

Dependence	Value					
D/d	1.91	2.5	2.55	3.33	3.67	4.88
$p_2/200$	0.85	0.36	0.32	0.094	0.09	0.03

It is possible to obtain very high excess pressures, up to 31% of pressure in front of the nozzle, which are sufficient for selective acid treatment and selective hydraulic fracturing. For example, with a pressure drop equal to 20.0 MPa and a diameter of 4.5 mm, the overpressure can be 6.2 MPa. These works can be carried out in production cores with a minimum strength and large diameters, i.e. where it's not even possible to install packers. The process of selective acid treatment is as follows (Figure 3).

Sandblasting hammer drill 2 is lowered into the well, setting against the selected treatment interval, and hydraulic clamps are used for strict retention of the perforator. The displacement of the latter eliminates the possibility of selective treatment.

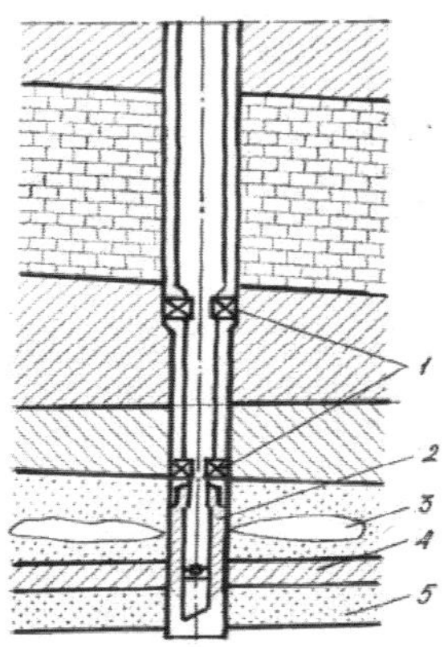

Figure 3. Selective acid treatment: 1 - fixator; 2 - punch; 3 - channel; 4 - low-permeability area; 5 - well permeable layer

After sandblasting and flushing the well from sand, without changing the position of the perforator, an acid solution is pumped into the pipes and filtered through its walls into the treated section of the formation. The part of the acid accumulated in the wellbore is forced into the formation by squeezing fluid through the annular space. The FG clamp prevents breakage, relieves tensile stresses in the pipe string during hydraulic sandblasting, and fixes it when centering the punch in a given treatment interval.

Sometimes acid treatment is unsuccessful, since the bottomhole zone is very clogged. In these cases, an acid gap is used. However, it is limited by high pressures required for forcing acid through the poorly permeable zone, or it requires special equipment (packers, anchors, wellhead fittings, high pressure pump units, etc.).

Particularly significant pressures, dangerous for the tightness of the production string can be observed in wells of great depths with dense, low-permeability reservoirs. In order to ensure that the pressure does not exceed the strength of the equipment, a multiple acid treatment method was used.

First, the filtered area is cleaned of sand plugs, then the tubing is lowered to the filter (Figure 4), and the wellhead is equipped with appropriate fittings designed for high pressure. After that, 3-4 m^3 of acid solution (Figure 4, a) and squeezing liquid (Figure 4, b) are pumped. Then, air or gas is pumped into the tubing and annulus (to the maximum pressure in the distribution box). In this state, the well is left under pressure for 20 hours (Figure 4, c), after which the well is drained to extract reaction products to the surface. Repeated hydro-acid treatment of the bottom-hole zone is carried out.

An analysis of the results of multiple (threefold) hydro-acid treatment shows (Figure 5) that after the first stage of treatment of the well, the initial injection pressure of the acid solution was reduced by 24%, and after the second treatment - by 55%. The production rate of well 172 increased from 3.8 to 6.3 t / day. Conventional hydroacid treatment did not give similar results.

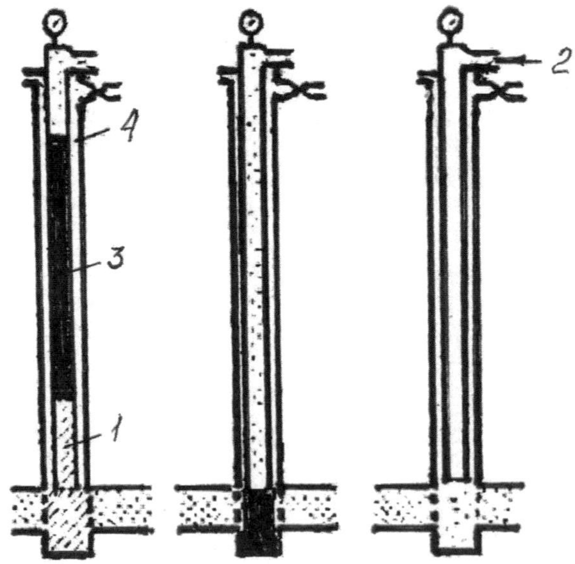

Figure 4. The technological scheme of one cycle of multiple hydro-acid treatment: 1 - water; 2 - air under pressure; 3 - oil; 4 - acid solution

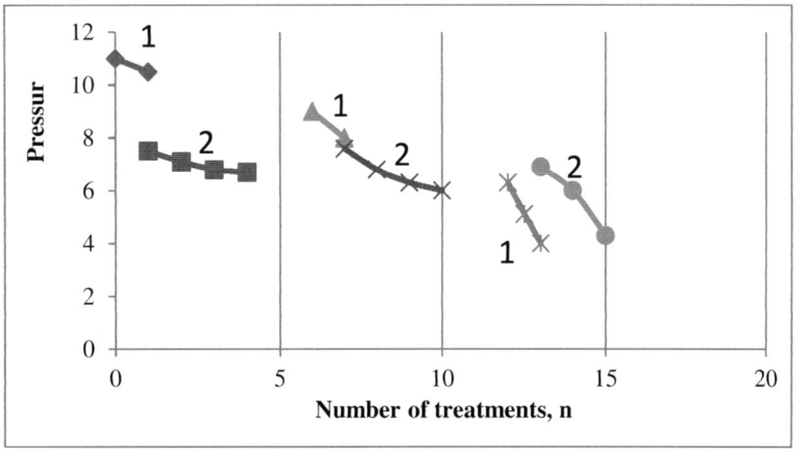

Figure 5. Pressure change during multiple (threefold) hydro-acid treatment in well 172: 1 - injection of acid solution into the well; 2 - the well is closed under excess pressure

Before injecting an acid solution, preventive measures are often taken to reduce wellhead pressure during treatment; one of them is an instant decrease in bottomhole pressure.

The effective penetration of an acid solution into the formation is determined by a number of factors:

- during the depletion of acid solution in reservoir conditions;
- the speed of its injection into the reservoir.

It should be noted that after complete depletion of acid solution, its further penetration into the formation is not beneficial and even harmful, since the depleted solution fills pore channels.

4. Conclusion

In order to increase the acid depletion time, i.e. to slow down the reaction rate, it is necessary to add special reagents to the solution. Syntanol DS-10 TU 2483-016-71150986-2012 (a non-ionic surfactant used as an effective surfactant) is a very effective reaction rate reducer. Adding it in an amount of 0.5% (by weight of the volume of the solution) one can reduce the reaction rate by 2.7 times.

It is much more difficult to increase the rate of acid solution injection into highly drained formations of low-production wells (up to 5–10 tons of fluid per day) with a low reservoir pressure due to the insufficient capacity of acid aggregates.

References

[1] Khabibullin M Ya and Suleimanov R I 2018 Selection of optimal design of a universal device for nonstationary pulse pumping of liquid in a reservoir pressure maintenance system *Chemical and Petroleum Engineering* **54(3-4)** 225-232 DOI: 10.1007/s10556-018-0467-2

[2] Khabibullin M Ya, Suleimanov R I, Sidorkin D I and Arslanov I G 2017 Parameters of damping of vibrations of tubing string in the operation of bottomhole pulse devices *Chemical and Petroleum Engineering* **53(5-6)** 378-384 DOI: 10.1007/s10556-017- 0350-6

[3] Khokhlov V I, Galimov Sh S, Devyatkova S G, Kotenev Yu A, Sultanov Sh Kh and Mukhametshin V Sh 2019 Justification of impact and planning of technology efficiency on the basis of limy-emulsion formulation in low-permeability highly-rugged reservoirs of Tyumen deposits *IOP Conference Series: Earth and Environmental Science (IPDME 2019 – International Workshop on Innovations and Prospects of Development of Mining Machinery and Electrical Engineering)* **378(1) 012114** 1–6 DOI: 10.1088/1755-1315/378/1/012114

[4] Korn G A and Korn T M 1984 *Mathematical Handbook for Scientists and Engineers: Definitions, Theorems, and Formulas for Reference and Review* (Moscow: Nauka)

[5] Welsh E 1987 Borehole Coupling in Porous Media *Ph. d. Thesis, Colorado School of Mines,*

Golden, Colo

[6] Barber A H, George C J, Stiles L H and Thomp-son B B 1983 Infill Drilling to Increase Reserves-Actual Experience in Nine Fields in Texas, Oklahoma and Illinois *J. Pet. Tech* 1530-1538

[7] Malyarenko A M, Bogdan V A, Kotenev Yu A, Mukhametshin V Sh, Umetbaev V G 2019 Wettability and formation conditions of reservoirs *IOP Conference Series: Earth and Environmental Science (IPDME 2019 – International Workshop on Innovations and Prospects of Development of Mining Machinery and Electrical Engineering)* **378(1) 012040** 1–6 DOI: 10.1088/1755-1315/378/1/012040

[8] Haoran Zh, Yongtu L and Xingyuan Zh 2017 Sensitivity analysis and optimal operation control for large-scale waterflooding pipeline network of oilfield *Journal of petroleum science and engineering* **154** 38-48

[9] Sun W and Mun-Hong H 2017 Forecasting and uncertainty quantification for naturally fractured reservoirs using a new data-space inversion procedure *European Assoc Geoscientists & Engineers Computational geosciences 15th Conference on the Mathematics of Oil Recovery (ECMOR) (Amsterdam, Netherlands)* **21(5-6)** 1443-1458

[10] Akhmetov R T, Kuleshova L S and Mukhametshin V V 2019 Application of the Brooks-Corey model in the conditions of lower cretaceous deposits in terrigenous reservoirs of Western Siberia *IOP Conference Series: Materials Science and Engineering (MEACS 2018 – International Conference on Mechanical Engineering, Automation and Control Systems)* **560(1) 012004** 1-4 DOI: 10.1088/1757-899X/560/1/012004

Improvement of formation fluid separation efficiency

M Ya Khabibullin

Ufa State Petroleum Technological University, Branch of the University in the City of Oktyabrsky, 54a, Devonskaya St., Oktyabrsky, Republic of Bashkortostan, 452607, Russian Federation

E-mail: m-hab@mail.ru

Abstract. Producers and researchers pay attention to the improvement of high-quality and productive separation of liquid heterogeneous systems in the oil and gas industry. To perform these operations, various designs of sedimentation tanks, separators, hydrocyclones, etc. based on gravitational and centrifugal separation of inhomogeneous masses of various densities are used. A further noticeable increase in the diameter of the rotor, for example, up to 1500 mm, is associated with the search for new design forms of separators. This is due to serious design studies and restructuring of the production base.

1. Introduction

Producers and researchers pay attention to the improvement of high-quality and productive separation of liquid heterogeneous systems in the oil and gas industry. To perform these operations, various designs of sedimentation tanks, separators, hydrocyclones, etc. based on gravitational and centrifugal separation of inhomogeneous masses of various densities are used.

2. Results and discussion

In liquid separators, the separation surface is formed by a package of conical plates. Their performance is determined by the performance index determined by formula [1, 2]:

$$\Sigma Q = \frac{2\omega^2 V}{g}, \tag{1}$$

where ω – angular frequency of rotation of the separator rotor, min^{-1}; V – volume limited by the working surface of the separation and its closing plane, m^3; g – gravity acceleration, m/s^2.

A natural way to increase ΣQ is to increase the size (diameter and height) and the number of plates. However, this leads to an unacceptable increase in the height of the rotor, worsening dynamic characteristics of the separators, and an uneven incease in the height of power supply of the package. It is known that modern relationships between the diameter and the height of rotors are dictated by many years of experience and supported by theoretical considerations regarding the optimal dynamic characteristics of separators. An increase in the number of plates is possible due to an increase in the height of the rotor, and a decrease in the gap between the plates for individual separated liquids. The volume limited to one plate of the separation packet in its constant form can be expressed as the product aR_T^3, where a is the proportionality coefficient; R_t is the largest radius of the plate.

$$\Sigma Q = \frac{2aR_T^3 z\,\omega^2}{g}, \tag{2}$$

where z number of plates in one bag.

IOP Conf. Series: Materials Science and Engineering **952** (2020) 012055 doi:10.1088/1757-899X/952/1/012055

Given that the working linear speed of the rotor points farthest from the axis of rotation is $\upsilon = \omega R$ (R – rotor radius), and conditionally assuming that the number of plates is proportional to R, we have:

$$\Sigma Q = \frac{2a_1 v^2 R^2}{g}, \quad (3)$$

where a_1 – a new coefficient of proportionality equal to the ratio of the total volume limited by all plates to R^4.

For further calculations, let us use the allowable value of the strength criterion of rapidly rotating nodes

$$[Ne] = \frac{[\sigma]}{\rho v^2}, \quad (4)$$

where $[\sigma]$ – allowable stress for the rotor material, MPa; ρ – density, kg/m^3; $[\sigma]/\rho = \gamma$ – specific strength of the material.

Then

$$\Sigma Q = 2\gamma a_1 R^2/\{g[Ne]\}, \quad (5)$$

The value [Ne] for a rotating steel cylindrical rotor filled with water can be taken equal to 2 [3,4]. In this case, we have

$$\Sigma Q = a_1 R^2/g, \quad (6)$$

According to the last expression, performance of the separator at a given linear speed of the rotor can be increased by increasing the radius of the plate and the rotor.

From the expression for [Ne] it follows that: $\vartheta^2 = (\omega R)^2 = \gamma[Ne] = const;$

$$\omega = \frac{1}{R}\sqrt{\gamma/[Ne]}, \quad (7)$$

The specific strength of the material is the square of the peripheral fracture rate of a rotating thin ring made of the same material. Denoting the linear safe speed through ϑ_d formula (7) at [Ne] = 2 can be written as

$$\omega = 0{,}71\frac{\vartheta_d}{R}. \quad (8)$$

At given values of density of its material and the liquid to be separated, the following relationships are given in [5] for a cylindrical rotor:

$$s/R = f([Ne]), \quad (9)$$

where S – rotor's wall thickness.

Based on these dependencies, we built the dependence of ratio ϑ_d/ϑ on the relation s/R (Figure 1).

3. Experiment

Let us use the example presented in [6]. It is required to design a rotor whose inner radius is 0.346 m, rotating with a frequency of 4500 rpm. Allowable stress [σ] = 450 MPa. The linear velocity of the rotor surface at a radius of 0.316 m is 163 m / s, while the safe speed is $\vartheta_d \leq$ 239.4 m/sec. The ratio $\vartheta_d/\vartheta = 1.47$.

This relation corresponds to the ratio s/R = 0.08; therefore, the wall thickness will be s = 27.7 mm.

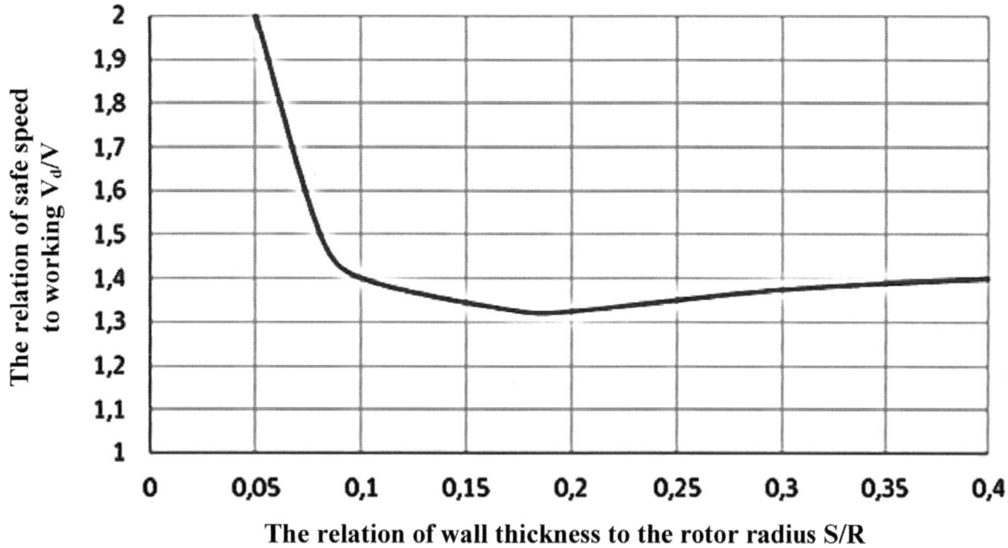

Figure 1. The dependence of the ratio v_d/v of safe speed to the working speed on the ratio S / R of the wall thickness to the radius of the rotor

As follows from Figure 1, the ratio s/R = 0.175 corresponds to a minimum of the ratio $\_\vartheta_d/\vartheta =$ 1.325. Therefore, at the same value ϑ it is possible to increase the speed by 1.47 / 1.325 = 1.1 times (i.e. by 10%), but the wall thickness will increase by 0.175 / 0.08 = 2.2 times (the rotor mass will also increase accordingly). Figure 1 shows that a further increase in operating speed (when 1.325 $\vartheta > \vartheta_d$) is excluded.

The reserve for increasing the rotor speed is possible by reducing the destructive speed. If you use the converted formula for the probability of non-destruction, taking the normal distribution of the strength of the material of the rotor, you can write [7]:

$$P = 1 - \frac{1}{\sqrt{2\pi}} \int_b^{\infty} \exp\left(-\frac{x^2}{2}\right) dx, \tag{10}$$

where $b = \frac{K_\vartheta^2 - 1}{\sqrt{\vartheta_s^2 K_\upsilon^4 + \vartheta_{s_i}^2}}$; P – rotor uptime; K_υ^4 – speed margin; $\vartheta_s = 0{,}19$, the coefficient of variation of specific strength of the rotor for steel; ϑ_{s_i} – coefficient of variation of the square of rotor speed, which is taken equal to zero.

Calculating values of b for different values of K_ϑ^4, let us find the values of probability of non-destruction of the rotor by the standard method [8] (table).

Table 1. The values of probability of non-destruction of the rotor

Parameters	Values						
K_υ	1.41	1.35	1.3	1.25	1.2	1.15	1.1
P	0.9957	-	0.9842	-	0.9463	0.8997	0.8186

Figure 2 shows the graph $P = f(K_\upsilon)$.

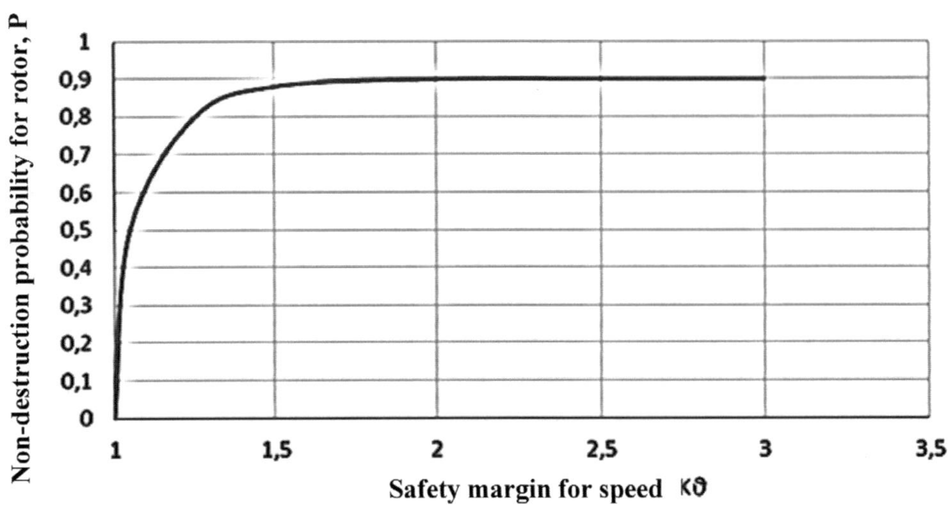

Figure 2. The dependence of probability of non-destruction of rotor P on the safety factor for speed K_υ

The reserve of an increase in speed of the rotor is possible by reducing the value of destructive speed and decreasing its reliability.

Sufficient safety is ensured at a speed margin of 1.41, which corresponds to the twofold safety margin. In this case, only four out of 1000 rotors can be destroyed [9, 10]. However, if the margin of speed is reduced to 1.1, the probability of operational safety will decrease to 0.8186, i.e., an increase in permissible speed by 31% leads to a significant increase in probability of destruction, since almost 20% of rotors can be destroyed. Figure 3 shows the dependences of the rotor diameter on its rotation speed on the basis of equality (7) for two values of P of the rotor failure-free operation. In Figure 3, dots denote parameters D = f (n) of modern separators. Most points are located near the bottom curve. The graph allows us to identify the border of the modern separator industry by values of the diameters of the rotors (shown by a dashed line).

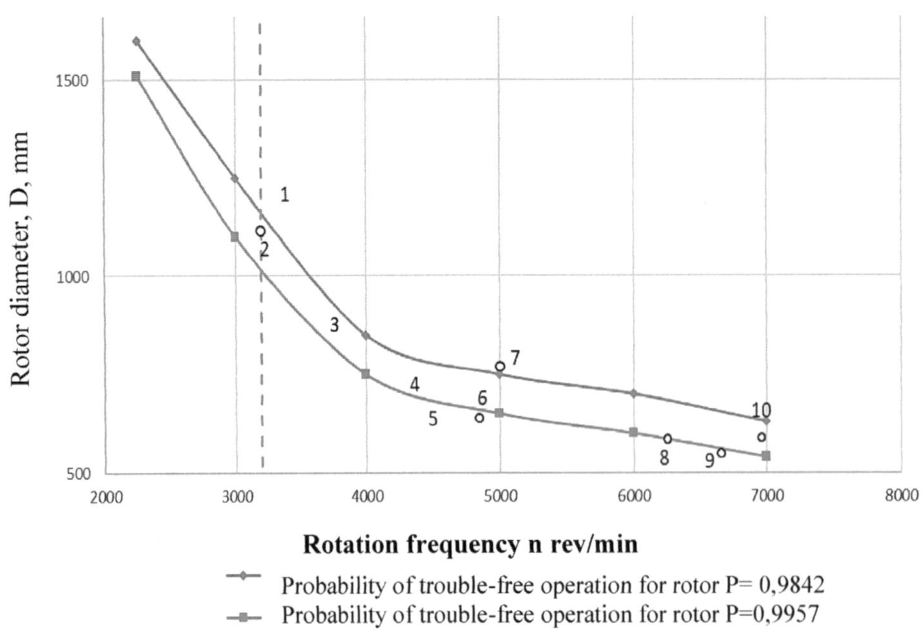

Figure 3. The dependence of diameter D of the rotor on speed n of various separators (the dashed line delineates the border of modern separator construction): 1 - SA-160-06-117; 2 - MDA-300; 3 - SDS-901 K-01; 4 - MRPX-418; 5 - SB-80-36-076; 6 - SDS-530K-1; 7 - SB-60-36-076; 8 - SDS-531K-01 .; 9 - SAMR5036M; 10 - SOS-501K-3

IOP Conf. Series: Materials Science and Engineering **952** (2020) 012055 doi:10.1088/1757-899X/952/1/012055

4. Conclusion

To increase speed of the rotor, it is necessary to reduce the value of destructive speed and decrease its reliability. A further noticeable increase in the diameter of the rotor, for example, up to 1500 mm, is associated with the search for new design forms of separators. This is due to serious design studies and restructuring of the production base.

References

[1] Khabibullin M Ya and Suleimanov R I 2018 Selection of optimal design of a universal device for nonstationary pulse pumping of liquid in a reservoir pressure maintenance system *Chemical and Petroleum Engineering* **54(3-4)** 225-232 DOI: 10.1007/s10556-018-0467-2

[2] Khabibullin M Ya, Suleimanov R I, Sidorkin D I and Arslanov I G 2017 Parameters of damping of vibrations of tubing string in the operation of bottomhole pulse devices *Chemical and Petroleum Engineering* **53(5-6)** 378-384 DOI: 10.1007/s10556-017- 0350-6.

[3] Polyakov V N, Chizhov A P, Kotenev Yu A and Mukhametshin V Sh 2019 Results of System Drilling Techniques and Completion of Oil and Gas Wells *IOP Conference Series: Earth and Environmental Science (IPDME 2019 – International Workshop on Innovations and Prospects of Development of Mining Machinery and Electrical Engineering)* **378(1) 012119** 1–7 DOI: 10.1088/1755-1315/378/1/012119

[4] Korn G A and Korn T M 1984 *Mathematical Handbook for Scientists and Engineers: Definitions, Theorems, and Formulas for Reference and Review (*Moscow: Nauka)

[5] Welsh E 1987 Borehole Coupling in Porous Media *Ph. d. Thesis, Colorado School of Mines, Golden, Colo*

[6] Barber A H, George C J, Stiles L H and Thomp-son B B 1983 Infill Drilling to Increase Reserves-Actual Experience in Nine Fields in Texas, Oklahoma and Illinois *J. Pet. Tech* 1530-1538

[7] Kuleshova L S, Kadyrov R R, Mukhametshin V V and Akhmetov R T 2019 Auxiliary equipment for downhole fittings of injection wells and water supply lines used to improve their performance in winter *IOP Conference Series: Materials Science and Engineering (MEACS 2018 – International Conference on Mechanical Engineering, Automation and Control Systems)* **560(1) 012071** 1-6 DOI: 10.1088/1757-899X/560/1/012071

[8] Haoran Zh, Yongtu L and Xingyuan Zh 2017 Sensitivity analysis and optimal operation control for large-scale waterflooding pipeline network of oilfield *Journal of petroleum science and engineering* **154** 38-48

[9] Sun W and Mun-Hong H 2017 Forecasting and uncertainty quantification for naturally fractured reservoirs using a new data-space inversion procedure *European Assoc Geoscientists & Engineers Computational geosciences 15th Conference on the Mathematics of Oil Recovery (ECMOR) (Amsterdam, Netherlands)* **21(5-6)** 1443-1458

[10] Malyarenko A M, Bogdan V A, Kotenev Yu A, Mukhametshin V Sh, Umetbaev V G 2019 Wettability and formation conditions of reservoirs *IOP Conference Series: Earth and Environmental Science (IPDME 2019 – International Workshop on Innovations and Prospects of Development of Mining Machinery and Electrical Engineering)* **378(1) 012040** 1–6 DOI: 10.1088/1755-1315/378/1/012040

IOP Conf. Series: Materials Science and Engineering **952** (2020) 012056 doi:10.1088/1757-899X/952/1/012056

Improvement of the efficiency of horizontal wells

R A Kozikhin[1], I G Fattakhov[1,2], L S Kuleshova[2], V E Andreev[3,4], A R Safiullina[2], A Kh Gabbasov[1], I G Andaryanov[1]

[1] PJSC TATNEFT named after V.D. Shashin, 75, Lenin St., Almetyevsk, Republic of Tatarstan, 423450, Russia
[2] Ufa State Petroleum Technological University, Branch of the University in the City of Oktyabrsky, 54a, Devonskaya St., Oktyabrsky, Republic of Bashkortostan, 452607, Russia
[3] Ufa State Petroleum Technological University, 1, Kosmonavtov st., Ufa, Republic of Bashkortostan, 450062, Russian Federation
[4] State autonomous scientific institution "Institute of strategic research of the Republic of Bashkortostan", 129/3, October avenue, Ufa, Republic of Bashkortostan, 450075, Russian Federation

E-mail: markl212@mail.ru

Abstract. The research purpose is to identify factors of premature flooding of wells, assess the influence of factors, simulate and compare results obtained with field data. An analysis of the reservoir properties, design and development indicators of horizontal wells showed that only 41% of cases of premature flooding are due to geological causes (heterogeneity of rock permeability). Technological reasons (high specific samplings) occur in 36% of cases, this problem is pronounced in wells where new technologies were implemented – division of the wellbore into sections using water and oil swell packers. In 23% of cases, premature flooding is due to wellbore drilling in the water-saturated zone.

1. Introduction

The accumulated experience of operation of horizontal wells has proven the effectiveness of vertical wells. In some cases, the development of individual fields or their individual sections by horizontal wells is the only possible way. In addition, the role of horizontal wells increases when developing hard-to-recover reserves, including from unconventional reservoirs. In addition to the indisputable advantages, horizontal wells have a number of problems whose solution can improve the efficiency of field development and increase the oil recovery coefficient. One of the key problems is the premature flooding which decreases the technological and economic efficiency of the well.

2. Materials and methods

The fluid inflow to the wellbore, regardless of its design, is proportional to fluid mobility (k/μ), filtration area (S), and formation depression (ΔP). The heterogeneity of any of these parameters leads to heterogeneity of the inflow. Advance production of individual intervals of the reservoir leads to an uneven rise in the oil-water contact, formation of watering cones and premature flooding of the well. Thus, to determine causes of flooding, it is necessary to determine the effect of fluid mobility (k/μ), a filtration area (S) and formation depression (ΔP) on the efficiency of horizontal wells.

3. Results and discussion

Influence of inflow heterogeneity on the efficiency of horizontal wells:

• *fluid mobility*

The fluid mobility parameter (k/μ) represents the ratio of rock permeability to fluid viscosity. Variability of permeability of reservoir rocks along the wellbore is the most common cause of heterogeneity of inflow during the development of traditional oil reserves.

From the intervals with the highest permeability, fluid sampling is more intensive (Figure 1, A), which leads to flooding of these wells. Viscosity of the fluid mixture decreases, which leads to an increase in its mobility and heterogeneity of the inflow. After a water breakthrough, the productivity of the local inflow interval can fully ensure the depression and block the intervals of oil inflow (Figure 1, B).

In case of development of unconventional reserves, such as super-viscous oil (UHV), the second component of fluid mobility - its viscosity - comes first. Since the viscosity of oil depends on temperature, during the development of the SHF by the method of steam gravity drainage (PGD), heterogeneity of the inflow can be caused by uneven distribution of the coolant along the horizontal trunk.

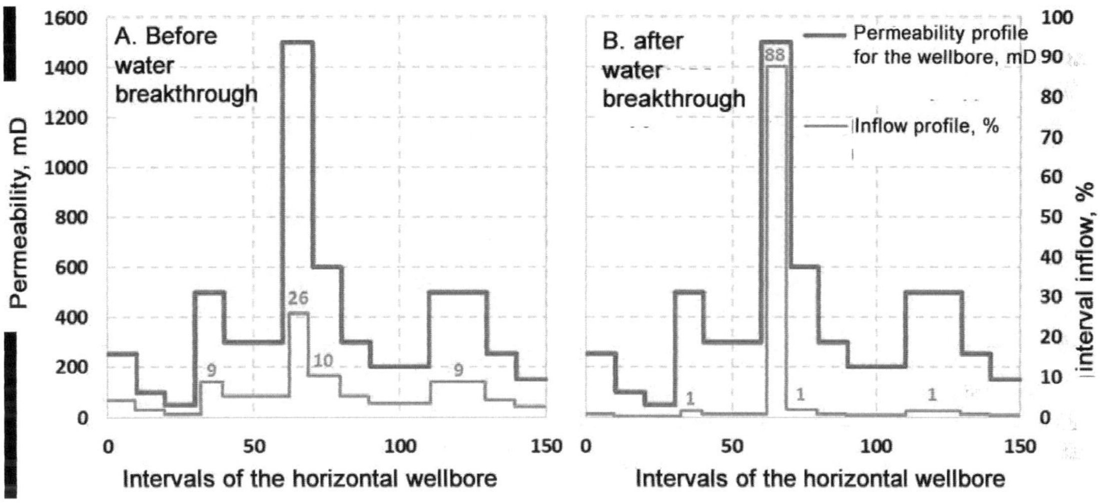

Figure 1. The schematic profile of the inflow along the horizontal wellbore with productive intervals of different permeabilities (before and after the water breakthrough)

• *Filtration area*

The filtration area is characterized by the length of the production interval - perforated formation thickness (h) for vertical wells or the length of the production interval (l) for horizontal wells. The unevenness of the inflow can be caused by the design of the well in case of sampling of productive intervals by packers or the use of filters.

Thus, wells with different lengths of operating intervals (equal fluid flow rates, horizontal wellbore lengths, the same FES) will differ in inflow uniformity. When the entire trunk is operating (Figure 2, A), uniform drainage of stocks occurs. In case of individual intervals (Figure 2, B), leading production and premature flooding occur.

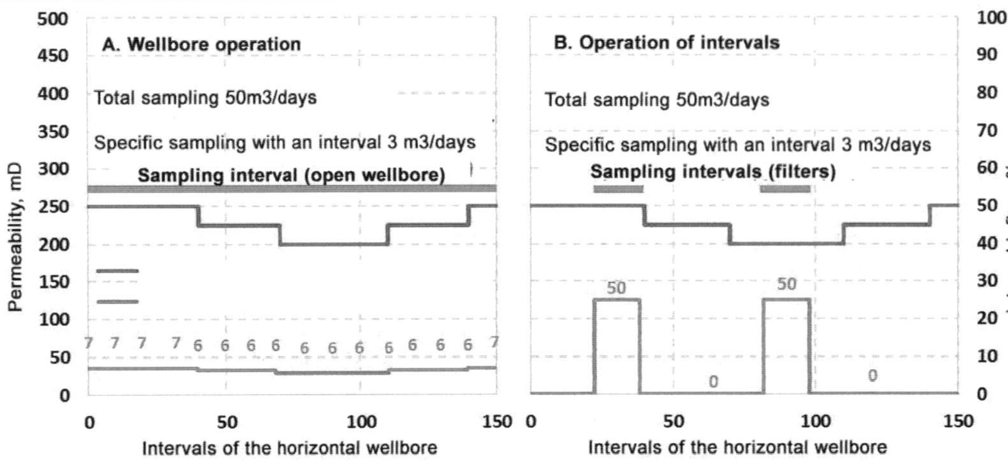

Figure 2. The schematic inflow profile along the horizontal wellbore with idle intervals

- *Reservoir depression*

Reservoir depression (ΔP). Heterogeneity of depression along the horizontal wellbore can be caused either by the opening of hydrodynamically unrelated formations with different levels of reservoir pressure, or by a change in the bottomhole pressure along the wellbore due to the heel-toe effect, which consists in the loss of friction pressure from the heel to the toe of the well and manifested as significant lengths of horizontal boreholes and high flow rates (Figure 3).

The mechanism of influence of inflow heterogeneity on the efficiency of horizontal wells

To determine the mechanism of influence of heterogeneity of the inflow on the efficiency of horizontal wells, the methods of numerical simulation were used. Heterogeneity of the inflow was described by two parameters: the ratio of the total length of working intervals and the total length of the horizontal wellbore, which characterizes the degree of involvement of reserves in the development or coverage of the formation by drainage. The ratio is calculated by formula:

$$l^* = \frac{\sum_{i=1}^{n} l_i}{L} * 100, \quad (1)$$

where l^* – share of working intervals, %; l_i – working interval length, m; L – total length of the drilled horizontal shaft.

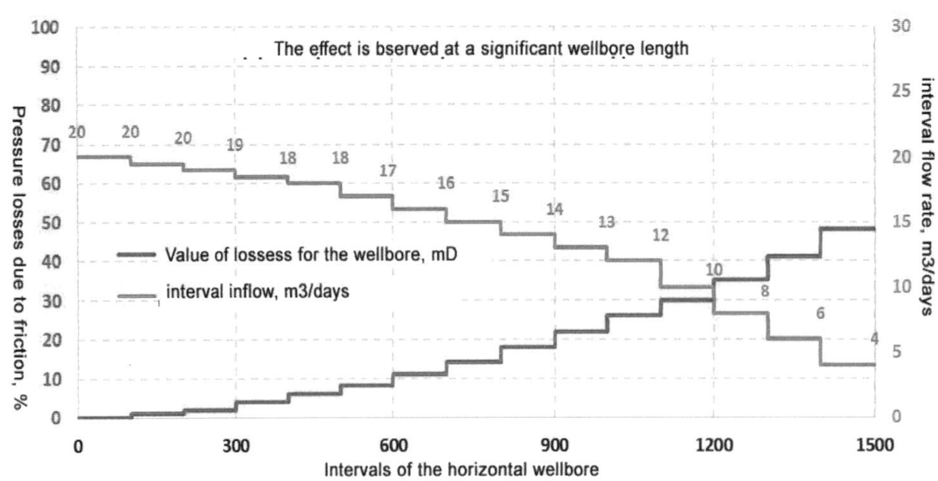

Figure 3. Estimated inflow profile along the horizontal wellbore of high extent with heel-toe effects

Specific fluid samplings along the horizontal wellbore are an indicator of the load on a reservoir section exposed by the horizontal wellbore. The physical meaning of the influence of this parameter on well operation can be described using the following example. The total well production rate of about 100 m^3/day can be ensured both by the interval of the productive formation 100 m long and by the interval of 10 m. In the first case, products from one part of the formation will be selected with specific samplings of 1 m^3 / day · m, and in the second case - 10 m^3 / day. Higher selection rates will lead to premature flooding.

Specific samplings are calculated by formula:

$$q_i^{\text{sp.}} = \frac{F_i * Q}{l_i}, \tag{2}$$

where $q_i^{\text{sp.}}$ – specific selection of the interval, m^3 /day · m; Q – total well production, m^3 / day; l_i – interval length, m; F_i – share of withdrawals from a single interval in the total well flow rate,% (based on the value of permeability of the interval).

The mathematical model was based on the following principles:

- the length of the horizontal wellbore is divided into intervals in terms of absolute permeability;
- each interval corresponds to the volume of drained reserves, determined during empirical modeling;
- the total fluid flow rate is equal to the sum of specific samplings for each interval;
- the share of samplings of each interval is calculated based on its length, phase permeability and depression;
- the water cut of each interval is a function of the initial oil and water saturation, phase permeability and accumulated oil and water samplings from this interval;
- the water cut is calculated as the weighted average water cut of the intervals according to the amount of samplings.

The model was adapted according to actual data on the operation of horizontal wells No. XXXX, YYYY, ZZZZ.

The analysis of geological data, as well as the results of geophysical and hydrodynamic studies showed that reservoir layers of the beaver horizon of the field have high permeability to ensure high productivity of all intervals of the horizontal wellbore.

The specific fluid samplings depend on the permeability of this section. Given this fact, as well as the design of the well, the distributions of specific samplings for wellbore No. XXXX, YYYY, ZZZZ were built. (figures 4, 6, 8). The simulation result is the calculated dynamics of the water cut of the wells presented in Figures 5, 7, 9.

Well No XXXX

With a fluid flow rate of about 50 m^3 / day and a total length of working intervals of 40 m, the average specific fluid samplings were 1.25 m^3 / day for one meter of the filter portion of the shank (Figure 4).

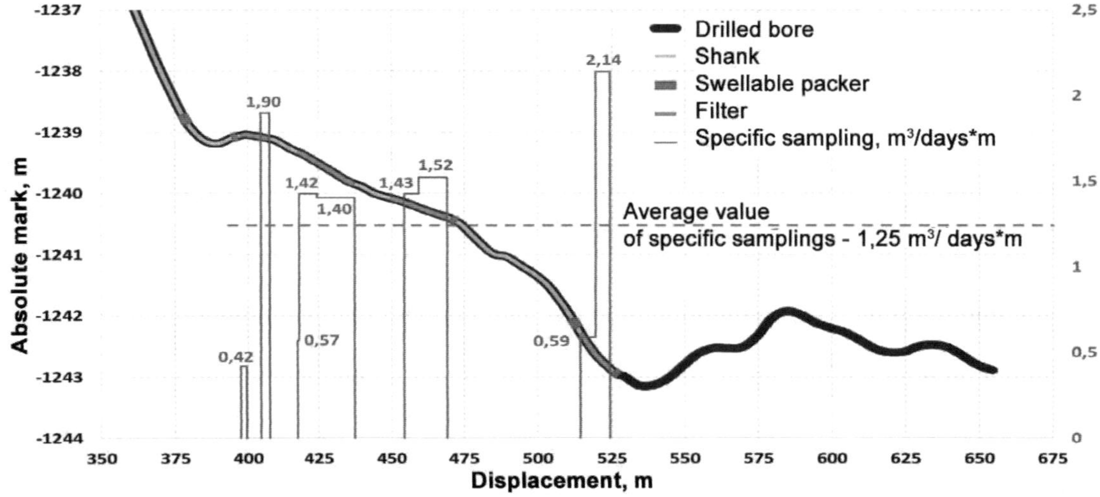

Figure 4. Distribution of specific samplings for well No. XXXX

Such a load on production intervals led to an early breakthrough of water in almost all intervals and complete watering of the well in the first year of operation (Figure 5).

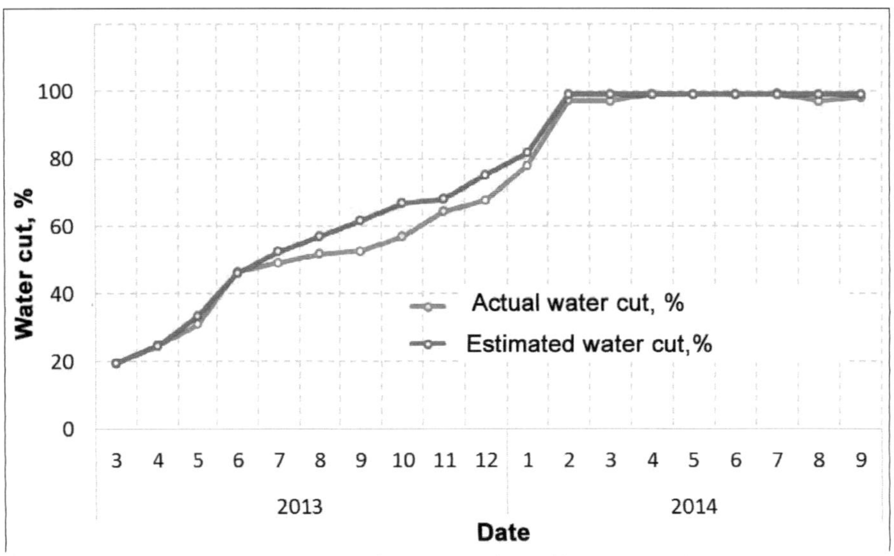

Figure 5. Actual and estimated dynamics of water cut in well No.XXXX

Well No YYYY

With a fluid flow rate of about 45 m^3 / day and a total length of working intervals of about 115 m, the average specific fluid samplings were 0.39 m^3 / day for one meter of the filter portion of the shank (Figure 6).

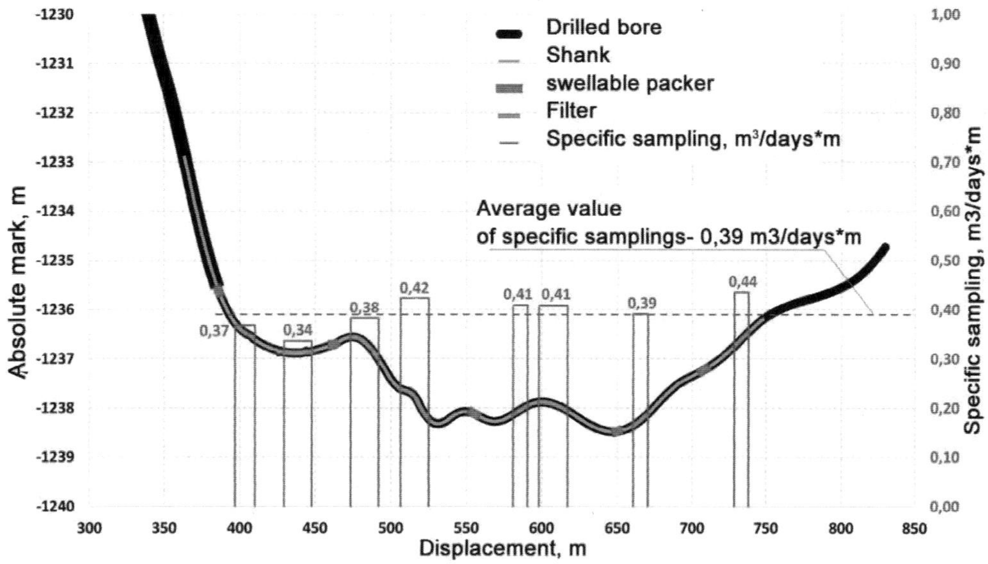

Figure 6. Distribution of specific samplings along well No. YYYY

For these geological conditions, such rates of sampling were the result of high rates of well watering (60% per year). A breakthrough of water occurred in individual highly permeable intervals of the horizontal trunk (Figure 7).

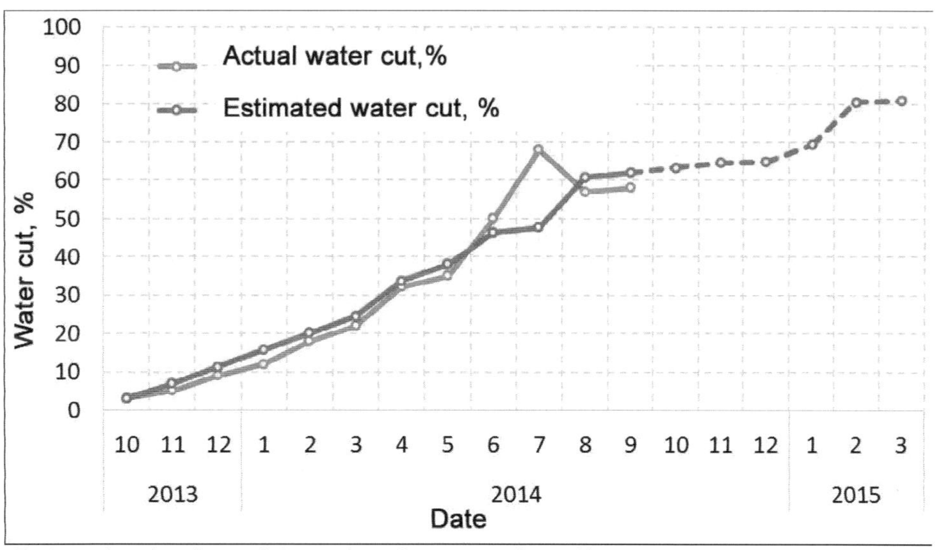

Figure 7. Actual and estimated dynamics of water cut in well No. YYYY

Well No ZZZZ

For well No. ZZZZ with a total fluid flow rate of about 40 m3 / day and a total length of working intervals of about 190 m, the average specific fluid samplings were 0.21 m^3 / day for one meter of the filter portion of the liner (Figure 8).

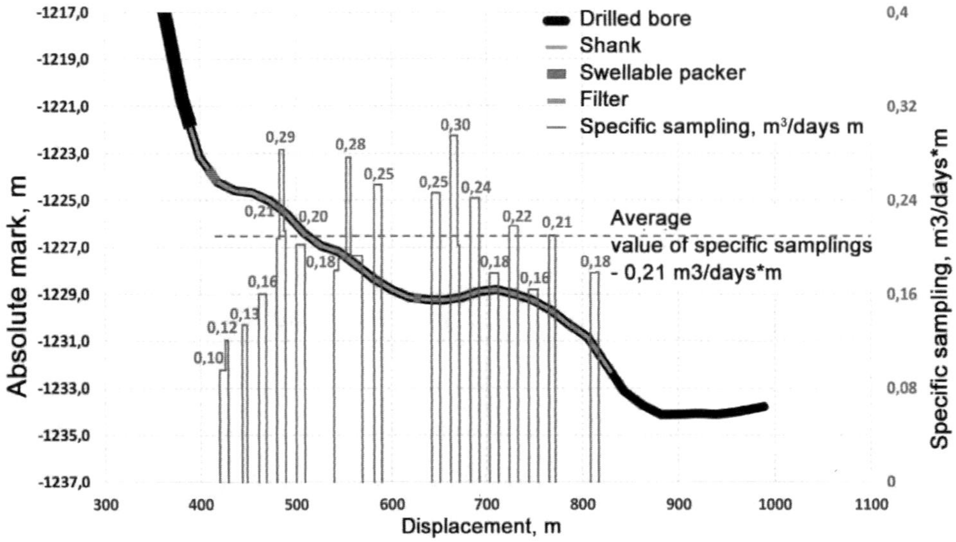

Figure 8. Distribution of specific samplings for wellbore No. ZZZZ

According to the actual data on the water cut and simulation results (Figure 9), the current value of specific samplings allows maintaining the rate of water cut at 10-12% per year.

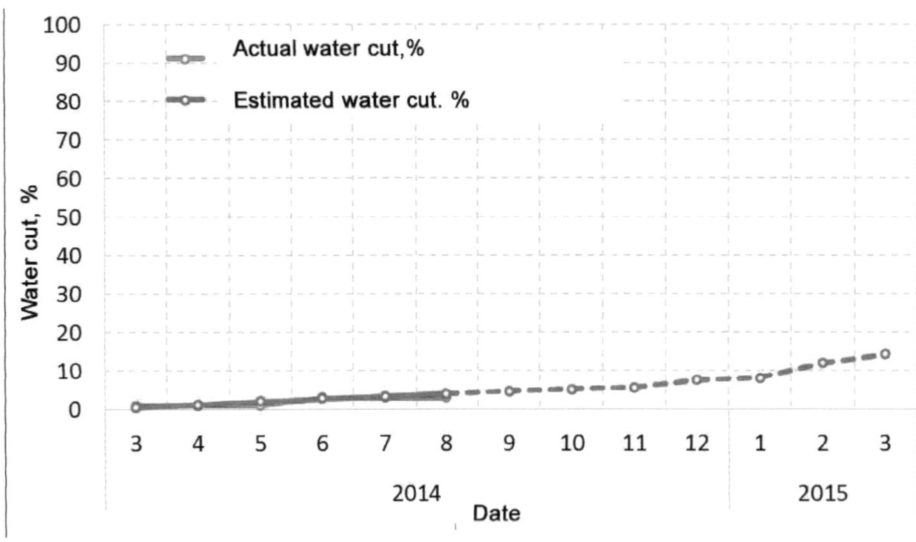

Figure 9. Actual and estimated dynamics of water cut in well No.ZZZZ

4. Conclusion

Thus, there is a direct relationship between the specific rates of sampling and the rate of watering. Obviously, the nature of this dependence is different for different fields depending on geological and technological characteristics. Thus, the reason for the flooding of horizontal wells XXXX and YYYY is high specific rates of sampling under the incomplete involvement of the reservoir.

The calculations showed that a similar pattern is observed in wells developing formations with a high degree of heterogeneity of reservoir properties. The rate of irrigation of products is a multiple of the specific rate of selection at the most permeable intervals. This reaffirms that another reason for premature flooding of the SGA products is the advance sampling of reserves at the most permeable intervals due to the uneven flow of fluid along the horizontal wellbore.

References

[1] Taipova V A, Shaidullin A A and Shamsutdinov M F 2017 Role of Horizontal Wells and Hydraulic Fracturing in Increasing the Efficiency of Oilfield Development Using the Example of Oil and Gas Production Department «Aznakayevskneft» Tatneft PJSC *Georesources* **19(3) part 1** 198–203

[2] Krivova N R, Reshetnikova D S, Fedorova K V and Kolesnik S V 2015 Efficiency improvement of low-permeable reservoirs development of the Western Siberia fields by the system of horizontal wells *Geology, Geophysics and Development of Oil and Gas Fields* **5** 52–57

[3] Sangyup Ch D, Furceribc P L, Saurabh M M and Poplawski-Ribeirob 2018 Oil prices and inflation dynamics: Evidence from advanced and developing economies *Journal of International Money and Finance* **82** 71-96

[4] Alkhelaiwi F T and Davies D R Inflow Control Devices: Application and Value Quantification of a Developing Technology *International Oil Conference and Exhibition in Mexico, 27-30 June, Veracruz, Mexico* 1–13 DOI: 10.2118/108700-MS

[5] Nurgaliev R .Z, Kozikhin R A, Fattakhov I G, Kuleshova L S and Gabbasov A Kh 2019 Prospects for the use of new technologies in assessing the impact of geological and technological risks *IOP Conference Series: Earth and Environmental Science (IPDME 2019 – International Workshop on Innovations and Prospects of Development of Mining Machinery and Electrical Engineering)* **378(1)** 1-9 DOI: 10.1088/1755-1315/378/1/012117

[6] Andreev V E, Chizhov A P, Chibisov A V and Mukhametshin V Sh 2019 Forecasting the use of enhanced oil recovery methods in oilfields of Bashkortostan *IOP Conference Series: Earth and Environmental Science (International Symposium «Earth sciences: history, contemporary issues and prospects»)* **350(1)** 1–6 DOI: 10.1088/1755-1315/350/1/012025

[7] Nurgaliev R Z, Kozikhin R A, Fattakhov I G and Kuleshova L S 2019 Application prospects for new technologies in geological and technological risk assessment *Mining Journal* **4(2261)** 36–40 DOI: 10.17580/gzh.2019.04.08

[8] Batalov S A, Andreev V E, Lobankov V M and Mukhametshin V Sh 2019 Numerical simulation of oil formation with regulated disturbances. Oil recovery quality simulation *Journal of Physics: Conference Series (ITBI 2019 – International Conference "Information Technologies in Business and Industry")* **1333(3)** 1-6 DOI: 10.1088/1742-6596/1333/3/032006

[9] Babichev I N, Fattakhov I G, Kuleshova L S, Zaripov L F and Morozov M A 2019 Potential for the wellbore zone development using the dynamic impact *IOP Conference Series: Earth and Environmental Science (IPDME 2019 – International Workshop on Innovations and Prospects of Development of Mining Machinery and Electrical Engineering)* **378(1)** 1-5 DOI: 10.1088/1755-1315/378/1/012111

[10] Batalov S A, Andreev V E, Lobankov V M and Mukhametshin V Sh 2019 Numerical simulation of the oil reservoir with regulated disturbances. Oil recovery stability simulation *Journal of Physics: Conference Series (ITBI 2019 – International Conference "Information Technologies in Business and Industry")* **1333(3)** 1-6 DOI:10.1088/1742-6596/1333/3/032007

IOP Conf. Series: Materials Science and Engineering **952** (2020) 012057 doi:10.1088/1757-899X/952/1/012057

The effect of axial loading on drill string deformation

L S Kuleshova, D I Akhmetshina, A A Fatkullin and A A Abdulmanov

Ufa State Petroleum Technological University, Branch of the University in the City of Oktyabrsky, 54a, Devonskaya St., Oktyabrsky, Republic of Bashkortostan, 452607, Russian Federation

E-mail: vsh@of.ugntu.ru

Abstract. A practical method for calculating the axial load of a string through the load on a hook is presented and the flotation capacity of elements for different well trajectories is provided. It was found that the traditional simplified method for calculating the buoyancy factor, which is used to calculate the axial load on a string and axial tension, can only be used in vertical wells with a freely suspended pipe string, since in this state buoyancy acts only on the lower part of the string. If the string is constrained by downhole tools such as a packer or an anchor, buoyancy cannot be considered a simple weight loss. In directional wells, buoyancy changes the axial load of the string and causes shear stresses in the cross section of the string. When calculating the effect of fluid on the string, it is necessary to take into account the sequence of operations and stresses from wellhead and downhole tools.

1. Introduction

In complex wells, such as high pressure-high temperature (HPHT) wells, deep wells and directional wells, it is important to predict and control the tubing axial load and axial deformations in order to ensure operational safety [1-4]. Errors made at the design stage can cause troublesome operations, even accidents. There are many factors affecting string deformations, including temperature, fluid pressure, tool weight, friction, well trajectory, string deformation, etc. [5–8]. The most fcommon one is the effect of fluid pressure on string deformations. The effect of fluid pressure on pipes is versatile. Firstly, the fluid pressure causes a sharp jump in the axial force at the bottom of the string and conical stages of the string due to the piston effect. Secondly, pressure of the fluid causes axial deformations due to the swelling effect. Thirdly, in directional wells, fluid pressure causes shear stresses in the cross section of the string, which increases bending deformations. Finally, fluid pressure causes a virtual axial load on the string, causing pipe bending, while pipe bending results in string shortening, contact loads and borehole friction.

Simple and practical methods for calculating the effect of fluid pressure can be found in various technical manuals. But the applicable conditions for the methods are not limited; therefore, when using them, mistakes are often made [9–12]. A method for calculating the axial force of a well string using the hook load to the bottom is presented. The change in fluid pressure and its effect on string deformations are described. The results will be of great engineering importance for complex strings.

2. Materials and methods

At any depth of the well, the internal pressure and external pressure of the fluid on the sting's surface are equal, and the buoyancy acting on the string is

$$F_f = \gamma_0 H\pi(b^2 - a^2) = \gamma_0 H A_s, \tag{1}$$

where $A_s = \pi(b^2 - a^2)$ – the cross-sectional area of the pipe wall.

Equation (1) shows that the buoyancy acting on the string is equal to the mass of fluid that it displaces and that is directed upward. Despite the fluid pressure applied to any point on the surface of the string, the actual buoyancy acts on the bottom of the string in a summation of forces. Let γ_s be the density of steel, the suspended weight (hook load) of the string is:

$$W = (\gamma_s - \gamma_0)HA_s. \tag{2}$$

Taking the wellhead as the origin of coordinates, the coordinate system is set along the depth of the well, and the axial force (positive tension, similarly here and below) of the entire string at any depth of the well is

$$F_x = W - \gamma_s x A_s = [\gamma_s(H - x) - \gamma_0 H]A_s. \tag{3}$$

Then

$$x = H, F_x|_{x=H} = -F_f = -\gamma_0 H A_s.$$

Namely, the buoyancy acts on the lower end of the string in vertical wells.

The total buoyancy is:

$$F_f = \gamma_0 H\pi b^2 - \gamma_i H\pi a^2 = \gamma_0 H A_0 - \gamma_i H A_i, \tag{4}$$

where γ_i – fluid density, $A_i = \pi a^2$ *and* $A_0 = \pi b^2$ –inner and outer cross-sectional areas of the string, respectively.

The string hook load is

$$W = \gamma_s H A_s - (\gamma_0 H A_0 - \gamma_i H A_i). \tag{5}$$

The axial load at any well depth is

$$F_x = \gamma_s H A_s - (\gamma_0 H A_0 - \gamma_i H A_i) - \gamma_s x A_s. \tag{6}$$

In many cases, there are packers at the bottom of the string, and sometimes the packer allows the axial movement of the lower end of the string, or a sliding joint is used to effect this movement. Only the case when the axial movement of the lower end of the column is impossible is studied. Buoyancy does act on the bottom of the string. Since the packer has been installed, the buoyancy effect is blocked by the packer, and the axial load of the string is not directly related to the fluid pressure under the packer. If the temperature and pressure do not change before and after the packer has been installed, equation (3) and equation (6) can be used. But this is not a common case. In fact, both temperature and pressure can change; therefore a new calculation method should be developed. Suppose that the hook load F0 is known, the distribution of the axial load of the string along the well depth is:

$$F_x = F_0 - x y_s A_s. \tag{7}$$

There is an unavoidable relationship between the hook load and the floating mass. If the hook load cannot be obtained directly, the axial load of the string can be calculated by the initial string length, temperature and pressure change before and after the packer is set. And the floating method could not be used in this process.

When using tapered strings, pressure drops occur inside and outside the string due to changes in the cross-sections. The resulting axial force which is called the piston force results in an abrupt change in the axial load of the string. In fact, the piston force is part of the buoyancy. With tapered strings, let A_{iu} and A_{ou} be for the inner and outer cross-sectional areas of the upper column, respectively, and A_{id} and A_{od}, be for the inner and outer cross-sectional areas of the bottom string, respectively. Therefore, the force of the piston at this point (upward is positive, similarly below) is equal to:

$$F_p = (A_{ou} - A_{od})P_0 - (A_{iu} - A_{id})P_i. \tag{8}$$

If a control valve is used, when the valve is closed, the fluid pressure above and below it will be different, and therefore the piston force is:

$$F_p = A_i(P_{id} - P_{iu}), \tag{9}$$

where P_{iu} and P_{id} – fluid pressure above and below the valve, respectively.

When a simple string is under the simple constraint, buoyancy is equal to the weight of the fluid displaced by the string. In fact, buoyancy is the resultant force of fluid pressure and should not be described as a displaced mass of fluid. In oil and gas wells, almost all strings can experience fluid pressure changes. These changes can be caused by surface production, formation pressure, thermal stress, different fluid densities, etc. Thus, when calculating the effect of fluid pressure on the stress and strain of the string, all factors and constraints must be taken into account. For example, if wellhead pressure or annular pressure is present, the string load derived from the fluid pressure should not be calculated based on the buoyancy.

According to the relationship between axial stress and deformation, the axial deformation of the string can be calculated. As a rule, the axial load of the string is not constant; therefore an integral method is required to determine axial deformations. In any well with depth x, intercepted by an infinitely small section dx, its axial elongation is equal to:

$$d(\Delta l) = \varepsilon_x dx = \frac{F_x}{A_s E} dx, \tag{10}$$

where ε_x – axial deformation; E – Young's steel modulus. This is the basic formula for calculating the axial elongation. Total axial deformation can be calculated by integration along the column. The calculation can be simplified using the axial force at the midpoint of the string.

Fluid pressure can cause axial deformations, a swelling effect. According to the basic theory of elasticity, the swelling effect can be calculated using the following formula:

This is the basic formula for calculating the axial elongation of the string. The full axial deformation is

$$\varepsilon_x = \frac{2\mu}{E} \cdot \frac{P_0 R^2 - P_i}{R^2 - 1}, \tag{11}$$

where μ – Poisson's ratio of steel, $R - OD/ID$ ratio. This formula is obtained in accordance with the plane stress hypothesis (the string can expand freely in the axial direction). The axial deformation caused by the swelling effect can be calculated using formula (11).

Bending is a common phenomenon of oil and gas well string deformation. The influence of fluid pressure on string bending is as follows: the axial load of the string and the virtual axial load. The effective axial load of string F_e (positive compression) at any well depth x is

$$F_e = -F_x + P_i A_i - P_0 A_0. \tag{12}$$

If the column is long enough, then the axial bending condition is:

$$F_e \geq F_{cr}, \tag{13}$$

where F_{cr} – critical bending force. The well size and the string limit are different, F_{cr} have different values. On the other hand, the bulge shape of the pillar is different. Column bending has two consequences: column shortening and friction.

Investigating the effect of buoyancy in deviated wells is a more laborious task than in vertical wells. In most cases, accurate calculation is impractical, therefore theoretical analysis is necessary. To trace this problem, let's take a curved section of the casing, Figure 1, a), where the internal fluid pressure is ignored.

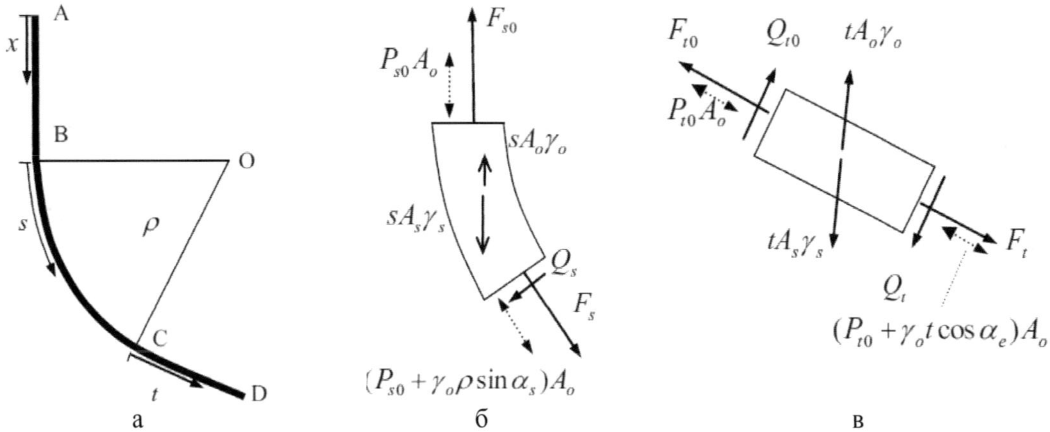

Figure 1. Distribution of forces on the curved section of the column: a - column diagram; b - flexible area; в - rejected part

For such a section, it is difficult to accurately calculate the effect of fluid on axial load and deformation, and this is not necessary. The simplified calculation is practical. In the process of simplification, attention should be paid to the concepts of fluid distribution and resultant forces. Suppose the hook load W is known. Then the axial distribution of the string force can be calculated as follows: divide the string into finite segments along the well path, then compute them from top to bottom using a recursive algorithm.

For a vertical segment of the AB string, fluid pressure is distributed symmetrically and the resulting force is zero in any axial micro-segment. The axial force is

$$F_x = W - \gamma_s x A_s. \tag{14}$$

The weight of the string is the only load. This formula can be used for the end of a vertical section.

If the curved section is close to an arc of a circle, then the calculation will be more convenient. As it shown in Figure 1, a), s — a curvilinear coordinate variable. Axial string force F_{s0} - the initial point of the curved coordinate, α_b - the angle between the axis of the string and the direction of gravity (slope); α_e - a slope of the string at the end of the arc segment$(s = S)$. Then the radius of curvature is:

$$\rho = \frac{S}{\alpha_e - \alpha_b},$$

The bend at any arc length is:

$$\alpha_s = \alpha_b + \frac{s}{\rho}.$$

Let P_{s0} be fluid pressure. Let us intersect a string segment of length s, as shown in Figure 1, b), then its forces from gravity and fluid pressure are buoyancy $sA_0\gamma_0$(upward) and gravity force $sA_s\gamma_s$ (downward). The resulting forces in the upper and lower sections of the hydrostatic pressure are$P_{s0}A_0$:

$$[P_{s0} + \gamma_0\rho(\sin\alpha_s - \sin\alpha_b)]A_0 = F_s A_0,$$

respectively, along the pipe axis. In fact, the forces at the top and bottom of the element are not fluid pressure, but axial forces $F_{s0} and\ F_s$ respectively. By equilibrium, they have the following relationship:

$$[F_{s0} + P_{s0}A_0 + sA_0\gamma_0 - sA_s\gamma_s]\cos\alpha_s - F_s - [P_{s0} + \gamma_0 p(\sin\alpha_s - \sin\alpha_b)]A_0 = 0,$$

from which the axial force Fs is:

$$F_s = (F_{s0} + P_{s0}A_0 + sA_0\gamma_0 - sA_s\gamma_s)\cos\alpha_s - [P_{s0} + \gamma_0 p(\sin\alpha_s - \sin\alpha_b)]A_0. \tag{15}$$

Similarly, the shearing force at a variable curvilinear coordinate *s*:

$$Q_s = (F_{s0} + P_{s0}A_0 + sA_0\gamma_0 - sA_s\gamma_s)\sin\alpha_s. \tag{16}$$

Thus, the fluid pressure can cause a shear force in the curved line, the force pointing in a direction like a radial arc. These two equations can be used up to the end of the arc section BC, and thus F_S and Q_S can be obtained at point C. If the curvature changes, the section should be divided into several parts; with each part the method described above should be used.

The method for analyzing the deflected section is similar, but simpler than the curved section, as shown in Figure 1, a and c). Where t is the rectilinear coordinates, the slope of the curved section is α_e. Let $F_{t0}(= F_s), Q_{t0}(= Q_s), P_{t0}$ for the axial force, shear force and fluid pressure at the starting point($t = 0$) respectively. Then the axial force and the shear force in any position t is equal to:

$$F_t = F_{t0} - tA_s\gamma_s \cos\alpha_e, \quad (17)$$

$$Q_t = Q_{t0} + t[A_0\gamma_0 - A_s\gamma_s] \sin\alpha_e. \quad (18)$$

Both efforts involve the bending of the string.

The resulting force of fluid pressure on the vertical section of the string is zero, and on the curved section of the column is:

$$F_{fc} = SA_0\gamma_0 + P_{s0}A_0 \cos\alpha_b - P_S A_0 \cos\alpha_e, \quad (19)$$

where

$$P_S = P_{s0} + \gamma_0\rho(\sin\alpha_e - \sin\alpha_b).$$

At $\alpha_b = 0$:

$$F_{fc} = SA_o\gamma_o + P_{s0}A_o - (P_{s0} + \gamma_o\rho \sin\alpha_e)A_o \cos\alpha_e. \quad (20)$$

$$F_f = LA_o\gamma_o \quad (21)$$

Aalthough in vertical wells the deformation state of the string differs from that in deviated wells, the buoyancy that the string receives is equal to the mass of fluid displaced by the string..

In the absence of relative movement between the pipe string and the packer, the fluid pressure at the lower end of the string is held by the packer, the buoyancy acting on the string is equal to:

$$F_{ft}^2 = TA_o\gamma_o + P_S A_o \cos\alpha_e - P_T A_o \cos\alpha_e.$$

If $P_T = P_S + \gamma_o T \cos\alpha_e$, we have:

$$F_{ft}^2 = TA_o\gamma_o \sin^2\alpha_e \quad (22)$$

The total buoyancy is

$$F_f = sA_o\gamma_o + P_{s0}A_o - (P_{s0} + \gamma_o p \sin\alpha_e)A_o \cos\alpha_e + TA_o\gamma_o \sin^2\alpha_e \quad (23)$$

When $\alpha_e = \frac{\pi}{2}$, the results of formula (23) and formula (21) will be the same, or will differ in the following cases.

The vertical section of the string does not receive the net buoyancy force from the fluid pressure; the effect of the fluid is transmitted to it in the form of an axial load from the string bottom. The internal forces are different in deviated and vertical wells. Firstly, the resulting buoyancy force is different, therefore the axial force distribution is different. Secondly, in deviated wells there is a shear force in the string, but in vertical wells, it is absent. The contact forces of the string with borehole walls are not taken into account. Internal fluid pressure is also ignored to reduce complexity.

3. Results and Discussion

In a thermal recovery well, casing prestressing is an effective method to improve safety of the casing. While cement is hardening, it is necessary to maintain pump pressure in the casing until a cement sheath has formed. As a result of increased pressure, the piston effect lengthens the string and the swelling effect shortens the casing, but the net effect is that the casing is elongated. After the wellhead has been set and the pump has been depressurized, the elongation becomes axial prestressing. The following parameters are given: casing outer diameter is 177.8 mm, wall thickness is 9.19 mm, casing length is 600 m, Young's elastic modulus is 1.98 × 105 MPa, Poisson's ratio is 0.3, pump pressure is 16 MPa. The calculation results are: The axial prestress of the casing by the pump-pressure is 26.3 MPa.

When testing high HPHT wells, permanent packers are often used. This packer allows the lower end of the string to slide freely, but separates the annular fluid pressure from the formation pressures below. At various stages, the axial force and deformation of the string are not linked.

Only the shut-off stage is taken into account. The internal valve locatd at the bottom of the tubing string is closed to separate the internal pressure from the formation pressure. The parameters are: the outer and inner diameters of the main pipe are 88.9 mm and 68.33 mm, the outer and inner diameters of the sliding pipe are 99.06 mm and 73.71 mm, the depth of the packer is 2800 m, the reservoir pressure is 60 MPa, the density of the fluid in the annular space is 1.5 g cm3, the density of the fluid in the string is 0.001 g / cm^3.

Calculation results: the piston effect shortens the string by 0.69 m, while the deflection effect shortens the string by 0.064 m. If the gate in the valve is at the wellhead, there will be a swelling effect that shortens the string by 0.66 m.

4. Conclusion

The influence of fluid pressure on the string depends on two factors: pressure distribution determined by the working medium and the operation stage and its spatial state and limitations. The methods for calculating string stresses and strains described in the technical manuals do not deal with these issues. When calculating stresses and strains in the string, a method and a formula must be carefully selected. It has been established that the traditional simplified method for calculating the buoyancy factor, which is used to calculate the axial force of the string and the axial elongation, can be used only for vertical wells with a freely suspended tubing, since in this case the buoyancy acts on the bottom of the string. If the string is bounded by downhole tools such as a packer or an anchor, the buoyancy cannot be considered in a traditional aspect. In a directional well, the buoyancy changes the axial force of the string and causes shear stress in the cross section. Examples from oilfield engineering show that when calculating the effect of fluid on the string, the sequence of operations and restrictions imposed by the wellhead and downhole tools should be taken into account.

References

[1] Nikolaev N I and Kozhevnikov E V 2014 Enhancing the cementing quality of the well with horizontal profile *Bulletin of PNRPU. Geology. Oil & Gas Engineering & Mining* **11** 29–36 DOI: 10.15593/2224-9923/2014.11.3

[2] Polyakov V N, Chizhov A P, Kotenev Yu A, and Mukhametshin V Sh 2019 Results of System Drilling Techniques and Completion of Oil and Gas Wells *IOP Conference Series: Earth and Environmental Science (IPDME 2019 – International Workshop on Innovations and Prospects of Development of Mining Machinery and Electrical Engineering)* **378 (1)** 1–7 DOI: 10.1088/1755-1315/378/1/012119

[3] Dmitriev A Yu 2008 *Fundamentals of drilling technology* Tomsk: TPU 216 p

[4] Mukhametshin V V 2018 Bottomhole formation zone treatment process modelling with the use geological and geophysical information *IOP Conference Series: Earth and Environmental Science (IPDME 2018 – International Conference on Innovations and Prospects of Development of Mining Machinery and Electrical Engineering)* **194(2)** 1–5 DOI: 10.1088/1755-1315/194/2/022024

[5] Zeigman Yu V, Mukhametshin V Sh, Khafizov A R and Kharina S B 2016 Prospects of Application of Multi-Functional Well Killing Fluids in Carbonate Reservoirs *SOCAR Proceedings* **3** 33–39 DOI: 10.5510/OGP20160300286

[6] Polyakov V N 1983 Requirements to the integrity and strength of the well in the process of completion of wells in the fields of Bashkortostan *Oil economy* **2** 27–28

[7] Kuleshova L S, Mukhametshin V V and Safiullina A R 2019 Applying information technologies in identifying the features of deposit identification under conditions of different oil-and gas provinces *Journal of Physics: Conference Series (ITBI 2019 – International Conference "Information Technologies in Business and Industry")* **1333(7)** 1–5 DOI: 10.1088/1742-6596/1333/7/072012

[8] Batalov S A, Andreev V E, Lobankov V M and Mukhametshin V Sh 2019 Numerical simulation of oil formation with regulated disturbances. Oil recovery quality simulation *Journal of Physics: Conference Series (ITBI 2019 – International Conference "Information Technologies in Business and Industry")* **1333(3)** 1–6 DOI: 10.1088/1742-6596/1333/3/032006

[9] Polyakov V N, Kuznetsov Yu S, Sagidullin I A, Shulgina N Yu, Dubrovskiy V S, Khusainov V M, Khaminov N I, Akhmetzyanov R G, Vildanov A A and Starov O E 2005 The solving of problems of tailing-in and wells operation in abnormal thermodynamic conditions *Oil Industry* **5** 104–110

[10] Andreev V E, Chizhov A P, Chibisov A V and Mukhametshin V Sh 2019 Forecasting the use of enhanced oil recovery methods in oilfields of Bashkortostan *IOP Conference Series: Earth and Environmental Science (International Symposium «Earth sciences: history, contemporary issues and prospects»)* **350 (1)** 1–6 DOI: 10.1088/1755-1315/350/1/012025

[11] Akhmetov R T, Mukhametshin V V, and Kuleshova L S 2019 Simulation of the absolute permeability based on the capillary pressure curves using the dumbbell model *Journal of Physics: Conference Series (ITBI 2019 – International Conference "Information Technologies in Business and Industry")* **1333 (3)** 1-8 DOI: 10.1088/1742-6596/1333/3/032001

[12] Akhmetov R T, Mukhametshin V V, and Kuleshova L S 2019 Simulation of the absolute permeability based on the capillary pressure curves using the dumbbell model *Journal of Physics: Conference Series (ITBI 2019 – International Conference "Information Technologies in Business and Industry")* **1333 (3)** 1-8 DOI: 10.1088/1742-6596/1333/3/032001

Technological features of pumping water into high-viscosity oil injection wells with the bottom hole pump

Sh G Mingulov[1], I Sh Mingulov[1] and V V Mukhametshin[2]

[1] Ufa State Petroleum Technological University, Branch of the University in the City of Oktyabrsky, 54a, Devonskaya St., Oktyabrsky, Republic of Bashkortostan, 452607, Russian Federation
[2] Ufa State Petroleum Technological University, 1, Kosmonavtov st., Ufa, Republic of Bashkortostan, 450062, Russian Federation

E-mail: vv@of.ugntu.ru

Abstract. The authors laid bare a correlation between the proportion of mechanical contaminations and oil products in discharge waters and the average oil viscosity in the field. Film thickness depends on the oil viscosity. More viscous oil forms thicker and heavier films on the surface of suspended solids. The use of electrical centrifugal pumps for fossil water injection into the well allows using the heat produced by the pump to warm the water. Standard measurement instruments show that the change of injected water temperature at the head of the well pump-over unit is between 1.1 and 8.4 °C. The average temperature increase is 3.6 °C.

1. Introduction

One of the ways to improve the efficiency of high-viscosity oil pool development is maintaining the bottom-hole pressure by injecting water into the bed [1–5]. However, the use of this development method under specific geological and field conditions requires checking the specifics of the process [6–10].

The prepared bottom water injected into the bed may contain some amounts of suspended particles consisting of both the oil phase and mineral substances [11, 12]. Oil film covers solid particles thus reducing the overall density of the suspension and complicating its settling-put in the preparation units. This is largely due to the fact that the hydrocarbon phase adhesion to solid surfaces significantly exceeds the water phase adhesion.

Solids that enter the well-bore from the bed get covered in oil film of a certain thickness as they move. Its thickness depends primarily on the physical and chemical properties of the oil phase.

Since the density of solid particles covered in the film depends on the phase components, i.e. The oil products and mineral substances, we analyzed the correlations between phase proportions and produced oil viscosity.

2. Materials and Methods

We analyzed the impact of oil product and mechanical contaminations in waters discharged from medium-viscosity oil at several water discharge units of Tuymazaneft NGDU, Chekmagushneft, and Krasnokholmskneft.

3. Results and Discussion

We determined that film thickness (oil volume in suspension) depends on the oil viscosity. More viscous oil forms thicker and heavier films on the surface of suspended solids.

Figure 1 shows the correlation between the proportion of oil products and suspended solids in discharge waters and the average oil viscosity in the field well groups. The correlation can be described with the $Vo/Vs = f(\mu)$ function where Vo is the oil product concentration in discharge waters in mg/l, Vs is the concentration of solids in discharge waters in mg/l; μ is the average viscosity in mPas.

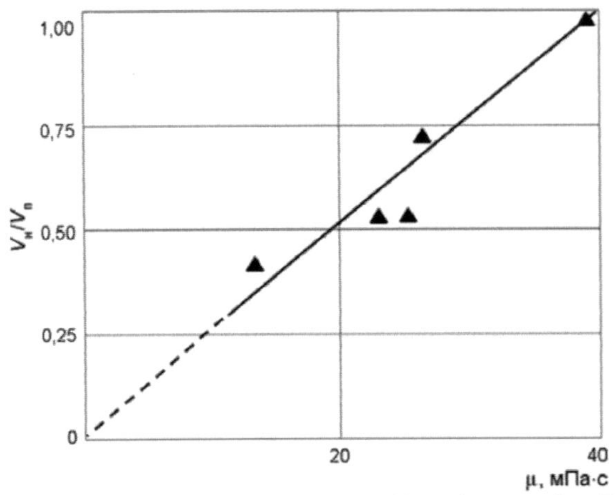

Figure 1. The correlation between the proportion of oil and water in discharge waters and the average oil viscosity in the field well group.

With high confidence, $R^2 = 0.78$ and this function is described by the following equation: $y = 0.0253x$.

Therefore, the surface of solids gets covered in the oil phase whose thickness depends on the viscosity of the produced oil. The content of oil products in discharge waters from high-viscosity oil fields increases along with the thickness of the film covering suspended solids. The quality of oil and discharge water preparation in preparation units and settling pits.

The presence of the film on solids significantly increases the clogging rate of the pore volume in the bed bottom hole area and reduces the injection capacity of the well. The coagulation of solid particles with the oil film and the increase of the clogging substance due to the presence of such film increases the contamination rates of the bottom-hole area (BHA) of the injection well and the pore volume clogging.

Thus, one of the ways to improve the efficiency of injection capacity recovery and BHA treatment intervals is the heating of injected water to reduce the viscosity of the oil film on the surface of the solids [13]. This allows reducing the volume of the colmatant by the dissolution of paraffinic hydrocarbons and a partial oil filtration in the pore volume.

There is a well pump-over (WPO) technology that uses the heat of the fluid as it passes through the downhole motor and the step decks. The dissipation of energy in the fluid fed to the downhole motor is achieved due to the stator winding resistance and the motor magnetic resistance. Besides, fluid heating is achieved through friction in the main bearings of impellers [14].

Figure 2 shows the basic diagram for WPO operation in an injection well when injecting the water from the upper water bed into the pay bed.

IOP Conf. Series: Materials Science and Engineering **952** (2020) 012058 doi:10.1088/1757-899X/952/1/012058

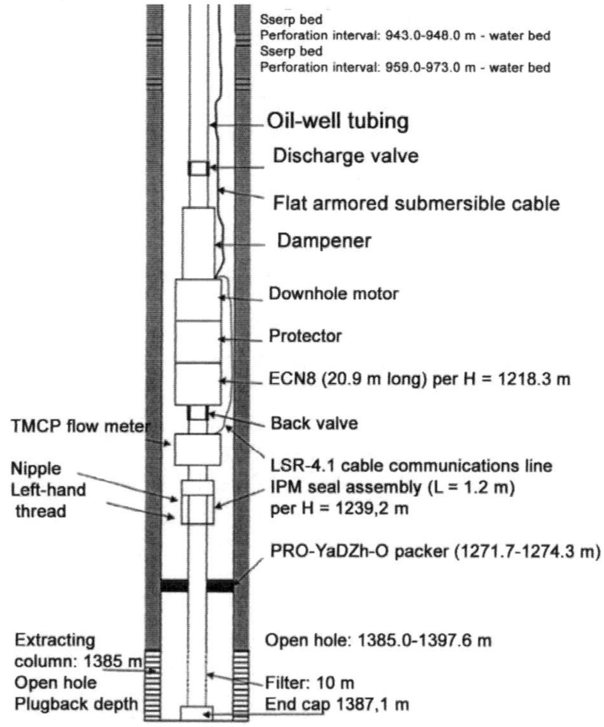

Figure 2. Water injection diagram for well 104 at Yubileynoye field

The electric centrifugal pump unit (ECPU) is tripped on the flow column into the well in the upside-down position. The water is injected into the bed through a pipe that passes the packer. At the pump head, there is a telemetry unit with a flowmeter installed that transmits information to the ground surface via the power cable of the electric motor. This information includes the flowrate and temperature of the water injected.

The fluid passes between the oil-well tubing and the extracting column, enters the pump suction, and is then fed into the pay bed via the packer pipe.

The water is injected from the upper water bed, the Sserp bed, that has the highest productivity factor, a broad external boundary, and a decent thickness of the water layer. Before hoisting down the pump, a packer is installed in the well. The packer has a pipe with a left-hand thread on the upper end. Besides, the upper end of the pipe features a cylinder that encapsulates the pump nipple. The ECPU is lowered as close to the pay bed as possible to prevent the heat leaking from the fluid pump to the borehole environment.

The goal of BHA heating is to measure the temperature change value at the WPO head and the impact of water heating on the operational parameters of the well.

To solve this problem, we researched injection well 104 at the Yubileynoye field of Tuymayaneft NGDU. The commercial oil beds in Yubileynoye include C1rd+C1bb of Bobrikovian-Radaevian horizon, CII benches of the Tournai stage, and D3fm2 packs of the Famennian stage. The water is injected in well 104 From Sserp bed to Dfams bed using the well pump-over unit.

According to the readings from the telemetry sensors installed on the well pump over (WPO) units in the operating wells of Mustafinskoye, Usen-Ivanovskoye, Solontsovskoye, and Abdulovskoye fields, the temperature variation interval is between 1.1 and 8.4°C.

Table 1 shows the data on the changes in water temperature at the well pump-over heads. The experiments showed that the average temperature increase is 36°C.

Table 1. Telemetry sensor readings at well pump-over units

Well No.	Field	Temperature increase at the WPO head in °C
127	Mustafinskoye	1.1
10MUF	Mustafinskoye	8.4
1784	Usen-Ivanovskoye	2.3
37УСТ	Solontsovskoye	5.0
201PTM	Abdulovskoye	1.5

It is especially relevant to assess the impact of temperature increase on the operation mode of the well equipped with a pump-over unit. To do that, we analyzed the dynamics of the operational characteristics of wells where water is injected from the Sserp bed wells.

To compare the operational characteristics, we selected five injection wells in the Ardatovskoye field where the Dfams bed was injected with water from the Sserp bed intake wells. Table 2 shows the geological and physical characteristics of the injection facility of well 104 at the Yubileynoye field and 5 wells in the Ardatovskoye field. We can see that the field development targets compared have similar geological and physical parameters.

Table 2. Geological and physical parameters of the Dfams bed of the mid-Famennian sub-stage at Yubileynoye and Ardatovskoye fields.

Parameter	Development targets	
	Yubileynoye field	Ardatovskoye field
	Mid-Famennian sub-stage	
	Dfams	
Average total thickness in m	13.5	12.1
Average effective oil height in m	12.8	7.1
Porosity factor	0.03	0.04
Oil saturation factor of the bed, in decimal quantities	0.90	0.90
Permeability, um^2	0.010	0.011
Sand fraction in decimal quantitiesед.	0.96	0.85
Stratification	1.6	2.6
Initial bottom-hole temperature in °C	25	25
Initial bottom-hole pressure in MPa	13.4	13.4
Oil viscosity under reservoir conditions in MPa s	72.38	36.4
Oil density under reservoir conditions in t/m^3	0.918	0.897
Oil density under surface conditions in t/m^3	0.924	0.908
Formation volume factor in decimal quantities	1.010	1.032
Sulfur content in oil in %	4.0	3.2
Paraffin content in oil in %	4.3	4.0
Bubble point pressure in MPa	3.0	5.2
Gas factor in m^3/t	4.2	12.3

Figure 3 shows the dynamics of operational parameters within the first year of operation for well 104 and the average for the 5 injection wells of the Ardatovskoye field where water is injected from intake wells. The changes in the injection capacity of well 104 can be described by the $Q_з = f(t)$ function. The injection capacity drop rate is described by the exponential equation $y = 89.268e^{-0.024x}$ with the accuracy value of $R^2 = 0.8709$. The average injection capacity drop rate for the 5 injection wells of the Ardatovskoye field where water is injected from intake wells can be described by the exponential equation $y = 84.644e^{-0.04x}$ with the accuracy value of $R^2 = 0.9707$.

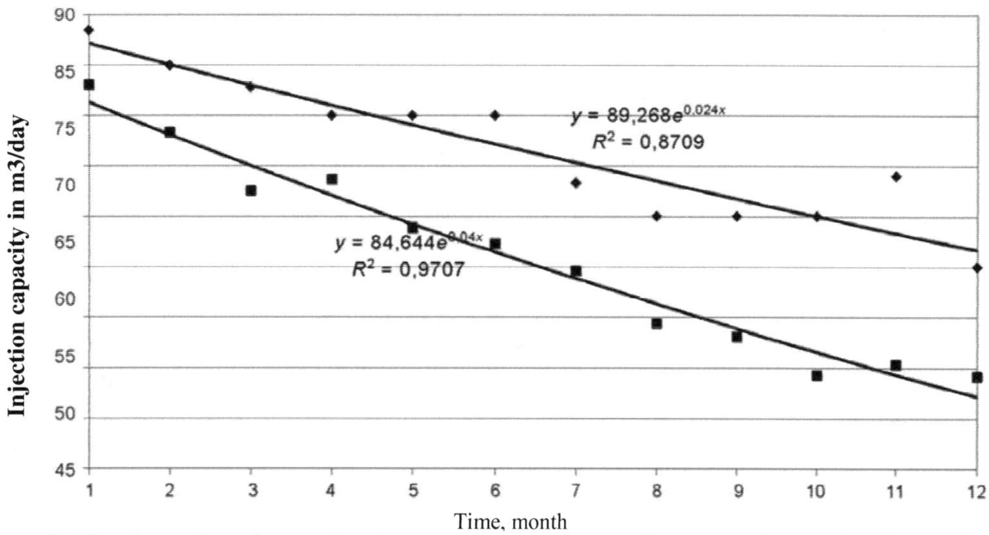

Figure 3. The dynamics of operational parameters within the first year of operation for well 104 at Yubileynoye field and the average for the 5 injection wells at the Ardatovskoye field. ◆ – well. 104 of Yubileynoye field, WPO; ■ – average for the 5 wells of Ardatovskoye field, water intake well injection

After about 12 months, the drop rate of the WPO well is lower than that of the water intake wells. The injection capacity drop of well 104 amounted to about 20 m^3/day, while for other, non-WPO wells, the average drop rate amounted to 30 m^3/day. The injection pressure in the wells in question was not considered because it varied insignificantly and within the measurement error. The use of ECPU allows injecting water under high pressure, which is sometimes impossible to maintain in the water injection system pipelines.

It happens when the increased injection pressure is only necessary for specific wells or the injection is performed at the remote parts of the field that lack the preparation infrastructure and produced water transportation.

4. Conclusion

1. The authors laid bare a correlation between the proportion of mechanical contaminations and oil products in discharge waters and the average oil viscosity in the field. Film thickness depends on the oil viscosity. More viscous oil forms thicker and heavier films on the surface of suspended solids.

2. The use of ECPU for fossil water injection into the well allows using the heat produced by the pump to warm the water.

3. Standard measurement instruments show that the change of injected water temperature at the head of the WPO unit is between 1.1 and 8.4 °C. The average temperature increase is 3.6 °C.

4. After about 12 months, the injection capacity drop rate of the WPO well is lower than that of the water intake wells. The injection capacity drop at a WPO well amounted to about 20 m^3/day, while for other, non-WPO wells, the average drop rate amounted to 30 m^3/day.

References

[1] Muslimov R Kh 2005 *Modern methods of oil recovery increasing: design, optimization and performance evaluation* (Kazan': FEN)

[2] Mukhametshin V V, and Kuleshova L S 2020 On uncertainty level reduction in managing waterflooding of the deposits with hard to extract reserves *Bulletin of the Tomsk Polytechnic University. Geo Assets Engineering* **331(5)** 140–146 DOI 10.18799/24131830/2020/5/2644

[3] Economides J M and Nolte K I 2000 *Reservoir stimulation* (West Sussex, England: John Wiley and Sons)

[4] Stenkin A V, Kotenev Yu A, Mukhametshin V Sh and Sultanov Sh Kh 2019 Use of low-mineralized water for displacing oil from clay productive field formations *IOP Conference Series: Materials Science and Engineering (MEACS 2018 – International Conference on Mechanical Engineering, Automation and Control Systems)* **560(1) 012202** 1-5 DOI: 10.1088/1757-899X/560/1/012202

[5] Valeev A S, Kotenev Yu A, Mukhametshin V Sh and Sultanov Sh Kh Substantiation of the recovery of residual oil from low-productive and heterogeneous formations in Western Siberia by improving the waterflood system using gas and water-gas impacts *IOP Conference Series: Materials Science and Engineering (MEACS 2018 – International Conference on Mechanical Engineering, Automation and Control Systems)* **560(1)** pp 1–6 DOI: 10.1088/1757-899X/560/1/012204

[6] Downey M W, Garvin J, Lagomarsino R C, and Nicklin D F 2011 Quick Look Determination of Oil-in-Place in Oil Shale Resource Plays *AAPG Annual Convention and Exhibition (Houston, Texas, USA, 10-13 April 2011)*

[7] Yakupov R F, Mukhametshin V Sh, Khakimzyanov I N and Trofimov V E 2019 Optimization of reserve production from water oil zones of D3ps horizon of Shkapovsky oil field by means of horizontal wells *Georesursy* **21(3)** 55-61. DOI: 10.18599/grs.2019.3.55-61

[8] Miller G A 2007 Some Perspectives on Various Methods of Oil Shale Extraction, Piceance Basin, Colorado *27th Oil Shale Symposium Colorado School of Mines Golden (Colorado, 15-19 October 2007)*

[9] Yakupov R F, Mukhametshin V Sh, and Tyncherov K T 2018 Filtration model of oil coning in a bottom water-drive reservoir *Periodico Tche Quimica* **15 (30)** 725-733

[10] Andreev A V, Mukhametshin V Sh, Kotenev Yu A 2016 Deposit Productivity Forecast in Carbonate Reservoirs with Hard to Recover Reserves *SOCAR Proceedings* **3** 40-45 DOI: 10.5510/OGP20160300287

[11] Usova L N, Zaydullina E M, and Fakhretdinov R R 2007 *Preliminary discharge of associated water at the Tuimazinsky oil field* (SPb.: Nedra)

[12] Niyazov R S 1984 Additional treatment of waste water used in the PPD system *Oil Industry* **4** 60-62

[13] Mingulov Sh G, Gabdrakhimov N M, and Yusupov I R 2001 Prospects for thermal treatment of oil-bearing reservoirs with deep dissipators *Oil and gas-2001: problems of production, transport, and processing: Intercollegiate collection of scientific papers* Ufa 22-24

[14] Mingulov Sh G 2013 *Recovery of injection well pick-up* (Moscow: Nedra)

IOP Conf. Series: Materials Science and Engineering **952** (2020) 012059 doi:10.1088/1757-899X/952/1/012059

On the influence of the amount and viscosity of oil on the content of solid particles in the water injected into the reservoir

Sh G Mingulov and I Sh Mingulov

Ufa State Petroleum Technological University, Branch of the University in the City of Oktyabrsky, 54a, Devonskaya St., Oktyabrsky, Republic of Bashkortostan, 452607, Russian Federation

E-mail: vsh@of.ugntu.ru

Abstract. Based on the studies carried out, a statistical dependence is obtained for calculating the amount of oil carried over by the discharged water from the viscosity of the extracted oil and the daily loading of the apparatus by liquid, and also a statistical relationship is established between the content of oil and mechanical impurities in the discharged water and the dependence of this ratio from the viscosity of oil.

1. Introduction

As known, water injection into reservoir is one of the ways to improve the efficiency of oil field development [1–8].

However, the exploitation of oil fields using the reservoir pressure maintenance (RPM) method by injecting reservoir water discharged from the fields is inevitably associated with clogging of the bottomhole zone of injection wells with mechanical impurities. In the end, the injectivity of the well may decrease to almost zero if no measures are taken to restore it.

2. Materials and methods

In most cases, the reservoir water of the RPM system is supplied to the receiving pumps of cluster stations from preliminary (track) discharge units, and the quality of this water in terms of the content of suspended solids (SS) and oil depends on the efficiency of the apparatus. Below are the results of the analysis of the amount of SS and oil in the discharged water at the discharge units of a number of Bashneft-Dobycha OOO facilities.

3. Results and Discussion

In [9] it was shown that the loading (productivity) of the apparatus by liquid affects the residual content of petroleum products in the discharged water. It is quite obvious that their amount in water will also depend on the time the liquid stays in the apparatus and on the physical properties of the oil itself. Therefore, when calculating, it is necessary to take into account the specific loading of the preliminary water discharge unit (PWDU), determined by the formula:

$$Q_{sp} = V / Q_{liq}, \text{ day} \tag{1}$$

where V is the useful volume of the discharge apparatus filled with liquid, m^3.

IOP Conf. Series: Materials Science and Engineering **952** (2020) 012059 doi:10.1088/1757-899X/952/1/012059

Formula (1) characterizes the residence time of a unit of liquid volume in the apparatus.

In accordance with the data on the operation of a number of pipe water separators (PWS) and the PWDU, statistical dependences of the amount of petroleum products in the discharged water on the specific load of the apparatus were built.

Figure 1 shows as an illustration a graph for the conditions of water discharge at the fields of Chekmagushneft NGDU. The graph shows that the amount of oil carried away by water increases with an increase in the specific load of liquid discharge units.

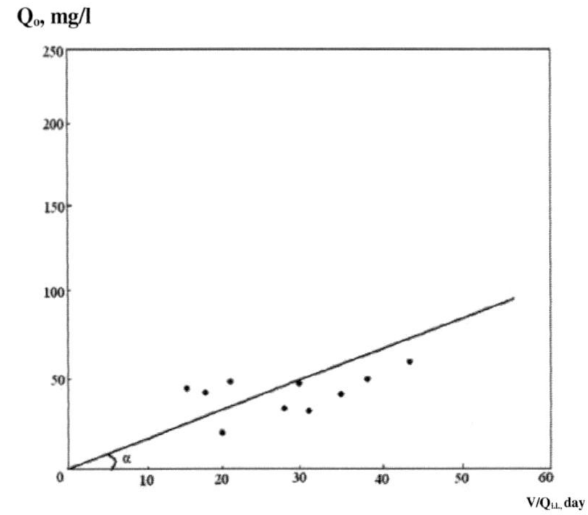

Figure 1. The graph of the dependence of the amount of oil carried away by disharged water from the loading of the apparatus (Chekmagushneft NGDU, $\mu_{av} = 41.8$ mPa · s)

The analysis also showed that the relative amount of oil carryover is different for different regions. The largest carryover value takes place at Krasnokholmskneft NGDU due to the fact that the average viscosity of degassed oil under standard conditions for this region is the highest and is about 45 mPa · s.

Figure 2 shows a graph of the dependence of the relative daily amount of oil phase carryover ($Q_l \cdot q_o / V$) on the oil viscosity. A linear dependence of the daily oil carryover from the apparatus on the oil viscosity is noted ($R^2 = 0.8$):

$$Q_l \cdot q_o / V = 1.08 \cdot 10^{-2} \mu_{av}, \quad (2)$$

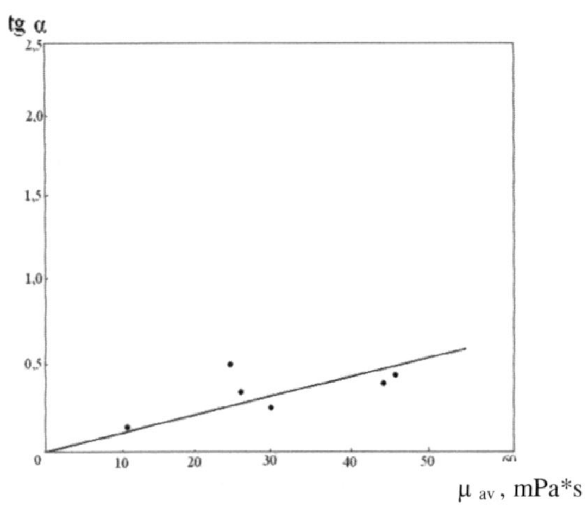

Figure 2. Graph of the dependence of the relative amount of carried over oil on the average value of

its viscosity

Values are deferred on the ordinate axis

$$tg\alpha = \frac{q_o}{V/Q_l} = \frac{q_o \cdot Q_l}{V}.$$

According to formula (2), it is possible to roughly calculate the amount of oil carried over by discharged water at known values of the loading of the apparatus, its volume and average viscosity of degassed oil produced in the region where the apparatus is located.

In the discharged water, along with petroleum products, there are suspended solid particles that are carried out from the reservoirs or entering the extracted liquid from other sources, for example, metal corrosion products, etc. Acceptable concentrations of both oil in water and suspended particles are usually 40–50 mg / l.

Statistical dependence between the amount of oil (q_o) and suspended particles ($q_{mech.\ imp.}$) during the year of operation was obtained for a number of preliminary water discharge units of the Tuimazinsky oil field.

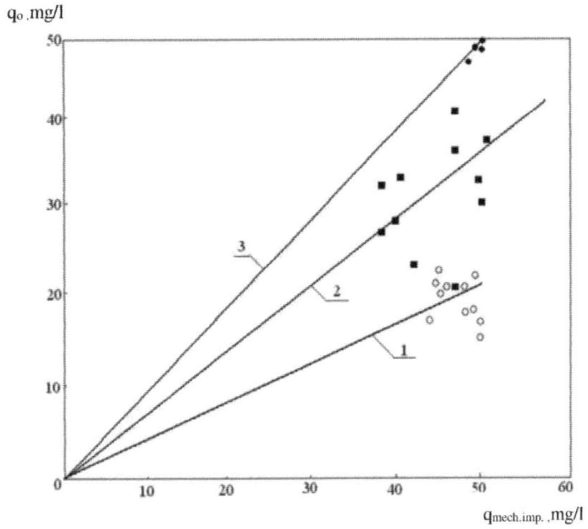

Figure 3. The relation between the amount of oil carried over by water (q_o) and mechanical impurities ($q_{mech.\ imp.}$) for water discharge units: 1 - TVO-20 (D); 2 - BKNS-28 (D); 3 - "Ardatovka" PDU

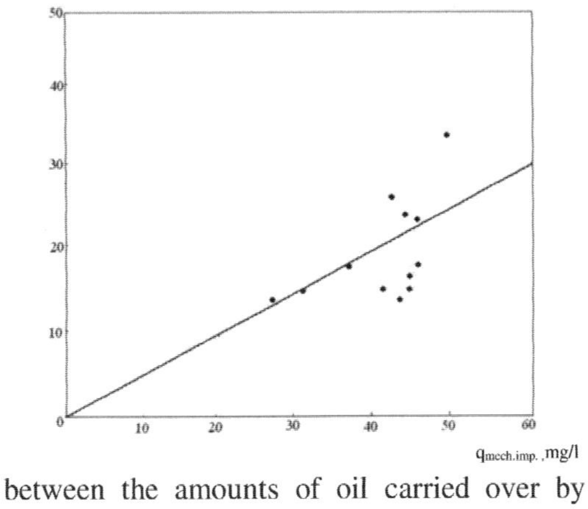

Figure 4. The relation between the amounts of oil carried over by water ($q_o$and mechanical impurities ($q_{mech.imp.}$) for the TVO-29 unit

IOP Conf. Series: Materials Science and Engineering **952** (2020) 012059 doi:10.1088/1757-899X/952/1/012059

Figures 3 and 4 show data for TVO-20 (D), BKNS-28 (D), "Ardatovka" PDU and TVO-29. During the year, the $q_o/q_{mech.imp.}$ ratio for each unit fluctuates. Lines drawn through the groups of points make it possible to determine the averaged $q_o/q_{mech.imp.}$ratios for each discharge unit. The slopes of these lines are different for each unit. For a number of units, a graph of the dependence of the $q_o/q_{mech.imp.}$ ratio on the viscosity of oil in the apparatus is built (Figure 5).

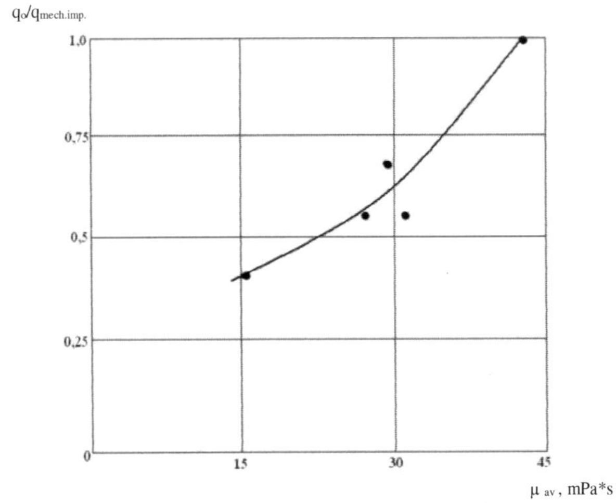

Figure 5. Graph of dependence of the ratio of oil and suspended particles in the discharged water on the oil viscosity

The resulting relation in the viscosity range of 15–45 mPa · s is described by empirical dependence ($R^2 = 0.7$):

$$q_o/q_{mech.imp.} = 0.2\ \exp(0{,}04\mu_{av}). \qquad (3)$$

Formula (3) allows calculating the amount of suspended particles in the discharged water for a given production region by the amount of oil carried over.

The results obtained show a close relationship between the amounts of oil and mechanical impurities carried over. This indicates that the oil phase, carried over by water, is mainly located on solid particles, enveloping them due to adhesion forces.

Thus, a solid particle covered with a layer of oil loses the possibility of sedimentation due to the fact that the oil part tends, on the contrary, to float. It is obvious that there is a certain thickness of oil film on a solid particle, at which the average density becomes equal to the density of water and the solid particle loses its ability to float or settle.

In practice, such water has a darkish tint, it is called "black water" and is a big problem in water treatment [10].

The research results show a significant deterioration in the quality of discharged water with an increase in the viscosity of the extracted oil. Together with this, an increase in viscosity, ceteris paribus, leads to a decrease in the degree of destruction of the emulsion at the inlet to the apparatus and to an increase in the amount of residual water in the oil discharged from the apparatus.

Therefore, when preparing oils of high viscosity, the use of demulsifiers alone is not effective enough to ensure the quality of the discharge [11–13]. At the same time, more effective measures are needed to prevent, first of all, intensive emulsification of watered oil in wells equipped with electric centrifugal pumps.

4. Conclusion

1. A statistical dependence is obtained for calculating the amount of oil carried over by the discharged water from the viscosity of the extracted oil and the daily loading of the apparatus by liquid.

2. A statistical relation between the content of oil and mechanical impurities in the discharged water and the dependence of this ratio from the viscosity of oil are established.

References

[1] Chudinova D Yu, Kotenev Yu A, Sultanov Sh Kh, and Mukhametshin V Sh 2018 The neural network for grouping wells of different facies zones of productive layer *IOP Conference Series: Earth and Environmental Science (IPDME 2018 – International Conference on Innovations and Prospects of Development of Mining Machinery and Electrical Engineering)* **194(8)** 1-6 DOI: 10.1088/1755-1315/194/8/082008

[2] Behrmann L, Huber K, McDonald B, Couët B, Dees J, Folse R, Handren P, Schmidt J, and Snider P 1996 Quo Vadis, Extreme Overbalance? *Oilfield Review* 18-33

[3] Mukhametshin V V, and Kuleshova L S 2020 On uncertainty level reduction in managing waterflooding of the deposits with hard to extract reserves *Bulletin of the Tomsk Polytechnic University. Geo Assets Engineering* **331(5)** 140–146 DOI 10.18799/24131830/2020/5/2644

[4] Yakupov R F, Gimazov A A, Mukhametshin V Sh, and Makaev R I 2018 Analytical method for estimating efficiency of oil recovery technology in case of bottom water-drive reservoir, verified on the hydrodynamic model *Oil Industry* **6** 66-69 DOI: 10.24887/0028-2448-2018-6-66-69

[5] Akhmetov R T, Mukhametshin V V, and Andreev A V 2017 A quantitative assessment method of the productive formation wettability indicator according to the data of geophysical surveys *SPE Russian Petroleum Technology Conference (16-18 October 2017, Moscow)* 12 p DOI: 10.2118/187907-MS

[6] Lestz R S, Clarke J N, Plattner D, and Byrd A C 2002 Perforating for Stimulation: An Engineered Solution *SPE Drilling & Completion* **17(1)** 8 p DOI: 10.2118/76812-PA

[7] Yakupov R F, Mukhametshin V Sh, Zeigman Yu V, Chervyakova A N, and Valeev M D 2017 Metamorphic aureole development technique in terms of Tuymazinskoye oil field *Oil Industry* **10** 36–40 DOI: 10.24887/0028-2448-2017-10-36-40

[8] Kadyrov R R, Nizaev R Kh, Yartiev A F, and Mukhametshin V V 2017 A novel water shut-off technique for horizontal wells at fields with hard-to-recover oil reserves *Oil Industry* **5** 44–47 DOI: 10.24887/0028-2448-2017-5-44-47

[9] Usova L N, Zaydullina E M, and Fakhretdinov R R 2007 *Preliminary discharge of associated water at the Tuimazinsky oil field* SPb.: Nedra 128 p

[10] Golubev M V, Minigalimov R Z, Usova L N, and Safonov V E 2007 Fundamentals of designing installations for preliminary water discharge in the production of watered oil *Electronic scientific journal "Oil and Gas Business"* **1** p 88

[11] Khaziev N N, Gazizov M G, Vildanov R G, and Golubev V F 1997 Pat. 2077918 Russian Federation CPC B01D 17/028 *Apparatus for dehydration of crude oil* proprietor Khaziev N N No. 94041531/25 field 17.11.1994 published 27.04.1997 bull. no. **12** 6 p

[12] Hutton A C 1987 Petrographic classification of oil shales *International Journal of Coal Geology* **8(3)** 203-231 DOI: 10.1016/0166-5162(87)90032-2

[13] Downey M W, Garvin J, Lagomarsino R C, and Nicklin D F 2011 Quick Look Determination of Oil-in-Place in Oil Shale Resource Plays *AAPG Annual Convention and Exhibition (Houston, Texas, USA, 10-13 April 2011)* 21 p

IOP Conf. Series: Materials Science and Engineering **952** (2020) 012060 doi:10.1088/1757-899X/952/1/012060

Restoring the intake capacity of injection wells by means of shock-wave exposure

Sh G Mingulov and V Sh Mukhametshin

Ufa State Petroleum Technological University, Branch of the University in the City of Oktyabrsky, 54a, Devonskaya St., Oktyabrsky, Republic of Bashkortostan, 452607, Russian Federation

E-mail: vsh@of.ugntu.ru

Abstract. New technology is proposed for restoring the intake capacity of injection wells, providing an opportunity to reduce the operation time, the amount of household waste, equipment used, the degree of environmental pollution and land loss in the absence of round trip operations.

1. Introduction

In order to maintain the design indicators of oil production during the development of oil fields, it is important to restore the intake capacity of injection wells in the conditions of using a reservoir pressure maintenance system (RPM) by produced water injecting [1-8]. Over time, the pore space of the bottom-hole zone of injection wells is clogged with oil and suspended particulate matters (SPM).

Technologies based on the creation of draw-down pressure are used to restore the intake capacity [9]. They differ in complexity of implementation and low technical and economic efficiency.

In [10], a new method of restoring the intake capacity, based on the creation of instantaneous draw-down pressures by relieving pressure at the wellhead. A rarefaction wave reaching the bottom causes a pressure pulse and the colmatant is detached from the rock. The induced inflow during repeated exposures brings the colmatant to the surface with an outpouring of liquid.

Hydro-swabbing, or the variable pressure method [11], is performed by periodically increasing the draw-down pressure with a liquid, preventing hydraulic fracturing, followed by the rapid opening of the wellhead. The effectiveness of the method is to create a draw-down pressure. Alternating significant pressure gradients destroy the structural bonds of emulsions and deposits in the pores of the bottom-hole zone, and high flow rates contribute to the removal of contaminants into the well-bore.

The basis of pulse-wave exposure on the reservoir can be considered a hydro-swabbing method, when applied, the wellhead pressure of the injected or poured liquid is sharply changed to create shock waves propagating through the column. In the case of shock wave transmission from the surface, the well fluid, most of which is water, due to its high density and low compressibility, is a favorable medium for the transmission of pressure drops over long distances.

It is generally accepted that for greater impact on the bottom-hole zone of the well, it is necessary to place equipment in the immediate vicinity of the processing object. The location of energy sources on the earth surface is recognized as non-optimal, primarily due to the refraction of waves in the earth stratum and their attenuation. This is true for the transmission of heat energy, for example, because the casing and tubing strings have high thermal conductivity and dissipate heat.

IOP Conf. Series: Materials Science and Engineering **952** (2020) 012060 doi:10.1088/1757-899X/952/1/012060

The tubing walls with considerable elasticity, contribute to a kind of "canalization" of the wavefront trajectory. This is true for infra-low frequency waves with low attenuation. Losses from the attenuation of Rayleigh waves and transverse waves, which are the main losses in the seismic method, are excluded.

The speed of the shock wave in the column is close to the sound speed. Pipes, having a large mass and, therefore, high inertia, do not have time to deform due to pressure drops during the passage of a shock wave. Therefore, gusts of the column and destruction of the cement stone are unlikely and were not observed during the testing process.

Shock-wave exposure forms processes in the oil reservoir that cannot be explained only by the influence of pressure and temperature. The shock wave affects the molecular chains, resulting in local overloads that can exceed the strength of the chemical bond.

The following options for creating impacts at the wellhead with subsequent energy transfer using a shock wave to the bottom have been developed [12]:

- propagation of molecular-wave vibrations along the metal of the pipe string, contributing to the separation of asphalt-resin-paraffin deposits (ARPD);
- use of an emitter at the wellhead that is mechanically connected to the tubing system to transfer energy to the perforation interval;
- implementation of a water hammer in the upper part of the column, its transmission along the liquid column and deviation in the perforation zone;
- alternation of periodic pressure pulses with well flushing;
- formation of a standing shock wave by alternating pressure and rarefaction at the wellhead;
- the formation of pressure drops between the near-wellbore area of the reservoir and the cavity of the wells with a gradual and uniform cleanup of the well-bore zone along the entire length of the perforated interval;
- the formation of pressure drops between the near-wellbore area of the reservoir and the cavity of the wells by periodic opening in time with the wellhead for the outflow of well-bore fluid and its filling;
- the formation of pressure drops between the near-wellbore area of the reservoir and the cavity of the wells by periodic programmable wellhead opening time;
- the amplification of the shock wave by using hydro-pneumatic accumulators for energy storage;
- pulse pumping of the working agent is alternated with exposure and dynamic spout;
- generating an elastic shock wave at the wellhead by pulsed injection of a metered volume of liquid and maintaining vibrations of the liquid column;
- the formation of pressure drops between the near-wellbore area of the reservoir and the cavity of the wells by periodic opening in time with the speed that provides a sharp drop in pressure at the wellhead to atmospheric, and closing – at a speed 2-5 times slower than the speed of its opening;
- synchronizing the closing and opening of the wellhead in phase with periodic decreases and increases in pressure in the bottom-hole zone during the formation of a shock wave, installing a check valve above the packer to allow flushing of the well between pulse-wave processing operations;
- pulsed injection through the annulus and spout through the tubing during the formation of a shock wave in wells with low reservoir pressure.

Most of these solutions offer to increase the pressure in the well and creating hydraulic shocks in the fluid column by periodically switching the wellhead equipment. In this case, the column of borehole fluid is subjected to a shock impact that extends to the bottom and can affect the filtration properties of the near-wellbore area of the formation.

2. Materials and methods

In comparison with other methods, the treatment of wells by pulsed-wave action from the surface without the use of lifting operations is characterized by low material and time costs. No expensive equipment is required. Concerning the special equipment, only modified wellhead strapping is used, in some cases a shock pulse generator and a pumping unit. The duration of treatment is usually from 3 to

8 hours.

Water impacts that are consistent in frequency with the speed of the shock wave and the depth of the well are amplified by resonance. Attenuation losses for infra-low frequency waves are on the order of 10-12 % per kilometer of the well length.

Technologies based on the listed patents provide an opportunity for directional processing of inter-layers that are not uniform in permeability to align the inflow or injectability, as well as selective processing of watered inter-layers. Pulse-wave processing helps to reduce the water content of the bottom-hole zone (BHZ).

The power of the shock wave, that is generated when the wellhead is opened, is proportional to the rate of liquid outflow, which, in turn, depends on the difference in down-hole and atmospheric pressure and the cross-section of the channel for liquid outflow.

According to the calculation, initially, the outflow rate will be 141 m/s. Practically, this speed is lower because it depends on the hydraulic resistance of the discharge hose. Due to the decrease in pressure, the spout rate decreases, which reduces the power of the water shock wave. On the supposition that the difference in the velocity of the liquid outflow during the formation of the water hammer was 20 m/s. Then the power of the shock wave formed at the well-head will be 1200 kW.

In the process of moving through the tubing, the energy of the pressure wave movement is significantly reduced due to hydraulic friction, which depends mainly on the column cross-section, resulting in the loss of pressure of the moving liquid. When the sump is reached, the kinetic energy of the liquid is converted into the potential energy of its elastic rarefaction, i.e. the pressure wave creates a new water hammer, that directly affects the near-wellbore area of the reservoir. The duration of the rarefaction stage does not depend on the force of the water hammer, but is determined only by the time of propagation of the shock wave along the column.

The water hammer generated by the spout of the well fluid develops a rarefaction wave, not a compression wave. Therefore, the spout helps to maintain the high-speed pressure of the shock wave.

The figure shows the wellhead pressure waveforms obtained using the PDMTI-60 pressure sensor and «Software for Vellemen PC scope, PCS64i» program.

3. Results and Discussion

Figure 1 *a* shows that when the well is depressurized, the wellhead pressure drops to atmospheric pressure, and then, when the well is closed, it is almost restored. After a period of time required for the wave to pass to the sump and back, a pressure pulse is formed at the wellhead, which in amplitude can be greater than the differential that generates the wave. The passage of a single pulse is accompanied by damped pressure fluctuations.

Figure 1*b* shows repeated pulses generated during regular depressurization of the wellhead. It is important to open the wellhead at the moment of the maximum amplitude of the reflected pulse to obtain the greatest pressure drop moving through the well.

Figure *c* shows the formation of a wave in a well with a high permeability of the bottom-hole zone. High attenuation is associated with a significant transfer of wave energy to the reservoir.

If there is the pulse-wave exposure from the wellhead, under the condition of sufficient pressure in the cavity wells, the repetition rate of the pulse-wave exposure is 12-20 times per minute and is determined by the speed of the shock wave and the depth of the well.

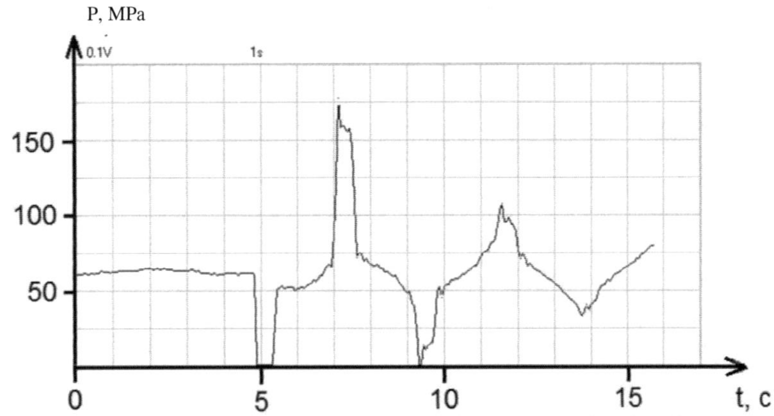

Figure 1. Wellhead pressure diagrams: a - for a short single spout; b - for short repeated spouts; c - for a short single spout in a well with high permeability of the near-wellbore area of the reservoir

The coverage area of the hydraulic shock is limited to rocks adjacent to the well since the force of the hydrodynamic impulse quickly fades as it moves away due to the high elastic capacity of the system and hydraulic resistances in the pores of the rocks. The depression effect caused by a rarefaction shock wave is nondurable due to the high speed of the wave movement. Cracks in the formation do not have time to deform. The rarefaction wave, moving along the well cavity from the wellhead to the bottom and back, creates differences in the bottom-hole zone. When processing, for example, an injection well, the wave does not penetrate into the depth of the formation, the shock wave energy is transmitted only to the surface layer of rock in the perforation zone, the loose crust of deposits is torn off from the surface of the bottom-hole zone, the formation is not processed in depth, although for most injection wells such processing is sufficient and gives satisfactory results.

Processing of pores and cracks in the depth of the formation is possible with intensive fluid movement through the bottom-hole zone. Considering that the impact of elastic waves separates rock particles from the walls of the cracks due to vibrations near the center of equilibrium, but does not move in space, a powerful alternating action is required that has energy sufficient to carry the bridging agent from the formation into the well - the "washing" effect.

The shock wave exposure from the wellhead compares favorably with the implosion technology in that when using the latter, after each single shock exposure, it is necessary to lift the column to replace the membrane or the destroyed capsule or create a vacuum in the container opposite the perforation without lifting it to the surface. From the wellhead, it is possible to repeat the pulse-wave exposure with a frequency determined by the time of bringing the source of hydraulic impulses to its original state.

4. Conclusion

When processing wells due to the non-use of lifting operations, the advantages of the technology are as follows:

- due to the short duration of the operation, there is no need to arrange household premises. Consequently, the amount of household waste is reduced;
- the absence of down-lifting operations can dramatically reduce the amount of equipment used in the well, which pollutes the soil and atmosphere;
- since most technologies use 2-3 cars, there is little air pollution and ground damage during transportation of equipment and people.

IOP Conf. Series: Materials Science and Engineering **952** (2020) 012060 doi:10.1088/1757-899X/952/1/012060

References

[1] Miller G A 2007 Some Perspectives on Various Methods of Oil Shale Extraction, Piceance Basin, Colorado *27th Oil Shale Symposium Colorado School of Mines Golden (Colorado, 15-19 October 2007)*

[2] Mukhametshin V V, and Kuleshova L S 2020 On uncertainty level reduction in managing waterflooding of the deposits with hard to extract reserves *Bulletin of the Tomsk Polytechnic University. Geo Assets Engineering* **331(5)** 140–146 DOI 10.18799/24131830/2020/5/2644

[3] Kuleshova L S, Mukhametshin V V and Safiullina A R 2019 Applying information technologies in identifying the features of deposit identification under conditions of different oil-and gas provinces *Journal of Physics: Conference Series (ITBI 2019 – International Conference "Information Technologies in Business and Industry")* **1333 (7)** 1–5 DOI: 10.1088/1742-6596/1333/7/072012

[4] Kimos A 2011 Engineering Natural Gas from U.S. Shale Deposits, available at: http://buildipedia.com/aec-pros/engineering-news/extracting-natural-gas-from-shale

[5] Akhmetov R T, Mukhametshin V V and Kuleshova L S 2019 Simulation of the absolute permeability based on the capillary pressure curves using the dumbbell model *Journal of Physics: Conference Series (ITBI 2019 – International Conference "Information Technologies in Business and Industry")* **1333 (3)** 1-8 DOI: 10.1088/1742-6596/1333/3/032001

[6] Kuleshova L S, Mukhametshin V V 2019 Elimination of uncertainties in predicting well interaction using indirect geological field information *IOP Conference Series: Earth and Environmental Science (IPDME 2019 – International Workshop on Innovations and Prospects of Development of Mining Machinery and Electrical Engineering)* **378 (1) (012115)** 1-8 DOI: 10.1088/1755-1315/378/1/012115

[7] Mukhametshin V V 2018 Rationale for trends in increasing oil reserves depletion in Western Siberia cretaceous deposits based on targets identification *Bulletin of the Tomsk Polytechnic University. Geo Assets Engineering* **329(5)** 117–124

[8] Economides J M, and Nolte K I 2000 *Reservoir stimulation* West Sussex, England: John Wiley and Sons 856 p

[9] Veres S P, Demchenko V A, Drozdov V N, Slidenko V M, Semenenko I A, Verba Yu V 1999 Pat. 2138617 Russian Federation CPC E21B 37/00, 28/00 *Device for cleaning of bottom-hole zone of bed* proprietor CJSC Scientific-Research Center "Yuganskneftegaz" No. 98119208/03 field 26.10.1998 published 27.09.1999 bull. no. **27** 1–10

[10] Nosov P I, Senochkin P D, Nurislamov N B, Zakiev M G, Minnullin R M 2000 Pat. 2159326 Russian Federation CPC E21B 43/25 *Process and gear for development and clearing of face zone of wells by pulse drainage* proprietor CJSC RITEK-implementation, LLC Scientific and Production company "Piter" No. 99126449/03 field 15.12.1999 published 20.11.2000 bull. no. **32** 1–16

[11] Gabdrakhmanov N Kh, Mingulov Sh G, Galiullin T S, Malets O N, and Yakupov R F 2000 Production of hard-to-recover residual oil at the Tuimazinsky oil field in the late development stage *Actual problems of oil production at the fields of NGDU "Tuymazaneft": collection of scientific papers* Ufa 14-16

[12] Shipulin A V, and Mingulov Sh G 2011 *Technologies of wave action on oil reservoirs* (Saint-Petersburg: Nedra)

IOP Conf. Series: Materials Science and Engineering **952** (2020) 012061 doi:10.1088/1757-899X/952/1/012061

Bottomhole zone treatment by creating pressure pulses at the wellhead of low-viscosity oil pools

Sh G Mingulov and I Sh Mingulov

Ufa State Petroleum Technological University, Branch of the University in the City of Oktyabrsky, 54a, Devonskaya St., Oktyabrsky, Republic of Bashkortostan, 452607, Russian Federation

E-mail: vsh@of.ugntu.ru

Abstract. It is shown that when a borehole fluid surface is impacted by a compressor and receiver, the wave propagates to the bottom hole zone, where its amplitude increases due to the addition of the reflected wave. It was found that if it is necessary to obtain high speeds, it is more expedient to pump down of the liquid using more accessible hydraulic equipment, such as a pumping unit. It is noted that the use of simple and inexpensive supporting equipment allows creating a fluid flow at the wellhead with the required speed.

1. Introduction

To increase the efficiency of oil field development are widely used methods of stimulating production [1-8].

Many years of experience in applying various technologies to reduce flow coefficients in the bottom hole formation zone (BHFZ) have shown high efficiency of hydraulic fracturing (HF). With successful HF is possible multiple increases in the filtration rate in the BHFZ [9, 10].

Despite its high efficiency, the HF method requires significant material costs for technical, technological, and geophysical support of the event.

To increase the size of existing cracks and perforation channels, it is advisable to use their "pacing" by periodically alternating the growth and reduction of the movement speed and pump pressure of the fracturing fluid. With a pressure surge in the treated area of the reservoir, the fracturing fluid penetrates the reservoir channel, acting as a wedge that pushes the rock apart and forms fractures in it. With repeated hydraulic preloading of the rock sample, the stress limits naturally decrease. In the case of existing microfractures in rocks, their further development occurs at significantly lower fracturing fluid pressures [11].

2. Materials and methods

It is possible to create pressure pulses at the bottom hole directly from the wellhead. A hydraulic shock pressure pulse is created using wellhead equipment without performing descent operations. Short-term pressure applications on the wellhead, creating a resonant movement of the compacted mass of the borehole fluid column, lead to compression of the underlying strata and their movement, therefore, to an increase in the moving mass, which, when the sump is reached, creates a pressure pulse that exceeds the wellhead in amplitude.

The energy of a hydro shock is determined by its duration, which depends on the depth and column stiffness. The impact of power depends on the pressure rate change in the perforation zone, which in turn is proportional to the flow rate at the time of stopping. The pressure build-up at the time of the hydraulic shock corresponds to a certain flow rate, but the duration of the hydraulic shock, and therefore its total energy increases with well depth. To preserve the casing integrity of the string and the hydrated cement behind it, it is necessary to limit the power of the hydraulic shock, which means the maximum pressure exerted on the column walls.

The advantages of the hydraulic shock method of impact on the well bore zone of the formation in comparison with the constant pressure used in HF are that:

- the hydraulic shock method creates a new network of fractures, which opens up additional drain in the bottom hole zone and provides a significant increase in the inflow or capacity of formations;
- the hydraulic shock use during HF contributes to the creation of large short-term pressures in the perforation zone, which cannot be created with a stationary flow of liquid due to pressure restrictions on pumping units and tubing strings;
- pulsed pressure creates such fractures that, due to the irreversibility of the rock deformation process, do not completely close under the influence of rock pressure without the use of a propping agent.

The borehole fluid can be set in motion by the impact of pressure gas (air or nitrogen) from the wellhead or by the liquid pumped by the pumping unit. The pumping unit design does not allow to sharply increase the flow rate of the injected fluid. For example, using the CA-320 unit, it is necessary to "gas" after opening the fluid supply valve to the well to increase the flow rate of the injected fluid as quickly as possible.

3. Results and Discussion

Figure 1 shows diagrams of pressure changes in the bottom-hole zone of the well during gas injection (see figure 1, *a)* or fluid injection (see figure 1, *b*). The diagrams were made using an ACM-4s submersible pressure gauge.

In both cases, the rate of pressure rise in the bottom–hole zone is 2.25-5.0 MPa/min. With such a low-pressure rise in the bottom-hole zone, wave processes that affect the productivity of the well bore zone are impossible. It is necessary to use special wellhead equipment that allows you to set the required speed of the injected medium. For example, in [12] it is proposed to give the necessary speed to the well fluid due to the discharge of the carbonic gas cylinder.

When the blast effect of gas impacts the surface of the borehole fluid, the wave propagates to the bottom-hole zone, where its amplitude increases due to addition with the reflected wave. Figure 2 shows the wave interference when the settling pond is reached. The diagram was obtained using a submersible pressure gauge in the settling pond of a well with a depth of 1270 m. In this case, a gas cylinder with a gas pressure of 4 MPa is discharged at the wellhead. The diagram shows that the pressure in the bottom-hole zone due to the addition of direct and reflected waves increases to 6.1 MPa. A similar effect can be obtained when pumping fluid into the well, which is more important since the compressor in the oil fields is more scarce equipment than the pumping unit. An example of a technological scheme for creating a high-speed pumping down is shown in figure 3.

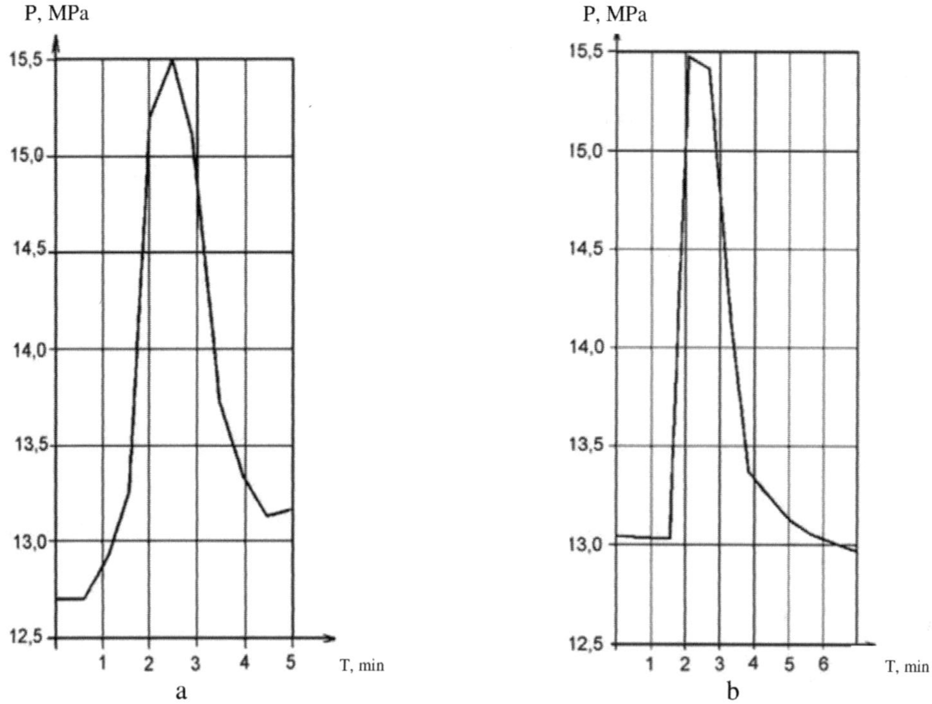

Figure 1. Diagrams of pressure changes in the bottom-hole zone when gas (a) or fluid (b) is injected into the well

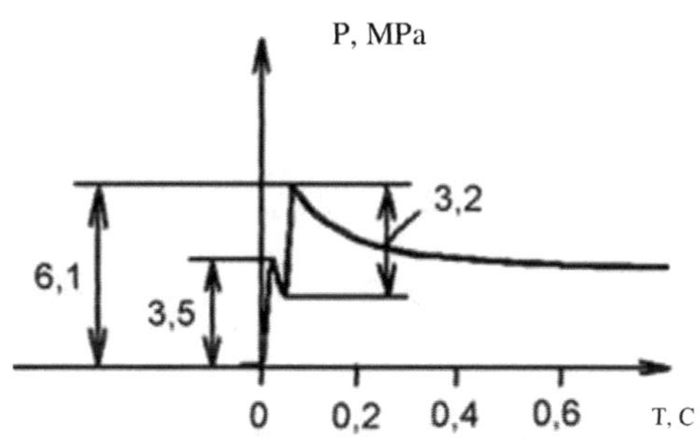

Figure 2. Diagram of pressure changes in the bottom-hole zone when a gas cylinder is discharged into the well

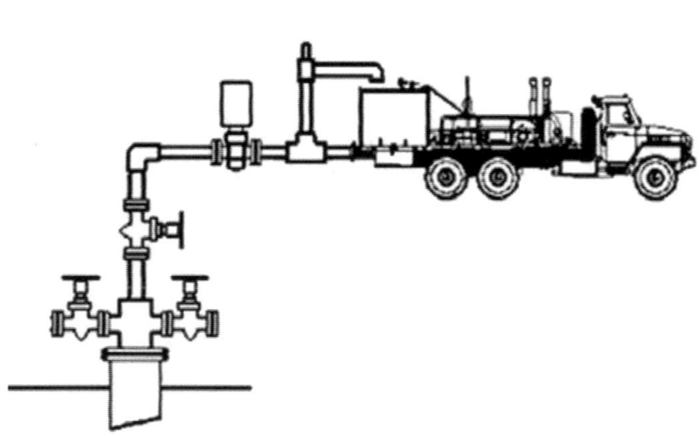

Figure 3. Example of implementation of the technology of wave action by a fluid

Between the output of the pumping unit and its gauge tank, a safety valve SPRV-4 is installed, which has an actuating area: 5.3–9.0; 8.5–12.4; 12.4–14.1; 14.0–16.0 MPa. First, the quick-acting valve connecting the pumping unit, and the wellhead is closed.

When the unit starts working, the pressure assembly at the outlet increases to the operating value of the safety valve, then the valve opens, and the liquid is poured into the gauge tank. The safety valve pressure is maintained at the outlet of the pumping unit. When the quick-acting valve is opened, the pressure at the outlet of the unit decreases, the safety valve closes, and all the fluid energy from the pump running at full capacity is transferred to the oil-well tubing.

The use of a pumping unit with high-speed and safety valves allows you to create a high speed of fluid injection, for example, for intensive cleaning of the bottom-hole zone of the injection well.

4. Conclusion

1. When the borehole fluid surface is impacted by a compressor and receiver, the wave propagates to the bottom hole zone, where its amplitude increases due to the addition of the reflected wave.

2. It is necessary to obtain high speeds, it is more expedient to pump down of the liquid using more accessible hydraulic equipment, such as a pumping unit.

The use of simple and inexpensive supporting equipment allows creating a fluid flow at the wellhead with the required speed.

References

[1] Economides J M, and Nolte K I 2000 *Reservoir stimulation* (West Sussex, England: John Wiley and Sons)

[2] Khuzina L B, Mukhametshin V S, Tyncherov K T and Shaikhutdinova A F 2018 On the choice of the oscillators' installation site *International Journal of Civil Engineering and Technology* **9 (9)** 1952-1959

[3] Mukhametshin V V, and Kuleshova L S 2020 On uncertainty level reduction in managing waterflooding of the deposits with hard to extract reserves *Bulletin of the Tomsk Polytechnic University. Geo Assets Engineering* **331(5)** 140–146 DOI 10.18799/24131830/2020/5/2644

[4] Zeigman Yu V, Mukhametshin V Sh, Khafizov A R, Kharina S B, Abutalipova E M, and Avrenyuk A N 2017 Peculiarities of selecting well-killng fluids composition for difficult conditions *Oil Industry* **1** 66–69

[5] Alvarado V, Stirpe M, La Roque C, Ponce R, and Farias M 2002 Streamline Simulation for Enhanced-Oil Recovery: Review and Laboratory Tests *Proceedings of INGEPET, EXPL-4-VA-84*

[6] Kuleshova L S, Mukhametshin V V and Safiullina A R 2019 Applying information technologies in identifying the features of deposit identification under conditions of different oil-and gas provinces *Journal of Physics: Conference Series (ITBI 2019 – International Conference "Information Technologies in Business and Industry")* **1333 (7)** 1–5 DOI: 10.1088/1742-6596/1333/7/072012

[7] Batalov S A, Andreev V E, Lobankov V M and Mukhametshin V Sh 2019 Numerical simulation of oil formation with regulated disturbances. Oil recovery quality simulation *Journal of Physics: Conference Series (ITBI 2019 – International Conference "Information Technologies in Business and Industry")* **1333(3)** 1-6 DOI: 10.1088/1742-6596/1333/3/032006

[8] Akhmetov R T, Mukhametshin V V and Kuleshova L S 2019 Simulation of the absolute permeability based on the capillary pressure curves using the dumbbell model *Journal of Physics: Conference Series (ITBI 2019 – International Conference "Information Technologies in Business and Industry")* **1333 (3)** 1-8 DOI: 10.1088/1742-6596/1333/3/032001

[9] Bou-Mikael S, Asmadi F, Marwoto D, and Cease C 2000 Minas Surfactant Field Trial Tests Two Newly Designed Surfactants with High EOR Potential *SPE Asia Pacific Oil and Gas Conference and Exhibition (Brisbane, Australia, 16-18 October 2000)* 12 p DOI: 10.2118/64288-MS

[10] Briones M J, Zambrano A, and Zerpa C 2002 Study of Gas-Condensate Well Productivity in Santa Barbara Field, Venezuela, by Well Test Analysis *SPE Annual Technical Conference and Exhibition (San Antonio, Texas, 29 September-2 October 2002)* 9 p DOI: 10.2118/77538-MS
[11] Shipulin A V, Mingulov Sh G, and Gorbatov V N 2006 Methods of creating hydrostrikes in a well *Actual problems of oil and gas business: collection of scientific papers* Ufa: USPTU **2** 177–181
[12] Balashkand M I, Andreev Yu N, Kaznin V A 1990 Treatment of bottom-hole zone of a formation by pulse pressure *Oil Industy* **8** 71-73

IOP Conf. Series: Materials Science and Engineering **952** (2020) 012062 doi:10.1088/1757-899X/952/1/012062

Dynamics control of a drilling tool by installing a downhole hydromechanical compensator of pressure fluctuations of drilling fluid into the bottom hole assembly

T N Minnivaleev, I G Arslanov, L S Kuleshova and A N Minnivaleev

Ufa State Petroleum Technological University, Branch of the University in the City of Oktyabrsky, 54a, Devonskaya St., Oktyabrsky, Republic of Bashkortostan, 452607, Russian Federation

E-mail: timxn@yandex.ru

Abstract. The fluctuation processes of the drilling tool and the pressure of the drilling fluid resulted from numerous factors of both downhole and wellhead nature lead to many negative consequences. The article describes the impact of non-uniformity of the pressure of the drilling fluid on the drilling process, provides an analytical study of the developed downhole hydromechanical compensator of pressure fluctuations of the drilling fluid installed into the bottom hole assembly. The authors obtained an analytical dependence of the degree of decrease in the amplitude of the pressure fluctuations of the drilling fluid in the compensator on the number of damping stages installed in it. The article describes the results of pilot tests of the compensator, indicating the reliability of the theoretical provisions and the high efficiency of the compensator.

1. Introduction

With the increasing requirements for the economic efficiency of the drilling process, it is important and urgent to improve the bottom hole assembly to increase the durability of drilling equipment, reduce the costs of tripping operations and increase the technical and economic performance of drilling as a whole. One of the directions of increasing drilling efficiency is to reduce the dynamics of the drilling tool during well drilling and thereby increase the mechanical drilling speed and durability of the drilling tool.

2. Materials and methods

Solving the hydrodynamic task, we used the basic laws of theoretical mechanics and hydromechanics as applied to the adopted technological scheme for installing the compensator into the drilling tool assembly, as well as empirical methods.

3. The contents of the method and evaluation of its effectiveness

The modern mud pumps supply pulsed fluid due to their design features. To reduce feed irregularities, they are made multi-cylinder, and the cranks are positioned at an angle to each other: in double-cylinder double-acting pumps the angle is 90° and in three-cylinder single-acting pumps the angle is 120°. A further increase in the number of cylinders does not lead to a significant reduction in feed non-uniformity, while it complicates the design of the pumps.

The non-uniformity of the pump supply depends on the number of cylinders, the principle of operation, the angle of the crank displacement, the operation of the valve mechanism, etc. Figures 1 and 2 graphically show the flow of a double-cylinder double-acting pump and a three-cylinder single-acting pump.

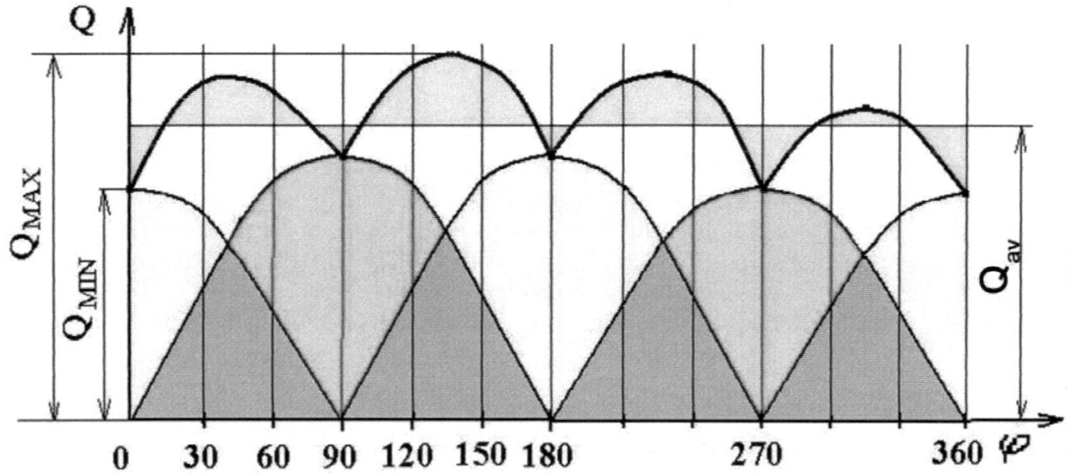

Figure 1. Supply work of a double-cylinder double-acting pump: Q - pump supply; Q_{max} - maximum pump flow; Q_{min} - minimum pump flow; Q_{av} - average pump flow; φ is the rotation angle of the crank.

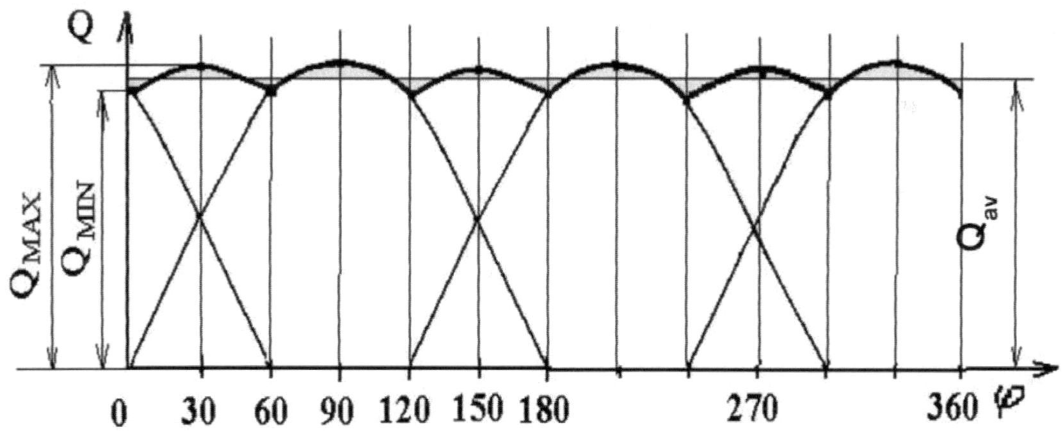

Figure 2. Supply work of a three-cylinder single-acting pump: Q - pump flow; Q_{max} - maximum pump flow; Q_{min} - minimum pump flow; Q_{av} - average pump flow; φ is the rotation angle of the crank.

In some cases, drilling conditions require the simultaneous use of two mud pumps. It can lead to the appearance of a pressure beating that occurs when two periodic fluctuations fold, for example, harmonic fluctuations with the same amplitudes and slightly different frequencies. The difference in the tension of the V-belt transmission connecting the electric motors to the pumps, as well as the relative slippage on the pulleys can explain the difference in the frequencies of the pressure fluctuations generated by the mud pumps. As a result, at the moments when these fluctuations occur sinphasically, the total amplitude will be maximum, and at the moment when they are in antiphase, they mutually cancel out.

The non-uniformity of the pressure of the drilling fluid while drilling leads to increased wear of the drilling equipment, a decrease in the motor resource of downhole motors, longitudinal and torsional vibrations of the drill string, as well as many negative consequences in the well [1-6]. It is important to note that due to forced tool fluctuations, up to 25-30% of the total power developed by the downhole motor is lost. Energy commensurate with the energy used to destroy the rock face can be spent only on the vibration of the drill collar, even without considering the friction forces. With an increase in the coefficient of dynamism, the longitudinal fluctuations of the bit and the drill string increase accordingly. It can result in short-term detachments of the bit from the bottom, and the bit can collectively be hanging above the bottom up to 50-60% of the total drilling time. Essentially, a reduction in the contact time of the bit with the bottom leads to a decrease in the time of rock destruction and results in a decrease in the mechanical drilling speed.

The pneumatic compensators installed on the flow line of the mud pumps do not adequately damp the fluid pressure fluctuations. The work [5] is one of the works testifying to the low efficiency of the currently used means of damping pressure fluctuations and flow rate of the flushing fluid. It presents the results of experimental studies conducted by the authors in the Bukhara Drilling Administration, located in the Republic of Uzbekistan. We measured the amplitude-frequency characteristics of forced pressure and flow fluctuations obtained under various operating modes of the UNV-600A mud pump. We found that the stationary pump operating mode and an average pressure of 5 MPa results in the impulse of pressure fluctuations in the hydraulic system with a frequency of 18 Hz and an amplitude of 2 MPa.

The study of the influence of pressure fluctuations and drilling fluid flow on the downhole equipment performance revealed the need to install a downhole pressure fluctuation compensator in the BHA. For this purpose, we have developed a downhole hydromechanical compensator, and give its description below.

Figure 3 shows a structural diagram of a multi-stage hydromechanical flushing fluid compensator, consisting of a housing 1 with partitions 2, 3, 4 installed inside it, forming closed chambers with pistons 5, 6, 7 connected by channels 8, 9, 10 with annular space. The installed bushings 11, 12, 13 allows limiting the stroke of the pistons. Springs 14 were installed between the pistons and baffles. Parts in the housing are installed with a sub 15. The housing is equipped with connecting threads 16 and 17. We provide seals 18 to prevent overflow between the pistons and baffles, and the seal 19 to prevent overflow between the body and baffles.

The pulsating fluid enters the compensator, presses on the pistons 5, 6, 7 and thereby compresses the spring in the sub-piston space connected to the annulus through the channels 8, 9, 10. Thereby, the amplitude of the fluctuations in the pressure of the flushing fluid in the cavity of the drilling tool decreases.

Table 1 presents the technical characteristics of the developed hydromechanical pressure compensator for drilling wells with bits with a diameter of 215.9 mm.

Based on the selected model of fluid flow interacting with the piston of a hydromechanical compensator, we performed an analytical solution of the hydrodynamic problem of determining the change in pressure and velocity of a fluid moving through the stages of the compensator for unsteady fluid flow, i.e. when the pressure $P_0$and speedV_0 of the fluid in the pipe change over time:

$$P_0 = P_0(t); \quad V_0 = V_0(t).$$

We conducted pilot field tests to check the performance of the prototype downhole compensator in real drilling conditions. We conducted the study at well No. 224 of the Yuzhno-Izmailovsky field of the Kutlumbekovsky district of the Orenburg region drilled by LLC MUN-TECHNOLOGY. The drilling was carried out by an ARB-100 drilling rig with a BRN-1 mud pump.

The the drilling tool assembly consists of:

- in 28-165 m, bit 295.3 SZ-GV, KS 293, D5-172, TsS 292, UBT 178 - 18 m, hydromechanical pressure compensator, UBT 178-80 m, SBT-114 * 9.19.
- in 165-741 m, bit 215.9 TD44, KS-214, D5-172, TsS 213, UBT 178-18 m, hydromechanical pressure compensator, UBT 178-80 m, SBT-114 * 9.19.

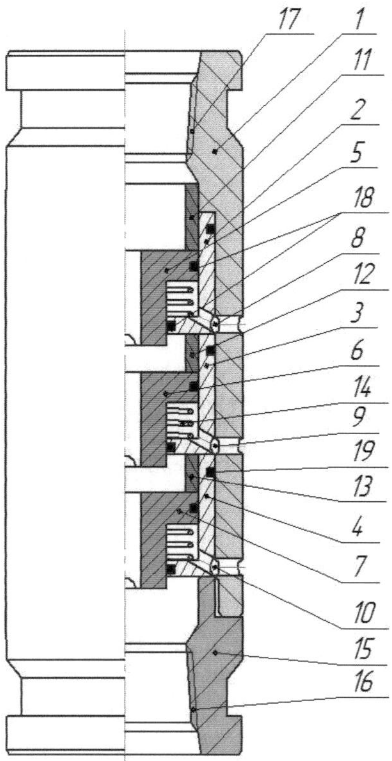

Figure 3. The structural scheme of the hydromechanical compensator: 1 - housing; 2, 3, 4 - partitions 5, 6, 7, 8, 9, 10 - channels; 11, 12, 13 - bushings; 14 - springs; 15 - sub; 16, 17 - connecting threads; 18, 19 - seals.

Table. 1. Technical characteristic of hydromechanical pressure compensator

Parameters of hydromechanical pressure compensator	For assembly with a 215.9 mm bit
Outer diameter, mm	178
Overall length, mm	914
Stages number	3
Piston diameter, mm	140
Piston bore diameter, mm	80
Connecting threads of the nipple and coupling	Z-147
The diameter of the channels in communication with the annulus, mm	10
Spring type	Screw cylindrical
Pressure drop, MPa	0.20–0.32

In the process of drilling the well, we periodically measured the fluctuations in the pressure of the flushing fluid with a pressure transducer measuring PDI-01-02 mounted on a high-pressure manifold.

As a result of the measurements, we obtained graphs of pressure changes in the circulation system of the drilling rig during drilling. Figure 4 presents some of them.

The upper part shows a graph of pressure changes during drilling without a compensator at an average pressure of 3.5–3.7 MPa. The graph shows that the amplitude of the pressure fluctuations reaches about 1.8 MPa. The bottom of the slide presents a graph of pressure changes when a hydromechanical compensator is installed in the drilling tool assembly. The graph shows that at an average pressure of 3.3 MPa, the amplitude range of the pressure fluctuations reaches 1–1.2 MPa.

We should also note that due to a more even bit working time, we obtained an increase in the mechanical drilling speed of 18%, unlike similar drilling without the use of a hydromechanical pressure compensator.

a

b

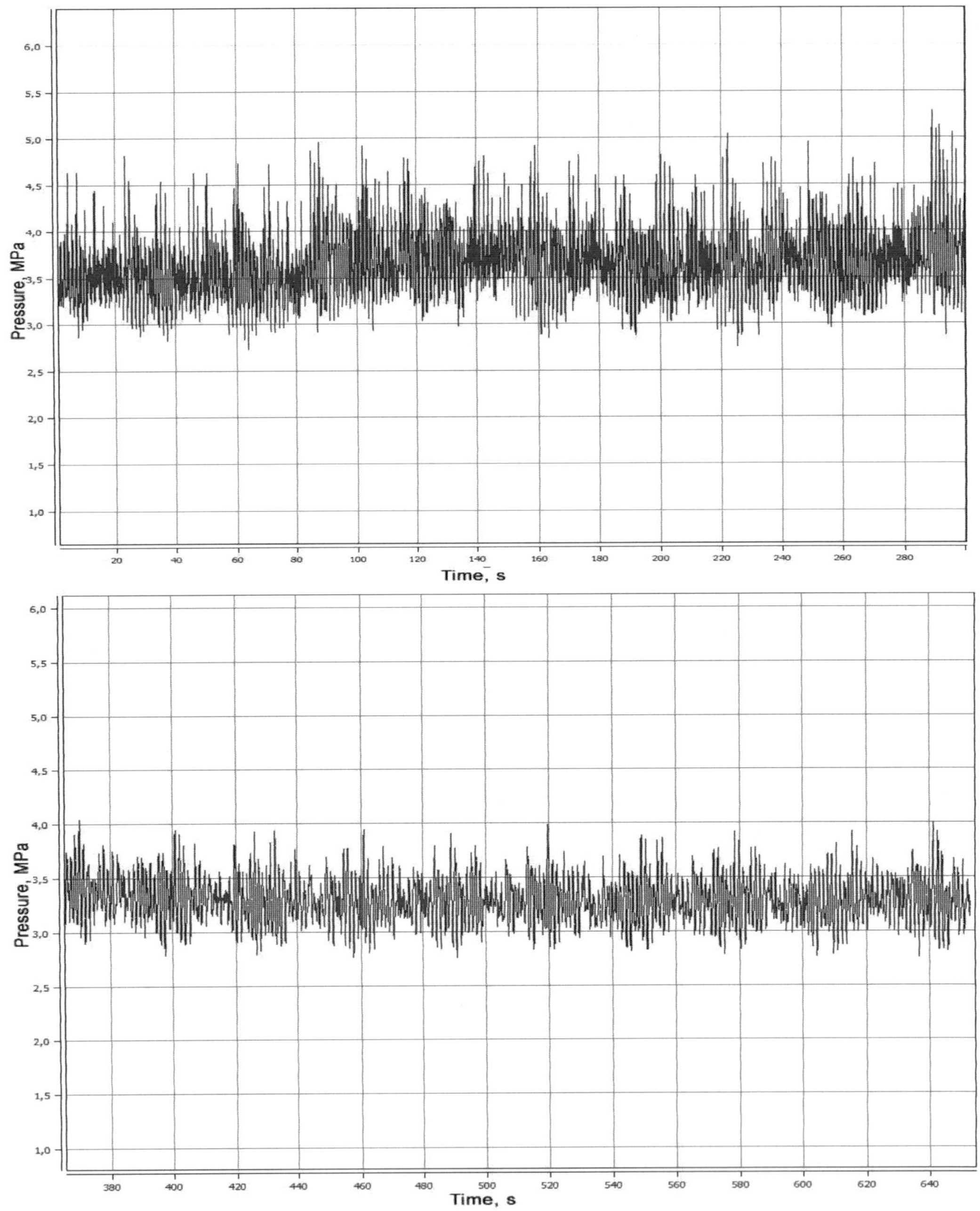

Figure 4. Graphs of pressure changes during drilling without a downhole hydromechanical compensator (a) and with a compensator (b).

4. Conclusion

Thus, despite the presence of regular pneumatic compensators for drilling pumps, pressure fluctuations of the drilling fluid in the drill string reach significant values and lead to various negative processes.

IOP Conf. Series: Materials Science and Engineering **952** (2020) 012062 doi:10.1088/1757-899X/952/1/012062

The developed downhole compensator made it possible to obtain a significant decrease in the amplitude of pressure fluctuations during the pilot tests, as well as an increase in the mechanical drilling speed that undoubtedly indicates a high efficiency of its operation.

References

[1] Minnivaleev T N 2018 Efficiency determination of amplitude decay of pressure fluctuation of circulating fluid with the help of a multi-stage piston-type downhole compensator *Advances in Engineering Research (AER) (International conference "Actual issues of mechanical engineering" (AIME 2018))* **157** 397-400 DOI: 10.2991/aime-18.2018.76

[2] Minnivaleev T N, Arslanov I G, Yagafarova K N 2018 Graphene aerogel is modern and promising material *IOP Conference Series: Earth and Environmental Science (IPDME 2018 – International Conference on Innovations and Prospects of Development of Mining Machinery and Electrical Engineering)* **194(4)** 1-5 DOI: 10.1088/1755-1315/194/4/042014

[3] Mukhametshin V Sh, Kotenev Yu A and Sultanov Sh Kh 2018 Assessment of the Need to Stimulate the Development of Hard-to-Recover Reserves in Carbonate Reservoirs *IOP Conference Series: Earth and Environmental Science (IPDME 2018 – International Conference on Innovations and Prospects of Development of Mining Machinery and Electrical Engineering)* **194(8)** 1-4 DOI: 10.1088/1755-1315/194/8/082027

[4] Batalov S A, Andreev V E, Lobankov V M and Mukhametshin V Sh 2019 Numerical simulation of oil formation with regulated disturbances. Oil recovery quality simulation *Journal of Physics: Conference Series (ITBI 2019 – International Conference "Information Technologies in Business and Industry")* **1333(3)** 1-6 DOI: 10.1088/1742-6596/1333/3/032006

[5] Mukhametshin V V 2018 Bottomhole formation zone treatment process modelling with the use geological and geophysical information *IOP Conference Series: Earth and Environmental Science (IPDME 2018 – International Conference on Innovations and Prospects of Development of Mining Machinery and Electrical Engineering)* **194(2)** 1-5 DOI: 10.1088/1755-1315/194/2/022024

[6] Kuleshova L S, Mukhametshin V V and Safiullina A R 2019 Applying information technologies in identifying the features of deposit identification under conditions of different oil-and gas provinces *Journal of Physics: Conference Series (ITBI 2019 – International Conference "Information Technologies in Business and Industry")* **1333(7)** 1-5 DOI: 10.1088/1742-6596/1333/7/072012

IOP Conf. Series: Materials Science and Engineering **952** (2020) 012063 doi:10.1088/1757-899X/952/1/012063

The results of applying the method of mathematical statistics to predict technological effectiveness of hydraulic fracturing

Al F Shakurova and Ay F Shakurova

Ufa State Petroleum Technological University, Branch of the University in the City of Oktyabrsky, 54a, Devonskaya St., Oktyabrsky, Republic of Bashkortostan, 452607, Russian Federation

E-mail: alsu0017@mail.ru

Abstract. The article discusses the apparatus of mathematical statistics for predicting the technological effectiveness of hydraulic fracturing. The authors made an attempt to improve the methods for processing experimental data in the conditions of a small sampling. The main method used for solving this problem was the second-order regression equation in the form of a polynomial. The comparison of calculated and actual data on technological results for hydraulic fracturing based on the solution of second-order polynomial regression equations for different fields showed a high level of convergence, which allows us to conclude that the proposed methodology for constructing a regression equation in the form of a power polynomial is acceptable.

1. Introduction

The most common way of processing experimental data is the regression analysis, which allows us to obtain a mathematical description of the technological process on the basis of experimental data in the form of an algebraic power polynomial. It is known that with an increase in the number of its members, reliability of the mathematical description of the technological process improves [1].

The practice of processing experimental data has shown that in most cases experimental results in the form of a tabular function are reflected with a sufficient approximation by a complete cubic polynomial. The third degree is sufficient and excessive, i.e. the number of terms of the polynomial can be reduced without significant loss of accuracy [2].

2. Materials and methods

As a research tool, we use the standard Microsoft Office software package (MS Excel). Functions that implement statistical methods of data processing and analysis are implemented in MS Excel in the form of special software - Analysis Package; for solving optimization problems, we used "Solution Search" [2].

3. Results and discussion

The apparatus of mathematical statistics has been widely used to process measurements of field data when evaluating the effectiveness of various methods of increasing oil recovery, as well as to predict the technological effect when planning hydraulic fracturing. The hydraulic fracturing

IOP Conf. Series: Materials Science and Engineering **952** (2020) 012063 doi:10.1088/1757-899X/952/1/012063

technology is sufficiently developed in relation to various geological, physical and geological and technical conditions [3].

However, the widespread practice of using this apparatus shows that in order to converge calculation results the actual data, a large sampling of measurements is required, which is not always possible in real production conditions.

According to [4, 5], the technological effect of hydraulic fracturing is influenced by geological, physical, and geological and technological parameters of the objects. Based on paired correlation dependencies, parameters that have a greater effect on the effectiveness of hydraulic fracturing was used. These parameters were used to forecast hydraulic fracturing results.

An attempt was made to improve the methods for processing experimental data in the conditions of a small sampling. As the main method for solving this problem was the second-order regression equation in the form of a polynomial.

For this purpose, the geological production results of hydraulic fracturing in the Bavlin field were analyzed for wells of Bobrikov and Devonian deposits in the Pavlovsk and East Leninogorsk areas of the Romashkin field. Based on the paired correlation dependencies, parameters that have a greater effect on the effectiveness of hydraulic fracturing were used. These parameters were used to predict the results of hydraulic fracturing.

The most common way of processing experimental data is the method of regression analysis, which allows us to obtain a mathematical description of the technological process on the basis of experimental data in the form of an algebraic power polynomial. It is known that with an increase in the number of its members, reliability of the mathematical description of the technological process improves.

The practice of processing experimental data has shown that in most cases the experimental results in the form of a tabular function are reflected with a sufficient approximation by a complete cubic polynomial. The third degree is not only sufficient, but also excessive, i.e. the number of members of the polynomial can be reduced without significant loss of accuracy.

Tabular functions were constructed for each hydraulic fracturing object.

Based on the tabular functions and using the MS Excel software, for each hydraulic fracturing object, second-order regression equations were constructed in the form of a polynomial:

$$Y=b_0 + b_1*x_1 + b_2*x_2 + b_3*x_3 + b_4*x_4 + b_5*x_5 + b_6*x_6 + b_7*x_7 + b_8*x_8 + +b_{12}*x_1*x_2 + b_{13}*x_1*x_3 + b_{14}*x_1*x_4 + \ldots + b_{67}*x_6*x_7 + b_{68}*x_6*x_8 + +b_{78}*x_7*x_8 + b_{11}*x_1^2 + b_{22}*x_2^2 + b_{33}*x_3^2 + \ldots\ldots\ldots + + b_{88}*x_8^2, \quad (1)$$

where Y – ΔQ_o (additional oil production), x_1 – K_p (porosity coefficient), x_2 – K_c (clay factor), x_3 – h_{perf} (perforation power), x_4 – h_{eff} (effective reservoir power), x_5 – K_{os} (oil saturation coefficient), x_6 – K_{per} (permeability coefficient), x_7 – V_f/V_{pr} (ratio of fracture fluid volume to proppant volume), x_8 – P_{hfp}– P_{lp} (difference between pressure of the hydraulic fracturing and layer pressure).

Let us study the components of the polynomial by means of a correlation analysis. Correlation is used to quantify the relationship of two data sets presented in a dimensionless form. The correlation coefficient of the sample is the covariance of two data sets divided by the product of their standard deviations [6].

The correlation analysis makes it possible to establish whether large values from one data set are associated with large values of another set (positive correlation), or, conversely, whether small values of one set are associated with large values of another one (negative correlation), or the data of two data sets are not connected in any way (the correlation is close to zero).

Then, using the "Analysis Package", the terms included in equation (1) was studied to assess the degree of their relationship and exclude identical ones from this equation.

To search for the coefficients of the regression equation, we used the"Solution Search" in "MS Excel".

"Solution Search" is a powerful data analysis tool in "MS Excel". It can determine at what values of the cells the formula in the target cell takes the desired value (minimum, maximum, or equal to any value). You can set restrictions for the solution search procedure, and it is not necessary that the same influencing cells are used. To calculate the set value, various mathematical search methods are used. You can set the mode in which the obtained values of the variables are automatically entered in the table. In addition, the results of the program can be presented in the form of a report.

The program "Solution Search" (in the original "Excel Solver") is an additional add-in of the table processor "MS Excel", which is designed to solve systems of equations, as well as linear and nonlinear optimization problems. The size of the problem that can be solved using the basic version of this program is limited by the following indicators:

• the number of unknowns is 200;
• the number of formula restrictions on the unknown is 100;
• the number of limit conditions for the unknown is 400.

Solution Search Options:

• Maximum time serves to limit the time allowed to find a solution. In this field you can enter the time in seconds, not exceeding 32,767 (approximately nine hours); a default value of 100 is acceptable for most simple tasks.

• The limiting number of iterations controls the time of solving the problem by limiting the number of computational cycles (iterations).

• Relative error determines the accuracy of calculations. The lower the value of this parameter, the higher the accuracy of the calculations.

• Tolerance is designed to specify the tolerance for deviation from the optimal solution, if the set of values of the influencing cell is limited to the set of integers. The larger the tolerance value, the less time it takes to find a solution.

In conclusion, using the methodology for finding the minimum standard deviation between the actual and calculated values (Table 1), the numerical values of the coefficients included in the regression equation (1) were determined with respect to each object of impact using a hydraulic fracturing technology.

As a result, the following second degree regression equations were obtained [7]:

- for Bobrikov deposits:

$$Y=1.943 + 19.499x_1 + 1.929x_2 - 12.92\,x_4 + 78.277x_5 - 73.183x_6 + 6.6x_7 + 54.217x_8 - 88.92x_1x_2 - 282.227x_1x_4 - 20.788x_1x_5 - 3.572x_1x_6 + 9.045x_1x_7 + 18.903x_1x_8 + 5.608x_2x_6 + 26.116x_4x_6 - 80.932x_5x_7 - 4.025x_5x_8 + 123.273x_7x_8 + 76.774x_1^2 - 24.542x_2^2 - 44.803x_4^2 + 6.606x_5^2 + 0.087x_6^2 + 1.214x_7^2 - 7.075x_8^2, \quad (2)$$

where $Y - \Delta Q_o$, $x_1 - K_p$, $x_2 - K_c$, $x_4 - h_{eff}$, $x_5 - K_{os}$, $x_6 - K_{per}$, $x_7 - V_f/V_{pr}$, $x_8 - P_{hfp} - P_{lp}$.

This equation (2) ensured that the calculated and actual values coincided with an accuracy of 1.94%.

- for Devonian sediments of the Bavlin deposit:

$$Y = -0.0104 - 0.157x_1 - 0.042x_2 - 0.017\,x_4 - 0.253x_5 - 0.451x_6 + 0.048x_7 - 0.773x_8 - 0.045x_1x_2 - 0.348x_1x_4 - 3.179x_1x_5 - 0.342x_1x_6 + 1.25x_1x_7 - 7.18x_1x_8 - 7.697x_2x_6 - 3.96x_4x_6 - 0.95x_5x_7 - 0.466x_5x_8 + 12.518x_7x_8 - 2.578x_1^2 + 0.954x_2^2 - 0.017x_4^2 + 5.774x_5^2 - 0.062x_6^2 + 0.881x_7^2 - 0.354x_8^2, \quad (3)$$

where $Y - \Delta Q_o$, $x_1 - K_p$, $x_2 - K_c$, $x_4 - h_{eff}$, $x_5 - K_{os}$, $x_6 - K_{per}$, $x_7 - V_f/V_{pr}$, $x_8 - P_{hfp} - P_{lp}$.

This equation (3) ensured that the calculated and actual values coincided with an accuracy of 0.95%.

- for Devonian deposits of Pavlovsk and East Leninogorsk areas:

$$Y = -4.482 - 42.359x_1 + 43.28x_2 + 36.319\,x_3 - 190.331x_5 - 304.513x_6 + 51.451x_7 - 968.722x_8 - 513.079x_1x_2 + 6.361x_1x_3 - 138.067x_1x_5 + 341.51x_1x_7 + 84.749x_1x_8 + 22.675x_2x_8 + 22674x_3x_6 + 108.98x_1^2 + 846.583x_2^2 - 121.647x_3^2 + 24.626x_5^2 + 0.019x_6^2 - 48.525x_7^2 - 2.435x_8^2, \quad (4)$$

where $Y - \Delta Q_o$, $x_1 - K_p$, $x_2 - K_c$, $x_4 - h_{eff}$, $x_5 - K_{os}$, $x_6 - K_{per}$, $x_7 - V_f/V_{pr}$, $x_8 - P_{hfp} - P_{lp}$.

This equation (4) ensured that the calculated and actual values coincided with an accuracy of 1.1%.

Table 1. Comparison of calculated values of additional oil production with actual field values obtained for Bobrikov deposits of the Bavlin field

No	Calculation	Fact	Deviation	Square deviation	Error, %
1	2956.2696	2956.3	-0.0303515	0.000921214	0.01
2	6402.1293	6402	0.12928633	0.016714955	0.02
3	4993.3307	4994	-0.66933333	0.448007102	0.13
4	4857.6724	4857.9	-0.22761772	0.051809824	0.05
5	1732.6263	1733	-0.37365975	0.139621607	0.22
6	3402.8769	3402	0.87693764	0.769019619	0.26
7	2371.8132	2373	-1.18675585	1.408389451	0.50
8	907.48939	907.6	-0.11060918	0.012234391	0.12
9	1603.2146	1602.6	0.61458128	0.377710147	0.38
10	5004.9362	5005.3	-0.36378511	0.132339604	0.07
11	9813.8538	9812.92	0.93383358	0.872045153	0.10
12	2924.4344	2924.3	0.13437208	0.018055856	0.05
13	3826.3922	3826.5	-0.10782055	0.011625272	0.03
			Total	**∑4.258494195**	**∑1.94%**

4. Conclusion

Thus, it was proved that even in the conditions of a small representative data sampling the power regression equation makes it possible to accurately predict the technological effect of hydraulic fracturing for the Bobrikov and Devonian deposits, provided that they are constructed taking into account specific geological, production and technological data corresponding to these deposits within each of these productive areas.

The comparison of calculated and actual technological results in hydraulic fracturing based on the solution of second-order polynomial regression equations for different fields showed a high level of convergence, which allows us to conclude that the proposed method for constructing a regression equation in the form of a power polynomial is acceptable.

References

[1] Mirzadzhanzade AKh and Stepanova G S 1977 *The mathematical theory of an experiment in the oil and gas ndustry* (Moscow: Nedra)

[2] Shashkov V B 2003 *Applied Regression Analysis. Multivariate Regression: Manual.* (Orenburg: OSU)

[3] Economides M J and Nolte K G 1989 Reservoir Stimulation *Prentice Hall, Eglewood Cliffs, New Jersey 07632*

[4] Shakurova Al F and. Shakurova Ay F 2018 The influence of hydraulic fracturing on the estimated ultimate recovery *IOP Conference Series: Earth and Environmental Science* 194(8) 082039 1-4.

[5] Shakurova A F 2010 The study of parameters affecting the efficiency of hydraulic fracturing in the conditions of the Romashkinskoye field *Scientific and technical journal "Oil field business"* 1 P. 17-19.

[6] Kuleshova L S, Mukhametshin V V and Safiullina A R 2019 Applying information technologies in identifying the features of deposit identification under conditions of different oil-and gas provinces *Journal of Physics: Conference Series (ITBI 2019 – International Conference "Information Technologies in Business and Industry")* **1333(7)** 1-5 DOI: 10.1088/1742-6596/1333/7/072012

[7] Shakurov Al F and Shakurova Ay F 2013 Determining the economic feasibility of applying hydraulic fracturing technology in the low-permeable formation zone *Electronic scientific journal "Oil and Gas Business"* **2** 254-261, available at: http://www.ogbus.ru/authors/Shakurova/Shakurova_3.pdf

Developing the efficiency of low-productivity oil deposits via internal flooding

N N Soloviev[1], V Sh Mukhametshin[2] and A R Safiullina[2]

[1] LLC Gazprom VNIIGAZ, Staraya Basmannaya St., 20, b. 8, Moscow, Russian Federation
[2] Ufa State Petroleum Technological University, Branch of the University in the City of Oktyabrsky, 54a, Devonskaya St., Oktyabrsky, Republic of Bashkortostan, 452607, Russian Federation

E-mail: vsh@of.ugntu.ru

Abstract. The article studies geological and technological parameters' impact on degree and nature of production wells. The wells influence the water injection into wells in the carbonate reservoirs of the Tournaisian stage of the Birsk saddle. The study presents an algorithm with substantiating technological solutions to increase the efficiency of developing low-productive deposits via internal flooding. It also considers the development of facilities for hard-to-recover reserves with high production costs and low profitability. This is made in order to use the resource base of the Volga-Ural oil and gas provinces more efficiently.

1. Introduction

To meet the country's needs for liquid hydrocarbons one should make a more complete and efficient use of the resource base of deposits with hard-to-recover reserves in the carbonate reservoirs of the Volga-Ural oil and gas province (VUOGP).

This resource base consists of two large groups:

- the first group includes relatively high-productive facilities of large and medium deposits developed for a long time with a high degree of development of recoverable reserves;
- the second group includes low-productivity small deposits not developed enough and drilled with a rare grid of wells. There is poorly organized water, and the degree of oil recovery from the bowels is several percent of geological reserves.

Scientists and industrialists are more interested in the second group due to the following reasons:

- significant total oil reserves and a low degree of involvement in development;
- location in a relatively positive climatic zone;
- confinement to territories with developed social and industrial infrastructures;
- provision of enterprises with human resources of high professional training;
- the ability to use wells drilled at other development sites, and the use of the created material base.

Thus, these objects are considered to have good perspective for good oil production in the region.

However, it is necessary to find ways and means of reducing costs of extracted products and ensure product profitability [1-6].

2. Materials and methods

One of such objects is massive deposits in the carbonate reservoirs of the Tournaisian stage of the Birsk saddle and objects similar to them in terms of geological and production characteristics, confined to other VUOGP tectonic elements.

Lithologically, the oil deposits are composed of gray and light gray biomorphic, biomorphic-detrital, clot-clumpy and pelitomorphic-clot limestones with insignificant interlayers of dolomites. In the structure of reservoir rocks, intergranular (granular) porosity plays a decisive role, similar to the rocks of the terrigenous series by the type of structure of the void space. Fracturing has a limited distribution, increasing the connectivity of voids in cavernous-porous areas. There are 3 structural types of collectors distinguished: 1 - pore; 2 - fissure-pore; 3 - fissure-cavernous-pore.

The first type of reservoir dominates sharply, determining the productivity of the deposits. The main reservoir rocks are clot-lumpy and detrital-biomorphic limestones, characterized by the highest porosity and permeability. Fracturing of limestone is expressed by the following average values of parameters: fracture porosity 0.02–0.2%, fracture permeability - up to $10 \cdot 10^{-3}$ μm^2 and bulk density of cracks - 20-50 m^{-1}.

The reservoir rocks are characterized by the following average values of parameters: porosity - 0.113; oil saturation - 0.80; permeability - $52.2 \cdot 10^{-3}$ $microns^2$; effective oil saturated thickness - 3.82 m; the thickness of oil saturated layers is 1 m.

According to the geological heterogeneity (estimated by the values of the heterogeneity parameter, which is the product of the standard deviation, variation, entropy and relative entropy of a particular geological parameter), objects can be characterized as relatively homogeneous in effective oil-saturated thickness, oil-saturated interlayers and porosity. The share of reservoir rocks in the total thickness of the reservoir with the values of dissection coefficient 3.8 is 0.21. The operating modes of the deposits are characterized as pressure-elastic with weak backing of bottom and marginal waters. The sealing of deposits near the surface of the oil-water contact due to the deposition of secondary calcite and viscous bitumen in a fractured-porous medium, the strong dissection of productive formations by dense impermeable differences and the lenticular structure of reservoir rocks. Thus, there appears a sharp drop in reservoir pressure and well flow rates in the absence of produced water in individual sections of the deposits indicating an elastic mode of operation. In view of this, in the absence of measures to maintain reservoir pressure, the operating mode of the reservoir subsequently switches to the dissolved gas mode.

The reservoir oils of the objects are viscous, have high values of density and viscosity at low saturation pressures and gas factors. Formation water is highly saline; the density is 1120–1180 kg / m^3. Calcium and sodium chlorides predominate in the salt composition. The first salinity reaches 83%. The concentration of calcium salts is slightly higher than magnesium salts. According to the classification of V.A.Sulin, they belong to the calcium chloride type, the chloride group, and the sodium water subgroup. The formula used for it is Palmer $S_1S_2A_2$.

Drilling of objects was conducted extremely unevenly and for a long time. At the same time, the analysis of the objects, the density of the well grid varies over a fairly wide range - from 16 to 240 ha per well in the external oil circuit. In some more productive areas, the density of the well network in the drilling zone reached 10 ha per well, when placing the wells on a uniform triangular and quadrangular grid, with a distance between the wells of 400 m.

Oil production is mainly conducted using natural energy of the reservoir - both due to elastic forces, and the weak backwater of plantar and marginal waters. The final oil recovery in areas developed under natural conditions is determined by the productivity and density of the well network and varies from 3.0 to 35.0% (oil recovery was predicted using field-statistical methods). Oil production rates decrease over time due to both a decrease in reservoir pressure and watering of produced products, and in some areas the reservoir pressure has decreased below the saturation pressure and a dissolved gas regime takes place [7].

In some areas with the most favorable geological characteristics, areal flooding was organized to make development process faster and increase the final oil recovery [8].

IOP Conf. Series: Materials Science and Engineering **952** (2020) 012064 doi:10.1088/1757-899X/952/1/012064

3. Results and Discussion

Water flooding analysis proves that along with an increase in the production rates of production wells surrounding injection wells, there is a lack of growth in production rates. In this case, the injection efficiency is determined by the geological structure of the formation and the distances between production and injection wells.

Improving the efficiency of object development considers their further drilling and areal and focal flooding organization in areas with favorable geological characteristics [9]. It is important to know the distance between the wells in areas with different geological characteristics. The producing wells in these places will respond to the injection of water into the injection.

The solution seems to be the use of oil displacement mode in the development of such facilities since the analysis shows, it will double the final oil recovery in a short period of time, as well as introduce similar facilities into development, the oil production of which is currently not profitable.

In this regard, there are areas with focal water flooding identified, and the change in the production rate of wells injection was studied. In this case it is very important to reduce extraneous "noise". Wells were excluded from the sample, in which there was any effect on the formation or bottomhole zone, which could be affected by injection from two or more injection wells, etc.

Wells were identified that responded to water injection and did not respond. Geological-field analysis showed that the likelihood of a response is largely determined by geological structure of the objects and a distance between the wells (F). Moreover, with water injection pressures less than 0.7 rock pressure, there is no increase of a reaction with an increase in injection volumes. It was revealed that with an increase in effective oil-saturated thicknesses (OS_{th}), average thicknesses of oil-saturated interlayers (OS_i), the number of oil-saturated interlayers (n), as well as average weighted porosity values (m) in both production and injection wells, the probability, time and intensity of response are growing. The response between the wells has a significant effect on response rates. It is important to note that about 75% of the wells experienced injection effects at following values: $OS_{th} > 6,5$ m; $OS_i > 2,5$ m; $m > 0,15$;

$F < 400$ m. The maximum values of the response time and intensity are confined to these intervals.

On the other hand, in areas with low values of parameters characterizing the geological and physical properties of the formation, a decrease in the distance between production and injection wells mitigates the negative impact of natural factors. This fact suggests the need for a differentiated approach to determining the value of F.

Nonparametric informational criteria revealed the best possible dependence [7, 8]

$$K = \frac{OS_{th}^2 \cdot m^2}{F},$$

where K is a complex parameter characterizing the geological and technological features of flooding as an element of the reservoir; OS_{th}, m is the average value of the effective oil-saturated thickness and porosity between production and injection wells.

Based on this, K values were established for the probability of response:

- 0% – ($K^0 < 1,4 \cdot 10^{-4}$ m);
- 25% – ($K^{25} = 2,4 \cdot 10^{-4}$ m);
- 50% – ($K^{50} = 3,4 \cdot 10^{-4}$ m);
- 75% – ($K^{75} = 4,4 \cdot 10^{-4}$ m);
- 100% – ($K^{100} > 5,4 \cdot 10^{-4}$ m),

due to the task, this allow scientifically substantiating the choice of wells in the drilled sections for transferring them to injection, transferring wells from other horizons to modify the object development process, planning the grid of wells and the waterflooding system at the developing facilities. In this case, it is necessary to ensure that F is less than $F^{100} = 1850\, OS_{th}^2 \cdot m^2$. If this is not possible for economic and technological reasons, you can use the other options presented.

4. Conclusion

The study and the results obtained suggest the following:

- carry out the substantiation of technological solutions to increase the efficiency of the development of low-productive deposits using internal flooding;
- consider putting into development facilities of hard-to-recover reserves with have high production costs and low profitability in order use the resource base of the Volga-Ural oil and gas province more efficiently.

References

[1] Yakupov R F, Mukhametshin V Sh, Khakimzyanov I N and Trofimov V E 2019 Optimization of reserve production from water oil zones of D3ps horizon of Shkapovsky oil field by means of horizontal wells *Georesursy* **21(3)** 55-61. DOI: 10.18599/grs.2019.3.55-61

[2] Kuleshova L S, Kadyrov R R, Mukhametshin V V and Akhmetov R T 2019 Auxiliary equipment for downhole fittings of injection wells and water supply lines used to improve their performance in winter *IOP Conference Series: Materials Science and Engineering (MEACS 2018 – International Conference on Mechanical Engineering, Automation and Control Systems)* **560(1) 012071** 1-6 DOI: 10.1088/1757-899X/560/1/012071

[3] Mukhametshin V V, and Kuleshova L S 2020 On uncertainty level reduction in managing waterflooding of the deposits with hard to extract reserves Bulletin of the Tomsk Polytechnic University. Geo Assets Engineering 331(5) 140–146 DOI 10.18799/24131830/2020/5/2644

[4] Stenkin A V, Kotenev Yu A, Mukhametshin V Sh and Sultanov Sh Kh 2019 Use of low-mineralized water for displacing oil from clay productive field formations *IOP Conference Series: Materials Science and Engineering (MEACS 2018 – International Conference on Mechanical Engineering, Automation and Control Systems)* **560(1) 012202** 1-5 DOI: 10.1088/1757-899X/560/1/012202

[5] Valeev A S, Kotenev Yu A, Mukhametshin V Sh and Sultanov Sh Kh Substantiation of the recovery of residual oil from low-productive and heterogeneous formations in Western Siberia by improving the waterflood system using gas and water-gas impacts *IOP Conference Series: Materials Science and Engineering (MEACS 2018 – International Conference on Mechanical Engineering, Automation and Control Systems)* **560(1)** pp 1–6 DOI: 10.1088/1757-899X/560/1/012204

[6] Andreev V E, Dubinskiy G S, Fedorov K M and Andreev A V 2014 Design and implementation of technology using acid "retarder" for carbonate reservoir treatment *Problems of Gathering, Treatment and Transportation of oil and oil products* **2** 5-14

[7] Mukhametshin V V and Kuleshova L S 2019 Justification of Low-Productive Oil Deposits Flooding Systems in the Conditions of Limited Information Amount *SOCAR Proceedings* **2** 16–22 DOI: 10.5510/OGP20190200384

[8] Akhmetov R T, Kuleshova L S and Mukhametshin V V 2019 Application of the Brooks-Corey model in the conditions of lower cretaceous deposits in terrigenous reservoirs of Western Siberia *IOP Conference Series: Materials Science and Engineering (MEACS 2018 – International Conference on Mechanical Engineering, Automation and Control Systems)* **560(1) 012004** 1-4 DOI: 10.1088/1757-899X/560/1/012004

[9] Kuleshova L S, Kadyrov R R, Mukhametshin V V and Safiullina A R 2019 Design changes of injection and supply wellhead fittings operating in winter conditions *IOP Conference Series: Materials Science and Engineering (MEACS 2018 – International Conference on Mechanical Engineering, Automation and Control Systems)* **560(1) 012072** 1-5 DOI: 10.1088/1757-899X/560/1/012072

IOP Conf. Series: Materials Science and Engineering **952** (2020) 012065 doi:10.1088/1757-899X/952/1/012065

Research deformation caused by hydraulic fracturing

A E Verisokin, A A Airapetova, L G Zhulina, A V Nazarenko and V A Tkachenko

Institute of Oil and Gas, North-Caucasus Federal University, 1, Pushkin St., Stavropol, 355009, Russia

E-mail: verisokin.aleksandr@mail.ru

Abstract. The paper discusses problems of tensile loadings affecting a soil layer when conducting hydraulic fracturing. The effect of tensile stress on rocks in a bottomhole formation zone during hydraulic fracturing is studied using numerical methods. A special analysis of the processes of cores deformation under reservoir conditions is carried out in the paper. The article provides formulas to calculate stress-strain state of rocks in bottomhole formation zone. Along with an exact numerical study of the distribution of stresses in bottomhole formation zone, some approximate analytical dependencies reflecting features of the stress-strain state of bottomhole formation zone are obtained with the aim of solving optimization problems. The paper presents the scheme of production layer vertical deformation which shows layer's roof and bottom displacement caused by increased pressure inrepression cone. Formulas describing the influence of deformation processes in bottomhole formation zone can also be used to study cyclic operation of underground gas storages.

The article studies [1] the processes of the influence of compressive loads on rocks in bottomhole formation zones of wells during withdrawal of formation fluid. It discusses the problem of the influence of tensile loads during hydraulic fracturing. The study of tensile stress was carried out on the basis of the method described in the article. The tensile stress occurs in the bottomhole formation zone during hydraulic fracturing [1].

Stretching of rocks under the influence of excessive loads developing in the productive formation layer during hydraulic fracturing may cause destruction of the rock during well operation.

The issue of deformation processes of rock samples in reservoir conditions requires special analysis. The stress-strain state of remote parts of the layer during fracturing is mainly determined by the process of one-dimensional stretching of rocks under the load created by power pumps. In the bottomhole zone of the well, i.e. in the zone a repression funnel, a complex non-uniform deformation of rocks occurs, which is dependent on many factors: features of pressure distribution in the repression funnel, ratio of elastic properties of layers and the surrounding rocks, thickness of a layer. Moreover, it is possible to establish a one-to-one correspondence between the change in formation pressure Δp at a given point of a layer and tensile effective stresses in this area with one-dimensional vertical deformation of the rock, which is a characteristic of remote zones of the layer. However, such dependence has more complex integral character in a near-wellbore zone of a well. This is due to the fact that, in accordance with the equilibrium equations for a saturated porous medium, the stress-strain state of a rock at a given point depends not only on the change in pore pressure in it, but also on the

IOP Conf. Series: Materials Science and Engineering **952** (2020) 012065 doi:10.1088/1757-899X/952/1/012065

features of its distribution throughout the repression funnel, i.e. from the function Δp(r) (r is the radius measured from the wellbore).

In such cases when the function Δp(r) changes insignificantly, or when the pore pressure is approximately constant over a large area of the reservoir, which is typical for areas remote from the well, complex integral dependence of the stress-strain state of a rock on the function Δp(r) takes the form of a fairly simple one-to-one relationship.

This makes it possible to determine the change in rock permeability k as a function of k(Δp), which is associated with the processes of rock deformation.

In the bottomhole formation zone of a well under the conditions of a sharp change in the function Δp(r), this approach is inaccurate, since the stress-strain state of a rock becomes axisymmetric and is characterized by dependence on many defining parameters. This happens due to the presence of a significant pore pressure gradient and the influence of formation top and bottom.

Thus, the reaction of a rock to a change in reservoir pressure by the same value Δp, which is expressed in a change in the permeability of a rock k, will be different in the near-wellbore zone and in remote areas of the formation.

Under the conditions of rock loading in the depth of the productive formation, the absence of lateral horizontal displacement of a rock is more consistent, which is due to the resistance of the rock mass located above and below the formation. Within the framework of the concept of elastic deformation of a saturated porous medium in the absence of lateral displacement, i.e. in the presence of rigid side walls, the lateral effective stresses will be equal to only a certain part of the vertical effective stress, which can be described by the value$v/(1-v)$=0.2–0.6 (where v is the Poisson's ratio of the rock).

Obviously, the change in the permeability of a rock during extension happens due to an increase in its pore volume. Within the framework of the concept of elastic deformation of a porous medium, the change in its elementary volume is proportional to the sum of the main effective stresses acting at a given point of the medium; the permeability of a rock should be a function of the sum of the main stresses.

Thus, in accordance with the indicated approach, the initial experimental dependence should be the dependence of permeability of a rock sample not on the pore pressure, but on the sum of the principal stresses developing in the sample when the pore pressure changes. This dependence can be used for the process of rock deformation both in the bottomhole zone of the well and in the depth of the formation.

It is possible to calculate the stress-strain state of a rock in the bottomhole zone of the well, caused by the repression funnel using the numerical methods [1]. In this case, the equations of elastic equilibrium of a saturated porous medium with an axisymmetric statement within the framework of the concept of effective stresses have the following form:

$$\sigma_{rr} + \tau_{rz,z} + (\sigma_r - \sigma_\varphi)/r = p_r(r), \tag{1}$$

$$\tau_{rz,r} + \sigma_{zz} + \tau_{rz}/r = 0, \tag{2}$$

$$\sigma_r = \lambda \cdot I + 2 \cdot \mu \cdot u_r, \tag{3}$$

$$\sigma_\varphi = \lambda \cdot I + 2 \cdot \mu \cdot u/r, \tag{4}$$

$$\sigma_z = \lambda \cdot I + 2 \cdot \mu \cdot \omega_z, \tag{5}$$

$$\tau_{rz} = \mu(u_z + \omega_r), \tag{6}$$

σ_r, σ_φ, σ_z – radial, vertical and angular stress components; τ, z, φ – corresponding coordinates;

λ=Eν/(1+ν)(1–2ν), μ=E/2(1+ν) – Lamé parameters, E – Young's modulus; ν – Poisson's ratio of the elastic medium; $I=u_r+w_z+u/r$, u and ω – are components of the radial and vertical displacement of the medium.

The p (r) function determines the distribution of pore pressure in the repression funnel in the reservoir. When it is outside of the reservoir it is zero. In general, the function p(r) should be defined as a solution to a separate equation describing the filtration flow in the bottomhole zone of the well; but to analyze the most significant features of the distribution, it is necessary to change the function. At the boundaries of elastic media with different properties, there should be conditions set for the equality of displacements and stresses of these media. Zero stresses are set at the outer boundary, far enough from the wellbore. In what follows, stress values are understood as additional stresses in the rock caused by the action of pore pressure p(r) in the repression funnel (the initial stress-strain state of the rock is taken as zero background).

Along with an accurate numerical study of the peculiarities of stress distribution in the wellbore zone, described by the system of equations (1)–(6), it is advisable to obtain some approximate analytical dependences reflecting features of the stress-strain state of the bottomhole zone to solve optimization problems.

For this, based on the analysis of the properties of exact one-dimensional solutions of the system of equations (1)–(6) and the analysis of some results of the numerical study of this system in the axisymmetric formulation, we have made the following assumptions. The approximate stress state of the rock can be represented as the sum of stresses developing in the horizontal plane and described by equations (1)–(6) in a one-dimensional version as a function of the radius r, and additional stresses caused by the vertical load on the formation from the side of its roof.

The one-dimensional solution σ_r(r), σ_φ(r) of equations (1)–(6) is rather difficult (in the form of integral dependences) expressed in terms of the function (r). However, as it follows from the analysis of these solutions, the sum of stresses $\sigma_r + \sigma_\varphi$in each point of the rock is expressed by the exact ratio in the following form:

$$\sigma_r + \sigma_\varphi = p(r)/(1 - v). \tag{7}$$

The analysis of numerical solution of the system of equations (1)–(6) shows that the vertical displacement of the formation top and bottom in the zone of a repression funnel during hydraulic fracturing is the same as it is schematically shown in Figure 1.

It should be noted that vertical displacement ω (r) in the central part of the repression funnel does not repeat the configuration of the pressure function p (r) with a sharp change in its gradient, but corresponds to the laws of elastic media with a smooth change in the magnitude of the bulge and, accordingly, a smooth change of effective vertical stress σ_z in the reservoir. This makes it possible to use an approximate description of the function σ_z as the simplest quadratic dependence:

$$\sigma_z(r) = \alpha_o \cdot \Delta p \cdot [1 - (r/R)^2] \tag{8}$$

where Δp is the pressure drop in the central part of the repression funnel.

As follows from the analysis of numerical calculations of the system of equations (1)–(6) [1, 5, 6], the coefficient t$\alpha_o$in formula (8) depends on determining parameters and can vary from 0 to 1, while the most probable range of values is 0.4–0.9. In general, it can be concluded that the proportion of vertical extension of production formation in the repression funnel in the principal stresses $\sum \sigma_i$ is the most significant at high extension of rocks (mainly rocks surrounding the pay zone) and a relatively small thickness of the pay zone. When the modulus of longitudinal elasticity (Young's modulus) of the production formation is approximately equal to or exceeds the stiffness of surrounding rocks, and the formation thickness is several meters, the parameter α_o tends to be 1. At high rock rigidity in the formation top and bottom this parameter decreases, rocks prevent vertical deformation of the formation in the center of the repression funnel. So, when the thickness of the formation is 7–8 m, and rigidity of surrounding rocks is 4 times higher than rigidity of the productive formation, the parameter $\alpha_o$decreases to 0.6–0.7. With an increase in the thickness of the reservoir, the parameter$\alpha_o$decreases.

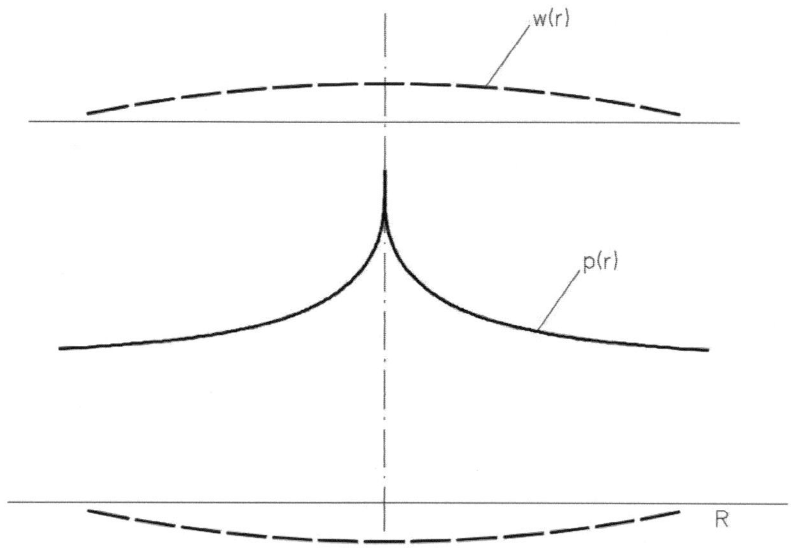

Figure 1. Scheme of deformation of the productive formation in the vertical direction: w(r) – displacement of the formation top and bottom caused by increased pressure in the repression funnel with a characteristic radius R

In the final assessment of the sum of principal stresses on rocks, it should be taken into account that its vertical extension in the absence (or small value) of lateral displacement increases the lateral stresses to a value proportional to the vertical load, with the proportionality coefficient v/(l–v). This means that [2v/(l–v)]–a_z(r) should be added to the sum of lateral stresses σ_r + $\sigma_\varphi$in formula (7) or, taking into account formula (8), it can be the termα_o[2v/(l–v)]–Δp–[l–(r/R)^2].

These estimates allow us to write down the final expression for the sum of the main effective stresses $\sum \sigma_i = \sigma_r + \sigma_\varphi + \sigma_z$ developing in the bottomhole formation zone and are caused by an increase in reservoir pressure in the repression funnel:

$$\sum \sigma_i = \frac{p(r)+\alpha\Delta p\ [1-(r/R)^2]}{1-v},\ \alpha = \alpha_0(1+v),\ p(r), \tag{9}$$

The dependence of rock permeability on effective stresses $k\sum\sigma_i$can be represented in a simple exponential form:

$$k(\sum\sigma_i) = k_0\exp(\sum\sigma_i/p^*), \tag{10}$$

where p^* is some characteristic value of stress (pore pressure), which determines the rate of change in the permeability of rocks during its extension (to reduce the number of designations used, we will assume that the parameter p* has already included the factor 1/(1–v) from formula (9).

The value of α_0 depends on many defining parameters and shows the degree of vertical stresses on rocks in the zone of the repression funnel during hydraulic fracturing.

Formulas describing the effect of deformation processes in the bottomhole formation zone of a well can also be used to assess the existing stresses in the rocks of underground gas storage during cyclic operation.

References

[1] Svalov A M 2008 Vliyanie deformatsii porody v prizaboinykh zonach skvazhin na vid indikatornykh krivykh *Neft. khoz-vo* **2** 63–65

[2] Bayuk E I, Tomashevskaia I S and Dobrynin, V M 1988 *Fizicheskie svoystva mineralov i gornykh porod pri vysokikh termodinamicheskikh parametrakh* Spravochnik (Moscow: Nedra) 255 p

[3] Zaitsev M V and Mihailov N N 2004 Vliyanie okoloskvazhinnoy zony na produktivnost' skvazhyny *Neft. khoz-vo* **1** 64–66

[4] Mishchenko I T, Bravicheva T B and Ermolaev A I *Vybor sposoba ekspluatatsii skvazhin neftyanykh mestorozhdeniy s trudno izvlekaemymi zapasami* 2005 (Moscow: FGUP Izd-vo Neft' i gas RGU nefti i gaz aim.) 448 p

[5] neftianykh plastov pri ikh sovmestnoy razrabotke *Neft. khoz-vo* **4** 85–87

[6] Svalov A M 2006 Dvumernyi analiz osobennostey raspredeleniya uprugikh napryazheniy v pristvol'nykh zonakh skvazhin *Gazovaiya promyshlennost'* **2** 34–36

[7] Vasilev V A and Verisokin A E 2016 Effektivnost' gidrorazryva plasta pri osvoenii i ekspluatatsii dobyvayushchikh skvazhin v treshchinnykh kollektorakh *Innovatsionnye tekhnologii v neftegazovoy otrasli Mater. II mezhdunarodnoy nauchno-prakticheskoy konf.* (Stavropol: OOO IDTESERA)

Backfilling for mines

L H Zagorodnyuk[1], A S Chernysh[2], V S Lesovik[1], A Yu Schekina[2], V D Ryzhikh[1], D A Sumskoy[1] and D A Sinebok[1]

[1]Department of building materials, products and structures, Belgorod State Technological University named after V.G. Shukhov, 46, Kostyukov Str., Belgorod, 308012, Russia
[2]Department of city cadastre and engineering surveys, Belgorod State Technological University named after V.G. Shukhov, 46, Kostyukov Str., Belgorod, 308012, Russia

E-mail: pr9nik2011@yandex.ru

Abstract. The article introduces results of studying the problem areas in filling masses of iron ore mining-and-processing enterprise. Some basic study of back filling after the blast hole drilling has been conducted by the authors. The samples picked from the disrupted back filling have been studied for strength. The analysis of the results evidences that all the samples meet the requirements of the strength class but there is a variation of the strength characteristics which evidences about the massif inhomogenuity. The results of chemical, X-ray and fluorescent analyses proved high inhomogenuity of solidified back filling masses in the massif. The microstructure of the back filling samples has inhomogeneous structure, with a great number of various pores. A set of studies showed lack of uniformity and consolidation of the filling solution. It resulted in the loss of strength and durability of the back filling area. The analysis of possible reasons for negative results has been given. The activities to improve the quality and durability of the back filling for mines have been suggested.

1. Introduction

Development of iron ore deposits and ore mining is a topical, prospective and expensive trend in mining works. This field encounters several problems: work labour input and arduous working conditions, environmental issues, insufficient study of the soil layers, obsolescence and physical deterioration of the facilities and others. Among these issues one of the most important is the quality of the back filling which must meet the required technological, strength, non-rigid, and physical characteristics. Low quality back filling can result in land fall in mines [1, 2]. Thus, improving the quality of the back filling is a topical issue at the mining-and-processing enterprises [3–7]. The quality of the back filling determines the possibility of further ore development, working conditions, ecological safety and others. A “dot” is used as a parting agent between the fractional and a whole part both in figures and tables.

2. Main body

Iron ore deposits development and ore mining is a topical prospective and an expensive trend in mining works. This field encounters several problems: work labour input and arduous working conditions, environmental issues, insufficient study of the soil layers, obsolescence and physical deterioration of the facilities and others. Among these issues one of the most important is the quality of the back filling which must meet the required technological, strength, non-rigid, and physical characteristics. Low quality back filling can result in land fall in mines [1, 2]. Thus, improving the quality of the back filling is a topical issue at the mining-and-processing enterprises [3–7]. The quality of the back filling determines the possibility of ore further development, working conditions, ecological safety and others. The article introduces results of studying the problem areas in filling masses of iron ore mining-and-processing enterprise. Some basic study of back filling after the blast hole drilling has been conducted by the authors.

Analysis of the back filling massif shows that all the samples have various external characteristics and peculiarities. First of all the strength of samples is different: some disintegrate by fingering others are rather strong. Sample humidity is different: from dry to wet. Great variety of the sample colors evidences poor quality of the back filling and internal strength in the massif layers. Figure 1 shows the photos of mine backfilling, different in colors and structure uniformity of the filling mass.

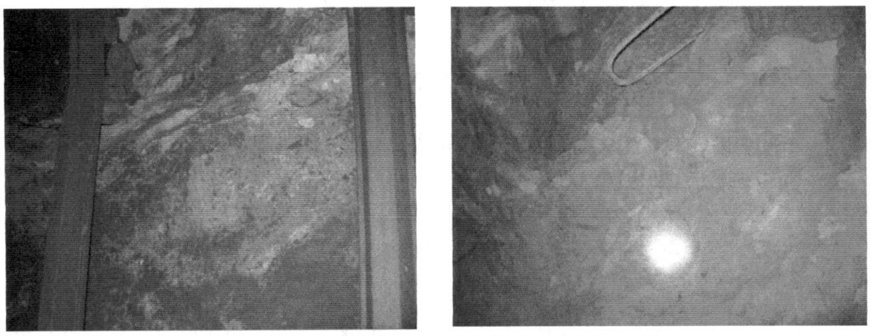

(a) (b)

Figure 1. Mine backfill

To determine the strength of the filling massif samples from different areas were taken 7 x 7 x 7 cm in size from the parts fallen from the filling massif. Split off lumps were 2–4 m^3 in volume, inhomogeneous in composition, which is seen as stripes of different color and strength, stripe thickness is 20–30 cm. Blocks were cut off from massifs 150 x 150 x 200 mm in size. Cubes were cut off from blocks, and the crushing strength limit for cubes has been determined. The conditionally determined type by the compressive strength varies within M21 to M100.

As a result of statistical treatment of the received values of strength the characteristic strength of the split lumps is 4.83 MPa, standard deviation S=1.41; variation coefficient V=0.29.

Samples were taken from the three lumps of the same massif. Such a variation in strength values and colors evidences undermixing the solution for the back filing, drawbacks in solution preparation, and transportation and mixture placing. The issues of quality and compatibility of the primary components for the mixture are studied separately [5, 8].

Besides, there were lustrous grains in samples, probably from the ore body. In order to study changes in properties of the mortar stone throughout the back filling massif samples were analyzed with the X-ray fluorescent spectrometer of ARL 9900 WorkStation series with in-built diffraction system.

The results of the chemical analysis allow determining indirectly the initial content of the Portland cement by the content of CaO in the original filling mixture. However, the CaO content in samples varies from 7.11 to 19.5 %, which makes variation nearly 3 times resulting in the inner massif lamination, unequal strength, internal stresses and further destruction.

The mineralogical composition of the representative sampling of the filling solution from the back filling has been studied. The equipment is DRON-2. Study parameters are: anode – Cu, step=0.05; E_{ksoz}=0.38; Speed=speed – 2; Maximum number of impacts=4630; S_{paeks}=4061; S_{total}=14674; K=27.7 %. Figure 2 shows the X-ray fragments of the back filling samples.

X-ray phase analysis of the solution showed that it contains cement hydration products: Portlandite $Ca(OH)_2$ (dÅ = 4.94; 2.978; 2.63; 1.708) and hydrated calcium silicates C-S-H(dÅ = 2.978; 2.784;

2.63; 2.188). Besides, there is a stage of silicone dioxide SiO_2 (dÅ = 4.26; 3.35), introduced into the system by quartzy sand.

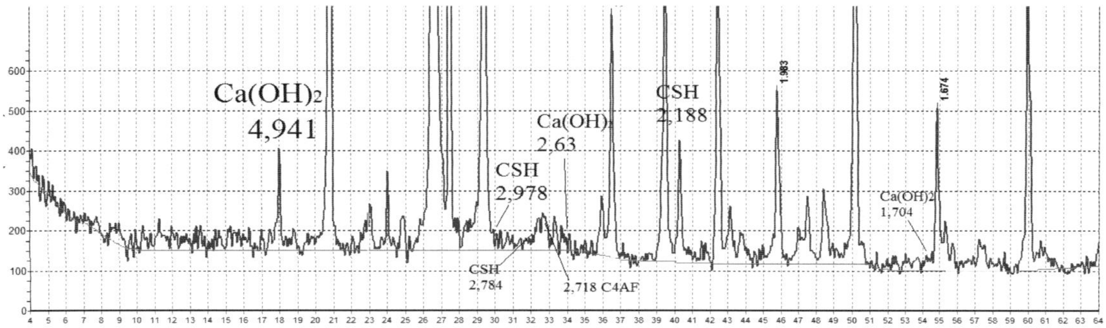

Figure 2. X-ray fragment of the filling mixture sample

It should be noted that the consolidated filling mixture has no diffraction maximums peculiar for concrete minerals which evidences about full cement hydration in the mixture. It is noted that the environment where the filling solidifies creates the best conditions: high humidity and constant temperature higher than 20 °C.

Thus, basing on the studied mineral composition of the samples of solidified filling a conclusion can be drawn that in filling massifs there are perfect conditions for concretion: high humidity and constant temperature higher than 20 °C that provides forming a solid, steady, durable massif. However, strength parameters of the filling mass first depend on a preliminary created mixture composition depending on the required designed strength.

Mechanical properties of the filling mass at a similar chemical and phase composition can alter depending on different geometrical parameters of new formations and pores, which characterize the structure of the solidified material [8-10]. To receive high-endurance stowage rock it is necessary to create volumetrically-homogenous, high-density structure with optimal dispersions of hydro silicates [11].

The hydration products of silicate minerals are in the form of tiny crystalline and sub microcrystalline (gellous) particles, and can be seen only by an electron microscope. An electron microscope of fine resolution TESCAN MIRA 3 LMU was used to study the microstructure of filling mixture samples. The results of the study are in figure 3.

Mixture microstructures are inhomogeneous, there is an entwinement of long-size lamina crystals with fine-crystalline new growths on the surface of the filler grains, and there is a non-uniform pore volumevoid space and numerous cavities, due to water evaporation. Forecasting the strength of such a sample we can say that such a mixture will have low strength characteristics.

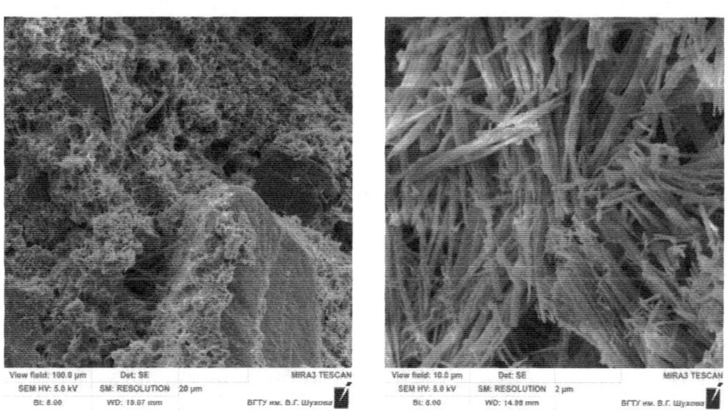

(a) (b)

Figure 3. Microphotographs of chip cleavage at various magnifications

For the generalized analysis we chose the following parameters: density, crushing strength, CaO content, Al_2O_3 content, sand SiO_2 content, SiO_2/CaO ratio (conventionally sand/concrete). Samples analysis, picked from the destroyed arrays, demonstrates a wide range of variations in density parameters (the difference in parameters is more than 42 %), the difference in parameters in crushing strength is from 1.35MPa to 14.59MPa, the ratio of Portland concrete to sand in the filling mass is from 3.34 to 11.2 (table 1).

Table 1. The results of a generalized analysis of samples from destroyed arrays

№ sample	Density, g/cm³	Crushing strength, MPa	mas. % CaO content	SiO_2	SiO_2/CaO
1	1.44	1.35	15.2	66.4	4.37
2	1.58	2.29	7.23	81.0	11.2.
4	1.48	1.86	7.11	78.0	11.0
5	1.45	2.1	7.6	78.0	10.3.
13/2	1.69	14.59	18.7	62.6	3.34
14/2	1.69	9.2	13.2	72.0	5.4
16/2	1.68	10.48	16.4	65.2	3.97
17/2	1.53	9.05	13.8	69.7	5.05
18/2	1.707	10.71	18.6	61.9	3.32

By SiO_2/CaO ratio the samples can be divided into 2 groups:

1. Additional layer, SiO_2/CaO ratio from 10.3 to 11.2, strength from 1.86 to 2.29 MPa, density 1.45–1.58 g/cm^3.
2. Base layer, SiO_2/CaO ratio from 3.32 to 5.45, strength from 1.35 to 14.59 MPa, density 1.2–1.707 g/cm^3.

Thus, basing on the results of the study we can say that the variation in CaO content evidences the mixture inhomogenuity when preparing basic materials or mortar mixes.

Density variations from 1.2 to 1.707 g/cm^3 can be caused by the inhomogeneous composition of the mixture at the stage of preparing basic materials, non-steady water-cement ration in the filling mixture.

Basing on the results of the works [5, 9–13] and research results of strength variations in massifs we can say about inhomogenuity of the filling mixture, and possible defective mixing of input components in the composite or wrong selection of the input products to form the filling.

To receive a complete picture of the reasons for variations in strength the study of basic materials for the filling mixture has been conducted. The results showed that the main input component of the filling massifs was sand with different characteristics. There is no constant composition of the filling mixture. The composition changes depending on the sand used. There is no coarse of mine stockpile at the enterprise. That was the reason for such a variation of strength of filling massifs, which is impermissible.

To solve this problem we have developed effective compositions with composite bindings and fillers to prepare filling mixtures with the substitution of Portland cement by the composite bindings. It will save up to 40 % of Portland cement and stabilize and improve strength characteristics of the filling massif up to 30 % of the guaranteed quality.

To solve such problems, a complex approach should be applied. So, to receive a quality and durable filling massif a number of other important issues should be solved. The first is possibility to use ashes, slag wastes, DMS, WMS, flotation in filling mixtures both as fillers and as components of the composite binding [1, 5, 6, 7, 13, 14].

To optimize the spending of the Portland cement and prepare quality filling mixtures a wet milling procedure is of special interest, when the binding is activated in the filling mixture [12. 15–18].

Dispersed reinforcement is also promising with technogenic raw material, introducing asbestos cement products wastes as fibro-optics, improving reinforcement of filling massifs and framing schemes of laying roofs for filling massifs.

Special attention should be paid to sand gradation for filling mixture in order to reduce water requirement of mixtures when preserving rheological properties. Introduction of large-size filler into the filling mixtures should also be considered. The technology of stowing operations is also important, exactly continuous feeding of the mixture along the pipeline.

Developing new justified systems of stowing complexes and technological lines of filling mixtures production is a step forward in developing further ore deposits with high economic effect.

Acknowledgments
The research has been conducted under the sponsorship of RFFP within the context of the scientific project №18-29-24113.

References
[1] Ageeva M S, Lesovik G A and Sopin D M 2014 *Filling mixtures based on technogenic raw materials* (Belgorod: Publ. House of BSTU named after V G Shukhov)
[2] German V I 2014 Rock failure prediction in mines by seismic monitoring data *J. of mining sci.* **50** 288–297
[3] Ageeva M S, Sopin D M, Ginzburg A V, Kalashnikov N V and Lesovik G A 2013 Developing composite bindings for filling mixtures *Bull. of BSTU named after V G Shukhov* **4** 43–47
[4] Ageeva M S, Lesovik G A, Kalashnikov N V and Mitrokhin A A 2014 Filling mixtures *Coll. of scie. papers* **9(2)** 14–19
[5] Lesovik V S, Zagorodnyuk L H and Chulkova I L 2014 Law affinity structures in material science *Fund. Res.* **3** 267–271
[6] Ptak M 2013 New solutions of mining waste management *Mining Sci.* **136(46)** 185–194
[7] Belagraa L 2016 Performance study of low environmental impact mortars based on mineral additions and cement resistant to sulfate (CRS) *Mining Sci.* **23** 65–76
[8] Golik V I 1999 Optimization of characteristic strength of consolidating mixtures when laying cavities *Mining inform. and analyt. Bull. (sci. and techn.l J.)* **3** 70–72
[9] Lesovik V S, Zagorodnuk L H, Tolmacheva M M et al 2014 Structure-formation of contact layers of composite materials *Life Sci. J.* **11** 948–953
[10] Agoshkov M I and Trebukov A L 1963 Effect of dry backfill on supporting strength of pillars *Sci. reports of NMRS A.A. Skochinsky Insti. for Mining* **18** 104–107
[11] Bronnikov D M and Zamesov N F 1971 *Production challenges of vast ore deposits of low angle-bending with systems with backfill* (Moscow: Nauka)
[12] Bronnikov D M, Zamesov N F and Kirichenko G S 1973 *Manufacturing science for underground development of ore deposits with filling* (Moscow: Science)
[13] Trebukhov A L 1981 *Applying consolidating stowing in underground ore mining* (M.: Nedra) 172 p
[14] Hayrutdinov M M and Shaymyardyanov I K 2009 Underground geotechnology with stowage to mined-out areas: disadvantages and improvement feasibilities *Mining inform. and analyt. Bull.* **1** 240–250
[15] Zagorodnyuk L H, Lasovik V S, SHamshurov A V and Belikov D A 2014 Composite bindings based on organo-mineral modifying agent for dry repair mortar *Bull. of BSTU named after V G Shukhov* **5** 25–31
[16] Zagorodnyuk L H, Sumskoy D A, Zolotykh S V and Kanaeva E V 2017 Obtaining binding compositions for thermal insulation solutions in the vortex jet mill *Bull. of BSTU named after V G Shukhov* **2** 25–35
[17] Zagorodnyuk L H, Lesovik V S, Sumskoy D A, Elistratkin M Yu and Makhortov D S 2017 Peculiarities of binding composition production in vortex jet mill *MEACS 2017 IOP Conf. Ser. Mater. Sci. and Engineer.* **327**
[18] Lesovik R V, Kazlitina O V, Ageeva M S, Sopin D M and Selukov M A 2017 Perspectives of using technogenic raw materials for filling mixtures. *Int. online congr. Fundamentals of binding materials science* pp 46–49

IOP Conf. Series: Materials Science and Engineering **952** (2020) 012067 doi:10.1088/1757-899X/952/1/012067

Well bottomhole cleaning device

L M Zaripova and M S Gabdrakhimov

Ufa State Petroleum Technological University, Branch of the University in the City of Oktyabrsky, 54a, Devonskaya St., Oktyabrsky, Republic of Bashkortostan, 452607, Russian Federation

E-mail: Lilyabert31@mail.ru

Abstract. Deterioration of bottomhole formation zone permeability is caused by mechanical pollution of porous medium, swelling of cement at its contact with water and accumulation of paraffin in low formation temperature deposits. In the process of oil production all extracted formation fluid: oil, water and gas pass through bottomhole zones of production wells and all water injected into formations – through the wellbore zone of injection wells. These processes occur at temperatures and pressures other than those at which these fluids (or gases) were originally on the surface or in the formation. As a result, different hydrocarbon components (asphalt-resin deposits) and different salts that fall out of the solutions as a result of thermodynamic equilibrium disorder can deposit in the wellbore zone, as in the filter. The choice of the impact method is based on the careful study of thermodynamic conditions and state of the wellbore zone, composition of rocks and fluids, as well as systematic study of accumulated production experience in this field.

1. Introduction

A productive formation containing gas, oil and formation fluid represents a complex hydrodynamic system in which physical and chemical processes are in a relatively equilibrium state prior to formation drilling.

During the operation of injection and production wells the permeability of rocks in the bottomhole zone decreases. Oil and gas formation was continually exposed to washing fluid impact at the stage of exploration works, as well as during drilling and throughout the whole period of operation up to formation development [1-5].

The main causes of pollution are as follows: penetration of killing or washing fluids into formations, swelling of clay particles in contact with fresh water, formation of water-oil emulsions, precipitation and deposition of asphalt-resin-paraffin components of oil or salts of produced water at the change of thermobaric conditions, accumulation of mechanical impurities and metal corrosion products in the wellbore zone, rock compaction as a result of hydrodynamic action on the wellbore zone during construction and operation of wells, fracture closing with the reduction of formation pressure. As a result of the above processes, the value of the colmataged net thickness can vary from 30 to 80% of the total penetrated thickness [6-11].

The choice of intensification method used in each case depends on many factors, in particular:

- type of collector;
- reservoir properties of the bottomhole formation zone;
- formation pressure;

- location of exposure intervals within the penetrated section;
- water-bearing intervals in the section;
- technical condition of the well;
- reasons for decrease of reservoir properties of the bottomhole formation zone – technical and financial abilities of the operating enterprise [12-14].

2. Materials and Methods

In order to influence the wellbore zone, a device is developed, the vibration exciter of which represents a ball.

The proposed device consists of a housing, a central washing channel and side washing channels arranged radially at a certain angle. The ball is arranged on the central channel to interact with the seat. The bottom of ball vibration excitation unit is made as ribs. The upper part of the device ends with an adapter. In the process of fluid circulation due to formation of vortices around the ball, the latter oscillates, which includes periodic separation of the ball from the seat formed by ribs. The ball is lifted due to irregularities 3-5 mm high under it, through which the fluid flows. The flow rate under the ball is higher than above and around it, resulting in a load-carrying capacity.

The periodic opening and closing of the central flushing channel results in double circulation of the fluid. When the central channel is closed, the entire flushing fluid flow is supplied by the wellbore zone. A batch jet blow occurs. When the ball is separated from the seat, the fluid is directed through the central channel into the annular space of a well.

The diameter of the valve chamber shall be 2-3 times higher the diameter of the central channel. In this case, in the lower position of the ball valve, the fluid flows freely to radial holes.

3. Results and Discussion

The amplitude of the fluid differential pressure oscillation within the device is determined by the Bernoulli equation.

The fluid entering the device flows out through three holes (exit area S_1, S_2, S_3) and the central hole of area S_4, which can be considered as a hole in a thin wall.

Let us write the Bernoulli equation for jets flowing out of the radial channels and the central opening:

$$\frac{aV^2}{2g} + \frac{p_0}{\rho g} = \frac{a_1 V_1^2}{2g} + \frac{p}{\rho g} + \xi_1 \frac{V_1^2}{2g}; \quad (1)$$

$$\frac{aV^2}{2g} + \frac{p_0}{\rho g} = \frac{a_2 V_2^2}{2g} + \frac{p}{\rho g} + \xi_2 \frac{V_2^2}{2g}; \quad (2)$$

$$\frac{aV^2}{2g} + \frac{p_0}{\rho g} = \frac{a_3 V_3^2}{2g} + \frac{p}{\rho g} + \xi_3 \frac{V_3^2}{2g}; \quad (3)$$

$$\frac{aV^2}{2g} + \frac{p_0}{\rho g} = \frac{a_4 V_4^2}{2g} + \frac{p}{\rho g} + \xi_4 \frac{V_4^2}{2g}, \quad (4)$$

where a_1, a_2, a_3, a_4 – Coriolis coefficients for fluid flow in the device, the first, second and third channels, in the central opening, respectively; V – flow rate inside the device; g – gravity acceleration; p_0 – pressure inside the device; ρ – density of the washing fluid; V_1, V_2, V_3, V_4 – jet velocity; p – fluid pressure in the bottomhole zone; $\xi_1, \xi_2, \xi_3, \xi_4$ – coefficients of overall resistance of channels and the opening.

The Coriolis coefficient for steady state flow is determined by the flow mode, which in turn depends on the Reynolds number:

$$Re = \frac{Ud}{\nu}, \quad (5)$$

where U – fluid flow rate; d – pipe diameter; ν – kinematic viscosity of the fluid.

The fluid motion can be considered to be established in view of the small size of the device and the high propagation velocity in the fluid. Considering that the velocity U depends on the volumetric flow rate Q and the internal diameter of the device d from the formula we obtain:

$$Re = \frac{4Qd}{\pi d^2 \nu} - \frac{Q}{d\nu}. \quad (6)$$

The calculations show that the fluid motion in the device is turbulent and the a_1, a_2, a_3, a_4 coefficients can be taken as one.

The resistances when flowing through the radial channels are mainly determined by local resistances. The resistance factor for the conical nozzle at an optimal angle β = 13.24 is 0.06 – the same as for the hole in the thin wall, so $\xi_1, \xi_2, \xi_3, \xi_4$ can be taken equal.

The fluid velocity in a bit is defined as follows:

$$U = \frac{Q}{\pi d^2/4} = \frac{4Q}{\pi d^2}. \tag{7}$$

The fluid flow rate through the central holes and channels will also make Q.

$$S_1V_1 + S_2V_2 + S_3V_3 + S_4V_4\varepsilon_4 = Q, \tag{8}$$

where ε_4 – jet compression ratio when flowing through the holes (ε_4=0.64).

Considering the channels to be the same, we get the following equation system:

$$\begin{gathered} \frac{V^2}{2g} = \frac{V_1^2}{2g}(1-\varepsilon) + \frac{p-p_0}{\rho g}, \\ \frac{V^2}{2g} = \frac{V_4^2}{2g}(1-\varepsilon) + \frac{p-p_0}{\rho g}, \\ 3S_1V_1 + \varepsilon_4 S_4 V_4 = Q, \\ V = \frac{4Q}{\pi d^2}. \end{gathered} \tag{9}$$

From which we get

$$V_1 = V_4 = \frac{Q}{3S_1+\varepsilon_4 S_4}, \tag{10}$$

$$p - p_0 = \rho Q^2\left(\left(\frac{4}{\pi d^2}\right)^2 - \left(\frac{1}{3S_1+\varepsilon_4 S_4}\right)^2\right)/2. \tag{11}$$

Thus, we determined the pressure drops in the device when the ball is raised. In the case where the ball closes the opening, the pressure drop can be determined by the same formula, considering $\varepsilon_4 = 0$:

$$(p - p_0)_{закр} = \rho Q^2\left(\left(\frac{4}{\pi d^2}\right)^2 - \left(\frac{1}{3S_1+\varepsilon_4 S_4}\right)^2\right)/2. \tag{12}$$

The pressure pulsation can be determined from the formulas:

$$\Delta p = \frac{\rho Q^2\left(\left(\frac{4}{\pi d^2}\right)^2 - \left(\frac{1}{3S_1+\varepsilon_4 S_4}\right)^2\right)}{2} = \frac{\rho Q(6S_1\varepsilon_4 S_4 + \varepsilon_4{}^2 S_4{}^2)}{18 S_1{}^2(3S_1+\varepsilon_4 S_4)}. \tag{13}$$

The jet pressure on the well wall generated by the channel is determined as follows. The enlargement in the jet section can be neglected. The pressure force on the well walls depends on the fluid density, the flow rate, the cross-sectional area and the angle Q formed by the incident jet with normal to the surface:

$$N = \rho S_1 V_1^2(1-\sin Q) = \rho S_1 \frac{Q}{3S_1+\varepsilon_4 S_4}(1-\sin Q).$$

4. Conclusion

The laboratory and field studies showed high performance of this device.

References

[1] Calabrese C, Mcbride-Wright M, Maitland G C and Trusler J P M 2019 Extension of Vibrating-Wire Viscometry to Electrically Conducting Fluids and Measurements of Viscosity and Density of Brines with Dissolved CO_2 at Reservoir Conditions *Journal of Chemical and Engineering Data* **64(9)** 3831-384

[2] Mukhametshin V V and Kuleshova L S 2019 Prediction of production well flow rates using survey data *IOP Conference Series: Earth and Environmental Science (IPDME 2019 – International Workshop on Innovations and Prospects of Development of Mining Machinery and Electrical Engineering)* **378(1) 012116** 1-7 DOI: 10.1088/1755-1315/378/1/012116

[3] Khokhlov V I, Galimov Sh S, Devyatkova S G, Kotenev Yu A, Sultanov Sh Kh and Mukhametshin V Sh 2019 Justification of impact and planning of technology efficiency on the basis of limy-emulsion formulation in low-permeability highly-rugged reservoirs of Tyumen

deposits *IOP Conference Series: Earth and Environmental Science (IPDME 2019 – International Workshop on Innovations and Prospects of Development of Mining Machinery and Electrical Engineering)* **378(1) 012114** 1–6 DOI: 10.1088/1755-1315/378/1/012114

[4] Fyrileiv O 2010 Effect of internal pressure on free spanning pipelines *Proceedings of the Biennial International Pipeline Conference IPC 3* 711-718

[5] Stepanova N V and Razumakov A A 2013 The 8 international forum on strategic technologies (IFOST 2013) **1** 240–242

[6] Suleimanov R I, Khabibullin M Ya and Suleimanov Re I 2019 Analysis of the reliability of the power cable of an electric-centrifugal pump unit *IOP Conference Series: Earth and Environmental Science (International Conference on Innovations and Prospects of Development of Mining Machinery and Electrical Engineering 2019, IPDME 2019)* **378(1)** 1-8

[7] Davydov A Yu 2018 Utilization of expert systems for screw pump sets with surface drive management *Advances in Engineering Research (AER) (International conference "Actual issues of mechanical engineering" (AIME 2018))* **157** 129-133

[8] Almuhametova E M, Kuzmina V V and Safiullina A R 2019 Scalimg Prevention on Wells of Tarasovskoe Field *Atlantis Highlights in Material Sciences and Technology (ISEES 2019 – International Symposium "Engineering and Earth Sciences: Applied and Fundamental Research" dedicated to the 85th anniversary of H.I. Ibragimov)* **1** 14–18 DOI: 10.2991/isees-19.2019.4

[9] Gilmanova A M 2019 Metrological support of geophysical equipment for acoustic logging *IOP Conference Series: Earth and Environmental Science (IPDME 2019 – International Workshop on Innovations and Prospects of Development of Mining Machinery and Electrical Engineering)* **378(1) 012080** 1–6 DOI: 10.1088/1755-1315/378/1/012080

[10] Zaripova L M and Gabdrakhimov M S 2019 A mathematical model of the pulsator for cleaning paraffin deposits of pipelines and downhole equipment *Journal of Physics: Conference Series* **1333(3)** 1-5

[11] Zaripova L M and Gabdrakhimov M S 2019 Restoration of intake capacity of injection well by vibrations *IOP Conference Series: Earth and Environmental Science (International Conference on Innovations and Prospects of Development of Mining Machinery and Electrical Engineering, IPDME 2019)* **378(1)** 1-4

[12] Zaripova L M and Gabdrakhimov M S 2019 Layout for drilling seismic wells by vibration method *IOP Conference Series: Earth and E nvironmental Science (2019 International Symposium on Earth Sciences: History, Contemporary Issues and Prospects, ESHCIP 2019)* **350(1)** 1-7

IOP Conf. Series: Materials Science and Engineering **952** (2020) 012068 doi:10.1088/1757-899X/952/1/012068

Analysis of the current issue of well bore and well bottom cleaning

L Z Zainagalina

Ufa State Petroleum Technological University, Branch of the University in the City of Oktyabrsky, 54a, Devonskaya St., Oktyabrsky, Republic of Bashkortostan, 452607, Russian Federation

E-mail: zlz11@mail.ru

Abstract. The paper considers the problem of insufficient cleaning of oil and gas wells from drill cuttings. The formation of slurry dunes is the main cause of complications and accidents in horizontal well drilling. The task related to the development of high-efficiency equipment for well bottomhole cleaning from drill cuttings is quite relevant. The existing techniques and technologies for borehole bottom cleaning from drill cuttings caused by rotary drilling were analyzed to find a way of improving the efficiency of borehole bottom cleaning. A new technical solution for cleaning vertical, directional and horizontal wells from drill cuttings was developed. The process functional diagram of at-bit sludge chopper during turbine drilling is presented to increase the cleaning efficiency of bottomhole and wellbore of vertical, directional and horizontal wells from drill cuttings during roller boring.

1. Introduction

Russia's energy strategy implies the annual increase of oil production. The increase of oil and gas production is inextricably connected with the increase of technical and economic indicators of well drilling. It is possible to reduce well construction time by increasing the drilling speed due to new types of rock-cutting tools, new types of washing fluids, mud circulating equipment: drill pumps, equipment for preparation and cleaning of washing fluids, as well as due to reasonable improvement of equipment operation. An essential reserve in improving the efficiency of drilling operations is the improvement of drilling techniques and technologies based on the introduction of efficient equipment for borehole bottom cleaning from drill cuttings. The degree of perfection of well circulation and the equipment for its implementation greatly affects successful, accident-free well drilling. The quality of borehole bottom cleaning from drill cuttings affects the cutting rate, well bore quality, penetration and resistance of a drill bit, a turbodrill, drilling equipment elements [7].

The bottomholes of oil and gas wells during drilling are usually contaminated with coarse cuttings thus deteriorating rock destruction and reducing bit operation parameters [1]. With the increase of the drilling depth at the bottomhole there remains an increasing amount of sediment consisting mainly of large-sized drill cuttings [3]. Cutting transport is complicated by the drilling of deep vertical, directional and horizontal wells, as well as by the drilling shallow wells – when rinsing fluid circulation does not create sufficient upward flow for cutting transport. The most pressing problem is the cleaning of the horizontal borehole from drill cuttings. At present, horizontal drilling during construction and reconstruction of wells is the most promising in oil and gas development industry [2, 5]. The development of oil and gas fields with the use of horizontal wells is a priority in the oil and gas production industry that involves hard-to-recover oil reserves (low-permeable heterogeneous formations and reservoirs, confined to water and gas-oil zones, oil basins of oil and gas deposits; deposits with high-viscosity oil, blind, peripheral and stagnant zones, with lens interlayers of different configuration, etc.) [4].

Insufficient cleaning from drill cuttings is the main cause of complications and accidents during horizontal well drilling, more than 80% of seizures occur due to well bore constraint during cuttings sinking. Due to complications and accidents caused by inefficient cleaning of wells with high slope angle, additional costs can reach up to 60% of the total costs of well construction [6]. Poor cleaning of wells from drill cuttings leads to such complications as the reduction of bit load, uncontrolled change of bore trajectory, difficulty with drilling tool orientation, change of the bottom-hole assembly operation mode, absorption of washing fluid, hydrodynamic seizure [9].

Compared to vertical wells the cleaning of horizontal bottomhole and well bore from drill cuttings is difficult due to radial orientation of gravity forces affecting mud particles and eccentric arrangement of a drill string in a well bore.

The effect of flooding increases due to complication of mud particles removal by washing fluid during drilling of direcitonal and horizontal wells, and the majority of technological and structural solutions connected with the transportation of drill cuttings to day surface is inefficient.

2. Materials and Methods

The existing techniques and technologies for borehole bottom cleaning from drill cuttings caused by rotary drilling were analyzed to find a way of improving the efficiency of borehole bottom cleaning.

F.A. Agzamov, T.O. Akbulatov, K.A.Kh. Alvan, T.E. Becker, A.I. Bulatov, M.S. Gabdrakhimov, N.A. Gukasov, B.I. Esman, G.P. Zozulya, V.I. Isaev, A.G. Kalinin, V.V. Kretsul, V.I. Krylov, E.G. Leonov, A.Kh. Mirzadzhanzade, J.J. Ozara, S. Okraji, G.A. Panfilov, A.N. Popov, V.I. Ryabchenko, B.Z. Sultanov, V.A. Fisher, R.I. Shishchenko, etc. made a great contribution to solving theoretical and practical issues of carrying out drill cuttings from wells with high slope angle and horizontal wells.

There are various techniques and technologies for improving the quality of well bore cleaning: improved quality composition of the drilling mud; increased flow rate of washing fluid; backflow; washing fluid flow separation devices; borehole washing with periodic lifting of a drill string by several stands and repeated drilling; introduction of devices with better hydrodynamic properties.

In order to increase the efficiency of bottomhole and borehole cleaning from drill cuttings, various devices built into a bit are used: ball, balance, acoustic vibrators. The quality of horizontal bores cleaning can be improved through radial oscillations of a drill string [8]. The use of vibration technologies also reduces the friction forces of a drill string against the well walls [2]. The disadvantage of vibrators in a drilling equipment is that in an extended horizontal section of a wellbore, there is a need for two or more vibration generators in a drill string to provide cleaning conditions and ensure self-synchronizing conditions in order to increase the efficiency of the hydraulic power of drill pumps consumed by the generators.

Various at-bit devices are used to improve the bottomhole cleaning: a drilling tool substitute equipped with a screw; screw helical finning drill pipes; jet pumps of different design. These devices improve the bottomhole cleaning to a certain extent, but do not remove coarse, solid mud particles, allow mud re-crushing, formation of mud dunes during horizontal drilling.

The effect of flooding increases due to complication of mud particles removal by washing fluid during drilling of direcitonal and horizontal wells, and the majority of technological and structural solutions connected with the transportation of drill cuttings to day surface is inefficient.

3. Results and Discussion

In horizontal drilling, all rock cuttings tend to fall on the bottom wall of a well. In order to prevent cuttings sinking and deposition, it is necessary that the rock cuttings are in a suspended state rather than in a transported state. However, increasing the supply and density of the drilling fluid results in the colmatation of well walls. According to literary data, rock cuttings with the size of not more than 1-2 mm can be transported in suspended state [10].

In order to increase the efficiency of well cleaning from drill cuttings, an at-bit device was developed, passing through which coarse rock cuttings are crushed to a size of 2 mm and less. The drum (grid) is mounted on a turbodrill shaft and is connected to a rotary shaft by means of springs. The rotary shaft of the turbodrill and the drum connected by springs form the oscillating system. The rotation of the turbodrill shaft creates torsional vibration of the drum due to the presence of springs, which provides grinding and screening of drill cuttings.

The device is installed during rotary drilling (rotary method, downhole motors) above the bit. The figure shows the process functional diagram of the at-bit sludge chopper during turbine drilling.

The main parts of the device include a housing 2, which is equipped with a screw 3 and a rotor 8, and a drum 5. The screw and the rotor are installed inside the drum, which has inlet and outlet channels 4, 9, holes 6, a baffle 7. At the top, the drum has a ribbed surface 12, which is integrated with the rotor end face. The device is equipped with a screw, which sucks the mud and sends it to the rotor, which, in turn, pushes the mud to the drum baffle 7. Thus, the mud is crushed, its fine particles are sieved through the drum, coarse particles are repeatedly crushed by the rotor end face.

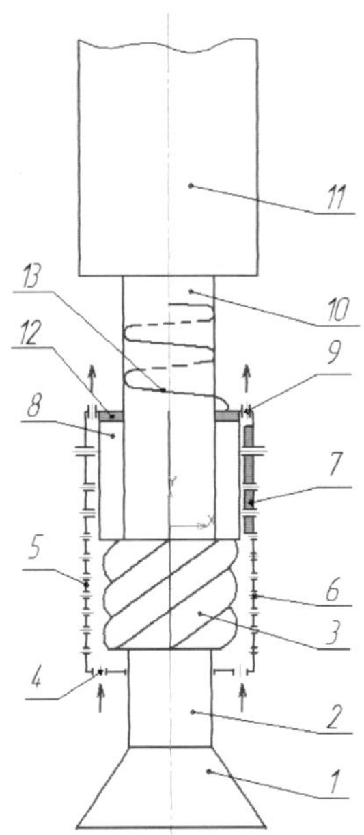

Figure 1. Process functional diagram of at-bit sludge chopper during turbine drilling
1 – drill bit; 2 – case; 3 – screw; 4 – inlet holes; 5 – drum; 6 – drum holes; 7 – baffle; 8 – rotor; 9 – outlet; 10 – turbodrill shaft; 11 – turbodrill; 12 – ribbed surface of the drum; 13 – torsion spring

4. Conclusion

The at-bit device ensuring crushing, sifting and withdrawal of drill cuttings from the bottomhole during drilling of vertical, directional and horizontal wells is developed. Fluid with drill cuttings is sent by the screw in axial and radial directions, at the same time drill cuttings are partially crushed due to shocks, fine particles are sifted through the grid (grid holes of 2-3 mm), coarse particles get into the rotor and are radially pushed onto the reflector, then drill cuttings are further crushed. Final grinding takes place when coarse particles get into the channels and are crushed by the rotor end face. Torsional vibrations of a drum (grid) under the drum inertia force, fluid thrown by the rotor onto eccentric reflector and spring forces contribute to drilling mud cleaning.

References

[1] Andreev V E, Chizhov A P, Chibisov A V and Mukhametshin V Sh 2019 Forecasting the use of enhanced oil recovery methods in oilfields of Bashkortostan *IOP Conference Series: Earth and Environmental Science (International Symposium «Earth sciences: history, contemporary issues and prospects»)* **350(1)** 1–6 DOI: 10.1088/1755-1315/350/1/012025

[2] Naganawa S, Sato R and Ishikawa M 2017 Cuttings-transport simulation combined with large-scale-flow-loop experimental results and logging-while-drilling data for hole-cleaning evaluation in directional drilling *SPE Drilling and Completion* **32(3)** 194-207

[3] Tombul H, Ozbayoglu A M and Ozbayoglu M E 2019 Computational intelligence models for PIV based particle (cuttings) direction and velocity estimation in multi-phase flows *Journal of Petroleum Science and Engineering* **172** 547-558

[4] Zeng C, Yan X, Zeng Z, Yang S 2018 The formation and broken of cuttings bed during reaming process in horizontal directional drilling *Tunnelling and Underground Space Technology* **76** 21-29

[5] Zainagalina L Z, Petrova L V and Petrov V A 2019 Analysis of the efficiency of telemetric systems for drilling wells *IOP Conference Series: Materials Science and Engineering (13th International Conference on Mechanical Engineering, Automation and Control Systems, MEACS 2018)* **560(1)** 012096 DOI: 10.1088/1757-899X/560/1/012096.

[6] Batalov S A, Andreev V E, Lobankov V M and Mukhametshin V Sh 2019 Numerical simulation of oil formation with regulated disturbances. Oil recovery quality simulation *Journal of Physics: Conference Series (ITBI 2019 – International Conference "Information Technologies in Business and Industry")* **1333(3)** 1-6 DOI: 10.1088/1742-6596/1333/3/032006

[7] Zainagalina L Z, Suleimanov R I, Gabdrakhimov M S and Khabibullin M Ya 2018 Determining oscillating sys-tem dynamic parameters of a near-bit junk pulper *Advances in Engi-neering Research (AER) (Inter-national conference "Actual is-sues of mechanical engineering" (AIME 2018))* **157** 642-645

[8] Ignatenko Y, Bocharov O, Gavrilov A and May R 2018 Steady-state cuttings transport simulation in horizontal borehole annulus *Proceedings of the International Conference on Offshore Mechanics and Arctic Engineering – OMAE 8*

[9] Khabibullin M Ya, Samigullina L Z and Zainagalina L Z 2018 Laboratory studies for the momentum of liquid optimal impact on hydro-carbon formation A*dvances in Engineering Research (AER) (International conference "Actual issues of mechanical engineering" (AIME 2018))* **157** 276-279

[10] Kuleshova L S, Mukhametshin V V and Safiullina A R 2019 Applying information technologies in identifying the features of deposit identification under conditions of different oil-and gas provinces *Journal of Physics: Conference Series (ITBI 2019 – International Conference "Information Technologies in Business and Industry")* **1333(7)** 1-5 DOI: 10.1088/1742-6596/1333/7/072012

IOP Conf. Series: Materials Science and Engineering **952** (2020) 012069 doi:10.1088/1757-899X/952/1/012069

Differential impact on wellbore zone based on hydrochloric-acid simulation

Yu V Zeigman[1], V V Mukhametshin[1], L S Kuleshova[2]

[1] Ufa State Petroleum Technological University, 1, Kosmonavtov st., Ufa, Republic of Bashkortostan, 450062, Russian Federation
[2] Ufa State Petroleum Technological University, Branch of the University in the City of Oktyabrsky, 54a, Devonskaya St., Oktyabrsky, Republic of Bashkortostan, 452607, Russian Federation

E-mail: vv@of.ugntu.ru

Abstract. On the basis of geologic and statistical modeling in high-viscosity oil deposits of the Tournaisian stage and using the obtained models the paper proposes to carry out forecasting, selection of wells, control and regulation of the impact process in order to reduce the number of inefficient operations and increase technical and economic indicators of fuel and energy complex enterprises.

1. Introduction

When carrying out, controlling and regulating the hydrochloric acid effect in the wellbore zone of the formation, it is important to know the forecast efficiency of this measure reflected by various criteria [1-8].

To solve this problem, in the conditions of high-viscosity oil deposits in carbonate reservoirs of the North-Western part of Bashkortostan, a generalization of the experience of conducting HAT was carried out with the construction of geological and statistical models using various volumes of field data.

2. Materials and methods

The following were used as independent variables: net oil thickness ($H_Э$), average thickness of oil-saturated sublayers ($H_П$) and their number (n), porosity factor ($M_Г$), share of reservoir rocks in the total reservoir thickness ($K_П$), time from the beginning of well operation to the HAT (t), maximum well flow rate before HAT ($Q_{H\,max}$), flow rate (Q_{H1}), watercut (f_1), accumulated oil production ($Q_{нак}$) of a well at the time of HAT, volume ($V_К$) and maximum pressure ($P_{зак}$) of acid injection into the formation. The following performance criteria were used as response functions: absolute ($Э_1$) and relative ($Э_2$) gain of oil flowrate; absolute ($Э_3$) and relative ($Э_4$) water production reduction; total gain of oil production during HAT ($Э_5$); relative increase of well productivity coefficient ($Э_6$).

This large number of performance criteria and different amounts of data is caused by the need to solve the tasks at various stages of development and in conditions of limited information on deposits (due to insufficient volumes of field studies due to organizational and financial reasons), as well as by the cases of changing the tactics and strategy of the enterprise in market conditions. This will allow a flexible response to changes in internal and external operating conditions [9-14].

3. Results and Discussion

The models were built using the step regression analysis.

Using the full amount of information (option 1), the following models were obtained:

$$Э_1 = 468 - 0{,}38t + 0{,}21Q_{H\,max} - 0{,}76Q_{H1} - 40{,}0f_1 + 8{,}18\cdot10^{-5}Q_{нак} + 15{,}6H_Э - 71{,}4H_П - 6{,}64M_Г - 81{,}1n + 7{,}03V_K;$$

$$Э_2 = 2254{,}1 + 0{,}76t - 1{,}62Q_{H1} + 38{,}8f_1 + 0{,}001Q_{нак} + 104{,}5H_Э - 407{,}4H_П - 73{,}0M_Г - 430{,}0n + 19{,}3V_K/H_Э + 8{,}50P_{зак};$$

$$Э_3 = -108{,}2 - 0{,}24t + 0{,}23Q_{H\,max} - 0{,}35Q_{H1} - 41{,}2f_1 - 3{,}54\cdot10^{-5}Q_{нак} - 11{,}1H_Э + 20{,}1H_П + 6{,}98M_Г + 36{,}6n - 4{,}06V_K;$$

$$Э_4 = -5677{,}3 - 2{,}78t + 0{,}024Q_{H\,max} + 2{,}57Q_{H1} - 603{,}0f_1 - 0{,}013Q_{нак} - 212{,}9H_Э + 658{,}2H_П + 206{,}9M_Г + 886{,}7n - 15{,}9V_K - 10{,}8P_{зак} + 1405{,}1K_П;$$

$$Э_5 = 2184{,}2 - 2{,}09t + 0{,}79Q_{H\,max} - 4{,}18Q_{H1} - 225{,}5f_1 + 0{,}001Q_{нак} + 67{,}8H_Э - 393{,}7H_П + 6{,}01M_Г - 430{,}7n + 56{,}3V_K;$$

$$Э_6 = 24{,}1 + 0{,}0001t + 0{,}23Q_{H\,max} - 0{,}046Q_{H1} + 5{,}31f_1 - 1{,}78\cdot10^{-5}Q_{нак} + 1{,}79H_Э - 1{,}94H_П - 2{,}12M_Г - 3{,}92n + 0{,}043V_K.$$

Using parameters reflecting the geological and physical properties of the formation at the point of its penetration by the well and the exposure technology (option 2), the following models were obtained:

$$Э_1 = -80{,}7 + 9{,}91H_Э - 33{,}9H_П + 3{,}74M_Г - 13{,}9n + 212{,}6V_K/H_Э;$$

$$Э_2 = 1036{,}3 + 4{,}00H_Э - 63{,}3H_П - 82{,}0M_Г + 16{,}7n + 412{,}1V_K/H_Э;$$

$$Э_3 = -151{,}0 - 6{,}04H_Э + 21{,}3H_П + 1{,}56M_Г + 37{,}7n - 1{,}47V_K;$$

$$Э_4 = -3837{,}5 - 79{,}7H_Э + 294{,}5H_П + 200{,}3M_Г + 358{,}8n - 37{,}3V_K;$$

$$Э_5 = 492{,}5 + 9{,}61H_Э - 246{,}3H_П + 27{,}0M_Г - 152{,}8n + 101{,}85V_K;$$

$$Э_5' = 0{,}035H_Э - 0{,}466H_П + 0{,}058M_Г - 0{,}264n + 0{,}616V_K;$$

$$Э_6 = 11{,}54 + 0{,}12H_Э - 1{,}21H_П - 0{,}37M_Г - 1{,}470n + 0{,}62V_K;$$

$$Э_6' = 0{,}056H_Э - 0{,}287H_П - 0{,}100M_Г - 0{,}317n + 0{,}469V_K.$$

When using parameters reflecting the technological features of wells and deposits, as well as the exposure technology (option 3), the following models were obtained:

$$Э_1 = 1{,}59 - 0{,}41t + 0{,}26Q_{H\,max} - 0{,}84Q_{H1} - 67{,}7f_1 + 0{,}00001Q_{нак} + 9{,}63V_K;$$

$$Э_2 = 255{,}4 - 0{,}65t - 2{,}38Q_{H1} - 405{,}1f_1 + 34{,}7V_K;$$

$$Э_3 = 39{,}9 - 0{,}21t - 0{,}08Q_{H1} - 15{,}0f_1 - 5{,}28V_K + 0{,}46Q_{H\,max};$$

$$Э_4 = -42{,}7 - 104{,}3V_K - 1{,}70t + 1{,}30Q_{H\,max} + 1{,}62Q_{H1} + 414{,}6f_1;$$

$$Э_5 = -132{,}7 + 69{,}0V_K - 1{,}81t + 1{,}28Q_{H\,max} - 4{,}24Q_{H1} - 249{,}7f_1;$$

$$Э_6 = 2{,}95 + 0{,}189V_K - 0{,}017t + 0{,}015Q_{H\,max} - 0{,}044Q_{H1} - 2{,}91f_1.$$

The relative errors according to option 1 vary from 19.4 to 48.9%, averaging 31.7%. Using a limited amount of information, the errors were naturally slightly increased, averaging between 2 and 40.4 %, and between 3 and 37.2 %, i.e. the dependencies obtained are rather quantitative and qualitative. In other words, they can be used in the selection of wells for hydrochloric acid treatment by comparing them with the values of any performance indicator (criterion). The increase of flowrate ($Э_1$, $Э_2$) and production ($Э_5$) of oil, reduction of watercut ($Э_3$,$Э_4$) and change of productivity coefficient ($Э_6$) for certain wells lead to significant errors, however, planning of these indicators by groups of wells (not less than 7-10) gives quite satisfactory results.

The analysis of the obtained models shows that in the vast majority (93%), the influence of geological and technological parameters on the treatment efficiency did not change using a limited amount of information.

With the increase of net oil thickness and decrease of the average thickness of oil-saturated sublayers, their number and porosity, the gain of oil flowrate and production, as well as productivity coefficients increase. At the same time, there is the increase of water cutting of well production.

In almost all parameters, the treatment efficiency decreases (with the exception of Э_6) with an increase in time from the moment of well commissioning to the moment of HAT. This fact is caused by the depletion of reservoir energy reserves and the decrease in the production reserve as oil reserves are developed.

With watercut increase (with the exception of Э_6), the treatment efficiency also decreases, which confirms the results of other researchers. With the increase of the maximum production rate of the well before HAT and the decrease of the oil production rate at the moment of impact, the efficiency of treatment increases.

It is particularly noteworthy that over time, as the water production increases, hydrochloric acid treatment leads to a greater increase in productivity. However, this does not indicate efficiency in terms of increasing production rates and additional oil production and reduction of water production. With the increase of well operation time and with the increase of watercut, the increase of oil production and watercut per unit of productivity coefficient change decreases.

With the increase of accumulated oil production at the time of impact, the increase in production rates and oil production increases, but at the same time the watercut also increases.

The increase in the amount of injected acid also leads to an increase in production rates, additional oil production and productivity ratio, but also the increase of water production.

The reduction of the maximum pressure of acid injection reduces the relative gain of oil flowrate, but also decreases the relative gain of water production. Based on the above, the selection of process impact indicators should take into account the fact that excessive acid injection and high injection pressure can lead to progressive watercut and the negative economic effect.

The analysis of the obtained results shows that a complex technological criterion of efficiency is needed, which would not only reflect the above efficiency criteria and various aspects of the hydrochloric acid effect, but also a compromise between the increase in the productivity coefficient, the gain in oil flowrates and production with the gain in water production. This criterion greatly simplifies the procedure for diagnosing, selecting wells and optimal process parameters of HAT.

As such a complex performance criterion, it is proposed to use the parameter Э_7 that is written in the form:

$$\text{Э}_7 = \frac{Q_{H2}}{Q_{H1}} \cdot \frac{f_1}{f_2} \cdot \tau,$$

where Q_{H1}, Q_{H2} – well flowrate before and after HAT, t/month; f_1, f_2 – water cutting of well production before and after HAT, %; τ – treatment, month.

The treatment duration was determined by the time during which the oil flowrate after treatment decreases to the oil flowrate before treatment. From a physical point of view, the efficiency criterion Э_7 characterizes the resource of oil production capacity of wells.

Using this complex performance criterion, the following models were obtained:

- option 1:

$$\text{Э}_7 = -3{,}85 - 0{,}06t - 0{,}02Q_{H\,max} - 0{,}04Q_{H1} + 15{,}21f_1 + 0{,}0001Q_{нак} - 6{,}89H_э + 11{,}2H_П - 1{,}33M_Г + 21{,}83n + 0{,}64V_К,$$

- option 2:

$$\text{Э}_7 = -55{,}2 - 6{,}11H_э + 15{,}4H_П - 1{,}39M_Г + 28{,}0n + 4{,}29V_К,$$

- option 3:

$$Э_7 = 16{,}8 - 0{,}02V_K - 0{,}05t - 0{,}008Q_{нак} - 0{,}05Q_{H1} + 8{,}61f_1.$$

The multiple correlation coefficients respectively make 0.895; 0.743; 0.820, and the relative errors – 20.6; 30.4; 26.0 %.

4. Conclusion

Thus, the performed geological and statistical modeling allows performing forecasting, selection of wells, control and regulation of the impact process in high-viscosity oil deposits of the Tournaisian stage in order to reduce the number of inefficient operations and increase technical and economic indicators of enterprises of the fuel and energy complex.

References

[1] Hognesen E J S, and Strand T A 2005 Waterflooding of preferential oil-wet carbonates: Oil recovery related to reservoir temperature and brine composition *SPE Europec/EAGE Annual Conference (Madrid, Spain, 13-16 June 2005)* 9 p DOI: 10.2118/94166-MS

[2] Kotenev Yu A, Mukhametshin V Sh, and Sultanov Sh Kh 2018 Energy-efficient technology for recovery of oil reserves with gas injection *IOP Conference Series: Earth and Environmental Science (IPDME 2018 – International Conference on Innovations and Prospects of Development of Mining Machinery and Electrical Engineering)* **194(8)** 1-6 DOI: 10.1088/1755-1315/194/8/082019

[3] Mukhametshin V V and Andreev V E 2017 Search and argumentation of decisions aimed at increasing the efficiency of bottom-hole zone stimulation in oil accumulations with challenged reserves *SPE Russian Petroleum Technology Conference (16-18 October 2017, Moscow)* 1–23 DOI: 10.2118/187785-MS

[4] Malyarenko A M, Bogdan V A, Kotenev Yu A, Mukhametshin V Sh, Umetbaev V G 2019 Wettability and formation conditions of reservoirs *IOP Conference Series: Earth and Environmental Science (IPDME 2019 – International Workshop on Innovations and Prospects of Development of Mining Machinery and Electrical Engineering)* **378(1) 012040** 1–6 DOI: 10.1088/1755-1315/378/1/012040

[5] Gomari K A R, Karoussi O, Hamouda A A 2006 Mechanistic Study of Interaction between Water and Carbonate Rocks for Enhancing Oil Recovery *SPE Europec/EAGE Annual Conference and Exhibition (Vienna, Austria, 12-15 June 2006)* 8 p DOI: 10.2118/99628-MS

[6] Valeev A S, Kotenev Yu A, and Mukhametshin V Sh 2018 Evaluation of Water-Alternating-Gas Efficiency when Using Wide Range of Gas Composition *IOP Conference Series: Earth and Environmental Science (IPDME 2018 – International Conference on Innovations and Prospects of Development of Mining Machinery and Electrical Engineering)* **194(8)** 1-8 DOI: 10.1088/1755-1315/194/8/082042

[7] Yakupov R F, and Mukhametshin V Sh 2013 Problem of efficiency of low-productivity carbonate reservoir development on example of Turnaisian stage of Tuymazinskoye field *Oil Industry* **12** 106–110

[8] Khokhlov V I, Galimov Sh S, Devyatkova S G, Kotenev Yu A, Sultanov Sh Kh and Mukhametshin V Sh 2019 Justification of impact and planning of technology efficiency on the basis of limy-emulsion formulation in low-permeability highly-rugged reservoirs of Tyumen deposits *IOP Conference Series: Earth and Environmental Science (IPDME 2019 – International Workshop on Innovations and Prospects of Development of Mining Machinery and Electrical Engineering)* **378(1) 012114** 1–6 DOI: 10.1088/1755-1315/378/1/012114

[9] Rabie A I, and Nasr-El-Din H A 2015 Effect of Acid Additives on the Reaction of Stimulating Fluids During Acidizing Treatments *SPE North Africa Technical Conference and Exhibition (Cairo, Egypt, 14-16 September 2015)* 27 p DOI: 10.2118/175827-MS

[10] Mukhametshin V Sh, Kotenev Yu A and Sultanov Sh Kh 2018 Assessment of the Need to Stimulate the Development of Hard-to-Recover Reserves in Carbonate Reservoirs *IOP Conference Series: Earth and Environmental Science (IPDME 2018 – International*

Conference on Innovations and Prospects of Development of Mining Machinery and Electrical Engineering) **194(8)** 1-4 DOI: 10.1088/1755-1315/194/8/082027

[11] Kadyrov R R, Nizaev R Kh, Yartiev A F, and Mukhametshin V V 2017 A novel water shut-off technique for horizontal wells at fields with hard-to-recover oil reserves *Oil Industry* **5** 44–47 DOI: 10.24887/0028-2448-2017-5-44-47

[12] Chudinova D Yu, Kotenev Yu A, Sultanov Sh Kh, and Mukhametshin V Sh 2018 The neural network for grouping wells of different facies zones of productive layer *IOP Conference Series: Earth and Environmental Science (IPDME 2018 – International Conference on Innovations and Prospects of Development of Mining Machinery and Electrical Engineering)* **194(8)** 1-6 DOI: 10.1088/1755-1315/194/8/082008

[13] Imqam A, Bai B, Wei M, Elue H, Muhammed F A 2016 Use of Hydrochloric Acid To Remove Filter-Cake Damage From Preformed Particle Gel During Conformance-Control Treatments *SPE Production & Operations* **31(3)** 11p DOI: 10.2118/172352-PA

[14] Yakupov R F, Gimazov A A, Mukhametshin V Sh, and Makaev R I 2018 Analytical method for estimating efficiency of oil recovery technology in case of bottom water-drive reservoir, verified on the hydrodynamic model *Oil Industry* **6** 66-69 DOI: 10.24887/0028-2448-2018-6-66-69

IOP Conf. Series: Materials Science and Engineering **952** (2020) 012070 doi:10.1088/1757-899X/952/1/012070

Improved rotary union in top-drive hydraulic power packs used on drilling rigs

R I Suleimanov and M Ya Khabibullin

Ufa State Petroleum Technological University, Branch of the University in the City of Oktyabrsky, 54a, Devonskaya St., Oktyabrsky, Republic of Bashkortostan, 452607, Russia

E-mail: rustamsul@rambler.ru

Abstract. The paper is concerned with a top-drive hydraulic power pack used on a drilling rig. The efficiency of hydraulic power packs can be ensured through a rotary union in a hose reel. An effective circulation of working fluid in the hose reel requires an improved rotary union, which allows the flow of working fluid to be redistributed with minimal energy and additional costs. As a result, the lifetime of the system is increased and the downtime of the rig is reduced.

1. Introduction
The top-drive drilling rig is recognized as one of the most significant advances in drilling technology since the advent of the rotary table. Unlike conventional drilling rigs, top-drive rigs are steadily demonstrating a higher drilling speed and safety level. Such drilling rigs are much less likely to be prone to drill pipe stickings. In addition, top-drive drilling allows the operator to drill to such horizons and formations that would be inaccessible with conventional rotary drilling [1, 2].

A top drive method, originally developed by American experts, dramatically changed the life of a drilling crew, greatly facilitating its work. The top drive provides torque to the drill string in the desired direction, in whatever position it is. In addition, the top drive drilling eliminates the need for additional manipulations and additional tools during drilling [3-5].

2. Materials and methods
Recently, the top drive has become the most popular method for drilling oil and gas wells, designed for quick and trouble-free drilling of vertical, directional and horizontal wells. Combining the functions of a swivel and a rotor, it is equipped with a set of devices for performing hoisting operations. The top drive is exploited in the oil industry in line with new safety rules, provided that a horizontal section is more than several hundred meters long [6, 7].

The paper discusses a fluid circulation system in the top-drive hydraulic power pack (Fig. 1). A technical challenge is to improve the reliability and durability of a hose reel, and simplify its maintenance.

IOP Conf. Series: Materials Science and Engineering **952** (2020) 012070 doi:10.1088/1757-899X/952/1/012070

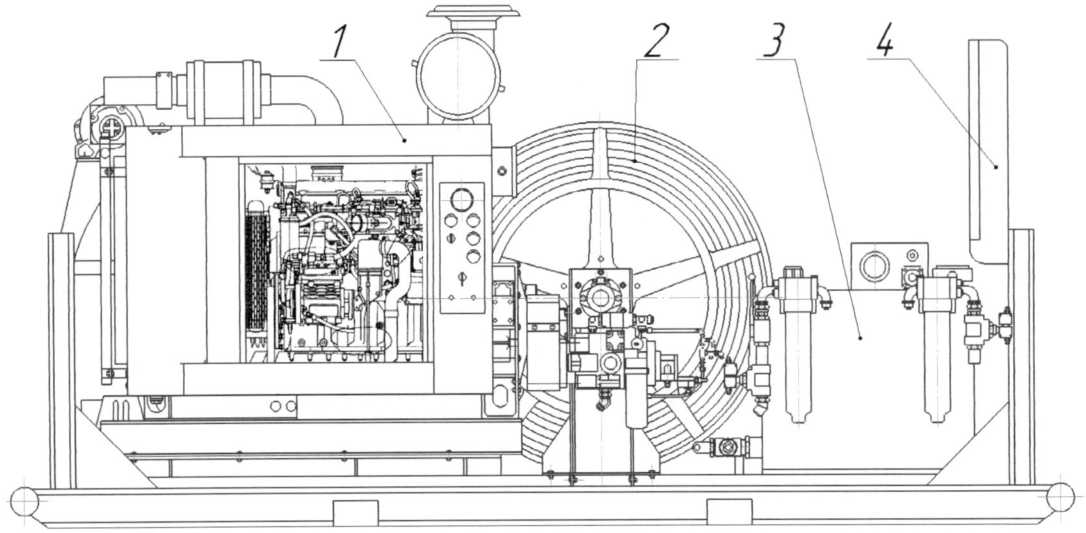

Figure 1. Top-drive hydraulic power pack: 1 – drive; 2 – hose reel; 3 – hydraulic oil reservoir; 4 – flow control

3. Results and discussion

The hose reel (Fig. 2) fixed in the top-drive hydraulic power pack is improved through quick disconnects removed from the assembly. They are located around three lines of the hydraulic system: the first line is circulation (supply) of hydraulic fluid to the hydraulic pump, the second line is circulation (return), the third line is drainage enabled by a triple-flow rotary union. The triple-flow rotary union is a mechanical device – a coupling through which hydraulic fluid is drawn from the fixed part of the hose reel to the rotating part of the hose reel. The triple-flow rotary union has three independent channels.

This technical solution can significantly increase its reliability, thereby reducing the cost of its maintenance and increase personnel safety.

Quick disconnects have a number of valve drawbacks, like cripplings of a shut-off valve spring, splinters getting into other items of the hydraulic system, freezing, leaks, deformation of the valve stem itself, foreign particles (dirt, sand, water, ice, etc.), and contaminated threaded matings [8].

These malfunctions can be eliminated either by replacing quick disconnects or repairing other equipment that has failed due to some destructed internal parts of quick disconnects, which leads to additional equipment downtimes and, consequently, financial losses for the entire enterprise.

The rotary union (Fig. 3) has a steel housing. The stem is made of 40X stainless steel. The metallized seal can withstand high working pressures in the system (P=3500 kPa). The bearing is mounted in the housing of the rotary union with an embedded part. It is possible to mount the rotary union on the equipment using a K-flange. The seal is made of teflon.

A sliding bearing is made of graphite, has an embedded element that stops the bearing from the torque in the housing. There is an annular groove on the rotor that indicats the wear of the seal – as soon as it appears from the housing, the seal must be changed [9].

The outer part of the triple-flow rotary union is attached to the adapter – the shaft (Fig. 4). The shaft is held by a flange-shaped cap that contains the bearing. The inner part of the triple-flow rotary union is mounted in the center of the hose reel, using four fastening bolts with a diameter of 10 mm, on a pentactinal inner base. Three ends of high-pressure hoses (HPH) are drawn to the pentactinal inner base, which, passing through the base, are sectionwise wound onto the hose reel [10].

The ends of the high-pressure hoses are crimped by special fittings screwed into threaded adapters at 45 ° and self-locking angle at 90 ° located at the base of the triple-flow rotary union.

The length of the HPH lines wound around the hose reel is 40-60 m, the inner diameter of which,

IOP Conf. Series: Materials Science and Engineering **952** (2020) 012070 doi:10.1088/1757-899X/952/1/012070

except for the drainage line, should be at least 32 mm. Finally, the diameter of the drainage circuit is at least 25 mm.

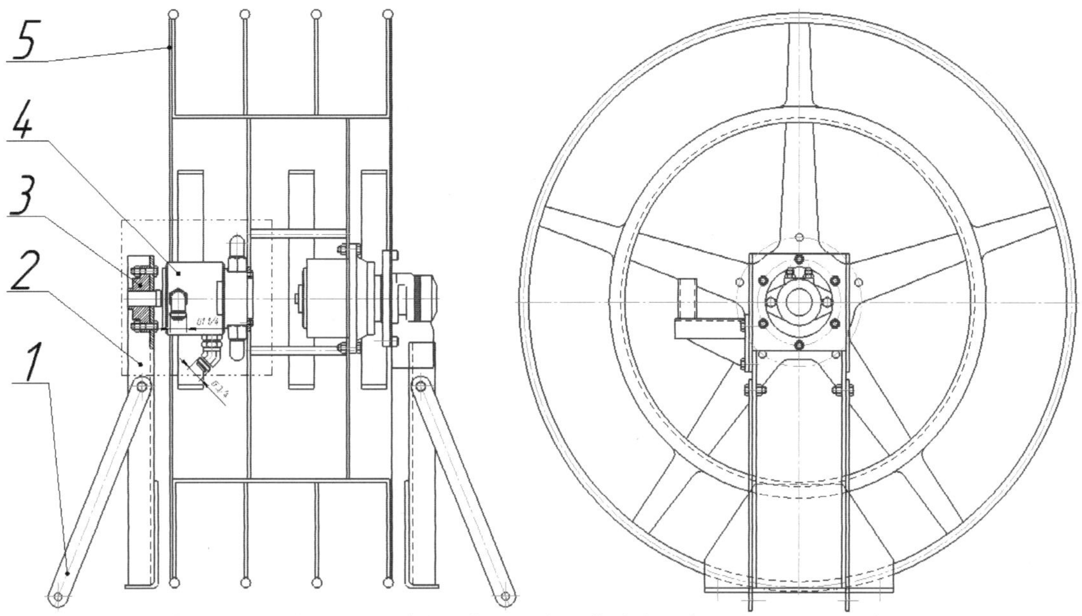

Figure 2. Hose reel: 1 – stand; 2 – frame; 3 – shaft bearing; 4 – rotary union; 5 – reel

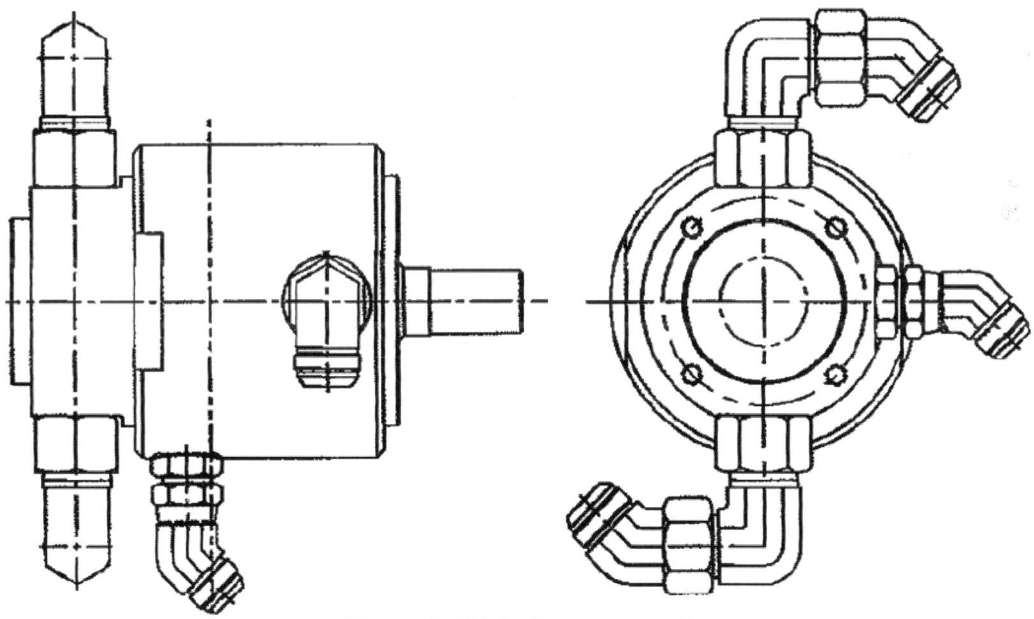

Figure 3. Triple-flow rotary union

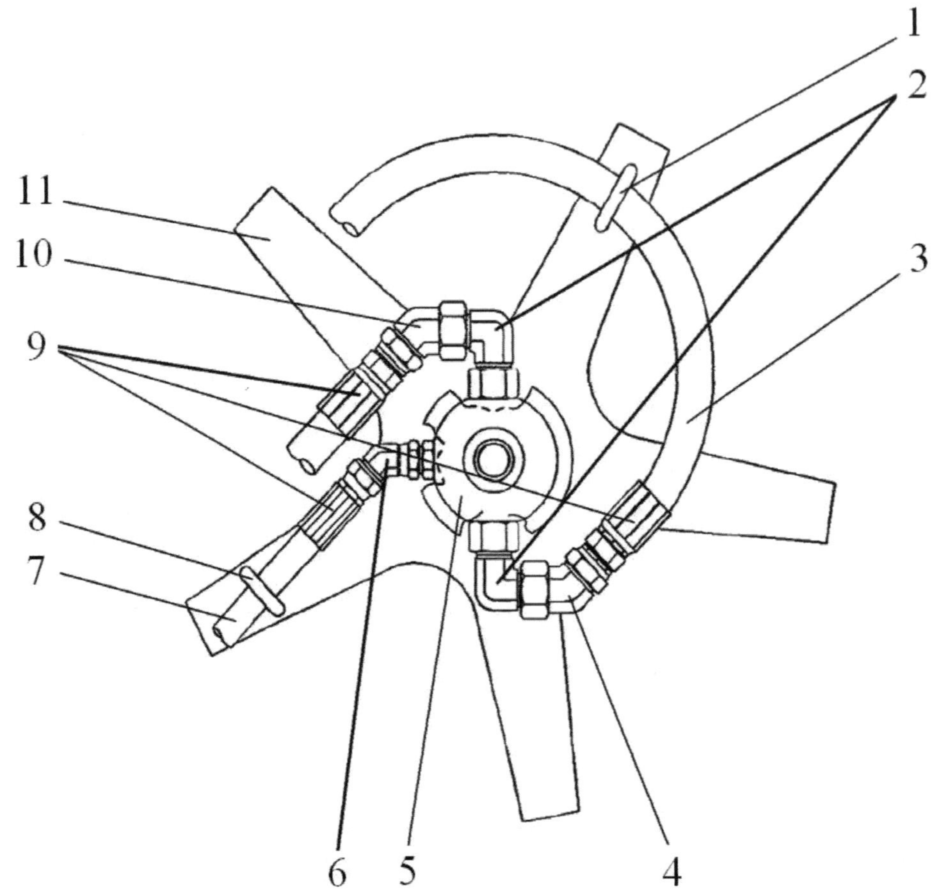

Figure 4. Pentactinal inner base assembled with triple-flow rotary union: 1 – retaining clip for a high-pressure hose (HPH); 2 – 90° self-locking circulation-line angle; 3 – circulation-line HPH; 4 – 45° circulation-line adaptive coupling; 5 – triple-flow rotary union; 6 – 45° drainage-line adaptive coupling; 7 – drainage-line HPH; 8 – retaining clip for HPH; 9 – fittings; 10 – 45° circulation-line adaptive coupling; 11 – pentactinal inner base

The heart of the rotary union is the mechanical seal (Fig. 5) that ensures its tightness. It is crucial to choose a proper mechanical seal to ensure the efficiency of the rotary joint, since in the event of a failure and leakage of the medium, the efficiency of the cooling or heating process as such is significantly reduced. The design of rotary unions provides a quick and easy replacement of mechanical seals, which is very important to reduce downtime and repair costs [11, 12].

Being an integral part of a swivel joint, mechanical seals should be selected based on the characteristics of the medium.

The load onto the mating surfaces of the seals is minimized and almost independent of the pressure of the medium, which significantly prolongs the lifetime of the seals and the union as a whole.

The optimum balance of the seals allows a thin lubricating film between the mating surfaces of the seals.

IOP Conf. Series: Materials Science and Engineering **952** (2020) 012070 doi:10.1088/1757-899X/952/1/012070

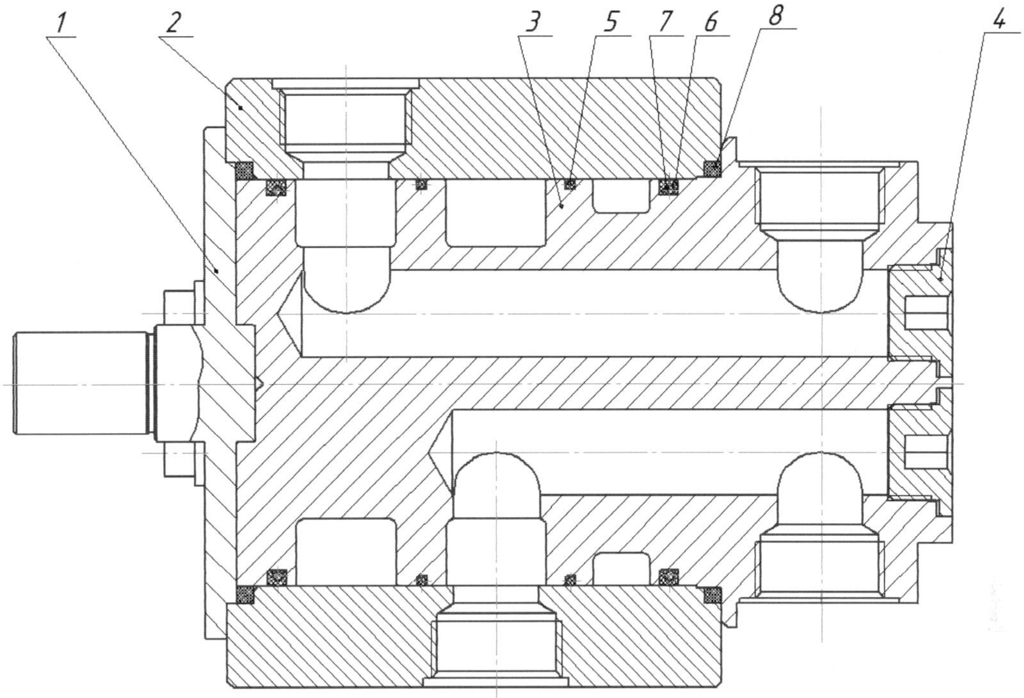

Figure 5. Triple-flow rotary union: 1 – cap; 2 – cylinder; 3 – rod; 4 – plug; 5 – retaining teflon ring; 6 – rubber seal; 7 – rubber ring; 8 – graphite ring

4. Conclusion

The technical effect implies that replacing an item of the top drive system extends the lifetime of the equipment, both of the item itself, the hose reel, and other elements located in the same line of the hydraulic system. The economic effect is achieved by reducing the cost of purchasing spare parts (quick disconnects), thereby eliminating equipment downtime during the operations performed by drilling crews. The advantages of the presented technical solution are that the load on the mating surfaces of the seals is minimized and almost independent of the pressure of the medium, which significantly prolongs the lifetime of the seals and the union as a whole.

References

[1] Sun Y, Shi Y, Wang Q and Yao Z 2018 Study on speed characteristics of hydraulic top drive under fluctuating load *Journal of Petroleum Science and Engineering* **167** 277-286

[2] Nassif A B, Said H and Lang D 2019 Harmonics and interharmonics of top and mudpump variable frequency drives in drilling rigs *Industrial and Commercial Power Systems Technical Conference 2019-May* 8733330

[3] Simonyants S L and Al Taee 2019 Stimulation of the drilling process with the top driven screw downhole motor *Journal of Mining Institute* **238** 438-442

[4] Khabibullin M Ya and Suleimanov R I 2019 Automatic packer reliability prediction under pulsed transient flooding of hydrocarbon reservoirs *IOP Conference Series: Materials Science and Engineering* **560(1)** 012024 DOI: 10.1088/1757-899X/560/1/012024.

[5] Suleimanov R I, Khabibullin M Ya and Suleimanov Re I 2019 Analysis of the reliability of the power cable of an electric-centrifugal pump unit *IOP Conference Series: Earth and Environmental Science* **378(1)** 012054 DOI: 10.1088/1755-1315/378/1/012054.

[6] Zainagalina L Z, Petrova L V and Petrov V A 2019 Analysis of the efficiency of telemetric systems for drilling wells *IOP Conference Series: Materials Science and Engineering* **560(1)** 012096 DOI: 10.1088/1757-899X/560/1/012096.

[7] Davydov A Yu and Yakhina I A 2019 Bit measuring systems *IOP Conference Series: Earth and Environmental Science* **378(1)** 012078 DOI: 10.1088/1755-1315/378/1/012078.

[8] Gabdrakhmanov N Kh, Davydova O V and Valeyev M D 2018 Studying gas separation at the inlet of the oil well bottom pump *IOP Conference Series: Earth and Environmental Science* **194(8)** 082011

[9] Zaripova L M and Gabdrakhimov M S 2019 Layout for drilling seismic wells by vibration method *IOP Conference Series: Earth and Environmental Science* **350(1)** 012048 DOI: 10.1088/1755-1315/350/1/012048.

[10] Batalov S A, Andreev V E, Lobankov V M and Mukhametshin V Sh 2019 Numerical simulation of the oil reservoir with regulated disturbances. Oil recovery stability simulation *Journal of Physics: Conference Series* **1333(3)** 032007 DOI:10.1088/1742-6596/1333/3/032007

[11] Kuleshova L S, Mukhametshin V V and Safiullina A R 2019 Applying information technologies in identifying the features of deposit identification under conditions of different oil-and gas provinces *Journal of Physics: Conference Series* **1333(7)** 072012 DOI: 10.1088/1742-6596/1333/7/072012

[12] Mahmud W M and Elmabrouk S K 2016 Rig selection and cost analysis; a comparison of top drive and rotary table drive rig systems *Proceedings of the International Conference on Industrial Engineering and Operations Management* pp 1029-1037

IOP Conf. Series: Materials Science and Engineering **952** (2020) 012071 doi:10.1088/1757-899X/952/1/012071

Degree and nature of shaft misalignments in pumping units assessed against vibration criteria

R N Suleymanov and O V Filimonov

Ufa State Petroleum Technological University, Branch of the University in the City of Oktyabrsky, 54a, Devonskaya St., Oktyabrsky, Republic of Bashkortostan, 452607, Russia

E-mail: rsulem@mail.ru

Abstract. The paper shows that the degree and type of shaft misalignments, along with imbalances, have a decisive impact on vibration levels in centrifugal pumps. An accurate assessment is impossible without a unit shutdown to measure shaft alignments with special devices. A method for determining misalignments is proposed to assess the degree of misalignment against vibration spectra, the so-called rotational harmonics.

1. Introduction

The most recognized techniques to backup centrifugal pumps involve detecting misalignment, shutting down a machine and measuring shaft alignment with special devices.

When actuating pumping units, the vibration level in many respects depends on whether the rotors constituting a unit are properly coupled, i.e. on their proper alignment. According to Fixtur-Lazer AB (Sweden), in 49% of cases, misaligned machinery causes premature breakdowns, and for pumping machinery this accounts to 60%! [3]. According to the laboratory of vibration diagnostics of USPTU branch, Oktyabrsky city, pumping units exploited at RPM OGPD, which have alignment capacity worse than the demanded specifications, make up about 90% of all examined units [1-3].

Shaft alignment is generally inspected once a unit is shut down exclusive of the stress applied to the elements of the unit during its operation. For RPM and OHP OGPD units, that is stress from the suction and discharge pipelines, insufficient rigidity of the foundation frame of the unit, etc. (dynamic misalignment), from various thermal expansion of the motor and pump (thermal misalignment). These phenomena are considered in detail in [4].

2. Materials and methods

There are four types of shaft alignments: axial, radial (or parallel), angular (or end), and combined. Misalignment can be estimated indirectly based on vibration spectra [2], however, the well-known criteria (based on vibration levels and rotational harmonics) are general [1].

To assess the effect of misalignment on the level of vibration and the possibility to diagnose it against vibration signatures (spectra), the authors conducted an experiment on a TsNS180 centrifugal pump, widely used in the industry, according to the research plan shown in Table 1.

Table 1. Types and degrees of misalignment

Type		Horizontal			Vertical		
		parallel	angular		parallel	angular	
		R^{11m}	R^{21m}		R^{21m}	R^{22m}	
		μm	°	μm	μm	°	μm
up	2	860	-	-	1190	-	-
up	2	670	0°8′57′′	495	530	-	-
up	2	285	0°3′37′′	200	360	0°2′	110
-	0	0	0	0	0	0	0
down	1	-380	-0°4′42′′	-260	-230	-0°3′31′′	-195
down	1	-750	-0°9′40′′	-535	-570	-0°6′09′′	-340
down	1	-900	-	-	-925	-0°10′08′′	-560

Note: 1. Approximation for angular misalignment was subject to a rear bearing offset about the axis of the pump (in μm).

3. Results and discussion

Unit No. 1 selected as a research object (CPS No. 14, RPM OGDU Arlanneft JSOC Bashneft) was mounted onto the capital foundation and comprised an asynchronous electric motor – 4AZMA 800/6000 and a centrifugal pump – TsNS 180-1200 coupled by an MZP toothed clutch with a shortened intermediate shaft in accordance with GOST 5006-83E. The misalignments were measured with a PTs-2 aligner equipped with dial gauges; the gaskets were calculated using a specialized Center calculator (Laboratory of vibration diagnostics of USPTU branch, Oktyabrsky city [5]). Vibration was measured with an AU014 vibration analyzer (DIAMECH, Moscow), while data processing was ensured by the TrendTest and Diamant software packages (DIAMECH).

.

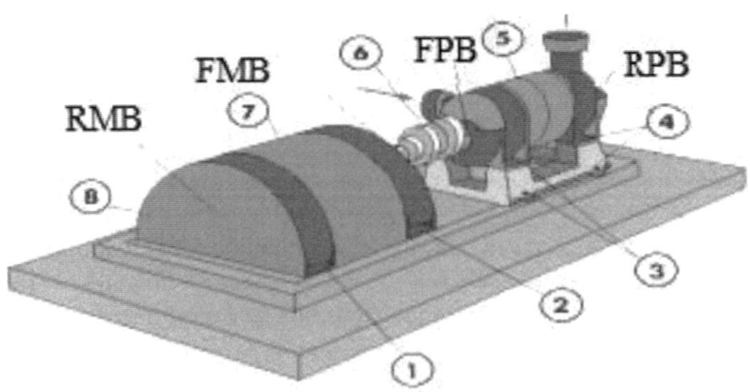

Figure 1. Schematic for measuring vibration level: RMB, FMB, FPB, RPB – rear motor bearing, front motor bearing, front pump bearing, rear pump bearing, respectively; 1, 2, 3, 4, 5, 6, 7, 8 – unit supports – measuring points of vibration on the lugs, frame and foundation.

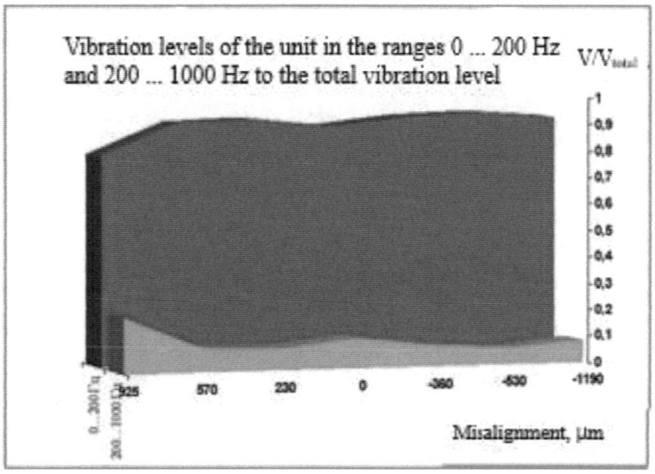

Figure 2. Levels of reduced vibration V/V_{total} in the frequency band 0 ... 200 Hz and 200 ... 1000 Hz

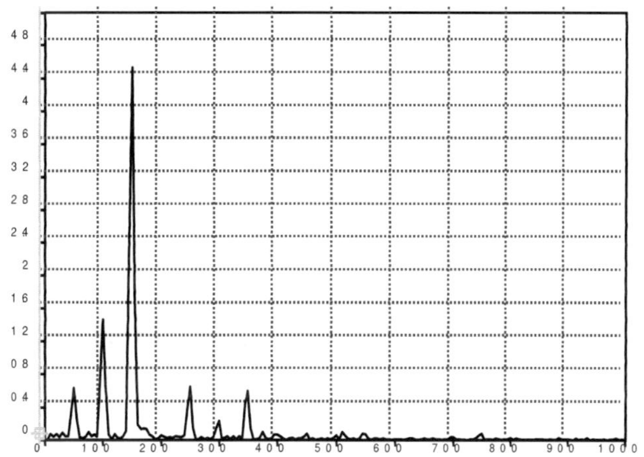

Figure 3. Malfunction spectrum "misalignment" of the shafts: about the X axis – frequency, about Y – vibration velocity (mm/s).

The vibration level was measured against the rms-vibration velocity in the frequency range 10 ... 1000 Hz on the bearings, in three orthogonal directions (Fig. 1), according to GOST ISO 10816-1-97.

Due to some bulkiness, vibration velocity measurements, as well as vibration-offset graphs, are not completely presented in the paper. Figure 4 shows a typical graph of the relationship between a dimensionless vibration level and the degree of offsets.

The target spectra were proved to contain the frequency band most sensitive to changing alignment values, namely the frequency range of 0..200 Hz (Fig. 2), i.e. the range to the fourth shaft rotational harmonic f0 (basically these are the first three shaft rotational harmonics f0, 2f0, 3f0 (Fig. 3), comprising ~ 90% of the energy of the entire frequency band.

Therefore, we decided to monitor the changes in the vibration level of the first three harmonics f0, 2f0, 3f0, as well as in the frequency band up to 150 Hz, inclusive, which are most affected by the misalignment.

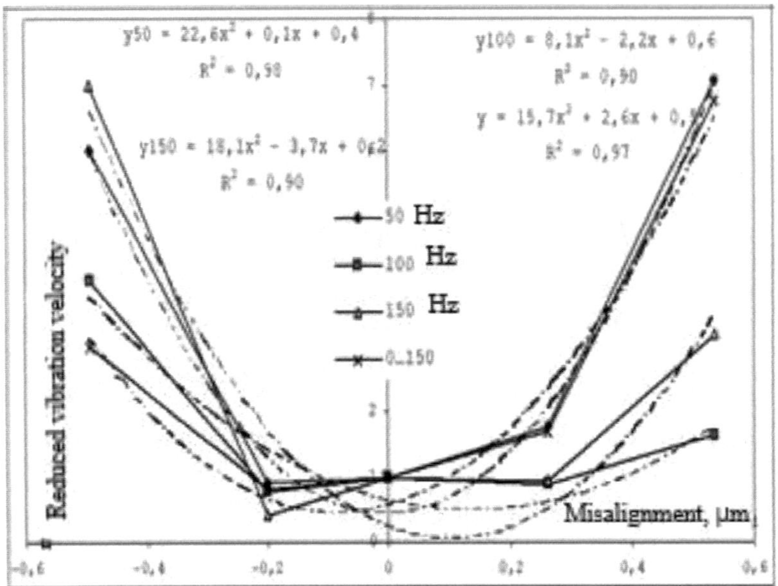

Figure 4. Vibration (reduced vibration velocity) vs misalignment: l, l50, l100 and l150 – respectively, total level, harmonic 50 Hz, 100 Hz and 150 Hz.

4. Diagnosis of misalignment by measuring the total level

For measurements, it is necessary to have the total vibration values in the band of 10 ... 1000 Hz. In the case under study, the total vector [4-5] in the band of 10 ... 150 Hz (p=4) V_{ij4}^{klm} is close to this value.

The misalignment for the total level is difficult to diagnose, only the following rules can be indicated (unlike [4], when processing measurement outputs, a polynomial approximation of the relationship between the rms-vibration velocity and the misalignment degree is accepted, namely, the quadratic dependence subject to parity parameter *m*):

- horizontal angular misalignment: there is a regular increase on all bearings and the whole unit, with a rate of 4 ... 9 1/mm^2 with an accuracy of over 0.85;
- vertical angular misalignment: there is a regular increase in radial vibrations at RMB and FMB, with a rate of 5 ... 19 1/mm^2 and 5 ... 9 1/mm^2 (FMB and RMB as a whole) with an accuracy of over 0.95;
- vertical radial misalignment R^{21m}: there is an increase with an accuracy of 0.9 with a rate of 2 1/mm^2 (RMB, axial and vertical).

In general, the above rules do not contradict generally accepted ideas. However, for a more accurate assessment (type, degree) of misalignments, it is necessary to analyze the variation spectra and establish the relationship between vibration growth at individual harmonics and misalignment signatures.

5. Diagnosis of alignment by vibration spectra

The scope of the studies, the cumbersome enumeration of the types of misalignment, the points of X, Y and Z vibration measurements at the corresponding rotational harmonics all suggest the formalization carried out in a pseudo-tensor format [6-9] R^{klm} (Table 2) and the corresponding vibration velocity values V_{ijp}^{klm}, where *k* is the direction of misalignment, respectively, horizontal and vertical, or 1 and 2; *l* is the type of misalignment, respectively, radial and angular or 1 and 2; *m* is the nature of misalignment, respectively, "up" and "down" (for vertical misalignment, for horizontal alignment the "up" position is taken to be the forward rotation of the coupling by 90 °, and "down" is the rotation of the coupling by 270 °) or 1 and 2; *i* is vibration measurement point (Fig. 1), respectively, RMB, FMB, FPB and RPB or 1, 2, 3 and 4; *j* is the direction of measurement, respectively, X, Y and Z, or 1, 2 and 3; *p* is the rotational harmonic or frequency band, respectively, 50, 100, 150 and 0 ... 150 Hz or 1, 2, 3 and 4 (rationale for these harmonics is provided below). Thus, the "vertical" "radial" "up" misalignment of 300 μm will be presented in the form R^{211} = 300 μm, and the corresponding vibration velocity of 12 mm/s at the RMB about the X axis at a frequency of 100 Hz – in the form V_{412}^{211} = 12 mm/s.

The *j* convolution gives the value of the vibration vector on the whole bearing over the shaft rotational harmonics. The X, Y and Z summation is performed:

$$V_{iop}^{klm} = \sqrt{\sum_{j=1}^{3} V_{ijp}^{klm} \cdot V_{ijp}^{klm}} \qquad (1)$$

The *i* convolution gives the value of the vibration vector on the whole unit. The summation is performed for the bearings:

$$V_{ojp}^{klm} = \sqrt{\sum_{i=1}^{4} V_{ijp}^{klm} \cdot V_{ijp}^{klm}} \qquad (2)$$

For example, *ijp* indices equal to 10p or 00p mean, respectively, the vibration vector for 1 bearing (RPB) as a whole – convolution solely for j index, or for the unit as a whole – convolution for *i* and *j* indices.

$$V_{oop}^{klm} = \sqrt{\sum_{j=1}^{3} V_{ojp}^{klm} \cdot V_{ojp}^{klm}} \qquad (3)$$

In addition, non-dimensionalization through normalization is preliminarily performed for the vibration level when misalignment is "zero" (m=0, which herein corresponds to the minimum radial misalignment achieved of no more than 30 μm and angular – no more than 0'23"):

$$\tilde{V}_{ijp}^{klm} = \frac{V_{ijp}^{klm}}{V_{ijp}^{klo}} \qquad (4)$$

Hereinafter, the reduced vibration velocity is being discussed and, hence, the "~" sign is omitted.

The diagnosis [10-12] requires a rotational spectrum and its first two harmonics (p=1, 2, 3). Then, the procedure for diagnosing the type and degree of misalignment will consist in applying the tensor V_{ijp}^{klm} to the mask (Table 3) constructed based on experimental data. Subject to the corresponding values of vibration velocity, the type and degree of misalignment is determined.

Moreover, the formalization of the comparison procedure allows such an assessment to be made in any algorithmic language or in Microsoft Excel. The degree of *P* forecast reliability is estimated in a usual statistics-related way [2–4] provided that the estimates for individual harmonics are independent. For example, in the case of "coincidence" F of the values V_{ijp}^{klm} for independent features, according to Bayes' theorem, the probability of misalignment is

$$P = P(r) \bullet \prod^{F} P_{ijp}^{kl}(r) / \left[P(r) \bullet \prod^{F} P_{ijp}^{kl}(r) + (1 - P(r)) \bullet \prod^{F} P_{ijp}^{kl}(o) \right] \qquad (5)$$

where P (r) is the frequency of "misalignment" occurring. Given that it is equal to 0.6 [5], then (1- P (r)) is equal to the frequency of misalignment (according to the probability theory, these events make up a complete group); $P_{ijp}^{klm}(r)$ and $P_{ijp}^{klm}(o)$ are the probability of increasing the vibration level at a point with "coordinates" k, l, i, j, p for a misaligned (*r*) and aligned (*o*) unit.

IOP Conf. Series: Materials Science and Engineering **952** (2020) 012071 doi:10.1088/1757-899X/952/1/012071

Table 2. Mask of different misalignments affecting the level of vibration

	R^{21m}				R^{22m}				R^{11m}				R^{12m}			
klp	1	2	3	4	1	2	3	4	1	2	3	4	1	2	3	4
11p	-	-	-	-	-	+3	±	-	-	-	-	-	+5	+2	+7	+5
12p	-	±	-	-	-	+21	-	-	-	-	-	-	-	+23	+10	+3
13p	-	-	-	-	±	±	-	-	-	-	-	-	+7	±	+8	+5
21p	-	-	-	-	-	+7	-	-	-	-	-	-	+9	-	+4	+9
22p	-	±	+3	-	-	+14	-	-	-	±	±	-	-	+13	+27	+5
23p	-	-	-	-	-	+28	-	±	-	+3	-	-	-	+20	-	+2
31p	-	-	-	-	+8	+7	-	+5	+2	-	-	-	+16	+5	±	+7
00p	-	-	-	-	-	+8	-	-	-	-	-	-	+5	+5	+7	+6

Note:

1. The "+" sign indicates a positive correlation with high accuracy (over 0.75), "-" indicates its absence, "±" indicates a positive correlation with low accuracy (less than 0.75).
2. The digits next to the "+" sign indicate the tempo (constant at degree 2 in the polynomial approximation function (Fig. 4), and generally characterize the effect of this type of misalignment on the value of vibration velocity.

6. Conclusion

1. Informative frequencies (frequency bands) have been established to make it possible to detect a "misalignment" malfunction.
2. A "mask" of relative vibration levels for rotational harmonics and a relevant comparison procedure have been developed to diagnose the type and degree of misalignment of the shafts.

References

[1] Suleymanov R N, Tincherov K T, Galeev A S, Filimonov O V and Selivanova M V 2019 Criteria for modeling the Gravity Clarifier *International Symposium "Engineering and Earth Sciences: Applied and Fundamental Research" (ISEES 2019) A. Atlantis Highlights in Material Sciences and Technology (AHMST)* **1** Available at: https://doi.org/10.2991/isees-19.2019.164.

[2] Galeev A S, Bikbulatova G I, Suleymanov R N, Filimonov O V, Sabanov S L and Boltneva Yu A 2019 Station for monitoring the state of a downhole rod pumping unit news of higher educational institutions *Oil and gas* **4** 82-91

[3] Galeev A S, Filimonov O V and Suleymanov R N 2019 Development of a laser beam to the line in measuring systems *IOP Conf. Series: Materials Science and Engineering* **560** 012125 doi:10.1088/1757-899X/560/1/0121251.

[4] Duda R O, Hart P and Stork D G 2001 *Pattern Classification,* 2nd ed (USA: John Wiley & Sons) 738 p

[5] Gilmanova A M 2019 Consideration of various factors influencing acoustic logging equipment production accuracy *IOP Conference Series: Materials Science and Engineering* **560(1)** 012063 DOI: 10.1088/1757-899X/560/1/012063.

[6] Gilmanova A M 2019 Metrological support of geophysical equipment for acoustic logging *IOP Conference Series: Earth and Environmental Science* **378(1)** 012080 DOI: 10.1088/1755-1315/378/1/012080.
[7] Derffel K 1969 *Statistics in analytical chemistry* (Moscow: Mir)
[8] Barri U A and Dzheyms A F 2015 *Mineral water processing technology* (Butterworth-Heinemann: Elsevier Ltd.) 512 p
[9] Radzhput S and Tkhakur N K 2016 *Geological control over gas hydrate and nontraditional* (Elsevier Inc.) 384 p
[10] Bikbulatova G I, Isaev A A and Boltneva Yu A 2018 Study of the influence of complicating factors on the efficiency of screw rod pumps *Bulletin of Tomsk Polytechnic University. Geo-Resource Engineering* **329(4)** 21-29
[11] Galeev A S, Bikbulatova G I, Boltneva Yu A, Kuzmin I A, Ermilov P P and Ryazantsev A O 2016 Experience of using the autonomous system of control and registration of technological parameters "ASKIR" *Oilfield business* **9** 43-47
[12] Bikbulatova G I, Ermilov P P, Boltneva Yu A and Sabanov S L 2015 Investigation of the effect of ultrasonic pulse radiation on threaded connections of pump and compressor pipes *Oil and Gas Territory* **6** 58-64

AUTHOR INDEX